S0-BWT-442

SPOROPOLLENIN

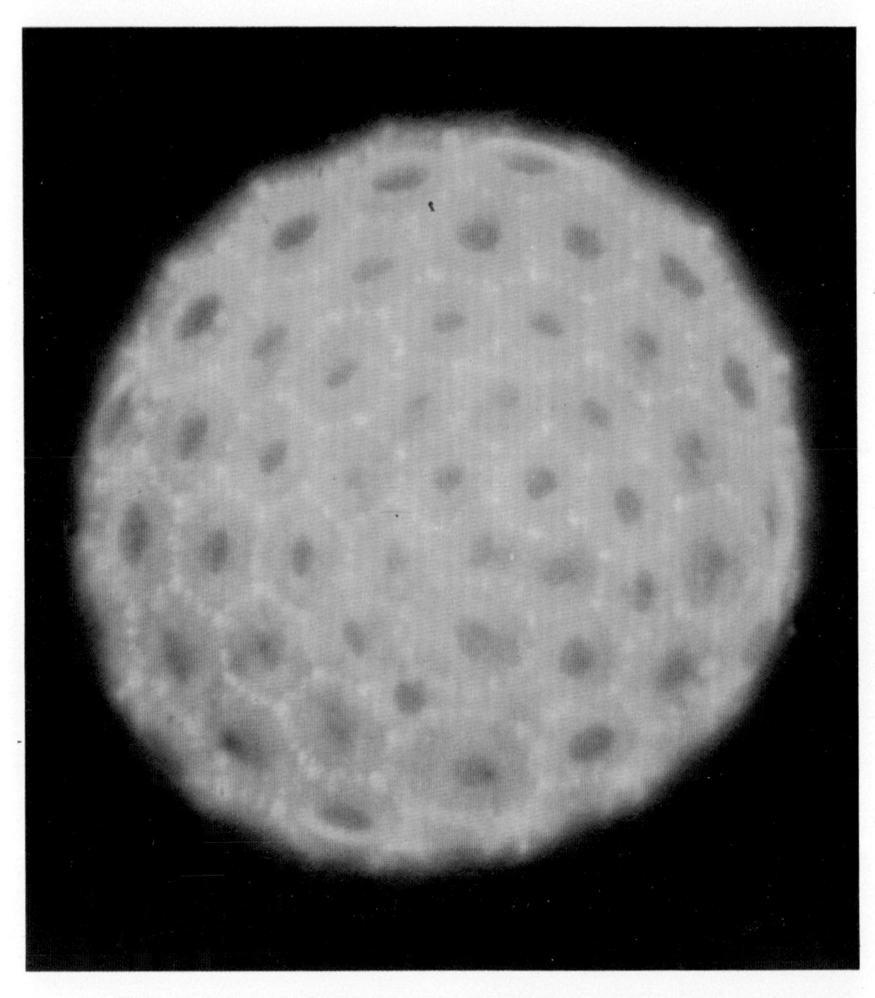

Ipomoea purpurea pollen grain.

Fluorescence micrograph of immature grain treated with highly dilute primuline solutions. The foot-layer and nexine-2 layers are becoming differentiated.

Symposium on Sporopollenin, Imperial College, 1970.

SPOROPOLLENIN

Proceedings of a Symposium held at the Geology Department, Imperial College, London, 23–25 September, 1970

Edited by

J. BROOKS, P. R. GRANT,
MARJORIE MUIR,
P. VAN GIJZEL *AND* G. SHAW

1971

ACADEMIC PRESS · LONDON · NEW YORK

ACADEMIC PRESS INC. (LONDON) LTD
Berkeley Square House
Berkeley Square,
London, W1X 6BA

U.S. Edition published by
ACADEMIC PRESS INC.
111 Fifth Avenue,
New York, New York 10003

Library of Congress Catalog Card Number: 71–149695
ISBN: 0–12–135750–3

PRINTED IN GREAT BRITAIN BY
William Clowes and Sons Limited, London, Beccles and Colchester

EDITORS' PREFACE

The concept of a meeting to discuss Sporopollenin began simply at an informal discussion held at the BP Research Centre in March, 1970. The participants at that meeting decided that it would be interesting to meet again in September, and to invite a few more people to contribute. It was estimated that perhaps about 20 people might attend. However, interest in the subject proved to be more widespread than initially supposed, and within a few weeks of its inception, the potential for a full international Symposium to be held under the aegis of the International Palynological Committee became apparent.

As we soon decided that the Symposium could provide a valuable publication, it became evident that supporting funds and accommodation would be necessary. It is with sincere thanks that we here acknowledge the generosity of the British Petroleum Company Limited, Shell Oil Company, and the Royal Society, Professor J. Sutton of the Geology Department and the authorities of Imperial College London for their kind hospitality and co-operation.

The staff of Academic Press are to be thanked for their aid and encouragement in producing the Proceedings Volume. We are also grateful to all the individual secretaries who laboured to produce the camera-ready copy; special thanks must go to Miss Wendy Coleman for her forbearance in making amendments and minor alterations to the scripts.

Any points in the discussion, adequately covered in papers, have been omitted from the transcript of the discussion.

We do not feel that it is necessary for us to make a formal introduction to the subject of Sporopollenin, as this is covered comprehensively in individual papers. The brief foreword by Professor Dr. R. Potonié and the summary by Professor Dr. F.P. Jonker are as effective as anything we could achieve here.

October 1970

FOREWORD

A Symposium on Sporopollenin was held on September 23rd - 25th, 1970 in the Geology Department, Imperial College, London. The meeting was excellently organised by Dr. M. MUIR and Mr. P. GRANT of that department, in co-operation with Dr. J. BROOKS, from the British Petroleum Company, Sunbury, Middlesex, and Dr. P. van GIJZEL, Afd. Fluorescence Microscopy, Faculty of Science of the Catholic University of Nijmegen, the Netherlands. We must thank the organisers very much for having provided a forum for discussion on the many different aspects of fossil and Recent spore-coats. Many of the papers published in the following pages are important to the further investigations of sporopollenin in pure science and in industry. Botanists long ago were aware that the substance of the coats of spores and pollen grains was very resistant. They called it cutine - a name which was also applied to the material which forms the outermost layers of the epidermal cells of the leaves of most land plants.

So that they could recognise this cutine, botanists proposed a series of microchemical reactions. Only when these reactions were positive was the substance called cutine, and indeed, these positive reactions also occurred with the spore coats of Carboniferous plants (R. POTONIÉ, 1915, 1920). From this, we also believed that these exines were not much altered compared with spores of Recent plants. This was the starting point for a more detailed chemical study of fossil spore coats.

About the year 1920, Dr. F. ZETZSCHE came to the Geologische Landesanstalt in Berlin and asked me to give him fossil material such as I had examined in my work in 1915 and 1920. We gave him then, amongst other things, spores taken from the Lower Carboniferous Brown-coal of the Moscow Basin, and further, the Australian Tasmanite, because at that time, we believed that it contained spores, and not algae as we now know. The summary formula given by ZETZSCHE and his collaborators then, were very interesting, but they told us little more than did the microchemical reactions. In every case, however, we could see that the composition of the Carboniferous spore coat substance had not changed fundamentally compared with Recent spores.

It was only with the work of G. SHAW & A. YEADON (1964) that the investigations took a new trend. These authors gave us structural formulae of Recent Sporopollenin, and it became possible to compare these formulae with the corresponding ones of Carboniferous spore coats of the Lycopsida (POTONIÉ & REHNELT, 1969), and we could see that coalification consisted of a very slow aromatisation of the Sporopollenin.

However, this is only one of the aspects which are treated with, and the reader must look to the following pages for many, varied investigations into the subject.

Professor Dr. R. Potonié,
KREFELD, Germany,
October, 1970.

CONTENTS

CONTENTS (continued)

SPOROPOLLENIN IN THE BIOLOGICAL CONTEXT

J. Heslop-Harrison, Herefordshire.

Introduction

The unifying theme bringing together chemists, botanists, geologists, pedologists and geochemists in this symposium is the concept of sporopollenin. The word itself we owe to Zetzsche (1932), whose work will be reviewed in some detail in later papers. The term, a compound of "sporonin" and "pollenin", also used by Zetzsche, was adopted as a collective appellation for the resistant wall materials found in the spores of pteridophytes and the pollen grains of gymnosperms and angiosperms. "Sporopollenin" therefore conveyed from the beginning a generalisation, namely that all these materials belong to the same family of compounds, chemically speaking. The early evidence for this conclusion came largely from similarities in the empirical formulae of spore and pollen-grain wall substances from different sources. Recently evidence of a more sophisticated kind, also to be reviewed later, has appeared to support this proposition. Dr Shaw and his colleagues have indeed extended the chemical generalisation to include the walls of the propagules of thallophyte groups, and this has been coupled with the suggestion that the "sporopollenins comprise a distinct and unique group of substances": unique presumably in the sense that they are chemically distinct from all other major classes of plant cell wall polymers.

In this introductory contribution I propose mainly to offer some thoughts on the functional roles of spore and pollen grain walls, and to say something about how plants deal with the business of handling sporopollenin as a wall material. However, because of the extraordinary interest of recent chemical findings, I begin with a few prefatory comments on this topic.

The Sporopollenin Concept

The conception of sporopollenin as a class of substances is a chemical one, and as such it must obviously be immune from restriction or re-definition on the basis of biological criteria. Yet without querying the validity of the chemical generalisations, it is legitimate enough to look critically at the biological implications .

I suspect botanists, while acknowledging a common function for the walls of spores, have been as much impressed by their diversity as by their community of character. I do not refer here wholly to structural diversity, which surely needs no emphasis, but to chemical variation as revealed by cytochemistry, including the cytochemistry of staining procedures, empirical and imprecise as this may have been. As Professor Faegri's findings showed (Faegri, 1935) the exine of the mature pollen grain has strata with markedly different staining properties, and developmental investigations have shown that these differ somewhat in the way that they are formed. Furthermore, the exine during its development shows progressive changes in its stainability and reaction to chemical fixation agents such as osmium tetroxide (Heslop-Harrison,

1968a). When major taxa are compared, differences in the stainability of spore walls are often found, and even within a group such as the Angiospermae exines show some variation. In a survey of the responses of the pollens of several species of flowering plants from a range of families to some half-dozen common histological stains, we found that almost every species could be characterised by the stainability of the exines of the mature, fresh freeze-sectioned pollen grains.

Some of the differences in stainability of fresh exines certainly result from variation in the amount and character of encrusting and impregnating materials, notably those added by the activity of the tapetum at the conclusion of wall development. This factor aside, however, it remains true that cleaned exine material from different sources very often does show cytochemical diversity. No doubt much of this is attributable to differences in degree of polymerisation and saturation, in amount of cross-linking, or in the proportion and order of monomers, factors insignificant enough so far as the broad chemical generalisation of the nature of exine materials is concerned. But the point is that there still remains a comparative chemistry of exines to be looked into, for such factors may not be without phylogenetic or adaptive significance.

I am aware from the abstract of Dr Chaloner's paper that he will be commenting on some of the phylogenetic issues, but I feel impelled to make some comment on the suggestion that sporopollenins comprise a unique group of compounds, peculiarly associated with the protection of propagules in all of the major plant groups. As many evolutionary morphologists of the first half of this century abundantly demonstrated, the development of the protected spore was a key event in the spread of life over the land surface of the earth. Bulky plants are by their very nature non-motile; and the abandonment of water as a medium for existence by organisms which could achieve photosynthesis on a massive scale on dry land only became possible with the development of the spore as an agent of dispersal. But the early spore-producers remained tied to water as a medium for fertilisation, and liberation from this limitation followed only with the telescoping of the haploid generation which turned the microspore into the pollen grain, the agent of gametic transfer. Protected unicellular or paucicellular propagules were not, of course, the exclusive property of vascular plants. The meiospores of bryophytes are invested with thick exines, and among thallophytes zygotes commonly bear heavy walls, constituting then zygospores or oospores. These latter structures represent quite a different phase of the life cycle, but their role is comparable with that of the archegoniate spore - namely, to attain dispersal and to maintain the species through adverse conditions, primarily of desiccation. Again we see that these are indispensable functions, even for species whose vegetative life is pursued in water or saturated atmospheres, when suitable habitats are scattered or subject to periodic drying, the conditions of life for many algae and fungi of the land surface.

Clearly, then, there is no particular difficulty in accepting that in all the groups where they occur wall-protected propagules - spores in the broad sense - have evolved in response to similar selective pressures. These are the conditions for parallel evolution, and the development of broadly similar structures in remote groups need accordingly offer no occasion for surprise. A common reliance on sporopollenin does, however, carry interesting implications. There may be good reason for accepting that the angiosperm pollen grain is the evolutionary derivative of the spore of some remote pteridophyte-like ancestor, but none whatever for linking it in any phylogenetic sense with the resting spores of algae, let alone the zygospores and oospores of fungi. Should the spore wall material of thallophyte, archegoniate and spermatophyte groups be chemically essentially identical, and should the compound concerned be indeed unique among plant wall materials, restricted entirely to propagules, the implication is that of all the compounds available to plants - biochemically so much more versatile than animals - only the one class fulfils the requirements for spore protection, and that selection has inexorably forced its adoption independently almost throughout the plant kingdom - even including the fungi, with their chemically specialised range of vegetative cell wall materials.

For green plants, the position would be different were the protective spore wall materials no more than specialised derivatives of ubiquitous cell wall materials present in somatic plant parts. The point has to be examined in relation to the function of the spore wall. Spore ecology tells us that one of the main protective functions of the wall must be against excessive desiccation, and I see no reason to doubt that the mature sporopollenin exine of the pollen grain of the vascular plant, in conjunction with its lipidic surface materials, does form a highly impermeable system. I am not at all impressed by accounts which seek to imply that the exine is as permeable as a sponge: at least in some species there undoubtedly are channels for movement of materials through it in early ontogenetic periods, but at maturity these seem always to be blocked by hydrophobic materials, and there is no evidence to show that at the time of dispersal the exine of the inter-apertural zones is permeable to water or water vapour (the role of the apertures themselves is discussed further below).

Now all the sub-aerial parts of terrestrial plants are liable to water stress, and control of water loss from the epidermis of the vegetative organs of green plants is accomplished by cutinisation and suberisation of cell walls and by the secretion of waxy films. I confess that for many years before the recent re-awakening of chemical interest in spore wall substances, I accepted without any particular question the likelihood that sporopollenin would prove to be chemically related to the cutins - notwithstanding the apparent differences, notably in resistance to chemical degradation. Although any substantial relationship has now apparently to be rejected, I would like to add a note here on some cytochemical observations we have recently been making on membranes of the anther other than those of the pollen grain. Two years ago, I reported the presence of an extratapetal membrane of acetolysis-resistant material investing the anther loculi of some species of Compositae (Heslop-Harrison, 1968a). The homologue of this membrane has now been found in various other families, and it is likely that the compound membrane system described by Banerjee (1967) from various grasses is derived by the association of the extratapetal membrane and the fenestrated membrane bearing the orbicules from the inner locular surface of the tapetal cells after removal of the cellulose wall and the cell contents. A structure homologous with the extratapetal membrane is being described in this symposium from the anther of *Pinus* by Dr Dickinson (p.42). In the Compositae, the thickening of the extratapetal membrane proceeds closely in step with the development of the spore walls, and the chemical similarity is testified to not only by resistance to solvents and degrading agents, but by response to many common staining procedures and by fluorescence properties. The point I wish to make is that initially the membrane is formed just in the manner of a cuticle, and is then indistinguishable from the outer cuticle of the anther wall. The staining properties change progressively during the ensuing period of thickening, but the general aspect remains that of a cuticle, and no new structure is ever formed. Are we truly confident that sporopollenin-like materials appear only in association with reproductive structures? Might it not now be worth extending the analytical methods recently applied to spore wall substances in a wider survey of the protective coatings of xerophytic vegetative organs?

Sporopollenin Sculpturing: Adaptive Roles

Turning to specific matters of spore wall function, it will surely be unnecessary here to stress the great range of structural variation. The morphological studies of Wodehouse and others, and of course the massive systematic works of Professor Erdtman and his collaborators, have thoroughly documented the remarkable diversity of exine structure in the flowering plants, and have provided all the evidence necessary to show that the higher plants have attained an extraordinary competence in the business of moulding and sculpturing sporopollenin in conformity with genetical control. What, then, is the significance of exine patterning? Evidently if we are not to assume that pollen wall sculpturings are no more than meaningless manifestations of a kind of morphogenetic virtuosity, we must accept that there is a high adaptive component, and that diversity has resulted from selective forces.

There are, of course, many clues to the adaptive values of spore wall structure. Basically, the wall must serve the protective function already discussed; further, it must be such as to facilitate, or at least permit, efficient germination. Beyond this, adaptations concerned with dispersal may be expected, at least in advanced groups. The different requirements may well be in some ways conflicting, and the product, accordingly, a compromise. Furthermore, as organic diversity so frequently illustrates, different solutions are likely to be available for the same kind of adaptive problem. In the taxonomic variability of exines, then, we may expect to see adaptive and fortuitous elements interwoven: adaptive, because the simple fact of survival must mean that all are functionally effective, and fortuitous, because there will always have been an element of chance in determining the pathways evolution has taken of the many available.

The apertures are among the more conspicuous features of the angiosperm pollen wall likely to show adaptive characteristics. The meiospores of some lower groups possess no particular specialisations for germination; they burst from internal pressure in an essentially random way. Alternatively, germination occurs through an opening produced at the triradiate scar marking the contact faces of the spore in the tetrad. In flowering plants we find distinctive aperture mechanisms, involving specialisations of both intine and exine. These form the preferred paths of exit for the pollen tube, but their function extends beyond simply providing an escape hatch. Recently, we have been looking at the detailed architecture of colpate and porate exines by scanning electron microscopy, and complementing the structural study with observations on what actually happens in the region of the aperture during final maturation of the grain and on imbibition and germination. The detailed findings will be reported elsewhere, but the one of particular note is that in the pollens we have examined the apertures often appear to act as regulatory devices concerned with controlling the ingress and egress of water and water vapour - harmomegathic in the sense of Wodehouse.

A common feature of colpate and porate exines is the presence of areas of the exine composed of overlapping plates or appressed granules of sporopollenin, not intimately fused with each other, but held in a loose framework by occasional anastomoses, or even simply by association with the underlying intine, with which there is often some interbedding. In monocolpate grains like that of lily, irregularly fissured plates cover the whole area of the aperture. In many of the tricolpate grains examined, the thickened, lamellated margin of each colpus grades into a zone of ribbon-like sporopollenin strands overlying the aperture proper. In the polyporate exines of the Malvaceae, the pores have no operculum, and the intine at the apertures is simply overlaid with a plug of granular sporopollenin. Where an operculum is present, as in the Caryophyllaceae, it is surrounded by a thinner annulus where the sporopollenin takes the form of overlapping platelets or amorphous granules. These features can be confirmed in many published transmission electron micrographs of apertures, and in particular the extensive studies of Roland (1967, 1968), although concerned rather with systematics than function, serve to show how widely the generalisation applies.

The significance of the presence of zones of lamellated or granulated sporopollenin at the apertures becomes clear when the grains are examined at different degrees of hydration, or immersed in solutions of different tonicity. The non-apertural exine is often itself notably elastic, as the quantitative studies of Banerjee *et al.* (1966) convincingly showed for *Sparganium*, but much of the accommodation to varying degrees of dilation of the vegetative cell is taken up not by uniform stretching and contraction over the entire surface, but by changes in the areas of the apertures. This is readily seen with monocolpate and tricolpate grains, where dehydration causes an infolding of the intine at the colpi, closing together the overlying plates or granules of sporopollenin, while imbibition brings about a gaping of the colpi, the sporopollenin particles being carried apart on the surface of the extruded intine, or left at the margins. The changes in porate exines are less conspicuous but no less meaningful. Here, changes in spore volume bring about disproportionately greater changes in aperture area, presumably because of the lesser strength of the exine over the pores or around the annulus. Again, the effect is to bring about a compacting of the sporopollenin as the vegetative cell contracts and a separation of the platelets or particles as it expands. The effect is entertainingly demonstrated by cucumber pollen; in hypotonic solutions, the opercula push out as the profiles of the underlying patches of intine rise; placed in a hypertonic solution, or dried again in air, they fit back into place. In passing, it may be noted that much of the variation in the compaction of sporopollenin at aperture sites seen in transmission electron micrographs of sectioned grains results from volume changes during fixation and processing.

The regulatory function of aperture mechanisms will now be clear. In effect, partial dehydration results in an increasing check to water loss, because with contraction of the vegetative cell the granular or lamellate sporopollenin at the aperture sites becomes compacted, and the surface of exposed, permeable intine reduced. In contrast, a positive feed-back system works during hydration: with each increase in volume, the surface of unprotected intine available for still further inward passage of water is increased.

We have also been impressed by the role of the surface materials of the exine in the control and regulation of water loss and uptake. In lily, the pollenkitt produced by the tapetum spreads preferentially over the residual cellulose of the primexine matrix between the meshes formed by the exine reticulum (Heslop-Harrison, 1968b). At the colpus, this material disperses irregularly; but after dehiscence, and as the grains become increasingly desiccated, the pollenkitt tends to form a globule over the colpus, and is sometimes trapped as the slit closes. Accordingly, not only does the permeable area of the colpus contract with loss of water, but also does it become occluded further by a lipid coating. When the mature pollen from an anther at dehiscence is brought into contact with water, the pollenkitt disperses immediately, presumably because the lipid is associated with surface proteins with detergent properties. Imbibition begins; the colpial area expands, and the cellulose surface of the intine is exposed for still further water uptake in the manner already described.

As regulatory devices, the apertural systems of pollen grains are simple enough, and not comparable in sophistication with the stomatal mechanism. However, that they can be shown to have regulatory functions at all is sufficient to convey the conviction that these features of the exine have adaptive value, and are likely therefore to have evolved under the pressure of selection. It is certainly reassuring to be able to attribute functional significance to apertural mechanisms; yet there are still more questions to be answered before the role of apertures in the total ecology of pollen can be regarded as being adequately understood. Why, for example, is the multiaperturate condition so common? The tube requires only one exit: why should so many pollens be garnished with several?

The triaperturate state is in some sense a reflection of symmetries imposed in the tetrad, as shown by Wodehouse (1935), but it is not sufficient to point to this relationship as causal in an evolutionary sense. The elaborate structural features of the three apertures of the pollen grains of the Compositae are far from being mere products of pressure, as the tri-radiate scars of fern spores may be; their location may reflect cell interactions, but their form arises from morphogenetic events which have the marks of purposefulness, if the impropriety of language may be forgiven. Nor can it convincingly be argued that the tetrad symmetries permit only a trimerous condition so that there is no choice for the spore but to produce its major wall features in triplicate. There are other forms of tetrad than the tetrahedral, and furthermore there are major alliances with monocolpate and monoporate pollens, and others where far more than three apertures are present.

The obvious conclusion is that aperture number itself is an adaptive feature, and this in turn must mean that apertures have functions other than existing to provide possible exits for the tube. One function is assuredly connected with the release of enzymes and other proteins on the stigma before and during germination. It has long been known that various materials flow freely from moistened pollen grains, including not only several hydrolytic enzymes but also antigens and allergens. It is now known from cytochemical and immunological evidence that all are derived in large part from extracellular sites, and in the case of the proteins, particularly from the apertural intine (Knox and Heslop-Harrison, 1969, 1970; Knox, Heslop-Harrison and Reed, 1970).

The apertures are concerned, therefore, not only in providing a way out for the pollen tube, but in the discharge of materials. The fact that among the proteins emitted are lytic enzymes suggests that one function is connected with the digestion of substrate on the stigma surface and so with the early nutrition of the pollen tube. There are indications, however, that the bulk of the protein is not enzymic. Some of the substances released are concerned in the compatibility reactions which determine what matings will produce offspring and what will not; they are thus "recognition" substances.

These facts place the evolution of pollen apertures in a different light. Adaptation has not been related only to the actual event of germination, but to the control of the breeding system and the physiology of the emerging pollen tube. The plurality of apertures found in many diocotyledonous groups finds its explanation when each colpus or pore is recognised as a site for the storage and ultimate release of physiologically important materials, and not merely as one potential exit for the pollen tube. Even the character of the aperture mechanisms themselves, with their provision for the protection of the underlying thickened intine until the beginning of hydration and then for the rapid exposure of this layer, becomes more explicable once the role of the apertural intine as a bag of leachable proteins is appreciated.

I have already mentioned the likelihood that in the course of evolution different solutions may have been found to the same adaptive problem, and the point is nicely illustrated by the diversity of ways in which pollen walls are specialised for the conveyance of materials in the general manner just described. Whereas in dicotyledons the "charge" of wall-borne materials is commonly distributed over several apertures, in monocotyledons with single apertures it is concentrated at the one. But there is no reason to think that this necessarily entails any substantial reduction in the total carried per grain of comparable size, for the volume of intine concerned is proportionately much the same, the aperture area itself being greater, or the underlying intine thicker. In the grasses where but a single pore is present the wall-borne proteins are packed into a large boss of intine underlying the sporopollenin operculum and its surrounding annulus. In grains with no clear apertures or with ill-defined colpi, the proteins are distributed in the intine over its whole area, and in this case the overlying sporopollenin of the exine does not provide a continuous barrier. This is beautifully illustrated in <u>Crocus</u>

vernus and *Gladiolus gandavensis*; in both of these species the exine is very thin and traversed by passages, and in *Gladiolus* it has been possible to show that the antigens are lost directly through it from the underlying intine (Knox, 1971).

The presence of pollen wall proteins in all angiosperm groups - and indeed among gymnosperms also - shows that the adaptation is an ancient one, and indeed it may be argued that the possession of exoenzymes has been important since the earliest evolution of siphonaceous fertilisation and the origin of a male gametophyte which was in effect a parasite. It is also interesting to note that the conveyance of recognition proteins concerned with incompatibility reactions may also be extremely ancient in the evolutionary sense; both Whitehouse (1950) and Lewis and Crowe (1958) have argued that self-incompatibility has been established in the angiosperms almost from the time of their origin.

We are less informed about adaptive features of the non-apertural exine, but it seems very probable that these are mostly concerned with dispersal. In the earlier years of the century while natural history was still a respectable pursuit, pollen dispersal mechanisms attracted a substantial amount of attention. Regrettably, in the modern search for recondite explanations, the correlations exposed in this earlier work are often overlooked. One quite conspicuous distinction, documented especially by Wodehouse, is between the exines of anemophilous and entomophilous pollens. The general rule is that the heavily sculptured exines are associated with insect transport, and the smoothly streamlined, with movement by wind. The correlations prevail even within families; thus anemophilous species of the prepronderantly entomophilous Compositae have dramatically less well developed exine processes than related entomophilous species. But the relationship between exine structure and dispersal agency cannot be understood without reference to another correlate - namely, the development of pollenkitt. No-one who has studied the ingenious and beautiful experiments of Knoll (1930) can fail to be convinced that the amount and nature of this tapetum-produced material is intimately related to pollination mode. Absent, or present in only trivial amounts, in anemophilous species, pollenkitt is a most conspicuous adjunct of the pollen of entomophilous plants, providing not only much of the visual and olfactory stimulus needed for the guidance of insects, but also the means for affixing the grains to the vector's body.

Taken together, then, the correlations worked out by Wodehouse, Knoll and others leave no reasonable doubt that the exine and its superficial materials are deeply concerned in the vital matter of pollen dispersal. So far as the sculpturings of the sporopollenin of the sexine are concerned, we may go so far as to say that the primary adaptive function must be related to dispersal and nothing else, otherwise the presence of heavy sculpturing could not be correlated with insect pollination, as it almost always is, except by the most extraordinary of coincidences.

This is not to claim that we are in a position yet to specify in any detail what may be the functional role of spinules, muri, tubercules and all the other exine embellishments of Professor Erdtman's lexicon. We may be sure, for one thing, that the fortuitous element in biological variation will have played its part in the evolution of exine types, combined with the principle of compromise, since the selective pressures affecting those aspects of sexine form concerned with dispersal cannot have been independent of others, effective at the time of germination, related to aperture number, position and form. There is, nevertheless, an intriguing challenge here. The stage is set for another Darwin, armed not with a hand-lens but a Stereoscan, to work out the intimate details of vector-pollen relationship, just as did Darwin himself for the insect-flower relationship. Because of the impressive cost of his equipment, he will be saved from being dismissed as a mere natural historian.

Spore Wall Patterning: Morphogenesis

Patterned exines provide a supreme example of the morphogenetic capacity of a single cell, and the means by which complex form is imparted to so refractory a material as sporopollenin are obviously of considerable interest. Fine-structural studies of the last decade have given some insight into the mechanics of spore wall formation, and have offered some clues as to how the process is controlled.

The architectural principles upon which exine structure is based must now be regarded as being reasonably well understood, and once more we owe largely to the work of Professor Erdtman the fact that we are able to make generalisations. No more than a summary is required here. The colpi and pores form one of the main features of wall patterning, and some account of the structure of the exine at aperture sites has been given earlier. The non-apertural exine consists of an outer, sculptured layer, the sexine, and an inner, unsculptured layer, the nexine. An important structural feature of the sexine is a population of radially directed sporopollenin columns, the bacula, arising from the nexine. The distribution of the bacula may be patterned, or essentially random. Where there is a pattern, the bacula are arranged in rows or groups, which in turn are distributed in characteristic ways over the surface of the grain, frequently to form a network with meshes of greater or less regularity. The heads of the bacula may be free; or they may be linked laterally to form muri; or they may be partly or completely roofed over at one or more levels, giving the tectate exine. From the surface of the tectum or at intervals on the muri spines, tubercles and other embellishments may arise. The location of these surface features is often related quite precisely to the disposition of the underlying bacula (Heslop-Harrison, 1969; Waterkeyn, 1971). Although there are some exines that cannot be analysed very readily in these terms, it is only necessary to look through the volumes of Professor Erdtman's World Pollen Flora or Professor Nair's excellent survey, _Pollen Morphology of Angiosperms_, to realise how generally the structural principles do apply.

One must emphasise the special importance of the distribution of the apertures and bacula as determinants of surface pattern. These are clearly the cardinal features; and it is usually possible to see that one is keyed to the other, since the patterns formed by the bacula are often focused upon the apertures.

The critical period of time for the determination of the exine pattern follows immediately upon the meiotic cleavages, and the pattern-determining events occur while the spores are still enclosed in the meiotic tetrads. The fact was known to early cytologists, and especially must credit be given to Beer (1911), whose early optical-microscopic studies showed how the patterned precursor of the exine was formed around each spore within the callose tetrad wall.

More recently, the associated events have been followed cytochemically and fine-structurally (for fuller review, see Heslop-Harrison, 1971). The first wall secreted by the spore within the tetrad - the primexine - has a cellulose component (Heslop-Harrison, 1968a; Flynn, 1971); almost from the beginning, the bacula are defined within it as radially oriented columns. No sporopollenin is present in the early primexine, and at first the bacula can often be seen to have a lamellate structure (Dickinson and Heslop-Harrison, 1968; Dickinson, 1970). Acetolysis-resistant material is subsequently deposited in the probacula, and as mentioned already the staining properties of this "protosporopollenin" differ from those of the sporopollenin of the mature exine. The feet of the probacula become linked at this time by the deposition of the first layer of the nexine, the nexine 1. The inner layer, nexine 2, is laid down towards the end of the tetrad period, and thickening continues in the young spore after release from the tetrad.

The participation of lamellae or membranes in the early growth of the nexine layers has been demonstrated especially by Rowley and Southworth (1967) and Echlin and Godwin (1968). These lamellae take the form of ribbons or platelets; they appear to arise at or near the plasmalemma of the spore, and to pass outwards to be apposed successively upon the inner face of the thickening wall. They undoubtedly provide a surface for the deposition of sporopollenin, and resemble therefore the lamellae of the probacula. Indeed in *Lilium longiflorum* it seems that all layers of the exine arise initially in much the same way - by the formation of lamellae in the vicinity of the plasmalemma, disposed radially in the bacula and tangentially in the nexine (Dickinson and Heslop-Harrison, 1968; Heslop-Harrison and Dickinson, 1969).

Should the circumstances be the same in other species as in *Lilium longiflorum*, we have an important clue as to how the cell handles the business of moulding sporopollenin into three dimensional form. It would seem that this is achieved by the generation of surfaces at the plasmalemma with some kind of preferential affinity for sporopollenin precursors. The distribution and orientation of these surfaces provides the initial pattern; this is subsequently amplified by the apposition of further lamellae, and in the case of the sexine probably also by the direct deposition of sporopollenin precursors originating outside of the spore (Heslop-Harrison and Dickinson, 1969).

These findings offer a glimpse of the mechanics of pattern formation; how, then, is control exerted by the cell? The question has several aspects. There is the problem of how the location of the key features of exine pattern - apertures and bacula - is determined from within the cytoplasm of the spores; and at a more basic level lies the question of genetic control. When are the genes concerned with spore wall pattern transcribed? How is the "information" they transfer, first to RNA and then to protein, ultimately translated into observable structure? Obviously answers to these questions cannot yet be given; nevertheless, considerable progress has been made in the last few years in understanding cellular control of wall patterning.

One of the most dramatic features of the whole course of sporogenesis and associated wall formation is that from an initially essentially non-polarised system, the parent meiocyte, four cells are produced that often show a strongly polarised organisation, both in the features of the sporopollenin wall and in the disposition of organelles within the cytoplasm. The polarisation may be carried forward to be expressed later in the planes of cell division in the gametophyte, and indeed in the differentiation of daughter cells as gametes and somatic cells. Wodehouse (1935) and Drahowzal (1936) pointed to the crucial correlate. The cleavage planes of the two meiotic divisions create proximal and distal faces for each of the daughter spores in the tetrad, and so establish the first polarisation. Wodehouse in particular showed that the "contact geometry" (Godwin, 1968) in the tetrad is often faithfully reflected in the location of apertures in the mature spore, even to the extent that aberrant cleavages are later made apparent in the misplacement of apertures.

Clearly then, the orientation of the meiotic spindles is the first key factor concerned in the establishment of exine pattern. Significantly, the different modes of cleavage are associated with different types of spore wall symmetry. Monocotyledonous exines often show a kind of dorsiventral symmetry, reflecting the fact that the spores are formed as quadrants of a sphere: the cleavage following meiosis I divides the mother cell into two hemispheres giving the dyad, and the two walls formed after meiosis II halve the daughter cells by walls oriented at right angles to the first. Many dicotyledonous spores possess a trimerous symmetry, and this arises because cleavage is simultaneous. No wall is formed after meiosis I, so that the daughter nuclei pass into meiosis II in the same cytoplasm. Secondary spindles are formed so as to link the four nuclei, which become disposed at the vertices of a tetrahedron. The six walls formed across the phragmoplasts then separate the spores, leaving them in the characteristic tetrahedral disposition. As we have seen, we owe to Wodehouse the demonstration that the position of the apertures is then related to the three contact faces each spore has with its sibs.

While spindle orientation is thus concerned in establishing the spatial framework for spore wall morphogenesis, there remains the question of how the symmetries set up at the time of cleavage are transferred to the material of the wall. One significant step has been the electron-microscopic detection of cytoplasmic structures whose disposition is such as to suggest they might be concerned with pattern control. Associations of cytoplasmic membranes with the plasmalemma of the young spore have been noted during the early tetrad period, when indications of pattern are first appearing in the primexine (literature cited in Heslop-Harrison, 1971). Perhaps the most conspicuous example is the definition of future aperture sites by the apposition of plates of endoplasmic reticulum at a time when the spores are invested directly in the callose special wall, before the commencement of primexine growth. It is perhaps not too fanciful to think of these plates as forming a kind of stencil, demarcating areas of the plasmalemma over which the matrix material of the primexine will not be deposited. The siting of bacula, also, has been shown in some species to be related to profiles of the endoplasmic reticulum, perhaps most strikingly in *Zea mays* (Skvarla and Larson, 1966). In lily, the membranes are not conspicuous at the sites of the bacula, but these are marked off by specialisations of the plasmalemma with associated polyribosomes (Dickinson and Heslop-Harrison, 1968; Dickinson, 1970).

The mere spatial concordance of cytoplasmic structures and future features of the primexine pattern does not of course prove that one determines the other, although the time sequence is suggestive. Yet the observational evidence does suggest how cytoplasmic constituents may be involved, and also points to the possibility of experimental test.

Spindle poisons offer a means of disrupting the meiotic

cleavages, and centrifugation a way of moving the membranes and organelles of the spore during critical periods of wall morphogenesis. Experiments with lily species have shown that, in general, expectation is realised, and that the pattern of sporopollenin deposition in the exine can be modified in predictable ways through these agencies (Heslop-Harrison, 1971b).

When both meiotic divisions are blocked in cultured buds in consequence of colchicine treatment, a restitution nucleus is formed, and the whole mother cell then behaves as a single spherical spore in respect to exine formation. Apertures are not defined at all, or one irregular colpus is formed, placed at random. In the normal exine of lily, the bacula are distributed in a reticulum over the non-aperturate surface. Colchicine-blocked mother cells produce the reticulate pattern in later development, but it is often uniform over the whole surface, or modified only slightly at the margin of the irregular colpus. When the second meiotic division is blocked but not the first, the two halves of the dyad behave as spores. Again, the colpus may be wholly eliminated, but generally each daughter cell forms one irregular colpus. The reticulate distribution of the bacula is sometimes more or less normal, but frequently the pattern is disrupted, and the sporopollenin of the exine is laid down in irregular bosses. Still later treatment with colchicine results mainly in errors in the placement of the bacula.

There are parallels between the effects on exine patterning produced by displacing the cell contents mechanically and those resulting from colchicine treatment. Cleavage aberrations result from centrifugation at the time of the meiotic divisions, and again these are accompanied by misplacement of the colpi. The most striking effects result, however, from centrifugation in the early tetrad period. The colpus may be defined in an essentially normal manner, but then the area becomes covered with a secondary reticulum, often with a mesh size different from the first. It is as though whatever is preventing the establishment of bacula over the area of the colpus is shifted, so leaving a free surface for more bacula to be produced.

Although not all of the experimental results can be wholly explained, they are in the main consistent with the dual hypothesis that the meiotic spindles are causally concerned in establishing the main symmetries of the sporopollenin exine, and that the role of intermediary is played by cytoplasmic constituents unstable enough to be moved mechanically. The ultrastructural evidence indicates of course that the labile cytoplasmic components include membranes of the endoplasmic reticulum.

Nothing can yet be said about the way genetic control is exerted over the siting of the cytoplasmic membranes themselves, although there is abundant room for speculation (Heslop-Harrison, 1968c). What seems reasonably certain now is that the genes for exine patterning are transcribed not in the haploid spores themselves, but probably in the mother cell before the meiotic cleavages. This issue has been discussed recently by Godwin (1968), who has mentioned some of the earlier observations bearing upon it. New evidence has come from the experiments with lily, just outlined. When cleavage is disrupted either by chemical or mechanical means, enucleate fragments are occasionally set off in the tetrad. At least some of these retain the capacity for forming a normally stratified exine, bearing the reticulate pattern of bacula. This must mean that the intervention of the spore nucleus is not required for these aspects of pattern formation. Taken with corresponding evidence from aberrant cleavages in hybrids, this leaves no reasonable doubt that the cytoplasm already carries the basic "programme" for exine pattern before the actual event of cleavage. It may be noted also that after the complete failure of nuclear division following colchicine treatment the mother cell may proceed to form an exine with a normal reticulate pattern. This too suggests that the cytoplasm is already programmed for wall production before the divisions, and that the programme can be carried through whatever the fate of the nuclei.

Conclusion

My aim in the foregoing has been to provide a brief sketch of certain functional features of spore and pollen grain walls and of the present state of knowledge concerning exine ontogeny. That we are still ill-informed about many aspects will be clear enough. Some of the outstanding problems lie primarily within the domain of the botanist, but in the context of this meeting it is certainly appropriate to stress that progress in many areas is going to be made most effectively by collaborative effort of chemist and biologist.

There is patently need for further chemical work on the characterisation of spore wall materials from all classes of plants - all the more now that we have a conception of what to expect from the work of Dr Shaw and his colleagues. But the chemical work must be backed by cytochemical and biochemical investigations directed at clarifying the nature of sporopollenin precursors and identifying their sources. The biosynthesis still remains an enigma, although with the suggestion that carotenoids and carotenoid esters may be involved a promising lead is available.

Many of the outstanding ontogenetic problems cannot be tackled without better knowledge of the biochemistry of sporopollenin. What is the nature of the lamellae upon which the earliest deposition of sporopollenin takes place during exine development? Why, in any event, is sporopollenin laid down only upon certain preferred extra-cellular surfaces? Is the accretion of sporopollenin in the growing exine enzyme dependent or not?

If one effect of this meeting turns out to be a better co-ordination of work on problems like these by all those with an interest in spores and spore wall materials, then this in itself will have justified the effort that has gone into organising it.

References

BANERJEE, U.C., 1967. Grana Palynol. 7, 365-377.

BANERJEE, U.C., ROWLEY, J.R. and ALESSIO, M.L., 1965. J. Palynol.

1, 70-89.

BEER, R., 1911. Ann. Bot. (Lond.) 25, 199-214.

DICKINSON, H.G., 1970. Cytobiologie 1, 437-449.

DICKINSON, H.G. and HESLOP-HARRISON, J., 1968. Nature (London) 220, 926-927.

DRAHOWZAL, G., 1936. Österr. bot. Zeitschr. 85, 241-269.

FAEGRI, K., 1956. Bot. Rev. 22, 639-664.

ECHLIN, P., and GODWIN, H., 1968. J. Cell Sci. 3, 175-186.

FLYNN, J., 1971. In Pollen: Physiology and Development, ed. J. Heslop-Harrison. Butterworths. In press.

GODWIN, H., 1968. New Phytol. 67, 667-676.

HESLOP-HARRISON, J., 1968a. Can. J. Bot. 46, 1185-1192.

HESLOP-HARRISON, J., 1968b. Science (N.Y.) 161, 230-237.

HESLOP-HARRISON, J., 1968c. Dev. Biol. Suppl. 2, 118-150.

HESLOP-HARRISON, J., 1969a. Can. J. Bot. 47, 541-542.

HESLOP-HARRISON, J., 1969b. Cytobios 2, 177-186.

HESLOP-HARRISON, J., 1971a. In Pollen: Development and Physiology, ed. J. Heslop-Harrison. Butterworths. In press.

HESLOP-HARRISON, J., 1971b. S.E.B. Symposium No. 25. In press.

HESLOP-HARRISON, J., and DICKINSON, H.G., 1969. Planta 84, 199-214.

KNOLL, F., 1930. Zeitschr. Bot. 23, 609-675.

KNOX, R.B., and HESLOP-HARRISON, J., 1969. Nature (London) 223, 92-94.

KNOX, R.B., and HESLOP-HARRISON, J., 1970. J. Cell Sci. 6, 1-27.

KNOX, R.B., HESLOP-HARRISON, J., and REED, C., 1970. Nature (London) 255, 1066-1068.

LEWIS, D., and CROWE, L.L., 1958. Heredity 12, 233-256.

NAIR, P.K.K., 1970. Pollen Morphology of Angiosperms. Scholar Publishing House, Lucknow, India.

ROLAND, F., 1966. Pollen et Spores, 8, 409-419.

ROLAND, F., 1968. Pollen et Spores, 10, 479-519.

ROWLEY, J., and SOUTHWORTH, D., 1967. Nature (London), 213, 703-704.

SKVARLA, J.J., and LARSON, D.A., 1966. Amer. J. Bot., 53, 1112-1125.

WATERKEYN, L., 1971. In Pollen: Physiology and Development, ed. J. Heslop Harrison, Butterworths (in press).

WHITEHOUSE, H.K.L., 1950. Ann.Bot.N.S.(London), 14, 199-216.

WODEHOUSE, R.P., 1935. Pollen Grains. McGraw-Hill, New York.

ZETZSCHE, F., 1932. Sporopollenine, in Handbuch der Pflanzen Analyse, Vol. 3, ed. G. Klein, Vienna.

THE ROLE PLAYED BY SPOROPOLLENIN IN THE DEVELOPMENT OF POLLEN IN PINUS BANKSIANA

H. G. Dickinson, Department of Botany and Microbiology,
University College, London, England.

Abstract

The mechanisms are described by which the main sporopollenin containing components of the microsporangium and microspores in *Pinus banksiana* are formed.

The patterning of the pollen wall sexine is shown to be established by the positioning of large cytoplasmic vesicles at the cell surface. Further activity by vesicles, apparently produced by golgi bodies, gives height to the sexine. Elements of the layered nexine form at the cell membrane and are subsequently packed into the wall. The early microspore wall does contain some sporopollenin while contained within the tetrad, but massive deposition of this material takes place after the young grain is released into the thecal fluid.

There are conspicuous similarities between the orbicules and the peritapetal membrane, both in composition and the manner of their formation. The character of these structures has been investigated and their possible functions during microsporogenesis are discussed.

The significance is considered of the changes occurring in the cells of the tapetum during the stages when sporopollenin is actively deposited within the microsporangium.

4

Introduction

Elements of three major components of the microspore and microsporangium in _Pinus banksiana_ are composed of sporopollenin. This inert and resistant material forms the outer layers of the wall that invests the pollen grain, a large part of the orbicules (or Übisch bodies) and one layer of the peritapetal membrane that encases both the microspores and tapetum.

The determination of patterning in and the subsequent development of the pollen wall have been extensively studied in many angiosperms. Heslop Harrison (1963) and Skvarla and Larson (1966) showed that at a certain stage in development of the pollen of _Silene_ and _Zea_, initials of the probacula corresponded in position with elements of the endoplasmic reticulum lying underneath the plasma membrane. Heslop Harrison also demonstrated that the position of the colpus in _Silene_ was determined by a sheet of endoplasmic reticulum which remained appressed to the plasma membrane throughout the early stages of pollen wall development. Such a correspondence between the endoplasmic reticulum and initials of the early wall was not found in _Tradescantia_ by Horvat (1965). Recently, Echlin and Godwin (1969), working on _Helleborus_, have confirmed the earlier investigations with regard to the presence of a sheet of endoplasmic reticulum under the region destined to be the colpus, but found no association between elements of the endoplasmic reticulum and probacula.

Early stages in pollen wall development have also been

studied in *Lilium*, (Dickinson and Heslop Harrison, 1968, Heslop Harrison, 1968, and Dickinson 1970) and the striking reticulate patterning of the pollen wall in this plant has been shown to result from the positioning, perhaps by microtubules, of protoplasmic protusions that precede the bacula of the sexine. Most recently, Vazart (1970) has demonstrated that, in *Linum*, the positioning of the bacula is determined by mitochondria that become appressed to the inner face of the plasma membrane.

It is thus clear that the determination of patterning in the outer layer of the pollen wall may be effected by different mechanisms in different species. There are however, marked similarities between species in the mode of formation of the nexine layer of the wall, and in the manner in which the outer sexine layer continues to develop.

Rowley and Southworth (1967) demonstrated that the nexine in *Anthurium* was composed of flat lamellae or 'tapes' formed by the thickening, on both sides, of an electron-lucent layer of unit membrane dimension. Similar structures have now been reported in *Ipomoea* (Godwin, Echlin and Chapman, 1967) and in *Lilium* (Dickinson and Heslop Harrison 1968), where they were shown to form at the plasma membrane and move up into the developing nexine. The initials of the exine become acetolysis resistant before release of the microspores from the tetrad, demonstrating that they, even at this early stage in development, contain a considerable proportion of sporopollenin.

On release from the callose tetrad wall the young pollen grains expand and, in most species studied, simple growth of the sexine takes place, possibly as a result of the accretion of tapetally synthysised sporopollenin precursors (Heslop-Harrison and Dickinson 1969).

The ontogeny of the gymnospermous pollen grain wall has attracted comparatively little attention and the few investigations undertaken have been almost exclusively descriptive (see Erdtman 1965). The sacci characteristic of certain gymnosperm pollens have fared somewhat better. From the earliest studies of Muhletahler (1955) and Ueno (1958) these structures have been regarded as being formed by a distension of the pollen wall sexine. Recently Ting and Tseng (1965) have reported that the saccus of *Pinus balfouriana* is bounded by both the sexine and elements of the nexine. As Pettitt (1966) points out, if this is the case, the expansion that results in the formation of the saccus must take place between the inner face of the nexine and either the plasma membrane or intine, rather than between the nexine and sexine.

The development of the orbicules on the inward facing and sometimes radial surfaces of the tapetal cells during the dyad-tetrad stages of pollen development was first followed with the light microscope (Schnarf, 1923, Ubisch 1927, and Py, 1932). Reinvestigation of orbicule development with the electron microscope has revealed many further details of their fine structure and the mechanism by which they are formed. Rowley (1963), with the aid of electron stains and acetolysis techniques, has demonstrated that the material composing the orbicules in *Poa* is identical with that of the pollen grain exine. Recent investigations in *Helleborus* (Echlin & Godwin, 1968) and *Lilium* (Heslop Harrison and Dickinson 1969) have revealed the orbicules to be formed by the deposition of sporopollenin, or its immediate precursors, upon globuli that are extruded into the loculus by the tapetum.

The function of the orbicules has yet to be satisfactorily determined. Rowley (1963) regards them as the result of overproduction of sporopollenin but, when the observations of Banerjee (1967) on the tapetal membranes of grasses are also considered, a role in the dispersal of the pollen grains may be reasonably hypothysised (Heslop Harrison & Dickinson 1969).

The presence of an acetolysis-resistant wall investing both the tapetum and microspores was first demonstrated by Pettitt (1966) in gymnosperms. He considered this structure to be formed by the deposition, onto the locular wall, of tapetally synthesised precursors of sporopollenin.

More recently Heslop Harrison (1970) has shown such a wall to exist in certain *Compositae* and proposes that it forms a 'culture-sac' in which the microspores develop during the latter stages of sexine formation. Heslop Harrison correctly points out that this membrane should not be identified with the 'Tapetal membrane' of grasses (Banerjee 1967) which is formed on the inner faces of the tapetal cells.

Materials and Methods

Sporangia were excised from male cones of *Pinus banksiana* and plunged into ice-cold 3% glutaric dialdehyde in 0.025 M phosphate buffer (pH 6.8) and fixed for 4 hours. Following several washes in buffer overnight and subsequent postfixation in 2% aqueous osmic acid for three hours, the material was dehydrated in ethyl alcohols, transferred to 1.2 epoxy propane and embedded in EPIKOTE resin (Shell, U.K.).

Thin sections for electron microscopy were cut on a Porter Blum MT 1 ultramicrotome and stained in 7% aqueous uranyl acetate followed by lead citrate (Reynolds 1963). Grids were examined in a Siemens Elmiskop 1 electron microscope operating at 60 kV.

For light microscopy, ribbons of thick sections were cut at 1.5 μm and mounted on gelatin coated slides. By means of the following treatments the distribution in the material of carbohydrates, lipids and carotenoids was investigated.

a) For Carbohydrates; oxidation in periodic acid followed by treatment with Schiffs reagent. (Feder and O'Brien, 1968).
b) For Lipids; staining with Sudan III (De Martino et al. 1968).
c) For Carotenoids; treatment with $SbCl_3$, HCl and H_2SO4. (Goodwin 1955).

Sections were also acetolysed using the mixtures described by Avetissian (1950).

Observations

The Pollen Wall

a) Development within the tetrad

Immediately prior to the development of the structural initials of the sexine small and frequent embayments form in the plasma membrane of the microspore (see Fig. 1). Despite their frequency these embayments seldom fuse to form a layer. Coincident with these developments, large numbers of vesicles become visible in the cytoplasm below the plasma membrane. Vesicles of similar size may also be observed in association with golgi bodies situated in the subjacent cytoplasm (See Fig. 1).

Subsequently, other vesicles begin to occupy this region. Deeper in the cytoplasm, these vesicles which can measure 0.1μm in diameter are frequently in continuity with large golgi bodies or cisternae of the endoplasmic reticulum (See Fig. 2). Once free, the vesicles move to the periphery of the cell and, when they make contact with the inner face of the plasma membrane the junction between the vesicles and the membrane undergoes a profound change (See Fig. 3). The membranous nature of these structures becomes no longer discernible and is replaced by an evenly staining matrix retaining the overall configuration of the components involved (See Fig. 3). At this stage, the inner faces of the vesicles fuse to form a highly convolute boundary that immediately commences to withdraw inwards, away from the callose wall. This boundary, which effectively forms a new plasma membrane occasionally contains elements of the vesicles (See Fig. 5). During the withdrawal, small vesicles, of a kind also seen in association with golgi bodies in the adjacent cytoplasm, accumulate under the new boundary of the protoplasm (See Fig. 4).

The space formed between the early sexine and the protoplast contains a dispersed fibrillar matrix (See Fig. 6). In the regions subsequently occupied by the sacci of the grain, this space is very much enlarged and its contents more condensed. Small vesicles, of equivalent dimension to those present in the cytoplasm may often be seen within the developing saccus (See Fig. 8). Very little or no distortion of the callose tetrad wall occurs during these early stages of saccus formation (See Fig. 11).

If material at this stage is treated with periodic acid and Schiffs reagent (PAS) the region immediately under the saccus stains heavily. The sacci themselves do stain, as does the remainder of the early sexine, byt not as intensely as the cytoplasm subjacent to the saccus (See Fig. 9).

The diffuse matrix replacing the outer faces of the vesicles, now separated from the new plasma membrane, thickens and becomes progressively more electron opaque (See Fig. 6). Further, the vertical components of this wall also elongate considerably, often descending to the level of the protoplast. It is at this stage that the initials of the sexine become capable of surviving acetolysis.

Sexine does not develop over the whole protoplast at one time. It is first formed over the areas subsequently occupied by the sacci and later spreads to the remainder of the cell surface, with the single exception of the region destined to be the germinal pore. Here, the small embayments are formed but large vesicles do not accumulate under the plasma membrane.

Formation of the early sexine is followed by development of the layered nexine. In *Pinus banksiana* the elements of this wall are apparently produced from the protoplast surface. The first indication of nexine formation is a thickening of the inner face of the plasma membrane (See Fig. 13). In small regions this thickening condenses into a second layer, of unit membrane dimension, lying immediately under the plasma membrane (See Fig.14). The areas of plasma membrane thus isolated from the body of the protoplast then separate from the underlying membrane and become thickened on their inner faces with electron opaque material (See Fig. 15). Thus formed, the thickened elements of the nexine become appressed to the other layers of the wall.

Layers of nexine also develop under the young sacci (See Fig. 10). Although the contents of the sacci and the primexine react positively with PAS treatment at this stage, the intense staining of the cytoplasm subjacent to the sacci is no longer evident (See Fig. 12).

Small embayments are present in the plasma membrane during all stages in nexine formation and small vesicles, which may also be seen associated with golgi bodies, are conspicuous in the peripheral cytoplasm.

Dissolution of the callose tetrad wall takes place in the latter stages of, or immediately following nexine development and the young microspores are released into the thecal fluid.

b) Development following release from the tetrad

On release, the young microspores increase in volume by some 300%, the sacci inflate and intine synthesis commences.

The contents of the newly expanded sacci are no longer PAS-positive, but the sexine boundary continues to give a reaction (See Fig. 17). As the grains mature, a PAS-reaction becomes undetectable even here (See Fig. 18).

Rapid growth of the sexine also takes place following dissolution of the callose wall and the initials formed within the tetrad become massively enlarged. No changes in structure may be observed within the wall during this growth, but, in the final stages of development, the material composing the sexine displays reduced affinity for electron stains (See Fig. 19).

The Tapetal Orbicules

In *Pinus banksiana*, formation of tapetal orbicules commences while the meiocytes are undergoing meiosis II and continues until the disintegration of the tapetal cells. In the earliest stages or globuli, invested in a shallow coat of electron opaque material, situated within the gelatinous inward-facing or radial walls of the tapetum (See Fig. 20). No orbicules are formed in the outward facing walls of the tapetum (See Fig. 23).

Sections of orbicules at these and immediately subsequent stages stain with Sudan III and other lipid stains (See Fig. 29). No reaction was observed in this material with $SbCl_2$, concentrated HCl or H_2SO4.

Electron opaque material continues to accumulate upon the orbicules during development of the pollen grains until it reaches a depth of about 0.05μm. Material of identical appearance with that condensing onto the orbicules may also be observed dispersed in the gelatinous tapetal walls (See Fig. 22).

During development of the orbicules the tapetal cytoplasm undergoes a striking alteration, changing from the densely staining tissue characteristic of the earlier stages of microsporogenesis to a vesiculate, loosely packed protoplast, containing copious quantities of membranous cisternae, golgi bodies, mitochondria and ribosomes (See Fig. 21). Vesicles of a type associated with both golgi bodies and endoplasmic reticulum cisternae within the cytoplasm, are frequently released into the gelatinous tapetal walls (See Fig. 23). The content of these vesicles is not consistent, but generally stains grey with electron-stains. Very few globuli, of the kind that form the 'cores' of the orbicules, are present within the tapetal cytoplasm while electron-opaque material is being deposited upon the orbicules in the gelatinous walls (See Fig. 24).

When the microspores reach the tetrad stage, the tapetum becomes coenocytic; massive channels form between the cells and nuclear fragmentation is frequent (See Fig. 20). The tapetum finally disintegrates shortly after release of the young microspores from the tetrad (See Fig. 26), leaving the maturing grains resting upon the orbicules which, in turn, are supported by the peri-tapetal wall (See Fig. 27).

At these later stages of their development the orbicules become highly acetolysis resistant (see Fig. 31) and their affinity for lipid stains is lost. As with the material composing the sexine, the outer layer of the orbicules becomes progressively less electron opaque during maturation.

The Peritapetal Wall

Formation of the peritapetal wall commences, as with the orbicules, while the meiocytes are undergoing meiosis II. Also, like the orbicules, it is seemingly composed of two components, a grey staining substratum and an electron-opaque upper layer (See Fig. 20).

The situation of this wall is not immediately evident but, since it forms not only peripherally between the tapetum and the anther wall cells, but also frequently radially, between the tapetal cells (See Fig. 20), it may be assumed to develop either on or in the original tapetal wall.

Accumulation of electron-opaque material upon the grey-staining substratum commences during the tetrad stage of pollen development, and, as in the gelatinous radial and inward facing walls, material identical with that being deposited may be seen dispersed in the space between the tapetal protoplast and the peritapetal wall (See Fig.23). Many vesicles, generally with grey-staining contents are also present at this interface (See Fig.23). Vesicles, of a similar dimension and content may be seen associated with golgi bodies and cisternae of the endoplasmic reticulum within the tapetal cytoplasm (See Fig. 21).

Treatment of material at this stage with Sudan III results in the staining of the wall (See Fig. 28), a property that is lost in later stages. The peritapetal wall is also resistant to acetolysis throughout the latter part of its development (See Fig. 30).

Visible degeneration of the tapetal cells takes place when accumulation of electron opaque material on the wall has reached a depth of about 0.05μm (See Fig. 26). At this juncture electron-opaque bodies often become evident both in the mitochondrial membrane system (See Fig. 25) and, less frequently, within the cisternae of the endoplasmic reticulum.

As the pollen grains mature, tapetal breakdown becomes complete and the grey staining component is lost from the peritapetal wall. The protoplasts of the microsporangial wall cells retain their integrity during these stages and many vesicles are present at the inward facing plasma membranes of these cells (See Fig.26).

Thus, in the desiccated microsporangium, the peritapetal membrane appears as a simple, acetolysis-resistant membrane investing both pollen grains and the tapetal orbicules (See Fig.27).

Discussion

The Pollen Wall

a) Development within the tetrad

The deposition of material between the plasma membrane and the inner face of the callose tetrad wall prior to development of the structural components of the early sexine, appears to be a general feature of pollen wall formation. Echlin and Godwin (1969) describe it in *Helleborus* and report its secretion by golgi vesicles; it is formed in a similar manner in *Lilium* (Dickinson, 1970) and Lepouse (1970) has also recorded its development in *Abies*. In *Lilium* this material contributes towards the early sexine and the evidence from *Helleborus* indicates that condensation of this material plays an important part in the development of the bacula. It would therefore, seem reasonable that the embayments in *Pinus banksiana* also, may contain substances that are subsequently involved in the formation of structural initials of the sexine.

In this species there is little doubt that the specific patterning of the pollen grain sexine is derived from the form of the vesicles that become appressed to the inner face of the plasma membrane. Vesicles of a similar size exist in the cytoplasm during the major part of microsporogenesis, but it is only following the formation of the embayments that they develop in large numbers. Their production in such quantity is presumably affected by the large and complex golgi bodies present in the cytoplasm.

The content of the vesicles is not known. It is thus impossible to determine whether the changes that occur when the outer faces of the vesicles merge with the plasma membrane are the result of action by agents contained in the vesicles themselves, or in the fibrous material outside the protoplast. These changes apparently confer considerable rigidity to the early sexine, as the combined outer faces of the vesicles may now be termed, for very little distortion takes place when the inner layer commences its withdrawal.

The retreat of the protoplast is probably best studied in the region destined to be the saccus where it is most pronounced. The fine structural and histochemical evidence, taken together, indicate that the protoplast is being displaced by the insertion of PAS positive material secreted between the early sexine and the protoplast. This material is presumably transfered by the golgi vesicles observed in the proximity of the plasma membrane at this stage. Indeed, it also appears that some vesicles may pass intact through the plasma membrane. Conceivably the membranes of these vesicles may have become affected by the thickening processes that are involved in the synthesis of primexine from membranous initials.

The early sexine becomes acetolysis-resistant during, or immediately following the displacement of the plasma membrane, indicating that the precursors of sporopollenin have commenced to polymerise upon the structural elements. The source of these precursors is not evident and they clearly may be contained in the materials inserted between the young sexine and the protoplast, the vesicles, or even the small embayments which, by this stage, can no longer be distinguished.

When considered with the mechanism by which patterning is established in the pollen wall of an angiosperm, i.e. *Lilium*, the events described in *Pinus* do not, at first sight, appear to be directly comparable. However, features common to both mechanisms do exist. Fibrous material is secreted in both cases, structural elements of the two sexines are derived from membranous initials and both walls increase in height by a similar displacement of the plasma membrane. Perhaps most important is the fact that, in spite of their wide taxonomic separation, patterning of the pollen wall in both species results from the sequential activities of different types of golgi body and their associated vesicles.

In both *Pinus* and *Lilium* the golgi vesicles involved in sexine development require considerable organisation, in the former they require orientation towards the periphery of the cell (with the exception of the area destined to be the germinal pore) and, in the latter, along the lines of the muri. In *Lilium*, wefts of microtubules have been observed adjacent to the vesicles in question and have been proposed, tentatively, as the agents by which organisation is effected. Very few microtubules have been observed in association with the vesicles in *Pinus*. It is perhaps premature to dismiss these structures from this rôle, for there is recent evidence (Franke *et al*. 1969) that fixation of tissue at 0°C., as was the material used in this study, can seriously impair their preservation.

There are interesting differences between the elements that compose the nexines of *Lilium* and *Pinus*; in the angiosperm there is equal thickening on both sides of the electron-lucent layer but, in the gymnosperm, the major part of the thickening occurs on one side only.

The process by which these elements are formed in *Pinus* is not evident. Clearly it involves the synthesis of a second membrane under the plasma membrane. However, the mechanism by which small regions of the plasma membrane separate from this lower layer and become unilaterally thickened has yet to be determined. The source of the electron-opaque thickening is not known, although it could be carried by the golgi vesicles that accumulate at the cell surface at this time. It is, however, equally possible that these structures are involved in the formation of the second membranous layer and that the thickening is simply secreted through the plasma membrane.

The deposition of layers of the nexine between the saccus and the protoplast possibly seals off its contents from the young spore. No distortion of the callose boundary of the saccus occurs despite the displacement of the protoplast by the developing saccus, suggesting that it is not under appreciable pressure whilst held in the tetrad.

The fact that the early exine, while still in the tetrad, is capable of surviving acetolysis suggests that it is, in part, composed of sporopollenin. Since the callose wall has been shown to be impermeable to molecules as small as thymidine (Heslop-Harrison and Mackenzie, 1967) it is likely that the precursors of sporopollenin originate from within the microspore, rather than the tapetum.

b) Development following release from the tetrad

The massive growth of the sexine that follows the dissolution of the tetrad wall may be explained in terms of accretion of material, already present in the thecal fluid, onto the existing initials. Such a hypothysis is supported by the observation that material electron-microscopically identical with that composing the developing sexine is present in the walls of the tapetum, having presumably condensed onto the gelatinous matrix during preparation.

It seems unlikely that the expansion of the sacci that follows microspore release from the tetrad results from the removal of compression or from simple growth of the sexine. Also, since the saccus is by this time isolated from the protoplast by elements of the nexine, it seems improbable that material is again transferred from within the grain. It is thus reasonable to assume that an agent responsible for the expansion may come from the thecal fluid. If this is the case, two mechanisms may be proposed.

While there is no evidence that the sporopollenin of the mature sexine is semipermeable, the possibility that the immature PAS-positive boundary might have this quality cannot be overlooked.

If the saccal contents were hypertonic in relation to the thecal fluid, osmotic pressure generated within the saccus could result in its inflation. Otherwise, the PAS-positive material inserted into the saccus might be a polysaccharide capable of swelling on imbibition. The dissolution of the callose wall would permit the hydration and subsequent expansion of the material, and consequently the inflation of the saccus. When fully hydrated this material may be leached through the porous mature sexine, or dry down during the desiccation of the pollen grain.

The decreasing affinity for osmium stain appears to be a general feature of maturation in the sexine, orbicules and peritapetal wall. No precise chemical interpretation of this change may be made at this stage but, if an analogy may be drawn with the staining properties of lipids, the lack of affinity to osmium of the mature structures may result from the increased chemical saturation of materials contained within them.

The Tapetal Orbicules and the Peritapetal Wall

There is no direct evidence from the present study that the grey-staining component of the orbicules and the peritapetal wall is formed first. However, in both *Helleborus* (Echlin and Godwin 1968) and *Lilium* (Heslop-Harrison and Dickinson 1969) production of this material is the first indication of orbicule development, and it is perhaps reasonable to assume that this is also the case in *Pinus*. The staining properties of this substance identify it as a lipid. It was impossible to demonstrate the presence of carotenoids in either the orbicules or peritapetal wall.

Echlin and Godwin (1968) report that, in *Helleborus*, extrusion of globuli continues after deposition of electron-opaque material has commenced upon those already free in the tapetal walls. In *Pinus* these events do not overlap, for by the time accretion of electron-opaque material has commenced, very few grey-staining globuli are present in the tapetal cytoplasm.

The experiments of Rowley (1963) and the acetolysis tests performed in this investigation leave little doubt that the electron opaque layer of the peritapetal wall and orbicules contains sporopollenin. There is considerable cytoplasmic activity in the tapetal cells during the deposition of this material onto the lipid component and, while the character of the contents of the golgi bodies and vesicles is not known, the timing of their development and the situation of the vesicles released from the cell, strongly indicates their involvement in the secretion of sporopollenin. The discovery by Risueño *et al.* (1969) that small sporopollenin bodies are present within the tapetal cytoplasm at the terminal stages of its development lends support to such a suggestion.

Why the sporopollenin only condenses onto the lipid surfaces, rather than onto the tapetal plasma membrane is not clear, especially since it accretes upon membranous initials during pollen wall formation. Conceivably the membranes involved in exine development contain polymerising enzymes or surface characteristics that promote condensation of sporopollenin and, unless induced in such a manner, deposition will otherwise only occur upon the tapetally synthesised lipid.

The electron-opaque bodies observed in the mitochondria during the latter stages of tapetal fragmentation may be the structures that Heslop-Harrison (1963) identified with the sites of sporopollenin synthesis in his early work. Considering the state of the surrounding tissue at this time, it is perhaps more likely that these bodies result from the aggregation of mitochondrial constituents during degeneration of the organelles.

The peritapetal wall may play more than one role in microsporogenesis. While the lipid-sporopollenin composition of the wall in its early stages lends support to the suggestion of Heslop-Harrison (1969) that it forms an impermeable "culture-sac" in which the pollen grains develop, it is also evident that the removal of the lipid-layer in the later stages of microsporogenesis is a controlling factor in the desiccation of the microsporangium. The considerable cytoplasmic activity in the microsporangial wall cells during this period indicates that the agents responsible for the elimination of this layer originate from outside the loculus. The peritapetal wall provides a firm surface for the orbicules and mature grains in the desiccated microsporangium and thereby may assist in the dispersal of the pollen.

Little is known concerning the function of the orbicules. They have been regarded as aids to pollen dispersal (Heslop-Harrison and Dickinson, 1969) and even the result of overproduction of sporopollenin by the tapetum (Rowley 1963). Whatever the present role played by these bodies, their striking similarity both in formation and structure to the peritapetal wall points to the interesting possibility that the orbicules originally arose as by-products of development of this wall.

By the time formation of the peritapetal wall has commenced there is little evidence of any polarity in the tapetal cells. Massive channels are formed between protoplasts, vesicles are discharged over the whole cell surface and there is o stratification of the cytoplasm.

Thus to ensure the construction of the peritapetal wall on their outer faces, the tapetum would be compelled to discharge its various components over its entire surface. While those materials discharged towards the microsporangial wall cells would undoubtedly form the wall, those released over the radial and inward facing surfaces would not. If the lipid component is formed first, as evidence from other species indicates, it would presumable remain as droplets, dispersed in the gelatinous tapetal walls. During deposition of the second layer onto the peritapetal wall, the globuli would be coated with a thick layer of sporopollenin and become indistinguishable from the bodies now recognised as tapetal orbicules.

Acknowledgements

I am grateful to Professor P.R. Bell for his help and encouragement during this study, and to the S.R.C. for financial assistance.

REFERENCES

1. AVETISSIAN, N.M.: 1950. J. Bot. U.R.S.S. 35. p. 385-386
2. BANERJEE, U.C.: 1967. Grana. Palymol. 7. p. 365-377.
3. DE MARTINO, C., P.G. NATALI, C.B. BRUNI and L. ACCINI: 1968. Histochemie. 16. p. 350-360.
4. DICKINSON, H.G.: 1970. Cytobiologie B1. 4. p. 437-449.
5. DICKINSON, H.G. and J. HESLOP-HARRISON: 1968. Nature (Lond.) 213. p. 926-927.
6. ECHLIN, P. and H. GODWIN: 1968. J. Cell Sci. 3 p. 161-174.
7. ECHLIN, P. and H. GODWIN: 1969. J. Cell Sci. 5 p. 459-479.
8. ERDTMAN, G.: 1965. Pollen and Spore Morphology V. III. Almquist & Wiksell. Stockholm.
9. FEDER, N. and T.P. O'BRIEN: 1968. Amer. J. Bot. 55 (1) p. 123-142.
10. FRANKE, W., S. KRIEN and R. BROWN: 1969. Histochemie 19. 2 p. 162-165.
11. GODWIN, H., P. ECHLIN and B. CHAPMAN: 1967. Rev. Paleobot. and Palynol. 3. p. 181-195.
12. GOODWIN, T.W.: 1955. Modern Methods in Plant Analysis. V. 3. p. 272-311. Springer Verlag. (Berl.)
13. HESLOP HARRISON, J: 1963. Grana Palynol. 4. p. 7-24.
14. HESLOP HARRISON, J: 1968. Canad. J. Bot. 46. p. 1185-1192.
15. HESLOP HARRISON, J: 1969. Canad. J. Bot. 47. p. 541
16. HESLOP HARRISON, J.and H.G. DICKINSON: 1969. Planta (Berl.) 84. p. 199-214.

17. HESLOP HARRISON, J. and A. MACKENZIE: 1967. J. Cell Sci. 2. p. 387-400.

18. HORVAT, F.: 1965. Grana Palynol. 6. p. 416-434.

19. LEPOUSE,: 1970. C. Rend. Acad. Sci. (Paris) D. 270. p.2929.

20. MUHLETAHLER, K.: 1955. Planta (Berl.) 46. p. 1-13.

21. PETTIT, J.: 1966. Bull. Brit. Mus. (Nat.Hist.) Geology. 13 (4). p.223-257.

22. PY, G.: 1932. Rev. Gen. Bot. 44. p.316-73.

23. REYNOLDS, E.S.: 1963. J. Cell Biol. 17. p.208-212.

24. RISUENO, M., G. GIMINEZ-MARTIN, J.F. LOPEZ-SAEZ and R.GARCIA: 1969. Protoplasma 67. p.361- 74.

25. ROWLEY, J.R.: 1963. Grana Palynol. 4. p.25-36.

26. ROWLEY, J.R. and D. SOUTHWORTH: 1967. Nature (LOND) 213. p.703-04.

27. SCHNARF, K.: 1923. Öst. Bot. 2. 72. p.242-245.

28. SKVARLA, J. and C.W. LARSON: 1966. Am. J. Bot. 53. p.1112-1125.

29. TSING, W. and C. TSENG: 1965. Pollen et spores. 7 (1) p.10.

30. ÜBISCH, G.: 1927. Planta (Berl.) 3. p.490-495.

31. UENO, J.: 1958. J. Inst. Polytech. Osaka Cy. Univ. 9. p.163-186.

32. VAZART: 1970. C. Rend. Acad. Sci. (Paris) D. 270. p.3210-3212.

Legends

Figures 1-6

Fig. 1. Small embayments in the microspore protoplasts (arrows). Note the cytoplasmic vesicles and Golgi body (G).

Fig. 2. Cytoplasmic vesicles (arrows), produced by a Golgi body (G), merging with the plasma membrane.

Fig. 3. Initials of the sexine emerging at the junctions between vesicles and the plasma membrane (arrows).

Fig. 4. Golgi bodies (G) and their associated vesicles (arrows) under the developing sexine (S).

Fig. 5. Sexine initials during the early stages of cytoplasmic withdrawal. Note remains of the walls of the vesicles involved (arrows).

Fig. 6. Thickening of early sexine (arrows) prior to development of the nexine. Fibrillar material may be seen between the sexine and the protoplast.

FIGURES 1 - 6

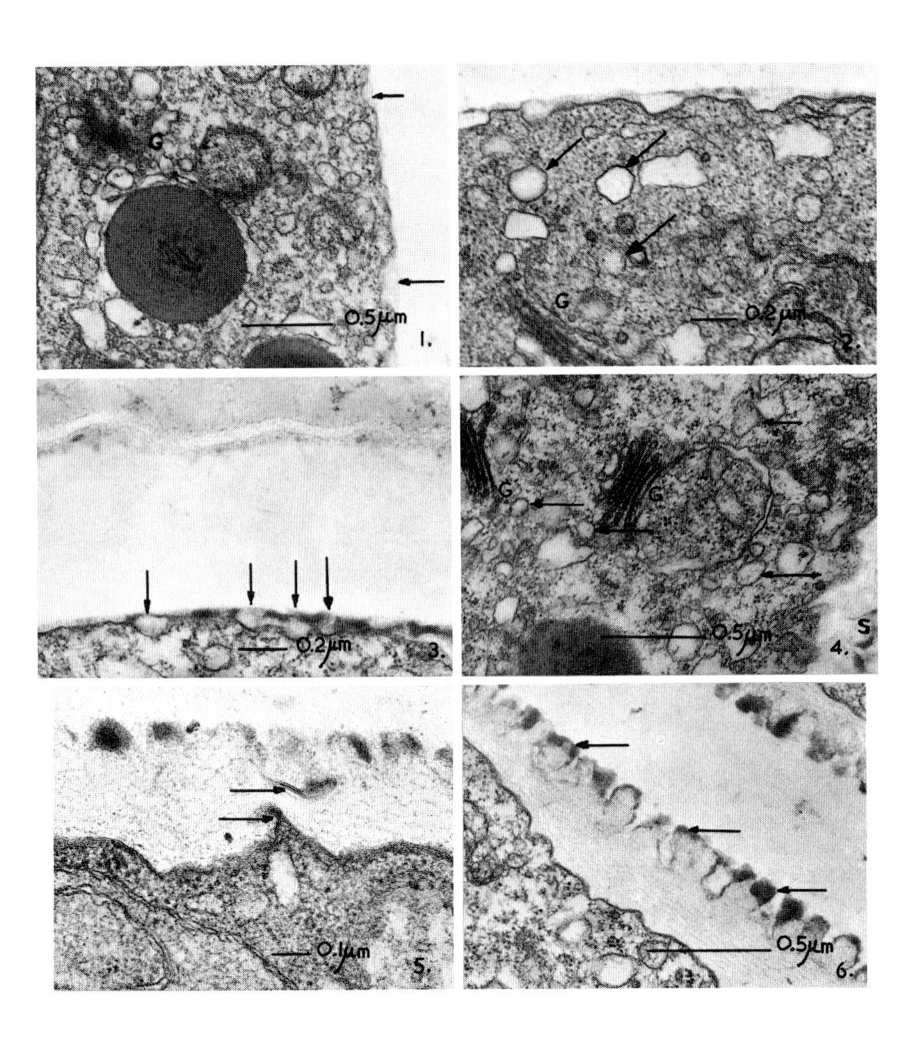

Legends

Figures 7-12

Fig. 7. Early stages in development of the saccus (S). An active Golgi body is visible in the subjacent cytoplasm.

Fig. 8. Detail of early saccus. Note the fibrillar content and membrane-bound vesicle (arrow).

Fig. 9. Material, at the same stage as shown in Figs. 7 and 8, after treatment with PAS. Intense staining occurs under the young sacci (arrows).

Fig. 10. Elements of the nexine (arrows) forming between the saccus (S) and the protoplast (P).

Fig. 11. Limit of microspore development within the tetrad. The sacci do not significantly distort the callose wall. (Phase Contrast).

Fig. 12. Material, at the same stage as shown in Fig. 11, after treatment with PAS. The contents of the saccus stain (S), as do the sexine and nexine (Xi and Xii).

FIGURES 7 - 12

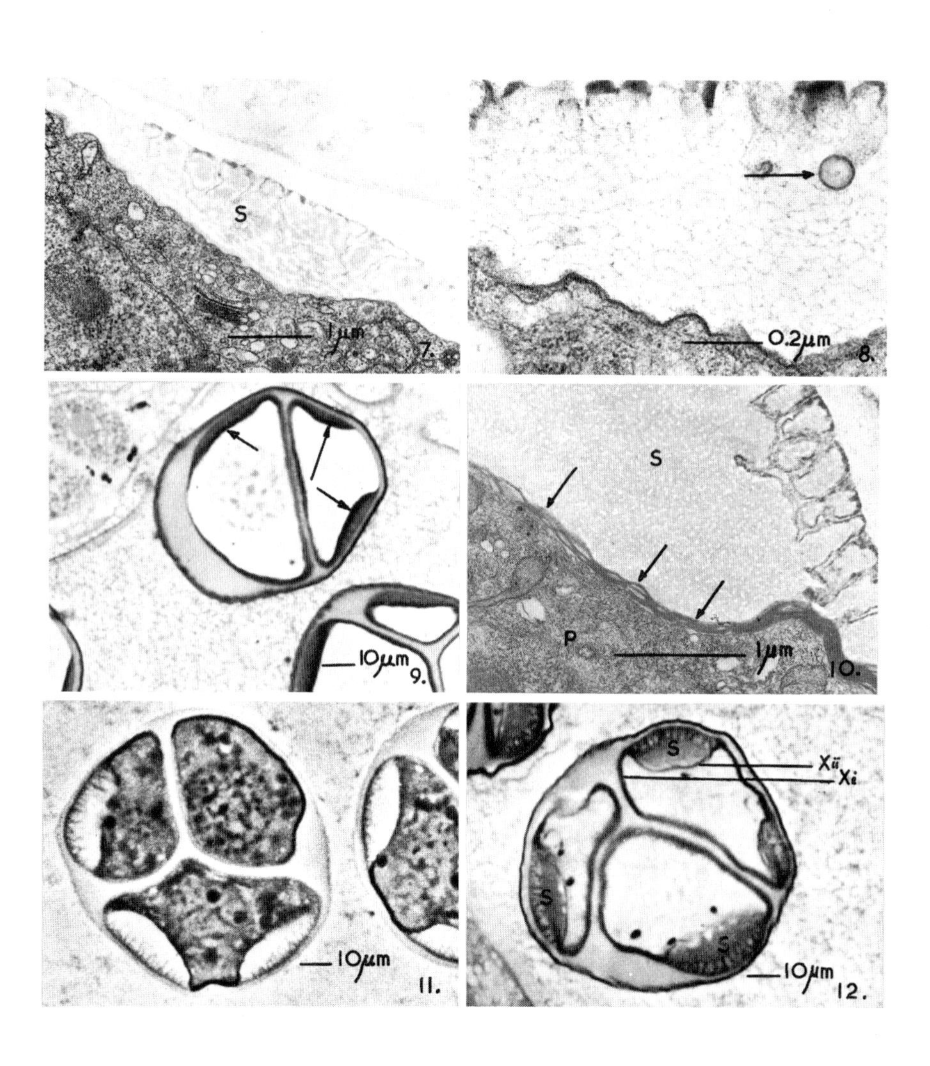

Legends

Figures 13-18

Fig. 13. Thickening of plasma membrane prior to nexine formation (arrows).

Fig. 14. The appearance, in small regions, of a second membrane under the plasma membrane (arrows).

Fig. 15. Separation of element of the nexine from the plasma membrane (arrow).

Fig. 16. Sexine (S) and nexine (N) layers following release of the microspore from the tetrad.

Fig. 17. Material, at stage also shown in Fig. 16, after treatment with PAS. Note the reaction of the sexine and nexine layers.

Fig. 18. Young pollen grain undergoing first mitosis. PAS. The sexine and nexine no longer react.

FIGURES 13 - 18

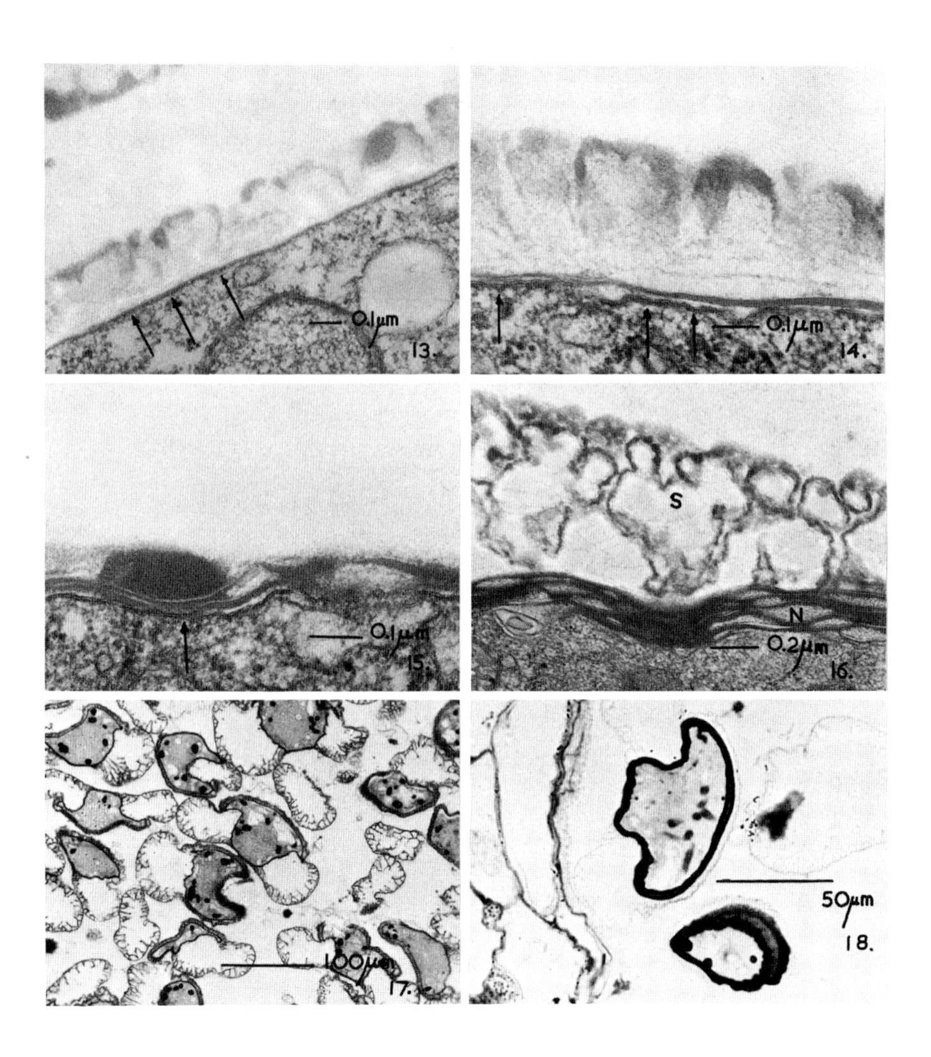

Legends

Figures 19-25

Fig. 19. Mature pollen wall showing sexine (S), nexine (N) and intine (I).

Fig. 20. Tapetal cells immediately prior to disintegration. Note the orbicules in the radial and inward-facing walls, nuclear fragmentation, channels between cells (C) and the peritapetal wall (arrows).

Fig. 21. Tapetal cytoplasm during sporopollenin synthesis. An active Golgi body is present (G), as are cisternae of the endoplasmic reticulum (ER).

Fig. 22. Deposition of sporopollenin coat upon young orbicules (arrows). Note the small vesicles produced by the tapetum (V) and electron-opaque material dispersed in the gelatinous walls.

Fig. 23. Vesicles at the interface between the peritapetal wall (arrows) and the tapetum.

Fig. 24. Orbicules in the radial and inward-facing walls of the tapetum. Small vesicles (arrows) are released from the protoplasts. The gelatinous walls contain dispersed electron-opaque material

Fig. 25. Mitochondria (M) within disintegrating tapetal cell. The electron-opaque bodies are clearly shown (arrows).

FIGURES 19 - 25

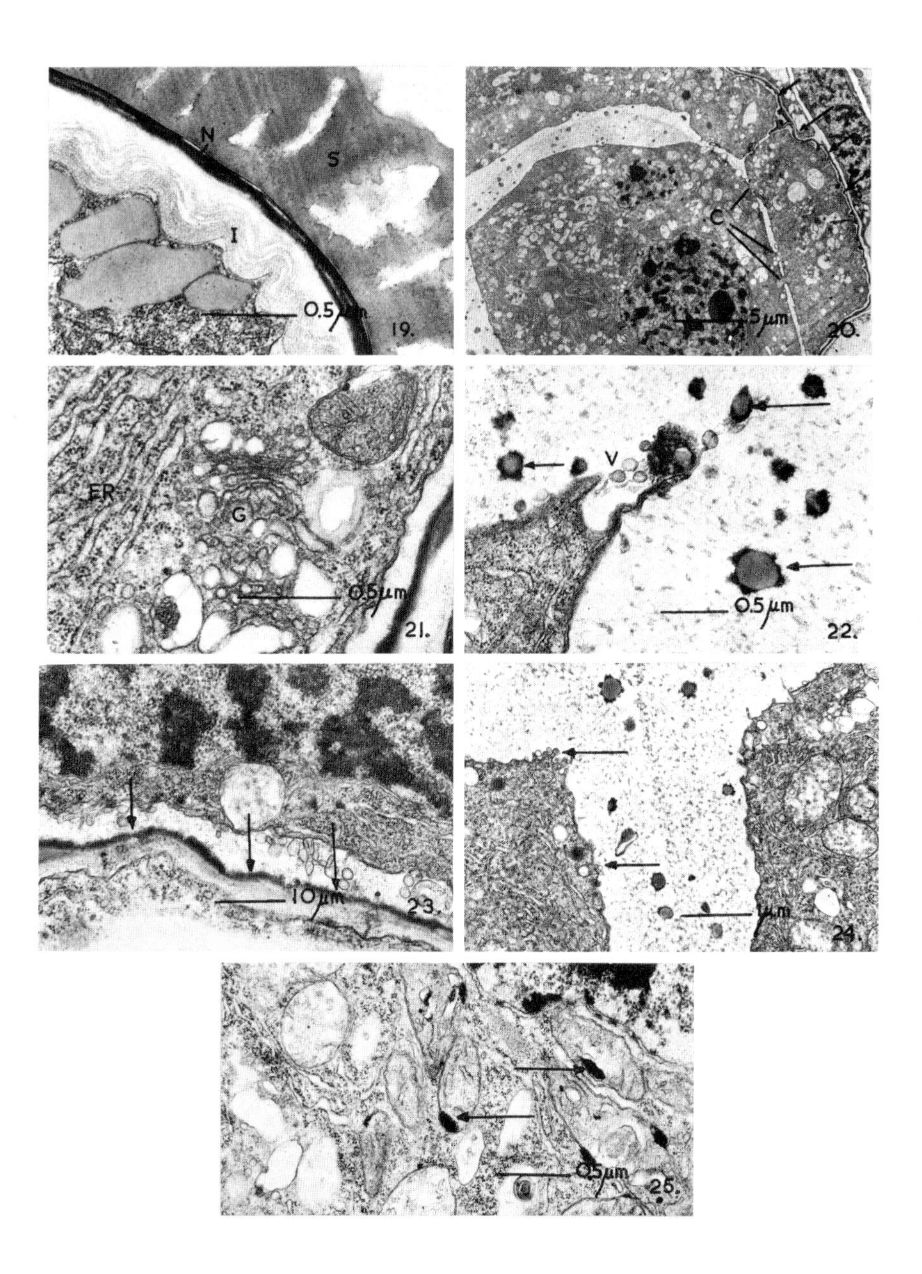

Legends

Figures 26-31

Fig. 26. Early stage of tapetal disintegration. Note the vesicular cytoplasm of the tapetal cell (T), the elimination in some regions of the lipid layer of the peritapetal wall (arrows) and the cytoplasmic activity of the microsporangial wall cell (M).

Fig. 27. Desiccated microsporangia. On either side of the dividing wall (W) orbicules can be seen (arrows) on the peritapetal wall. On one side (S) the orbicules support a pollen grain saccus.

Fig. 28. Sudan III stained section, showing microspore tetrads (T) and the peritapetal wall. Tapetal cells (C).

Fig. 29. Sudan III stained section, showing tetrads (T) and tapetal orbicules (arrows). Tapetal cells (C).

Fig. 30. Peritapetal wall acetolysed during the tetrad stage.

Fig. 31. Acetolysed peritapetal wall and orbicules (arrows), from desiccated microsporangium.

FIGURES 26 - 31

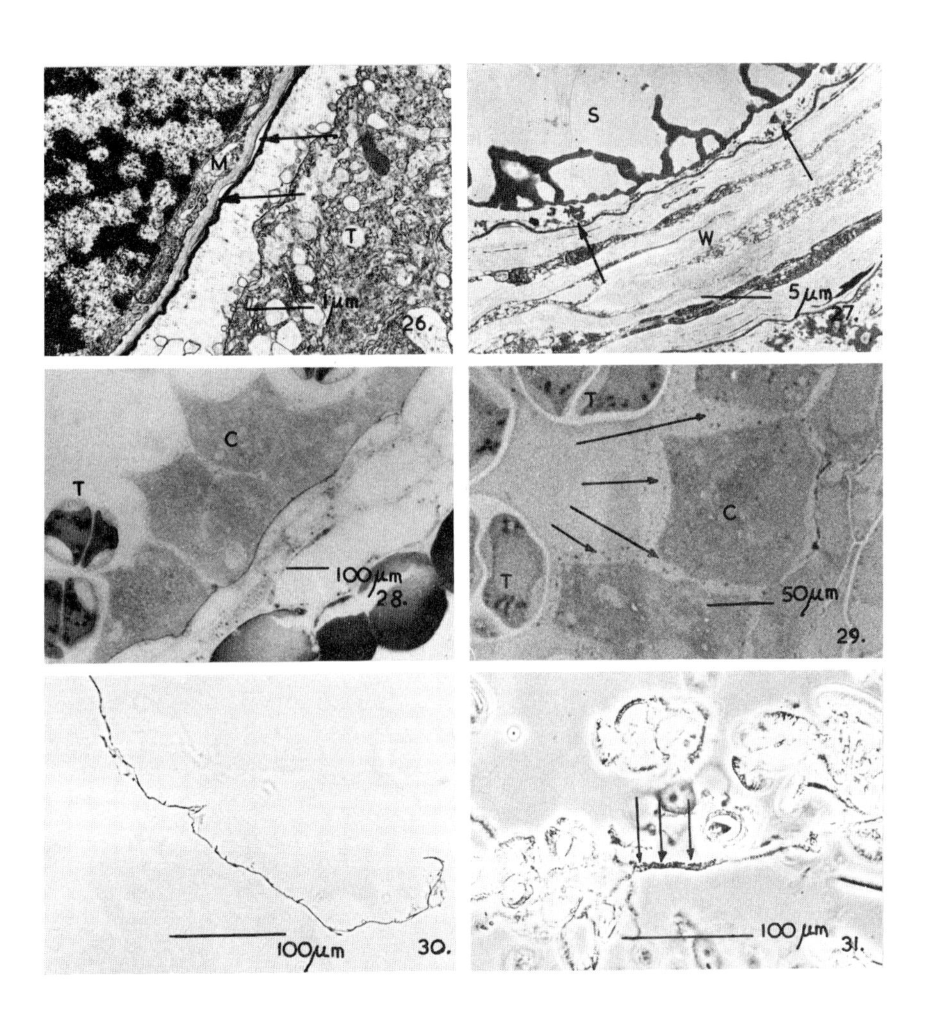

DISCUSSION ON PROFESSOR J. HESLOP-HARRISON and DR. H.G. DICKINSON'S PAPERS.

WILLEMSE: Dr. Dickinson: after the microspores break out from the callose walls, the tectum, baculae and footlayer all thicken considerably. Is it possible that this sporopollenin is coming from the tapetal cells?

DICKINSON: Yes. I would agree with that entirely. Once the spore is released from the tetrad, there is a considerable increase in size, caused presumably by accretion.

SIEVERS: Do I understand that Dr. Dickinson found vesicles which are extruded from the cytoplasm through the plasmamembrane and which are found as membrane-bound vesicles outside the plasmamembrane?

DICKINSON: Yes. There is an extra cellular lumen and there we see circular vesicles of the same dimensions and the same contents as are seen within the membrane, but undoubtedly outside the membrane.

SIEVERS: Then can you explain how the vesicles were extruded through the membrane?

DICKINSON: I just don't know, but would very much like to know. I do not understand how they can pass through a single membrane and leave it unstrained - it would probably be much easier to explain if there were two membranes.

HESLOP-HARRISON: We can see such vesicles in many examples, and we can see that they must come out of the cell. Therefore there must be some mechanism for discharging them. It could be that the process itself is an instantaneous one, something that we could never hope to catch with the electron microscope. Perhaps when the membrane of the vesicle contacts the membrane of the cell, the latter has its stability altered - it would be wrong to think of things passing through - and it is then transformed into a new stable level.

SIEVERS: If these extracellular structures are real structures, I could imagine that a vesicle or vesicle membrane is incorporated in the plasmalemma, and then 'drives off' a new vesicle.

HESLOP-HARRISON: This is well worth examining. I think part of our difficulty arises because we try to think of these structures as balloons passing through a solid wall. What we have, in fact, is a system with various parastable states, any one of which can be assumed, and presumably all that happens, is that we get a reduction of free energy, and whichever way this goes, the final product must be stable in order for us to see it with the electron microscope.

SHAW: I think that we must here be careful with terminology. Dr. Dickinson mentioned that he probably has some hemi-cellulose in the sporopollenin of *Pinus*. What he means is that the sporopollenin may have been associated with hemi-cellulose. Sporopollenin was originally chemically defined by Zetzsche and until we have a better definition, we must stick by this. Professor Heslop-Harrison also mentioned the possible relationship between cutin and sporopollenin: here again there seems to be a conflict between biologists and chemists who have rigid definitions of these substances. Cutin is readily hydrolysed, while sporopollenin is not.

LINSKENS: Does Professor Heslop-Harrison believe that sporopollenin polymerisation is enzymatic or non-enzymatic? And is there any way of synthesising sporopollenin in cell-free systems?

HESLOP-HARRISON: No. I do not think that enzymes have any part in the final assembly of sporopollenin (in the chemists' strict sense). This is also true of cellulose. Work by Ben-Hayyim and Ohad (*J. Cell Biol.* 25, 191-207, 1965) on the formation of cellulose by *Acetobacter* in culture showed that a diffusible "prefibrous" form of cellulose leaves the cell, and that the cellulose microfibrils arise by polymerisation remote from the cell by a process not involving enzymes. A further question was asked about orbicules: sporopollenin synthesis can take place at several sites inside the loculus, and sporopollenin precursors seem to be present in the tapetal fluid. This being so, why does not the sporopollenin synthesis take place everywhere? There must obviously be some surfaces which attract the deposition of the sporopollenin monomers and some which resist deposition. Surfaces unsuitable for deposition apparently include cellulose, hemicellulose and callose. The tapetum surfaces are mostly protected by polysaccharide. However, we have found that after centrifugation tapetal cells sometimes break up, and some of their contents are released from the polysaccharide matrix, and ball up to form simple spherical shapes (these have the smallest surface area). They now become suitable for the deposition of sporopollenin. This seems to be the origin of the unpatterned spherules I illustrated in my paper (see Heslop-Harrison, *S.E.B. Symposium* No. 24, in press). Where fragments of the meiocyte gather up to a spherical shape after disruption, patterned spherules are formed. It is important to distinguish between two quite different things: 1) the deposition of sporopollenin which can occur on any suitable surface; and 2) the capacity to produce a *patterned* distribution of sporopollenin, which is a function of the meiocyte and its products.

LINSKENS: Is the formation of sporopollenin aerobic?

HESLOP-HARRISON: Yes. Shaw's work clearly indicated that the polymerisation is oxidative.

MORPHOLOGICAL AND FLUORESCENCE MICROSCOPICAL INVESTIGATION ON SPOROPOLLENIN FORMATION AT PINUS SYLVESTRIS AND GASTERIA VERRUCOSA.

M. T. M. WILLEMSE, Botanical Laboratory, the Catholic University, NIJMEGEN, Driehuizerweg 200, The Netherlands.

Abstract.

Pollen wall-formation in *Pinus sylvestris* begins during the formation of the callose wall in the tetrad. A great number of Golgi vesicles, with a heterogeneous content, transport the primesexine outside the cell. At the same time the callose wall-formation is locally blocked at the places where the Golgi vesicles come out of the cytoplasm. Where the contact between plasmamembrane and callose wall persists, callose wall-formation continues. In this way, a template for the coming pollen wall pattern is determined inside the callose wall. Material for the tectum and bacules, the sexine, is derived from the Golgi vesicles. Material for the footlayer, the nexine 1 , is produced in the microspore and deposited mainly on the plasmamembrane.

The fluorescence features of the exine during microsporogenesis have been investigated on fresh pollen of *Pinus sylvestris* and *Gasteria verrucosa* . These pollen show an increase of intensity and a shift in the maximum of the single peak spectrum, probably caused by intine formation. The intensity and the spectral maxima of both pollen types do not have any agreement.

Introduction.

The great variety of pollen wall patterns in Gymnosperm and Angiosperm pollen is reflected in numerous studies on pollen wall-formation and on the nomenclature of pollen wall stratification. When on Pinus sylvestris pollen a further different mode of pollen wall-formation is shown, it is necessary to restrict our results to Pinus sylvestris.

In this study the pollen wall is divided into exine and intine. The exine is subdivided into sexine and nexine. The sexine includes the tectum and bacules. The nexine is subdivided into nexine 1 or footlayer, and nexine 2.

Material and methods.

For electron microscopy, microsporophylls of Pinus sylvestris were fixed for one hour in 1% OsO_4 in phosphate buffer pH 7,2 at $0^{\circ}C$. After washing they were stained for 15 minutes with 1% uranyl acetate in water, followed by another washing and then put for 15 minutes into 1% $KMnO_4$ in water. The specimens were embedded in Epon 812 and thin sections were stained with lead citrate. The material was examined by means of the Philips electron microscope EM 300.

Fresh material of Gasteria verrucosa and fresh material, stored for a long time at - 10° C, of Pinus sylvestris was squashed in water and used for microspectrophotometry. A Leitz microspectrophotometer and a photomultiplier with a Knott apparatus were used.

The objects were exposed to light with a wavelength of mainly 365 nm (UV-fluorescence).

Four characteristics of primary fluorescence were measured. First, the fluorescence spectrum from 415 to 625 nm was measured during a time of 90 seconds. All treated pollen showed a single peak spectrum. This peak was reached after about 30 seconds and has been recorded simultaneously. During ultra violet exposure a rapid fading of the fluorescence of the pollen walls occurs, even in the first 30 seconds, which can be measured in a decrease in fluorescence intensity and a change in the fluorescence spectrum (see also VAN GIJZEL, elsewhere in this Volume). This fading has been determined during the initial 30 seconds and its amount is given in percentages. Thirdly, the fluorescence intensity after 30 seconds at a fixed kilovoltage was measured and expressed in arbitary units. Finally, the spectral maximum reached after 6 seconds was determined. The average value and deviation were calculated from the results.

TABLE 1

	Megaspore wall:		Pollen wall:	
	Pinus sylvestris	*Pinus mugo*	*Pinus sylvestris*	*Pinus mugo*
Spectral maximum after 30 seconds:	478 ± 2	482 ± 2	486 ± 2	495 ± 1
% Fading after 30 seconds:	35 ± 4	27 ± 5	34 ± 9	34 ± 3

Morphological development of Pinus sylvestris pollen.

The early tetrad stage.

The early tetrad cell is surrounded by a thin electron dense wall, which is formed during the early leptotene stage between the plasma-membrane and the cellulose wall of the pollen mother-cell. Under it lies the thick and very electron transparent callose wall. The formation of this callose wall starts during the diplotene stage and becomes increasingly thicker until the early tetrad. The cell has four nuclei which are arranged opposite to each other near the plasmamembrane (fig. 1).

In the cytoplasm are plastids with a starch granule and without starch, many globular and baculiform mitochondria and lipid granules, strongly electron dense rounded organelles, sometimes surrounded with dark dots only visible after staining with lead, or sometimes connected with an electron transparent vesicle. Furthermore some Golgi bodies with little vesicles can be observed, the contents of the Golgi vesicles are electron transparent. Very little rough endoplasmatic reticulum (RER) is found, but many cisternae of the smooth endoplasmatic reticulum (SER) are present. In the cytoplasm between these organelles ribosomes appear as very little dark dots, polysomes are recognizable as groups of ribosomes. Occasionally microtubuli can be observed, oriented at right angles to the callose wall or the coming new cell plate. The first signs of cell plate formation are visible in the center of the cell as a dark lined structure composed of cisternae of the SER (fig.2).

The callose wall-formation.

During the beginning of cell plate formation the cytoplasm shows cisternae of SER and globular particles, sometimes with ribosomes attached to the membrane, which are probably derived from RER. They are situated near the nucleus on the side of the cell centre as well as near the existing callose wall (fig. 3 and 4). The globular particles with ribosomes lose their ribosomes and will turn into elements of SER.

The first indication of the cell plate is a line composed of elongated cisternae of SER. From these flat cisternae a more electron dense double membrane is formed, which surrounds a more electron transparent dot. The increase in size of such vesicles occurs by fusion with other cisternae of the SER.(fig. 5).

The cell plate becomes visible as a line of vesicles with electron transparent contents surrounded by a clear double membrane. Between these vesicles groups of microtubules are visible (fig 6).

After the growth of each vesicle the presence of callose is firstly indicated as a very light spot in the centre of the vesicle, which is surrounded by the plasmamembrane. Elements of SER remain in the vicinity of the future cell plate (fig.7).

Also the callose wall-formation takes place outside the plasmamembrane. The material for the callose wall, derived from SER cisternae passes the plasmamembrane. The callose wall grows centripetally by fusion of the callose containing cell plate vesicles. Golgi bodies are rather small and produce a small number of little Golgi vesicles. Possibly they may be involved also in callose wall formation (fig 8).

The tetrad.

The formation of the pollen wall starts during the tetrad stage. A thick callose wall has divided the cell into four daughter cells. The nucleus of each cell lies excentrically near the outercallose wall (fig 9). In the cytoplasm a remarkable number of Golgi vesicles with a heterogeneous structure is present. The Golgi vesicles are scattered throughout the entire cytoplasm (fig 10). On the contrary few vesicles are found in the cytoplasmic strip between nucleus and the exterior callose wall (fig 11).

The Golgi body.

The Golgi vesicles are produced by a fairly large Golgi body. The vesicles at the end of the membranes have a unit membrane and heterogeneous contents. The centre has electron dense granular and fibrillar material which is connected with thin fibrils and the membrane; an electron transparent zone surrounds this completely (fig 12). These contents of the Golgi vesicles are visible in the space between the plasmamembrane and the callose wall. The vesicles are also transported outside of the cell (fig. 13).

During excretion, the membrane of a Golgi vesicle makes contact with the plasmamembrane and the contents of the vesicle pour into the space between plasmamembrane and callose wall (Fig. 14 and 15). At the site where a vesicle is excreted, the plasmamembrane is locally pushed away from the callose wall. This separation will become more general, as more Golgi vesicles are **excreted,** although the contact between the callose wall and the plasmamembrane will be maintained at certain sites (fig. 16). In the cytoplasm as well as on the Golgi body itself, Golgi vesicles can fuse to form bigger vesicles (fig. 17).

The callose wall.

During the excretion of the Golgi vesicles, the callose wall undergoes some changes. Shortly after its formation and during its thickening, the plasmamembrane is situated against the straight callose wall (fig. 18). When the contents of the Golgi vesicles appear between plasmamembrane and callose wall, the plasmamembrane is no longer in contact with the callose wall. As more vesicles are poure out, the contact between the callose wall and the plasmamembrane becom looser. In places where the contact between these two structures persists, the plasmamembrane makes a cone-like fold (fig. 19 and 20) The inner side of the callose wall has lost its eveness; especially at the connection of the plasmamembrane with the callose wall, the latter develops a protrusion into the direction of the cell (fig. 21 and 16).

This means, that where Golgi vesicles are excreted and the plasmamembrane has lost contact with the callose wall, the formation of the callose wall is also blocked. Where contact between plasmamembrane and callose wall remains intact, the callose wall-formation will continue. During the excretion of material for the pollen wall , the production of material for the formation of the callose wall continues, but only there where the contact with the plasmamembrane persists. Thus protrusions are formed on the inner side of the callose wall as a template.

When the excretion of Golgi vesicles starts, there are several possibilities: a contact between callose wall and plasmamembrane persists either on both sides of one vesicle, or between two or more vesicles. If all the Golgi vesicles are juxtaposed, a very regular template in the callose wall may be formed.

Elements of SER are continuously present during this stage. The material for the callose wall is also excreted on the plasmamembrane, besides contact between callose wall and plasmamembrane is necessary.

The template is used immediately: where the callose wall-formation is blocked, material derived from the Golgi vesicles will precipitate there. This is the first appearance of the tectum (fig. 22). Thereupon material will precipitate onto the protrusions, thus resulting in the construction of the bacules (fig. 23).

The material for the tectum and the bacules originate from Golgi vesicles and more particularly from the central part which is composed of thin granular and fibrillar material. Therefore it may be called primesexine. The electron transparent content of the Golgi vesicles will remain in the extracellular space. The precipitation is a strictly gradual process. Meanwhile the granular and fibrillar Golgi material changes into a network structure.

Formation of the tectum and the bacules is a process connected with the products of Golgi vesicles in combination with callose wall-formati On the contrary the formation of the footlayer has another origin.

The late tetrad stage.

The cell of the late tetrad stage appears differently from the tetrad The nucleus is located now more centrally in the cell (fig. 24). The structure of the pollen wall is already visible. Also the wings become apparent. Most remarkable are the very electron dense thick tapes which are found outside the **cytoplasm** along the plasmamembrane, with long extensions going into the cytoplasm. The lipid granules are also very numerous. They are increasing in size. The cytoplasm contains many ribosomes and a few membranes of the endoplasmatic reticulum.

The origin and localisation of the footlayer.

The very homogeneous electron dense footlayer lies mainly parallel to the plasmamembrane, with long plate or tape like extensions coming out of the cytoplasm. Between the wings on the outside of the tetrad, where formation of tectum and bacules is lacking, a great number of these tapes can be observed (fig. 25).

The plasmamembrane runs parallel to the long tape penetrating deeply into the cytoplasm, but at the end this separation is not clear and there appears a junction between the plasmamembrane and the tape (fig

26 and 27). Where the footlayer runs parallel to the plasmamembrane, the formation of this material takes place on the plasmamembrane. In the cytoplasm sometimes small fragments of endoplasmatic reticulum are visible along the tapes (fig. 26 and 27). Here and there groups of bent, very thin electron dense membranes appear (fig. 28). Occasionally at the terminal portion a connection between lipid granules and the tape becomes visible (fig. 26).

On the outside, parallel to the plasmamembrane, the footlayer appears also as a strongly electron dense line. This footlayer is not deposited simultaneously over its entire length. Its thickness may also vary (fig. 29). With the appearance of the footlayer, the last connections between plasmamembrane and callose wall are broken and the callose wall-production is arrested. The structure of the pollen wall has become more accentuated. In electron transparent material the tectum and the bacules become visible, both intermingled with the network of primesexine. The footlayer is also connected with this network (fig.30).

Formation of the footlayer may be a function of membranes, either the plasmamembrane or the endoplasmatic reticulum . The big lipid granules may be a source of a precursor product. This origin of the footlayer on membranes is also shown by the long electron transparent tapes within the very thick layer, which are a remnant of the space between the two membranes of the endoplasmatic reticulum (fig. 28). The membrane origin of the footlayer differs also from the origin of tectum and bacules, which has been derived on their turn from Golgi bodies. The Golgi bodies disappear during the late tetrad (fig. 26).

The formation of the wings.

The construction of the pollen wall of *Pinus* shows local differences Two wings are formed. Tectum and bacules are lacking between the wings on the outside of the tetrad . This absence of tectum and bacules is a result of a small number of Golgi vesicles. On this side the position of the nucleus near the plasmamembrane leaves only a very thin layer of cytoplasm, thus reducing chances that there may be many Golgi vesicles.

A different situation exists on the sides where the wings are formed Here many Golgi vesicles are excreted. This preferential position may be related to the close vicinity of tapetal fluid and the position of the nucleus in the cell.

At the level of the wings the process of pollen wall-formation can be followed easily. Where the future site of the wings is situated, many Golgi vesicles are excreted in a very early stage. The cytoplasm is pushed back and the plasmamembrane has deeply bent invagi nations. The extension of the plasmamembrane is possibly due to uptake of membranes from the Golgi vesicles. Where connections of the plasmamembrane with the callose wall can be observed, it will remain intact. Long protrusions on the callose wall are formed (fig. 31). A tangential section shows the rounded protrusions. The pattern of the developing pollen wall is also distinguishable by the locali- sation of the protrusions, visible as light dots (fig. 32). At the onset the electron dense contents of the Golgi vesicles are distinguishable as granular and fibrillar materials. If much material is being exorcted, it becomes fine grained, while between the protrusions, during precipitation of the material of the tectum, a fine network of fibrillar material is formed (fig. 33). Finally, during the late tetrad stage, when tectum and bacules are clearly visible, all the material turns into a fine network (fig. 34). Tectum and bacules are formed by condensation of the material from the Golgi vesicles, probably in the fibrillar form. The switch from a granular and fibrillar form to a network is probably due to polymerisation. This network or the polymerized material condenses also against the callose wall protrusions in the tectum and the bacules (fig. 34). The footlayer is in contact with the fine network and it may be possible that some of this material takes part in the construction of the footlayer.

After the pollen wall has acquired its first configuration, the two accompanying walls, the thin cell wall and the callose wall are dissolved by substances present in the surrounding fluid. The site of breakdown is above the wings. A local dissolution is sufficient to let the microspore come out (fig. 33 and 34).

The tapetal cell.

Pollen wall development is very closely connected with the surrounding tapetal cells. For many Angiosperm species the tapetal cells are supposed to produce the sporopollenin. In Pinus sylvestris the tapetal cells excrete also electron dense substances during the formation of the pollen wall. During late telophase II little electron dense droplets are still visible outside the plasmamembrane of the cell.

The tapetal cell is surrounded by a thin cell wall and contains one or more nuclei. The cytoplasm has a great electron density, which is caused by a great number of ribosomes. Several organelles such as plastids with and without starch, mitochondria, vesicles, several Golgi bodies and endoplasmatic reticulum are present. During the late tetrad stage the little dark granules can be seen, next to globular structures with some electron dense material on the rim outside the ce: But they remain within a very thin dissolving cell wall.

There is also production of pollen wall like material, but no indication has been found, that it must considered to be the source of sporopollenin during the tetrad stage.

Survey of pollen wall-formation.

A survey of the pollen wall-formation is presented in the diagram of figure 35.

During the early tetrad stage callose wall-formation takes place. Fusion of predominantly SER cisternae gives rise to the formation of the plasmamembrane and they contain material for the callose wall.

Callose wall production in connection with SER is also reported by ANGOLD (1966) in Endymion. Such a production connected with endoplasmatic reticulum during formation of the sieve plate pores has been reported by ESAU, CHEADLE and RISLEY (1962).

During the tetrad stage a production of heterogeneous Golgi vesicles and an excretion of the contents of these vesicles occurs. The vesicles contain an electron transparent material, of which the nature is unknown, probably enzymes for polymerization and carbohydrates. The center of the Golgi vesicles contains the primesexine. At the same time callose wall production takes place. Where the plasmamembrane is in contact with the callose wall, the callose wall-formation remains and protrusions as template are formed in the callose wall. Tectum and bacules are constructed in this stage.

Golgi vesicles with heterogeneous contents have also been observed during cell wall-formation of roothairs of Zea mais by SIEVERS (1963). In pictures of pollen wall development of various Angiosperms they are also visible during formation of the primexine. The pollen wall pattern reproduced in the callose wall has been reported in Ipomoea by WATERKEYN and BIENFAIT (1968) and WATERKEYN (1970) .

During the late tetrad stage the material for the footlayer or nexine 1, is produced on membranes, coming from the microspore. Lipid granules possibly contain the precursors for the footlayer.

This production on membranes is also assumed in Podocarpus by VASIL and ALDRICH (1970).

The electron transparent line between the thick tapes has been observed in many organisms: during pollen wall-formation in

Ipomoea by GODWIN, ECHLIN and CHAPMAN (1967), in Tradescantia by MEPHAM and LANE (1969), in Allium by RISUENO, GIMENEZ-MARTIN, LOPEZ-SAEZ and RISUENO-GARCIA (1969) and also in oilglands of Arctium by SCHNEPF (1968) as well as in cell walls containing cork by SITTE (1962).

In the pine Pinus sylvestris , tectum and bacules have another origin as the footlayer. This means that the difference between sexine and nexine 1 is not determined only by morphological differences. Also the nexine 2 has the same origin as the nexine 1, but it appears first in the young microspore.

The young microspore.

After emergence, the cell increases its volume by vacuolisation. In the beginning the nucleus is located near the cell base , but proir to the first mitosis its position is in the center of the cell. The wings formed by the sexine, have very electron transparent conten (fig.36).

However, the pollen wall-formation is not finished now. First there is a thickening of tectum, bacules and footlayer (fig. 37). Thereafter the nexine 2 as well as thin electron dense tapes appear. When the cell nucleus lies in the center of the cell, the nexine production stops and the formation of intine starts (fig. 38).

The ripe pollen.

The ripe pollen of Pinus sylvestris contains a generative cell, two degenerated prothallial cells and a vegetative cell.

The intine is remarkable thick, especially above the generative cell. The morphological structure of this intine is complex: starting from the cell membrane above the generative cell, there are two lamellar layers separated by a more granular one. On the outer lamellar layer is a series of electron dense lobes, oriented toward the periphery, which is rich in granular material.

The chemical composition of these pollen wall, exine and intine, as reported by SHAW and YEADON (1966) is approximatly as follows: 10-15% cellulose, 10% hemicellulose, 10-15% lignine like substances and a lipid fraction of 55-65%, which probably consists of carotenoids and carotenoid esters, according to BROOKS and SHAW (1968). WATERKEYN (1964) and MARTENS, WATERKEYN and HUYSKENS (1967) described the presence of callose in the intine, laying mainly in the outer intine, and of pecto-cellulosic materials in the inner intine.

All these components may be responsible for the pollen wall fluorescence.

Pollen wall fluorescence.

The morphological changes during pollen wall development have been compared with the results of primary fluorescence determinations of the pollen wall.

During the development a difference in fluorescence in pollen occurs. In particular the pollen wall shows a very distinct primary fluorescence contrary to the fluorescence of the cytoplasm,which is very weak. The fluorescence of Pinus sylvestris pollen is yellowish during the tetrad stage and bluish when they are mature. The tetrad and the wings have a small intensity. The fluorescence colour in the tapetal cell is the same as in the microspores.

Measurements.

The fluorescence in four stages of development of Pinus sylvestris pollen has been determined, in the stages of: the tetrad, the young microspore with one nucleus, the microspore with prothallial cells and generative cell before drying out, and complete ripe and dried pollen. The results are given in figure 39.

During the development after the tetrad stage a shift in the spectral maximum occurs to bluish green around 490 nm. During the tetrad stage, Pinus sylvestris pollen have the same spectral maximum at 500 nm after 6 or 30 seconds. When in the following stages of development both maxima after 6 and 30 seconds are measured, the values differ. Statistical analysis has been made by means of the t-test. With the 5% level of significance it shows that, the values of the measurements after 6 and 30 seconds at these different stages of development give significant differences.

The values of the fluorescence maximum of the tapetal cell and the wing are situated both in the region of values for microspores. The fluorescence intensity of the microspore increases after the tetrad stage. Then the intensity of the tapetal cell and the wing is comparable to the intensity of the microspore. Between developing microspores and the wing there is only a minor difference in fading . The fading of the tapetal cell is very low. The fading of the tetrad is lower than in the following stages of development.

During exposure an immediate decrease in intensity and a shift in the maximum during a few seconds occurs. During microspore development there is also a shift of the maximum to the blue. The physical and chemical factors responsible for these changes in fading and the shift of the maximum, will not be discussed here. From the morphological point of view, the changes in the fluorescence maximum during microspore development are mainly caused by the changes in the composition of the pollen wall. During the tetrad stage, cellulosic or hemicellulosic substances are absent and the spectral maximum is determined by components of the sexine and nexine 1. After the tetrad stage the production of components for the intine starts, mainly with cellulosic and hemicellulosic substances. The callose appears somewhat later and probably causes a further decrease in wavelength of the fluorescence maximum. The increase in intensity and fading,after 30 seconds, may be caused also by the formation of the intine.

Chemical treatments.

The pollen were treated according to the method used by SHAW and YEADON (1966). The results of fluorescence determinations are given in figure 40.

During treatment with ether, ethanol and boiling water, but before treatment with potassium hydroxide, neither fresh ripe *Pinus pinaster* pollen nor fresh ripe *Pinus sylvestris* pollen show any change in wavelength. Only after 6% potassium hydroxide treatment there is a shift in the spectral maximum to the green after 30 seconds, reaching about the same value at ± 500 nm as found during the tetrad stage. During saponification some of the cellulosic and hemicellulosic and probably also callosic substances are destroyed. The potassium hydroxide solution after the treatment used in the experiments, shows a weak fluorescence by itself.

What is obtained after this treatment, shows a very close resemblance to the composition of the tetrad wall,as a result of the removal of some intine substances.

Cleaning of the *Pinus pinaster* pollen by rinsing with water proceeds more slowly than for *Pinus sylvestris*, due to the homogenisation of the pollen of the latter species before the treatment. Extraction of lipids during subsequent treatments does not result in any change. By uncertain reasons, after saponification the intensity of the fluorescence also increases.

Measurements of Gasteria microspores.

Not all pollen fluorescence analysis give the same results. Measurements on Gasteria verrucosa pollen, a membrer of the family of Liliaceae, show some differences in fluorescence with Pinus sylvestris pollen.

On Gasteria an easier description of the five different stages of development can be made: first the tetrad stage (1), then the young microspore without food reserves (2), followed by the microspore with big granules of food reserve (3), next comes the microspore with two round nuclei (4) and finally the ripe pollen with the elongated generative nucleus (5).

The morphology of the pollen wall during stage 1,3 and 5 shows the following differences (fig. 41): The developing pollen wall, inside the tetrad, has an electron dense band between the bacules. The young microspore has a pollen wall which **contains** some electron dense material between the bacules. A thick cap of intine lies on one side. The ripe pollen has the same structure, the only difference is a very thin layer of intine which now surrounds the whole plasmamembrane. The thick intine cap remains in the colporal region. On the pollen wall some pollen glue is visible.

The primary fluorescence of the ripe Gasteria pollen is light blue and also the tetrad appears to be bluish. The tapetal cells show little fluorescent dots, the globules.

The results of the primary fluorescence of the pollen wall are given in figure 42.

Also Gasteria microspores show a shift in the spectral maximum during development. However, it starts with a blue colour during the tetrad stage, which in the following stages shifts to the green. In ripe pollen again there is a return to the blue, possibly in relation to the pollen glue. After the second stage , a difference in the spectral maximum after 6 and 30 seconds occurs. Statistical analysis by means of the t-test and the 5 % level of significance, show that **such** differences appear to be significant for the three last stage

Although starting with different colour, Gasteria produces **a** shift after the first stages **of microspore** development **as do** Pinus sylvestris **microspores.** The shift in the maximum during development of Gasteria microspores and the shift after 6 and 30 seconds seems to be caused by components **of** the intine.

The spectral maxima of the tapetal cells during two stages **corresponds with** values obtained with the microspores during these stages. Also an empty pollen grain shows the same value.

The intensity increases after the tetrad stage. Tapetal cells and empty pollen grains have the same values.

The fading of fluorescence of both the tetrad and the **tapetal** cell during the first stage are lower in comparison with the following stages. The fading of the empty **pollen** grain shows no difference. In comparison with Pinus sylvestris the values of intensity and fading are lower for Gasteria verrucosa.

These measurements demonstrate clearly the differences in fluorescence between Gasteria verrucosa and Pinus sylvestris pollen. Also with respect to changes during strong UV-exposure, both pollen show a great specificity, particularly the shift in the spectral maximum during pollen wall development.

The megaspore wall.

The pollen wall is not the only sporopollenin containing cell wall. THOMSON (1905) has compared the pollen wall with the "megaspore membrane" , the wall around the female part of the gymnosperm Cycas revoluta. His conclusion was based on morphological and histochemical observations.

A similar megaspore wall surrounds the macroprothallium in Pinus sylvestris (fig. 43). Histochemical staining of the megaspore wall shows a very small amount of cellulose, but a great similarity with the histochemical reaction of the pollen wall.

Submicroscopical research shows very electron dense granules around the syncytium of the developing macroprothallium (fig. 44 and 45). These granules are produced by the prothallial cells. The strongly electron dense material, situated deeper in the cytoplasm, lies very closely against mitochondria and plastids. In the periphery of the cytoplasm many electron dense granules occur (fig. 46). After excretion, these granules arrange themselves around the macroprothallial cells. This granular layer grows and the dark granules in the cytoplasm will disappear. At the same time Golgi vesicles appear and these vesicles are also excreted (fig.47). In an early stage the material of the wall is loosely embedded in electron transparent material. In addition, along the plasmamembrane are tiny pieces of electron dense material (fig. 48) which turn into a continuous, very electron dense layer (fig.49). Above this layer some little granules fuse. The ultimate result is a two-layered wall with an exterior network of granular structure and along the cytoplasm, a continuous layer of the same electron density (fig.50).

The primary fluorescence of the megaspore wall around a young prothallium without archegonia is blue. The whole spectrum has one peak. The results of measurements of the spectral maxima of pollen wall and megaspore wall after 30 seconds are given in table 1.

Measurements of the spectral maxima of Pinus sylvestris and Pinus mugo megaspore walls show that it is situated at a lower wavelength than in the pollen wall. The shift in the maximum is very low during exposure after 6 and 30 seconds. The same fading phenomena occur as in the pollen wall.

A special wall surrounds the female gametophyte. This wall is not similar in morphology or reaction to UV-exposure to the pollen wall and also differs in origin. The special wall around male and female gametophytes is a noteworthy feature.

Because this study has been made mainly on Pinus sylvestris, it only goes part of the way to simplify the very complex problems of pollen wall formation and the composition of sporopollenin.

Literature

ANGOLD, R.E. (1967) Rev. Paleobot. Palynol. 3, 205-212.

BROOKS, J and SHAW,G. (1968) Grana Palynologia 8, 227- 234-.

ESAU,K. CHEADLE,V.I. and RISLEY, E.B. (1962) Bot. Gazette 123, 233-243.

GODWIN,H. ECHLIN,P. and CHAPMAN,B. (1967) Rev. Paleobot. Palynol 3, 181-195.

MARTENS,P. WATERKEYN,L and HUYSKENS, M. (1967) Phytomorphology 17, 114-118.

MEPHAM,R.H. and LANE, G.R. (1969) Protoplasma 68,175-192.

RISUENO,M.C. GIMENEZ-MARTIN,G. LOPEZ-SAEZ,J.F. and RISUENO-GARCIA,M.I. (1969) Protoplasma 67, 361- 374.

SCHNEPF,E. (1969) Protoplasma 67, 185-194.

SIEVERS,A. (1963) Protoplasma 56, 188-192.

SHAW,G.J. and YEADON, A. (1966) J.Chem. Soc. C. 16-22.

SITTE, P. (1962) Protoplasma 54, 555-559.

VASIL,I.K. and ALDRICH,H.C. (1970) Protoplasma, in press.

WATERKEYN, L. (1964) In: Pollen Physiology and function . ed. H.F. LINSKENS North Holl. Publ. Co. Amsterdam. p. 52-58.

WATERKEYN,L. and BIENFAIT,A. (1968) C.R.Acad. Sc. Paris 267, 56-58.

WATERKEYN,L. (1970) Grana, in press.

Figure 1. Early tetrad stage with beginning cell plate formation.

Figure 2. Detail cytoplasm.

In the figures the following abbreviations are used: C = callose wall, In = intine, Pm = plasmamembrane, N = nucleus, A = plastid with starch P = plastid, M = mitochondria, L = lipidgranule, G = Golgi body, V = Golgi vesicle, S = Smooth endoplasmatic reticulum, r = ribosome, p = polysome, and m = microtubule.

The line in the edge of the figures corresponds to 1 micron, or if it is noted above the line, to 0,1 micron.

Figure 3 - 8. Callose wall-formation.

Cisterns of SER are near the nucleus (fig.3) and callose wall (fig.4), sometimes occupied with ribosomes. From SER cell plate vesicles are derived (fig.5, arrow). The vesicles are arranged in a line, sometime passed by microtubuli (fig.6). In the vesicles the callose appears as a light dot (fig.7, arrow). The callose wall grows centripetally. Golgi bodies produce some little vesicles (fig.8).

Figure 9. The tetrad. In the cytoplasm are many heterogeneous Golgi vesicles(fig. 10). Between nucleus and the exterior callose wall are few vesicles (fig.11).

Figure 12-17. The Golgi body.

The Golgi body produces vesicles with a heterogeneous contents(fig.12 arrow). The contents of the Golgi vesicles is transported outside the cell (fig.13,arrow). The membrane of the Golgi vesicle dissappears (fig.14,arrow),the whole contents is poured out (fig 15, arrow) and appears between plasmamembrane and callose wall (fig. 16,arrow). Sometimes fusion of Golgi vesicles takes place in the cytoplasm (fig.17).

Figure 18-23. The callose wall.
In the beginning the plasmamembrane lies parallel to the callose wall (fig.18) During the excretion the eveness of the callose wall dissapears (fig.19) and protrusions are coming visible (fig.20,arrow). The plasmamembrane makes a contact with the protrusions (fig.21,arrow). The material from the Golgi vesicles precipitates between the protrusions (fig.22, arrows), the tectum, thereafter against the protrusions, the bacules (fig.23,arrows).

Figure 24. Two cells in the late tetrad stage, the wings are visible, in the cytoplasm many lipid granules.

Figure 25-28. The footlayer.
Mainly between the wings on the outside of the tetrad, long tapes are coming out of the cytoplasm (fig.25). The tape lies outside the plasmamembrane, on some places a lipid granule lies close to the tape (fig.26) and parallel to it somewhat endoplasmatic reticulum is situated (fig.26 and 27, double arrow) . In the cytoplasm groups of electron dense membranes appear (fig.28,arrow). In the tapes remnants of the space between two membranes of the endoplasmatic reticulum are visible as white lines (fig.28).

Figure 29. The formation of the footlayer along the plasmamembrane is not at all places simultaneously. The footlayer appears gradually, produced outside the plasmamembrane (fig.30).

Figure 31-34. Formation of the wings.
A great quantity of granular and fibrillar material derived from Golgi vesicles is visible between plasmamembrane and callose wall (fig.31). In tangential section the protrusions are round (fig.32,arrows). The material derived from the Golgi vesicles becomes fine granular (fig.33 and turns in a network of fibrills (fig.34) This material precipitate against the protrusions (fig.34,arrow).

Figure 35. Survey of pollen wall-formation.

Figure 36. The young microspore, the cytoplasm has great vacuoles.

Figure 37. The formation of the nexine 2 starts with thin tapes (see arrow),followed by intine formation (fig.38).

Figure 39. Results of the measurements of primary fluorescence during development of _Pinus sylvestris_ pollen.

Figure 40. Results of the measurement of the spectral maximum and intensity of primary fluorescence during different successive treatmen of _Pinus sylvestris_ and _Pinus pinaster_ pollen.

Figure 41. The morphology of the pollen wall of _Gasteria verrucosa_ in three stages of development. In the last stage pollen glue is visible.

Figure 42. Results of the measurements of primary fluorescence during the development of Gasteria verrucosa microspores.

Figure 43-50. The megaspore wall-formation of Pinus sylvestris. The megaspore wall lies around the female prothallium surrounded with cells of the spongy tissue (fig.43,arrow). Around the prothallium granular wall material appears (fig.44). This material is also found in the perifery of the cell (fig.45) and lies deeper in the cytoplasm along mitochondria and plastids (fig.46). When all the material is excreted, many Golgi bodies appear in the cytoplasm (fig.47). Along the plasmamembrane appear tiny pieces of electron dense material (fig. 48) The tiny pieces form a continuous layer , above this the united little granules (fig.49). The complete wall has a continuous layer and a more granular outer layer (fig.50).

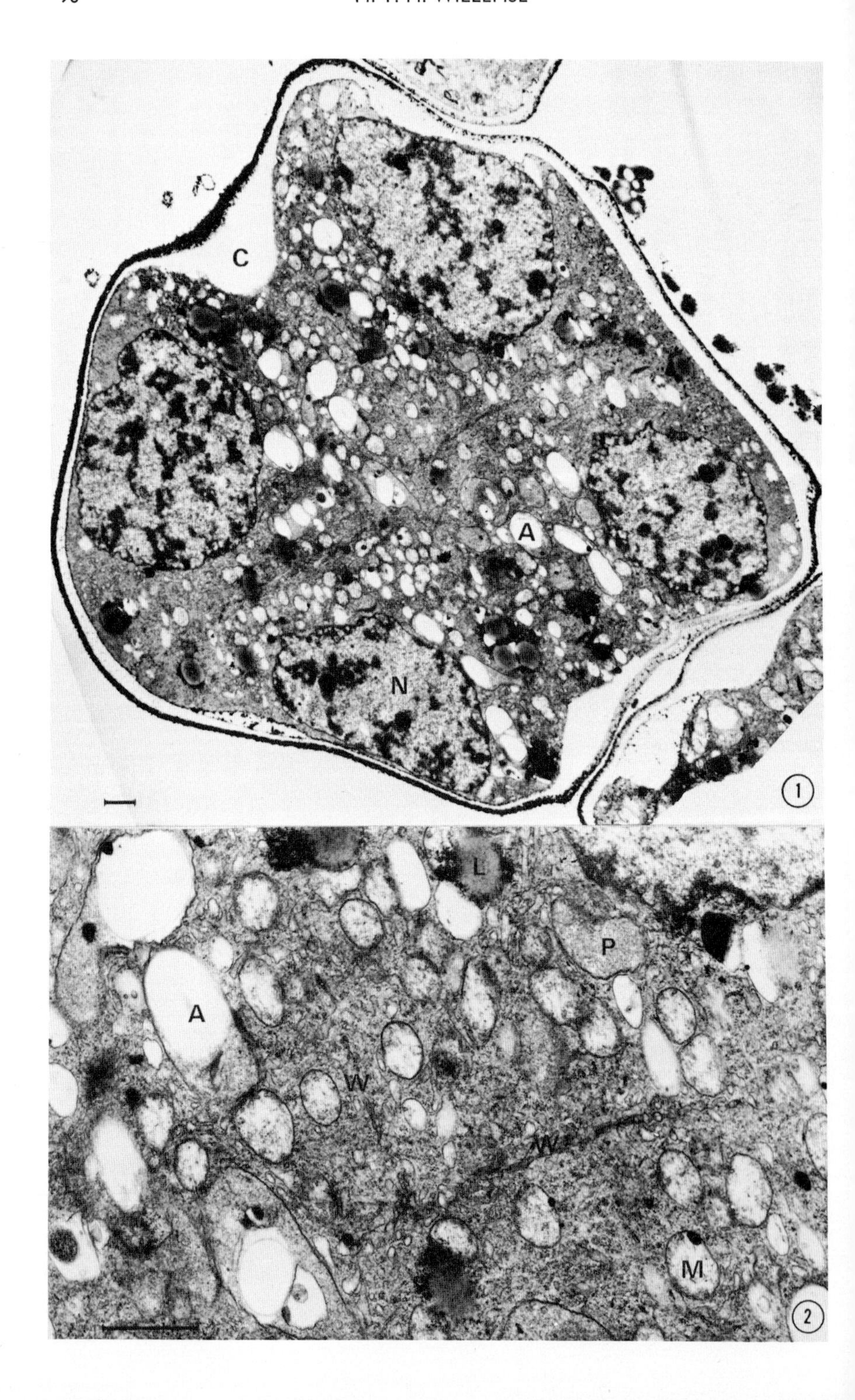
C
A
N
1
L
P
A
W
W
M
2

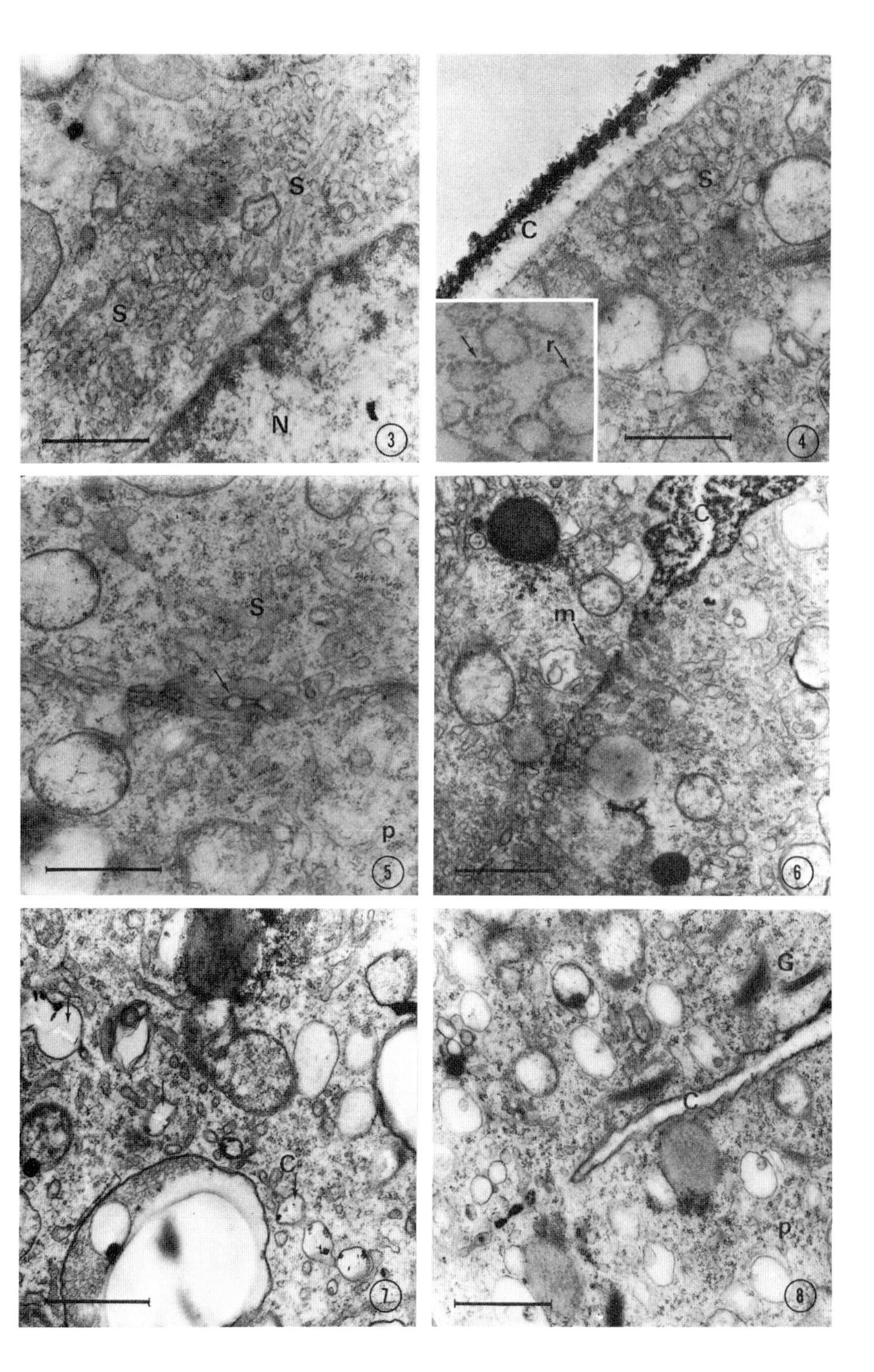
S
S
N
3
C
S
r
4
S
p
5
C
m
6
C
7
G
C
p
8

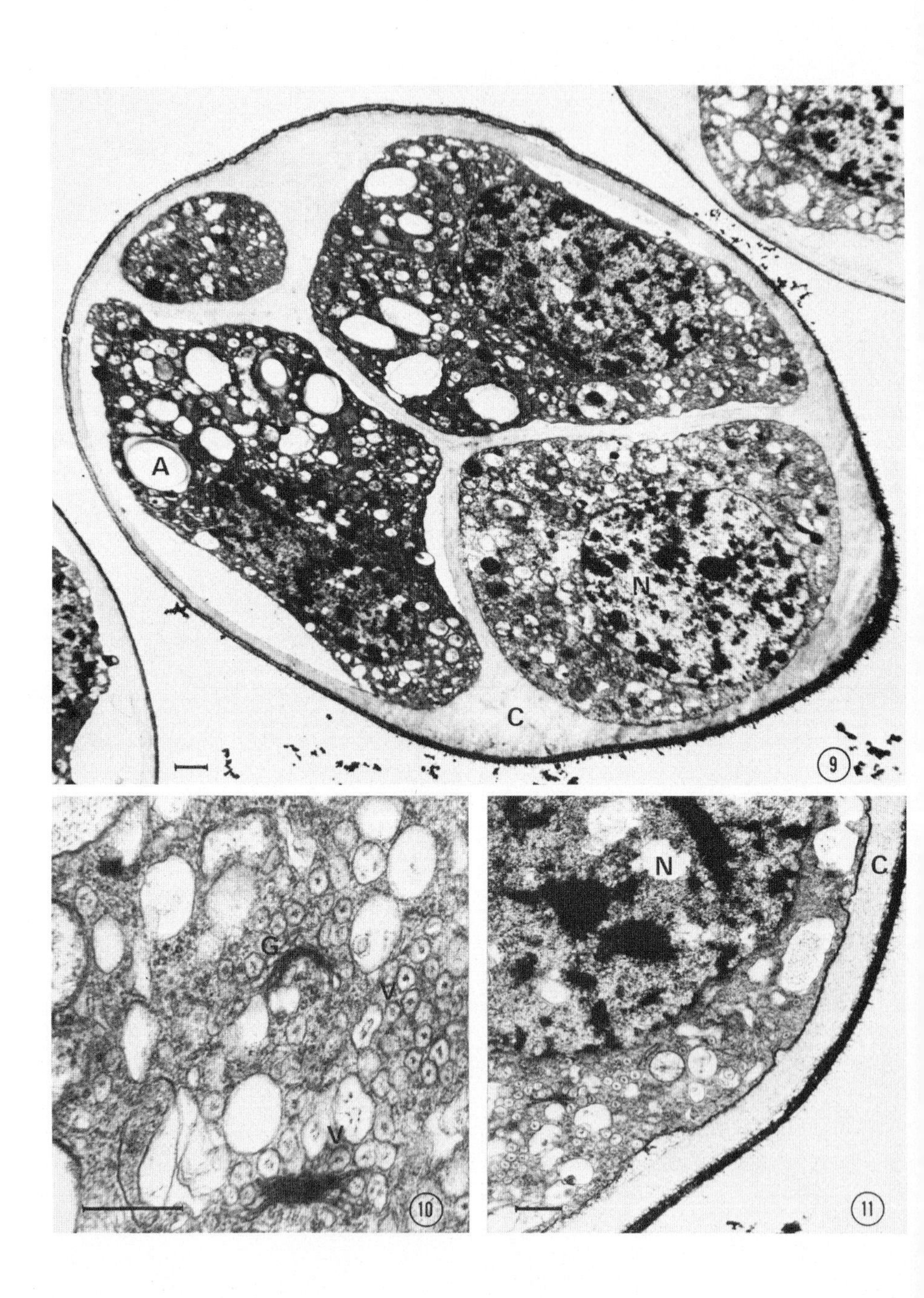
A
N
C
9
G
V
V
10
N
C
11

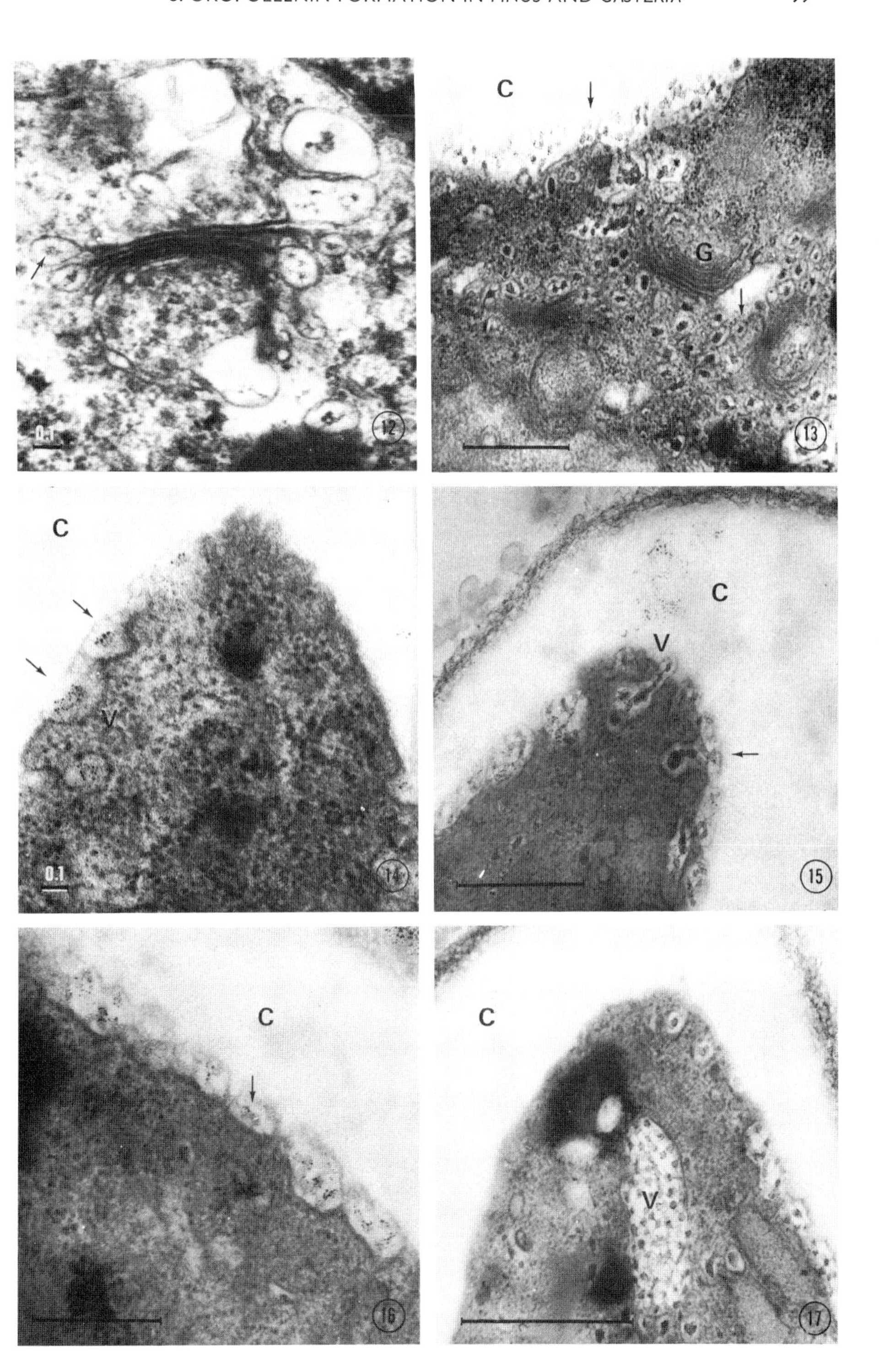
C
G
C
V
C
V
C
C
V
0.1
0.1
12
13
14
15
16
17

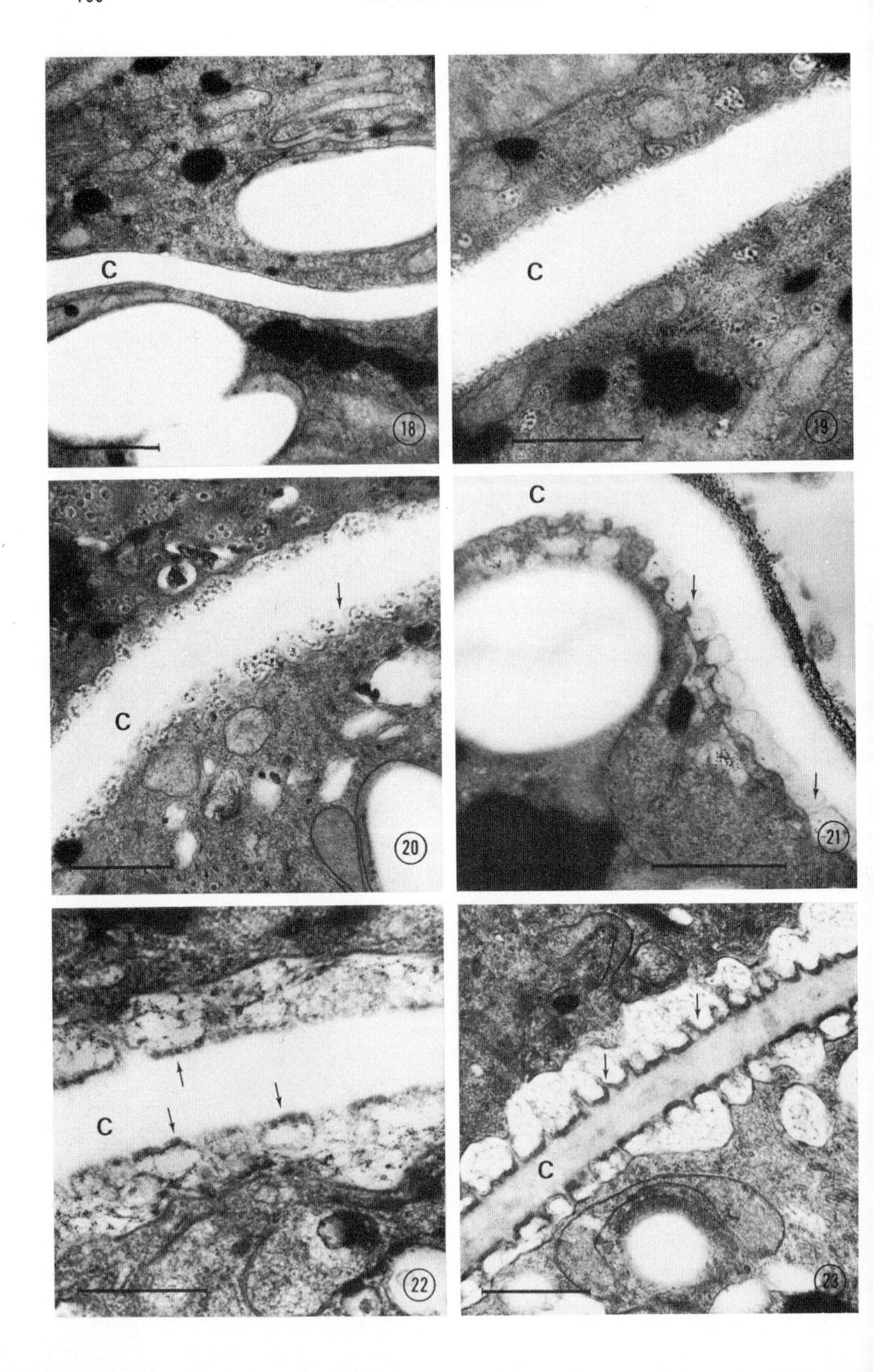
C
18
C
19
C
20
C
21
C
22
C
23

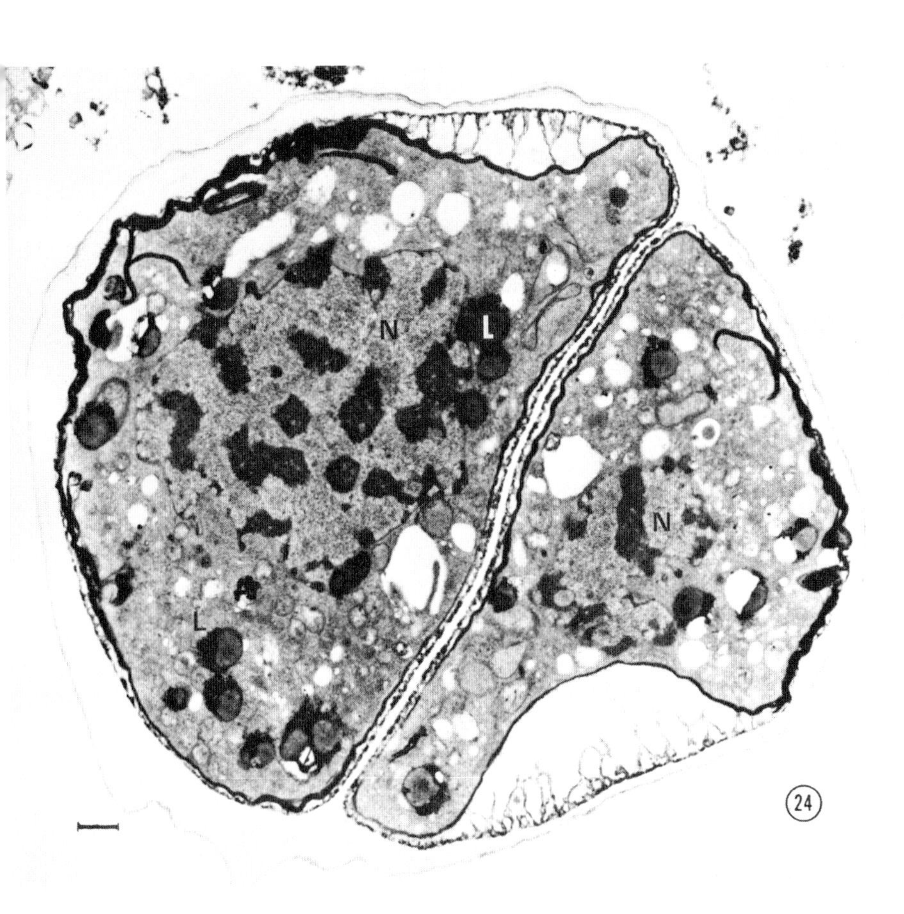
N
L
N
L
24

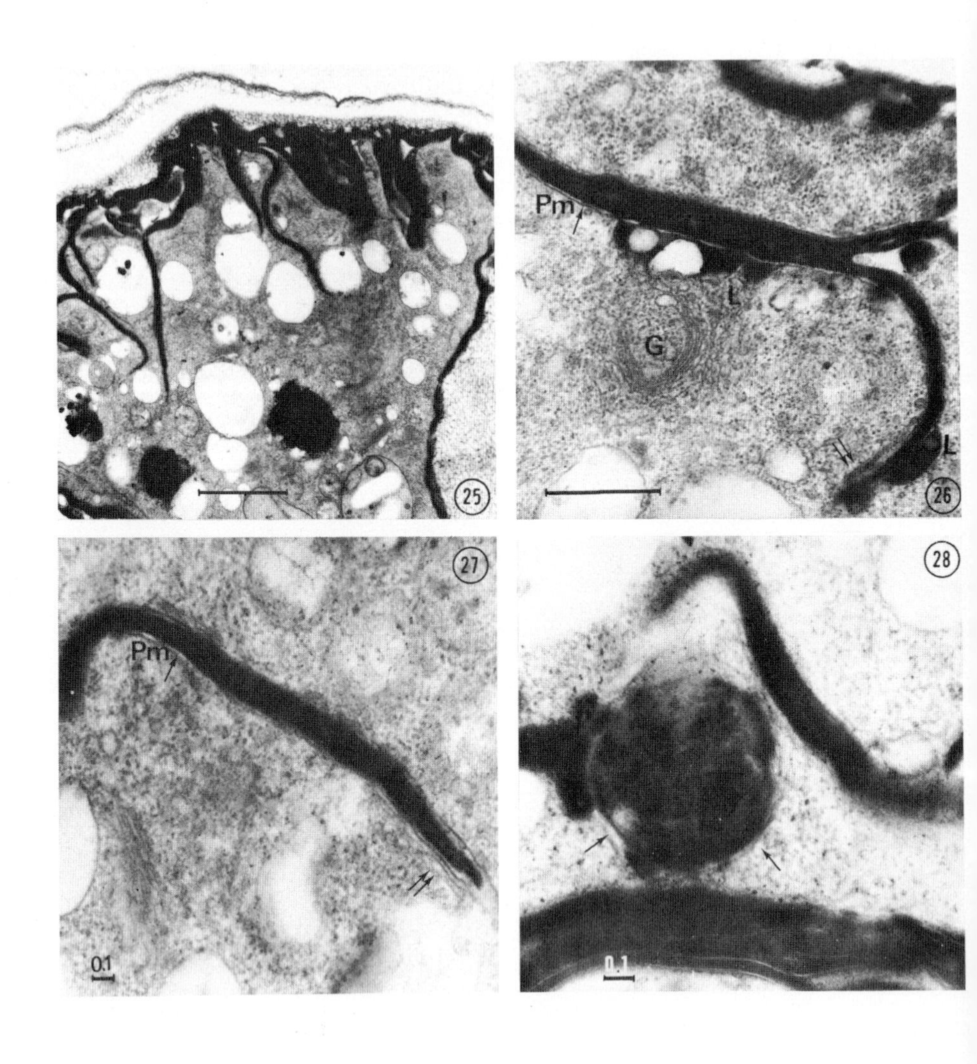
25
Pm
L
G
L
26
27
Pm
0.1
28
0.1

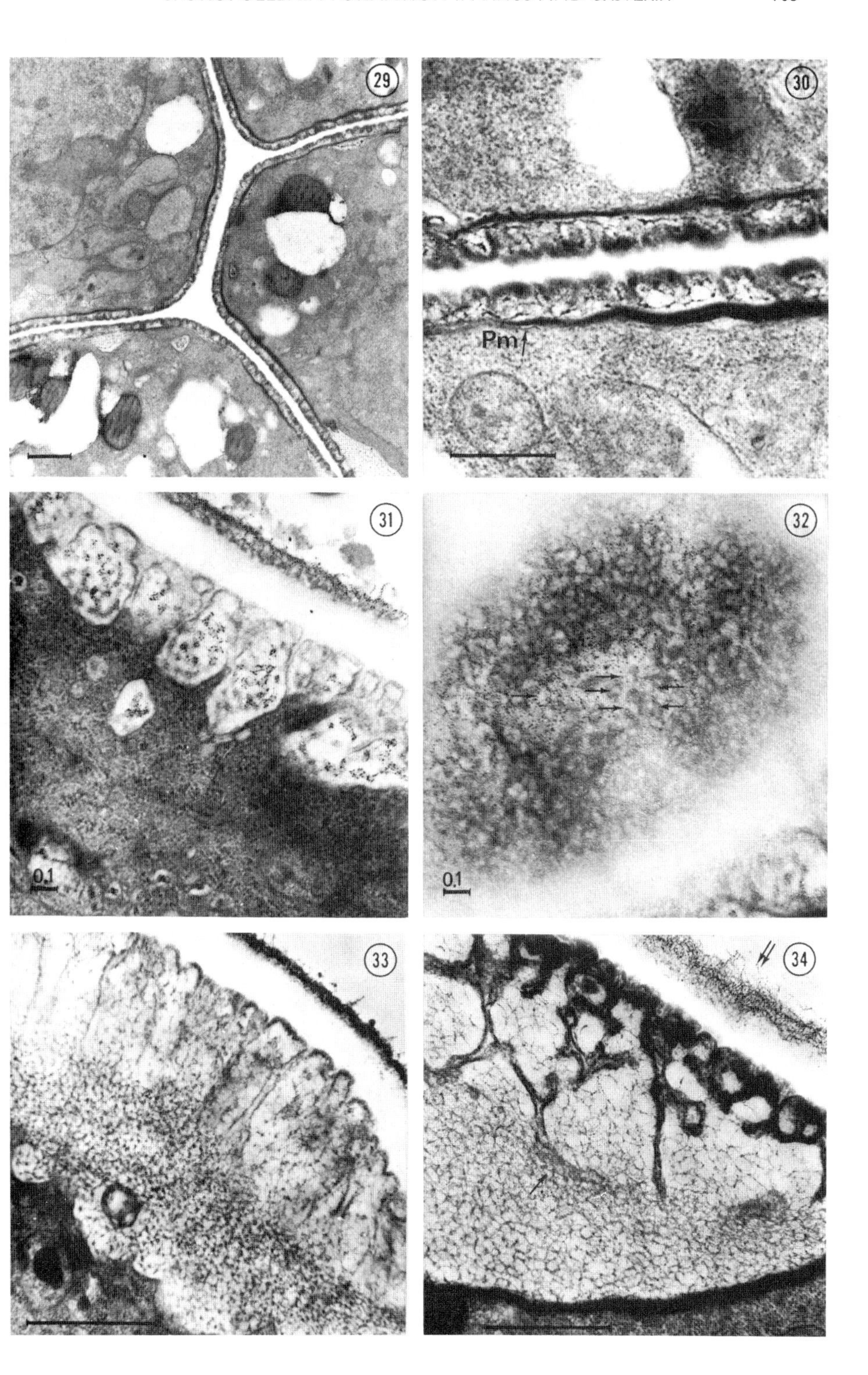
29
30
Pm
31
0.1
32
0.1
33
34

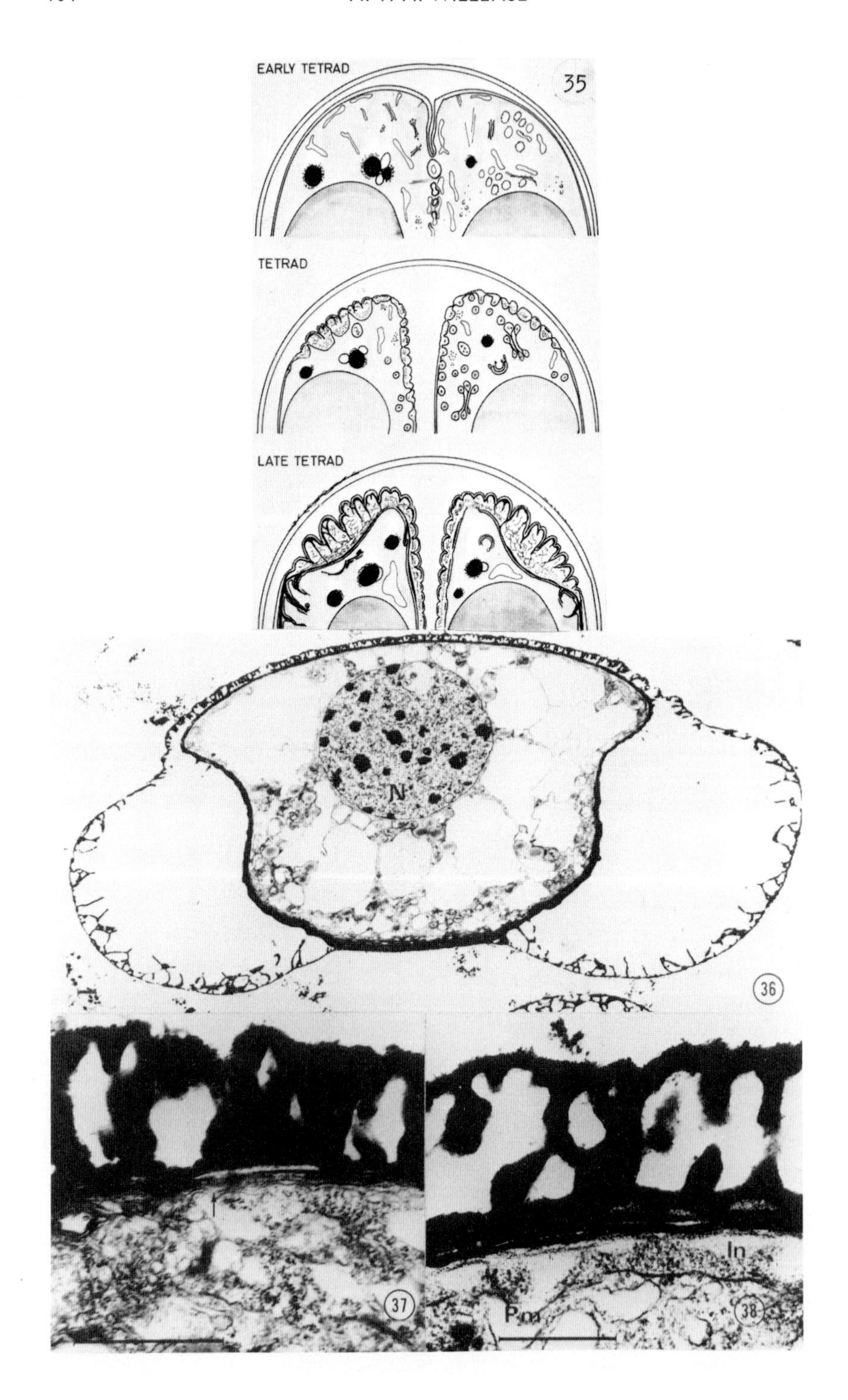
EARLY TETRAD
35
TETRAD
LATE TETRAD
N
36
37
In
Pm
38

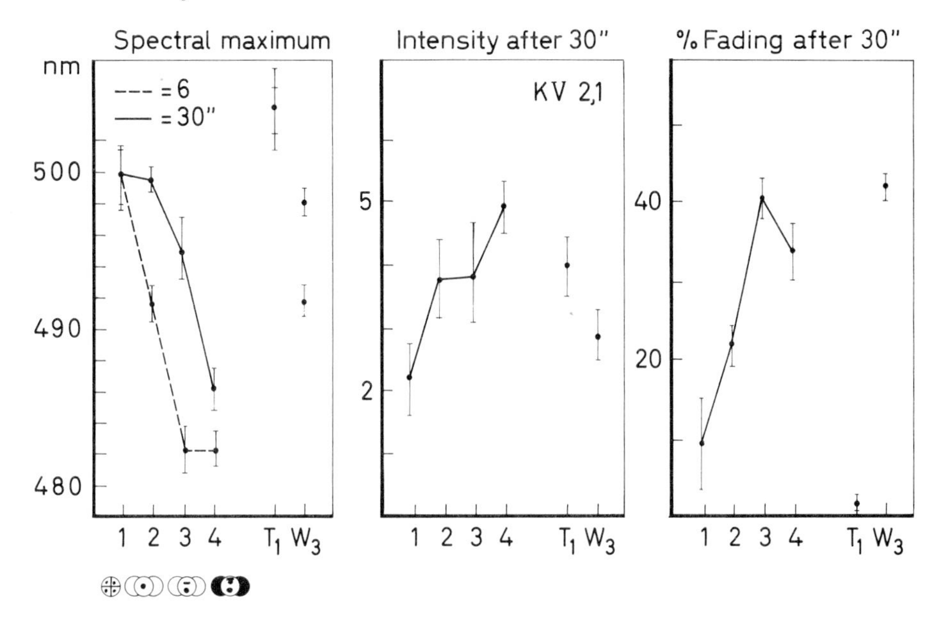

1,2 : Stage T: Tapetal cell W: Wing

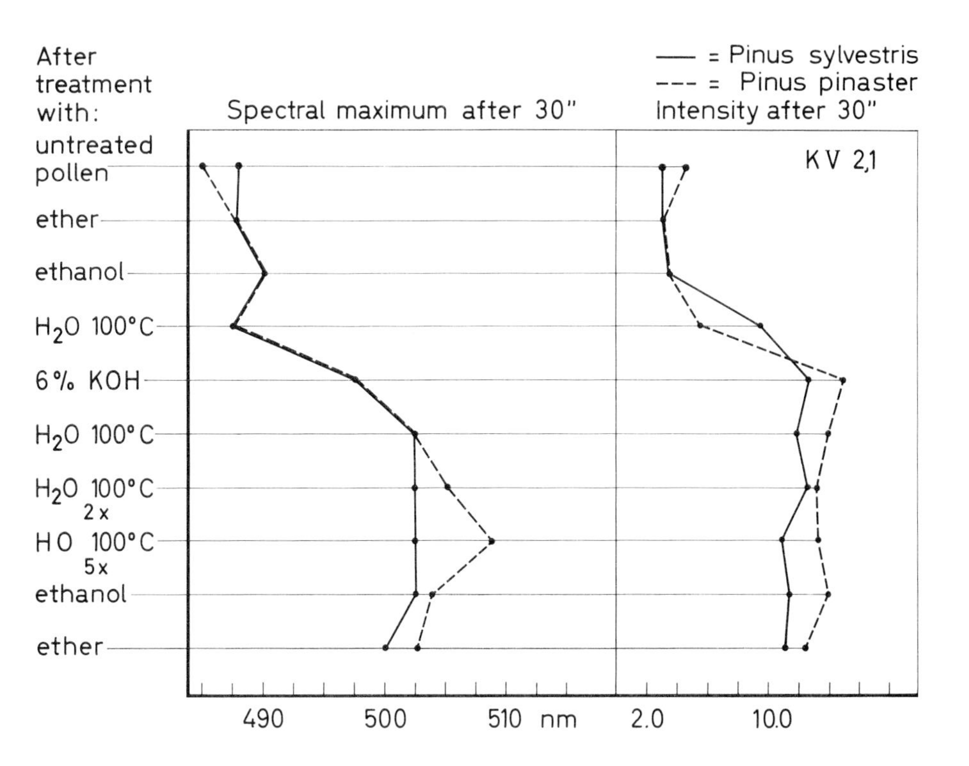

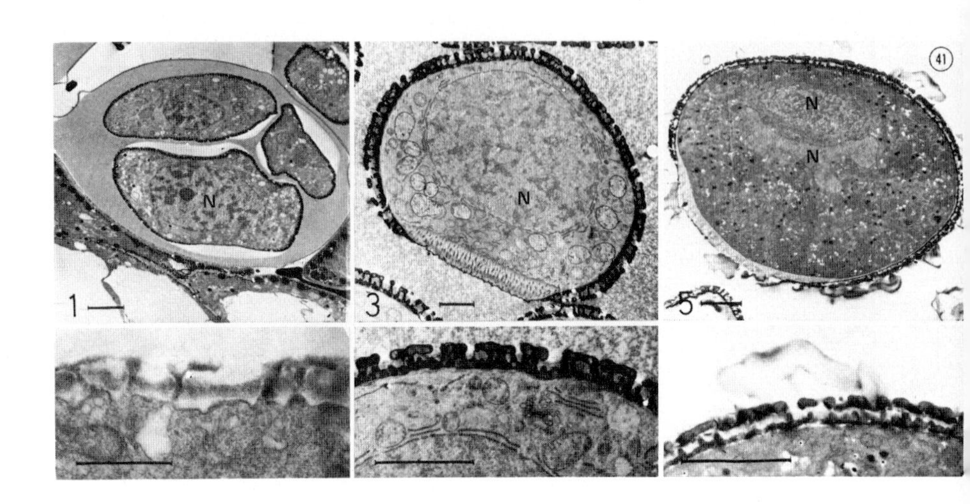
41
N
N
N
N
1
3
5

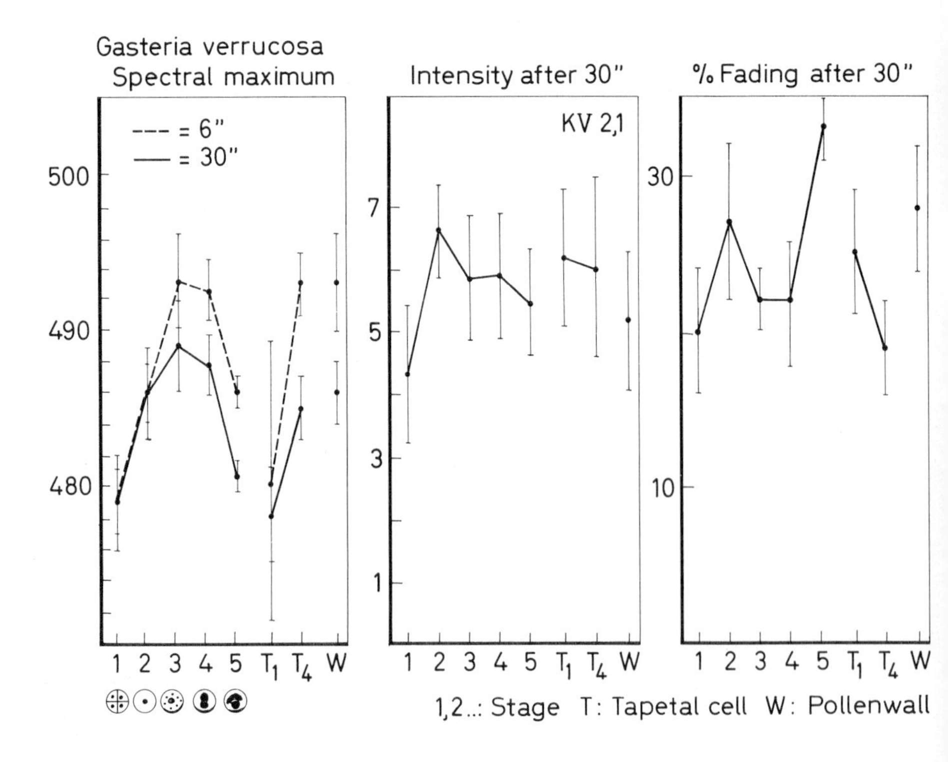
Gasteria verrucosa
Spectral maximum
--- = 6"
— = 30"
500
490
480
1 2 3 4 5 T1 T4 W
Intensity after 30"
KV 2,1
7
5
3
1
1 2 3 4 5 T1 T4 W
% Fading after 30"
30
10
1 2 3 4 5 T1 T4 W
1,2..: Stage T: Tapetal cell W: Pollenwall

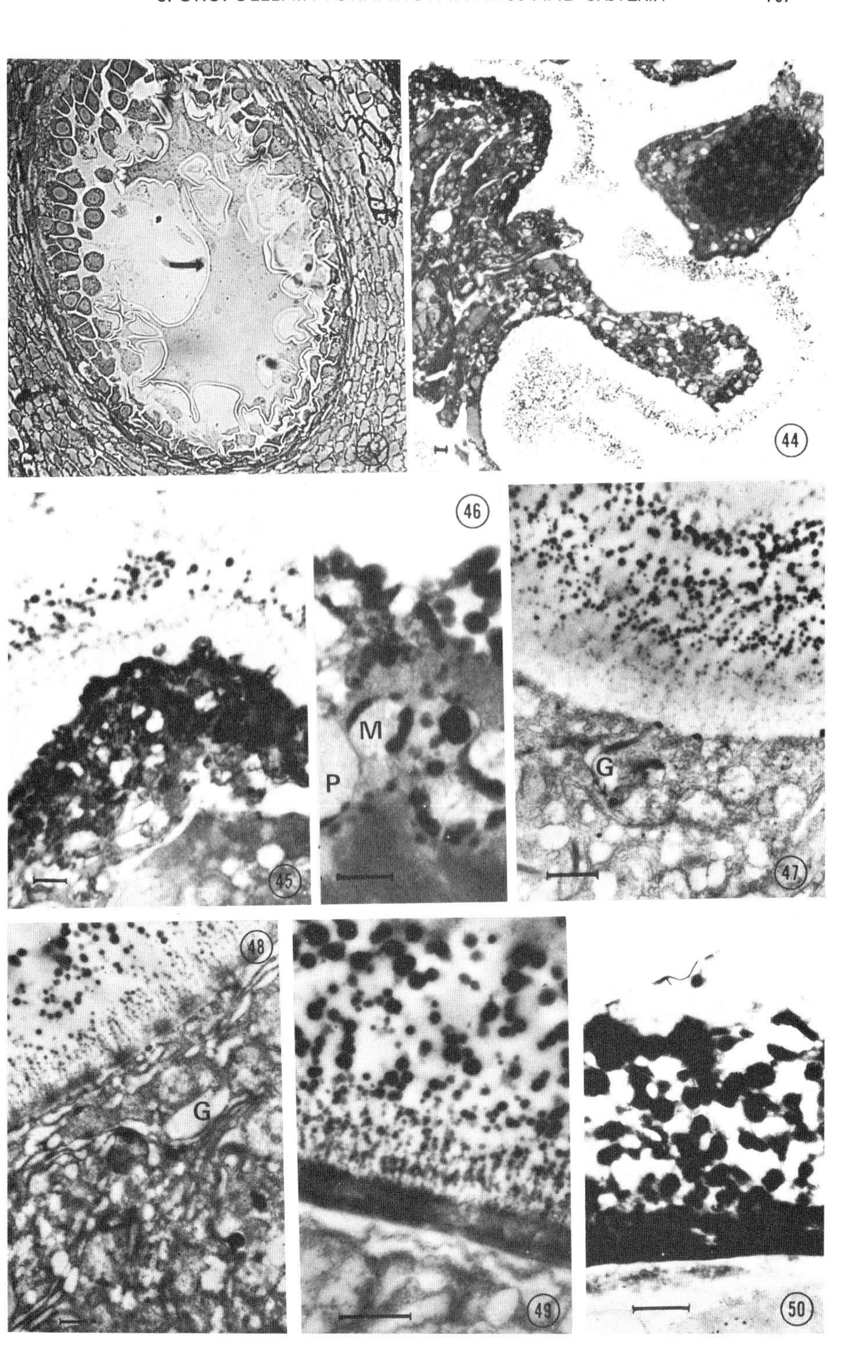
44
46
M
P
G
45
47
48
G
49
50

PRIMULINE INDUCED FLUORESCENCE OF THE FIRST EXINE ELEMENTS AND UBISCH BODIES IN *IPOMOEA* AND *LILIUM*

L. Waterkeyn & A. Bienfait

(Université Catholique de Louvain, Belgium)

Abstract

In order to make more apparent the first elements of the exine in *Ipomoea* and *Lilium*, a number of fluorochromes were tried. Good results were obtained with primuline. Microspores of both species and at diverse developmental stages were fixed by neutral formaldehyde, treated by highly diluted primuline solutions and observed in the fluorescence microscope. In *Ipomoea*, the delicate network formed by the probacula and spine rudiments of the primexine exhibits a specific and intense silver-white or yellowish-white fluorescence. The primexine matrix, in which these isolated elements are embedded, shows no fluorescence at all. At later stages, the foot-layer and nexine-2 fluoresce too, although not so intensively. In *Lilium*, the very first sexine patterning and the site of the colpus are clearly outlined, even in the coherent tetrad stage. The later formed foot-layer and the nexine-2 exhibit no appreciable fluorescence. Older pollen grains show a dull brown fluorescence. In both species there was also a bright and specific lightening of the Ubisch bodies.

Introduction

Fluorescence microscopy, using U.V. or deep blue excitation light, has only to a slight extent been used in palynological studies. Auto-fluorescence of pollen and spores, excited by short wave-length radiations, is a well known phenomenon. Whole pollen grains or isolated exine show characteristic fluorescence colours, which recently have been spectrophotometrically analyzed (van GIJZEL 1967).

Induced or secondary fluorescence, obtained after treatment by a fluorochrome, has sporadically been used, for example, to locate pollen and spores in sediments (SHELLHORN & HULL 1964). Fluorochromes also have been used for the study of pollen and pollen tube content (BHADURI & BHARYA 1962). Recently they have been used to evaluate pollen viability (HESLOP-HARRISON & HESLOP-HARRISON 1970).

In this connection, we may also cite the now common use of secondary fluorescence, evident in callose deposits, produced by aniline blue solutions, a method introduced by ARENS (1949). This last method, standardized by ESCHRICH & CURRIER (1964), has greatly improved the study of the callose deposits formed during microsporogenesis (special wall) and pollen germination (intine and pollen tube wall).

But, as far as we know, fluorescence methods have not been used in ontogenetic studies of spore and pollen walls. We have recently tried a number of fluorochromes in order to make more apparent the early formed exine and to follow the evolution of its structure. Good results were obtained with primuline. Coriphosphine also has given some interesting information. Both fluorochromes were initially tested on microspores, at diverse developmental stages, of *Ipomoea purpurea*. In a previous work, we have encountered some difficulty with this species in obtaining a clear view of the early pattern formed by the baculoid elements of the primexine (WATERKEYN & BIENFAIT 1968, 1970). The same species has been studied by BEER (1911), GODWIN *et al.* (1967) and ECHLIN (1968). Because of the interesting results obtained, we also tested *Lilium*, which has been intensively studied by others (HESLOP-HARRISON 1968 a, b,c; HESLOP-HARRISON & DICKINSON 1969).

Material and Methods

In addition to *Ipomoea purpurea* (L.) Roth., we have used *I. rubro-coerulea* (cv), *Lilium martagon* L., *L. regale* (cv) and *Oxalis acetosella* L. The anthers, from tetrad stage to mature pollen, were fixed by neutral phosphate buffered formaldehyde and stored at 4°C. Smears of anther contents were made in highly diluted neutral solutions of primuline (Edw. Gurr 64) or coriphosphine (Edw. Gurr 590) and sealed with vaseline. A Leitz-Ortholux fluorescence microscope was fitted with either UG1 (2mm) + BG38 or BG12 (5mm) excitation filters. The barrier filters were respectively K430 and K570. Photographic recording was made with a Leitz-Orthomat camera on Ilford FP4 (22DIN) negative emulsion or reversible daylight Agfacolor CT film (18DIN) for transparencies.

Observations

Ipomoea

At the tetrad stage, when the microspores are still embedded in the callosic special wall, the baculoid elements of the primexine are hardly visible. Contrary to the primexine matrix, which is specifically stained by toluidine blue (WATERKEYN & BIENFAIT 1968, 1970), these early elements have up to now not been adequately stained. In addition, the refractive cytoplasmic organelles (oil drops, starch) and the wafered structure of the spore chamber wall have a disturbing action on image formation. Nevertheless, some useful pictures may exceptionnaly be obtained by phase contrast on partly squeezed and flattened spores (Fig.1).

Now, if primuline is added to the smear, which is then observed in the fluorescence microscope, the very delicate network of probacula and spine rudiments shows an intense and specific fluorescence (silver-white in U.V. and yellowish-white in blue excitation light). In face view, these isolated elements are clearly outlined as many bright dots (Fig.2). These structures are of two sizes: large ones, corresponding to spine rudiments at the angles of the polygonal meshes, and little ones or probacula, occupying the sides of these meshes. They stand out against the black background in absence of any fluorescence of the primexine matrix and the faint dark blue or greenish-blue fluorescence of the cytoplasm. With very flattened spores it is possible to explore, in its entirety, by varying the fine focus adjustment, the whole primexine network, because out-of-focus images vanish immediately.

The callosic special wall also becomes fluorescent with primuline. This latter fluorescence at the primexine stage is always less than with aniline blue and is not cumbersome.

At the time the special wall dissolves, the baculoid elements of the primexine thicken and become very conspicuous. The nexine-1, or foot-layer, makes its appearance and from the beginning is less fluorescent than the probacula and pierced by numerous circular pores, each occupying the center of a polygonal mesh of the primexine (Fig.4). At a more advanced stage, when the nexine-2 is formed, the fluorescent spines and bacula of the sexine remain visible and stand out against the less fluorescent underlying layers (Fig.5 and 6).

From now on, the large spines and bacula of the sexine gradually lose their fluorescence and become highly refracting. The total fluorescence of the spore seems to diminish, but the pores retain a bright contour (Fig.7). When microspores are extremely flattened by pressure on the coverslip, they show various "moiré" patterns formed by the superposition of the pore system of both walls (Fig.8).

In control smears, made in the sole buffer solution, the probacula and spine rudiments present a weak pale blue primary fluorescence in U.V. light and no fluorescence in blue excitation. Older microspores show respectively blue and green autofluorescence.

Coriphosphine solutions have no particular action on the first stages of exine formation, but this fluorochrome confers a bright gold-yellow to orange-red fluorescence to older microspores and near mature pollen grains. In addition, the nuclei of the spores, as well as of the tapetal cells, acquire a proper green fluorescence. Coriphosphine, contrary to primuline, has no action on the callose layers of the special wall.

Essentially the same results were obtained with the two species of <u>Ipomoea</u>. Nevertheless, some specific differences with regard to the distribution of the large spines were noted and will be discussed below.

<u>Lilium</u>

As in <u>Ipomoea</u>, the first patterning of exine in <u>Lilium</u> is also clearly outlined by means of the primuline induced fluorescence. Primuline confers a bright fluorescence to the probacula of the primexine, when the spores are still grouped in tetrads. The site of the colpus (<u>Lilium</u> has monocolporate microspores) is readily visible and occupies the distal face.

As soon as the spores are released from the tetrad special wall, which also shows fluorescence, the gradual enlargement of the pila and their partial confluence may be easily followed. In face view, the sexine elements form a network of irregular meshes with beaded limits. These structures stand out against the dark background, for the thin foot-layer and the nexine-2 exhibit no appreciable fluorescence (Fig.10). At the distal side, the colpus area often bears grouped baculoid elements (Fig.9).

In older spores, the over all fluorescence is suddenly affected and turns from bright yellowish-white to dull brown. The primary fluorescence, observed in control slides, although weaker, presents also a characteristic colour change, from sky blue to dull light brown (in U.V. light).

Coriphosphine were also tried on diverse stages of exine formation in *Lilium*. This fluorochrome, as in the case of *Ipomoea*, is unable to display the very first sexine patterning, but it produces striking changes in colour during later developmental stages. During the growth period of the microspores, the sexine shows a rather yellowish-green hue, whereas older microspores and near mature pollen grains acquire a bright orange-red fluorescence.

Ubisch bodies

In both species, at the time the microspores are released from the tetrad wall, a multitude of bright dots can be observed in primuline treated slides. These dots are the well known "Ubisch bodies" or sporopollenin orbicules, which, just like sexine elements, show a specific fluorescence. These Ubisch bodies cover the walls of the tapetal cells turned towards the anther loculus. In *Ipomoea*, they are spheroidal and retain a bright yellowish-white fluorescence to the end, whereas with coriphosphine they become gradually orange to red.

In _Ipomoea_ _rubro-coerulea_, besides small spheroidal Ubisch bodies, we have observed large stellate bodies, looking like grouped spines (Fig.15). In _Lilium_, the Ubisch bodies densely cover the inner tapetal wall and the limits outlined by the anticlinal walls (Fig.14). At high magnification, they present the form of a perforated circular disk. They assume a more polygonal contour when closely packed.

During the later growth stages their fluorescence after primuline or coriphosphine presents the same colour change as shown by the sexine.

Some additional observations were performed on _Oxalis_ _acetosella_. Ubisch bodies of various _Oxalis_ species are large and present a typical morphology (CARNIEL 1967). Primuline and coriphosphine solutions induce a bright fluorescence at the level of these curious structures which can easily be located in the granular debris of tapetal cells (Fig.11 an 12). At high magnification they are circular disk-shaped and perforated bodies or large multiperforated plates (Fig. 13).

Discussion

Primuline is not a commonly used fluorochrome and was introduced in animal histology as a fluorescent vital stain (PICK 1935). Then, BISHOP & SMILES (1957) used it to distinguish living from dead spermatozoids. It was also utilized as a general cell wall fluorochrome in plant histology (STRUGGER 1939, ZIEGENSPECK 1949, HAITINGER 1959) and especially to distinguish lignified from non-lignified walls (LUHAN & TOTH 1951). Lastly, primuline has be used as a fluorochrome in starch studies (CZAJA 1960, STERLING 1965).

Coriphosphine, on the other hand, has more often been used and gives polychromatic fluorescence with plant material. Like primuline, it permits the recognition of lignified cell walls. In addition, it produces a greenish fluorescence of the nuclei (HAITINGER 1959).

Fluorescence techniques, in spite of their poor specificity and generally non-predictable results, offer nevertheless many advantages and would be a useful tool in solving some specific palynological problems. Some of these problems have attracted the attention of many investigators , in particular: the nature and structure of the primexine and the origin of control of its pattern; the exact nature and biosynthesis of sporopollenin; the origin and significance of the Ubisch bodies (ECHLIN & GODWIN 1968, GODWIN 1968, HESLOP-HARRISON 1968a,b,c, HESLOP-HARRISON & DICKINSON 1969).

In the present study, the enhanced fluorescence induced by primuline allows us to check in _Ipomoea_, with an accuracy never before attained, the fixed and specific distribution of probacula and spine rudiments. The two species of _Ipomoea_ used in the present study have almost identical pollen grains. But from a careful comparison between the distribution of the large spines it may be shown that in _I. purpurea_ two spines are regularly missed at the angles of some meshes of the primexine, whereas in _I. rubro-coerulea_ all the corners bear a spine. We have shown in a preceding paper, how the fixed distribution of these spines is related to the wall pattern shown by the callosic spore chamber (WATERKEYN & BIENFAIT 1970). Although in _I. rubro-coerulea_ all meshes of the primexine are strictly identical, the wall pattern of the spore chamber in this species is constructed just like that of _I. purpurea_: a hollow network of large hexagonal and pentagonal meshes, each with a central knob.

This corroborates our previous views concerning the possible rôle of the spore chamber wall. The wafered structure of this wall is not a passive imprint produced by the primexine pattern, for it is not a perfect negative of the latter. The callosic spore chamber wall possess its proper pattern and serves as a temporary matrix or mould for the primexine.

The primuline induced fluorescence constitutes a

new means which allows us to locate the colpus site and the pores, even at the coherent tetrad stage. But the method must now be tested on other spore material and compared with existing methods (STAINIER _et al_. 1967).

The observed colour and intensity changes, displayed by the fluorescence during growth stages of the microspores and Ubisch bodies, may be due to chemical and also physical changes of the sporopollenin fraction itself or may be related to the appearance of accompanying products (encrusting substances, pigmented lipids, carotenoids, "phytine"). In any case, it seems clear that exine and Ubisch bodies undergo a parallel evolution. This results from our present observations and from many others (HESLOP-HARRISON 1968c, HESLOP-HARRISON & DICKINSON 1969).

In both species of _Ipomoea_, and also in _Lilium_, the sexine structures display a difference in colour and intensity of fluorescence when compared with the nexine layers. Difference in autofluorescence colour between "ektexine" and "endexine" has already been noted (van GIJZEL 1967). Recently, SOUTHWORTH (1969) has shown a difference in U.V. absorption displayed by the sexine and nexine. This may corroborate the often admitted opinion, that sexine and nexine have different composition, structure and/or origin .

The enhanced fluorescence produced by fluorochromes, not only greatly improves the observation of pure morphological changes of the exine, but may also be helpful to follow and characterize the parallel chemical and physical evolution of spore walls.

References

Arens, K. 1949. Lilloa 18, 71.

Beer, R. 1911. Ann. Bot. 25, 199.

Bhaduri, P.N. & Bharya, P.K. 1962. Stain Techn. 37, 351.

Bishop, M.W.H. & Smiles, J. 1957. Nature 179, 308.

Carniel, K. 1967. Österr. bot. Z. 114, 490.

Czaja, A.T. 1960. Photogr. u. Forschung. 8, 110.

Echlin, P. 1968. Ber. dtsch. Bot. Gesell. 81, 461.

Echlin, P. & Godwin, H. 1868. J. Cell Sci. 3, 161.

Eschrich, W. & Currier, H.B. 1964. Stain Techn. 39, 303.

Gijzel, P. van. 1967. Leidse Geol. Med. 40, 263.

Godwin, H. 1968. New Phytol. 67, 667.

Godwin, H. Echlin, P. & Chapman, B. 1967. Rev. Palaeobotan. Palynol. 3, 181.

Haitinger, M. 1959. "Fluoreszenz-Mikroskopie". Akad. Verl. 2d ed. Leipzig.

Heslop-Harrison, J. 1968a. Science 161, 230.

Heslop-Harrison, J. 1968b. Canad. J. Bot., 46, 1185.

Heslop-Harrison, J. 1968c. New Phytol. 67, 779.

Heslop-Harrison, J. & Dickinson, H.G. 1969. Planta 84, 199.

Heslop-Harrison, J. & Heslop-Harrison, Y. 1970. Stain Technol. 45, 115.

Luhan, M. & Toth, A. 1951. Mikrosk. (Österr.) 6, 299.

Pick, J. 1935. Z. wiss. Mikrosk. 51, 338.

Shellhorn, S.J. & Hull, H.M. 1964. Nature 202, 315.

Southworth, D. 1969. Grana Palynol. 9, 5.

Stainier, F. Huard, D. & Bronckers, F. 1967. Pollen et Spores. 8, 367.

Sterling, C. 1965. Protoplasma 59, 180.

Strugger, S. 1939. Biol. Zentralbl. 59, 409.

Waterkeyn, L. & Bienfait, A. 1968. Compt. Rend. Ac. Sci. (Paris). Sér.D, 267, 56.

Waterkeyn, L. & Bienfait, A. 1970. Grana. 10 (in press).

Ziegenspeck, H. 1949. Mikrosk. (Österr.). Sonderb.1 ,71.

Explanation of the figures

All figures (except fig.1 an 11) are fluorescence micrographs of formaldehyde fixed material treated with highly diluted primuline solutions. Line scales represent ca. 10 μ on all figures, except fig.8 (100 μ).

Fig.1 *Ipomoea purpurea*. General view of the primexine network, formed by the probacula and spine rudiments (black dots), surrounded by the highly expanded spore chamber. Phase contrast.

Fig.2 Id. Flattened microspore, at the tetrad stage, showing specific fluorescence of the baculoid elements.

Fig.3 Id. Whole view of the primexine network at an early stage. Microspore mechanically extruded from the spore chamber.

Fig.4 Id. First appearance of the foot-layer (nexine-1) underneath the isolated elements of the sexine. Circular pores visible at center of each mesh.

Fig.5 Id. Still older microspore with more differentiated foot-layer and nexine-2.

Fig.6 _Ipomoea rubro-coerulea_. Face view of an exine fragment showing different fluorescence intensities in sexine and nexine.

Fig.7 Id. General fluorescence of a near mature pollen. Each pore possess a bright margin. Large spines are no longer fluorescent.

Fig.8 Id. Microspores, mechanically compressed between slide and cover-slip, showing the "moiré" pattern obtained by the superposed systems of pores of both walls.

Fig.9 and 10 _Lilium regale_. Two face views of differently oriented microspores. Colpus site (distal) partly covered by sexine elements. Note absence of fluorescence of the nexine layers.

Fig.11 and 12 *Oxalis acetosella*. Microspore and tapetal debris. Two views of the same area in bright field illumination and in fluorescence. In the latter, only the exine and Ubisch bodies display fluorescence.

Fig.13 Id. Single and compound Ubisch bodies at higher magnification.

Fig.14 *Lilium regale*. Face view of tapetal cell walls, oriented towards anther loculus, covered with Ubisch bodies.

Fig.15 *Ipomoea rubro-coerulea*. Several large stellate Ubisch bodies.

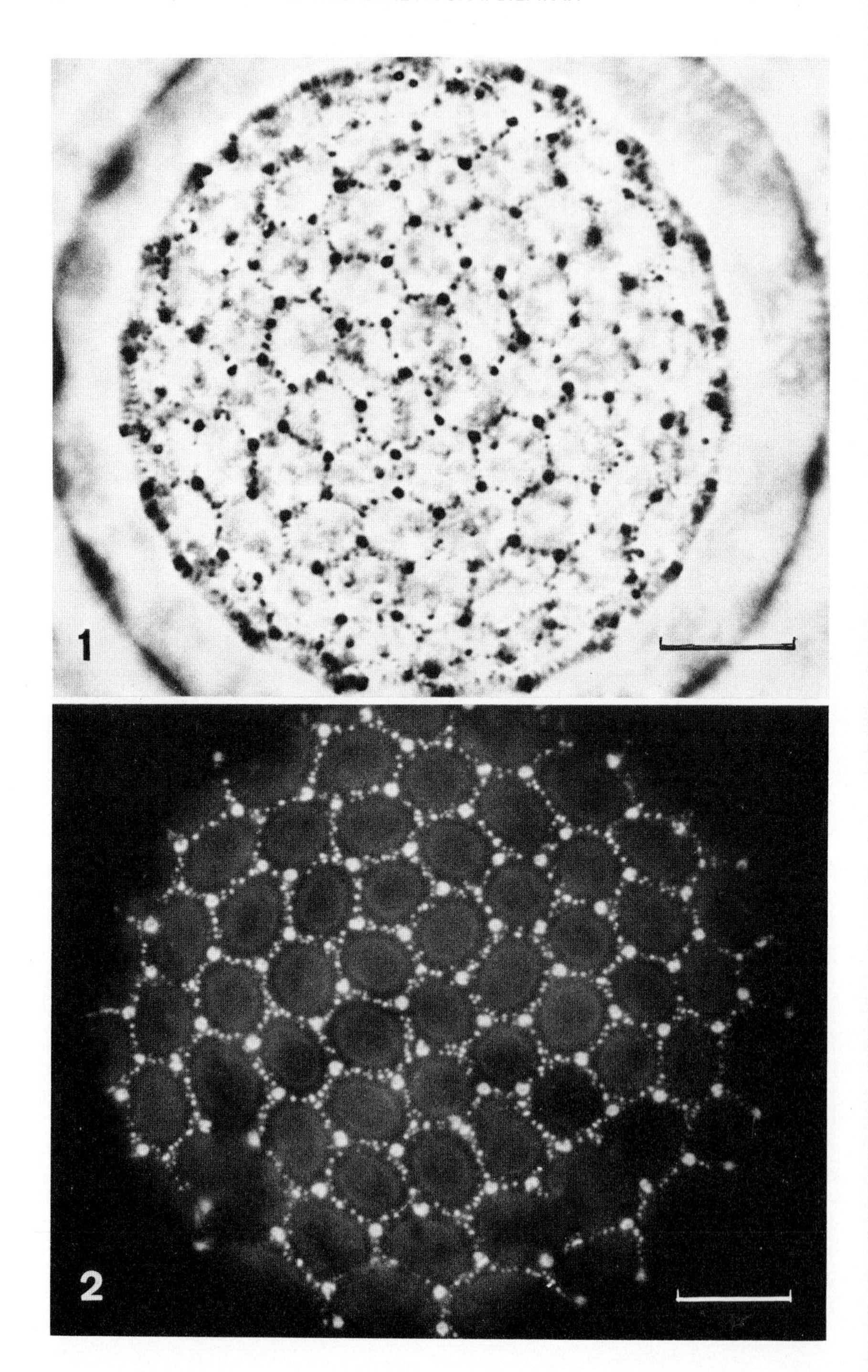
1
2

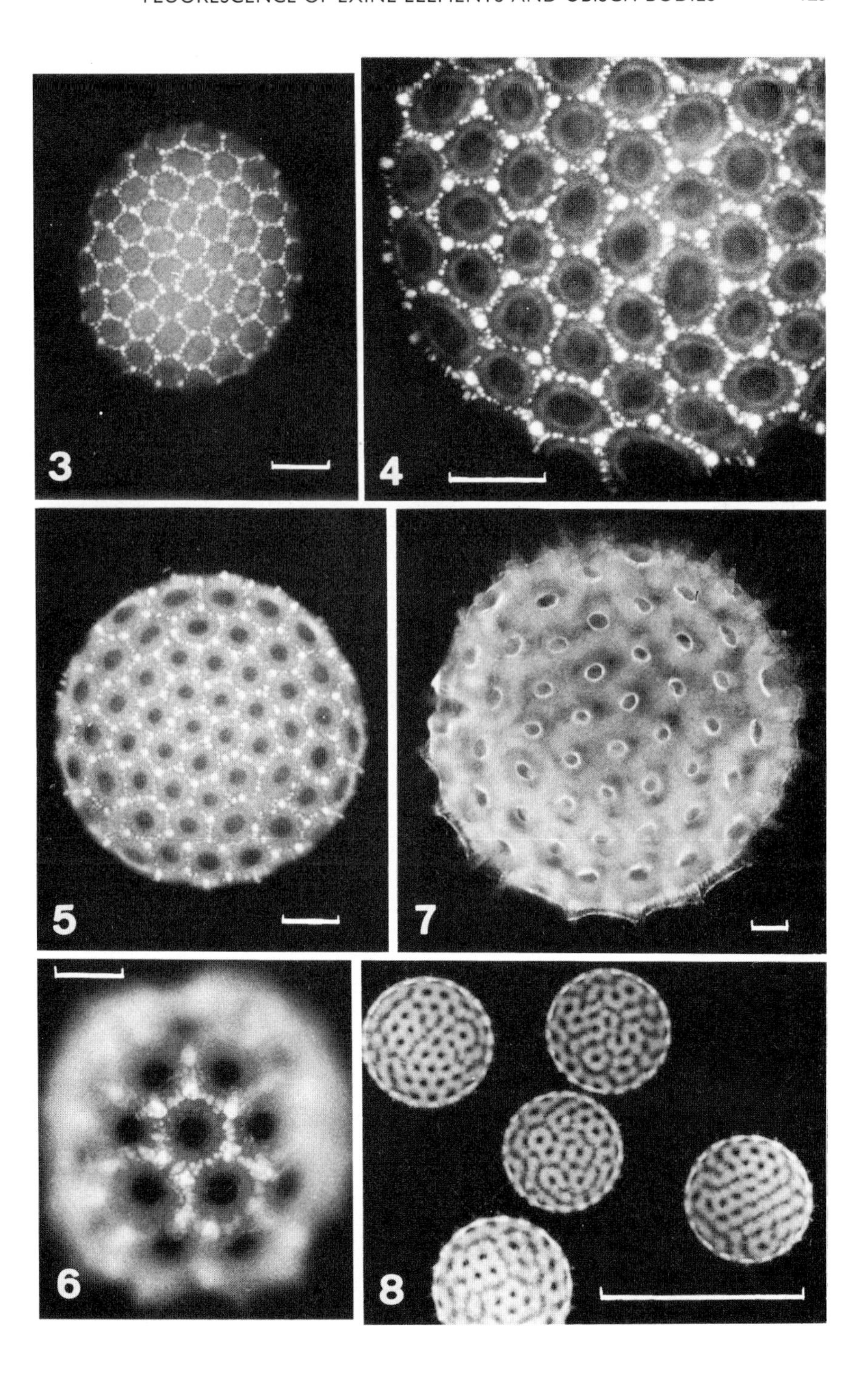
3
4
5
7
6
8

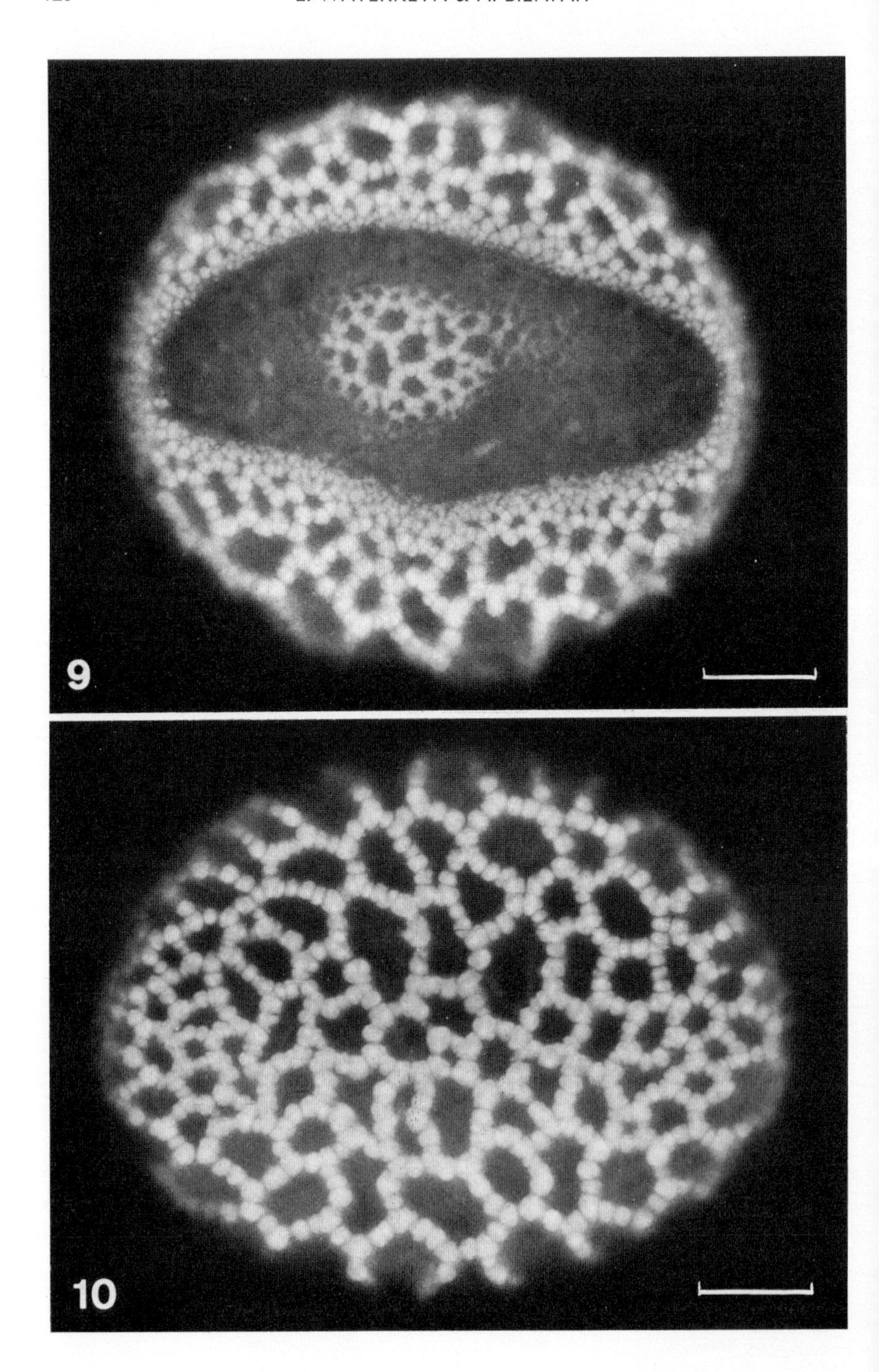
9
10

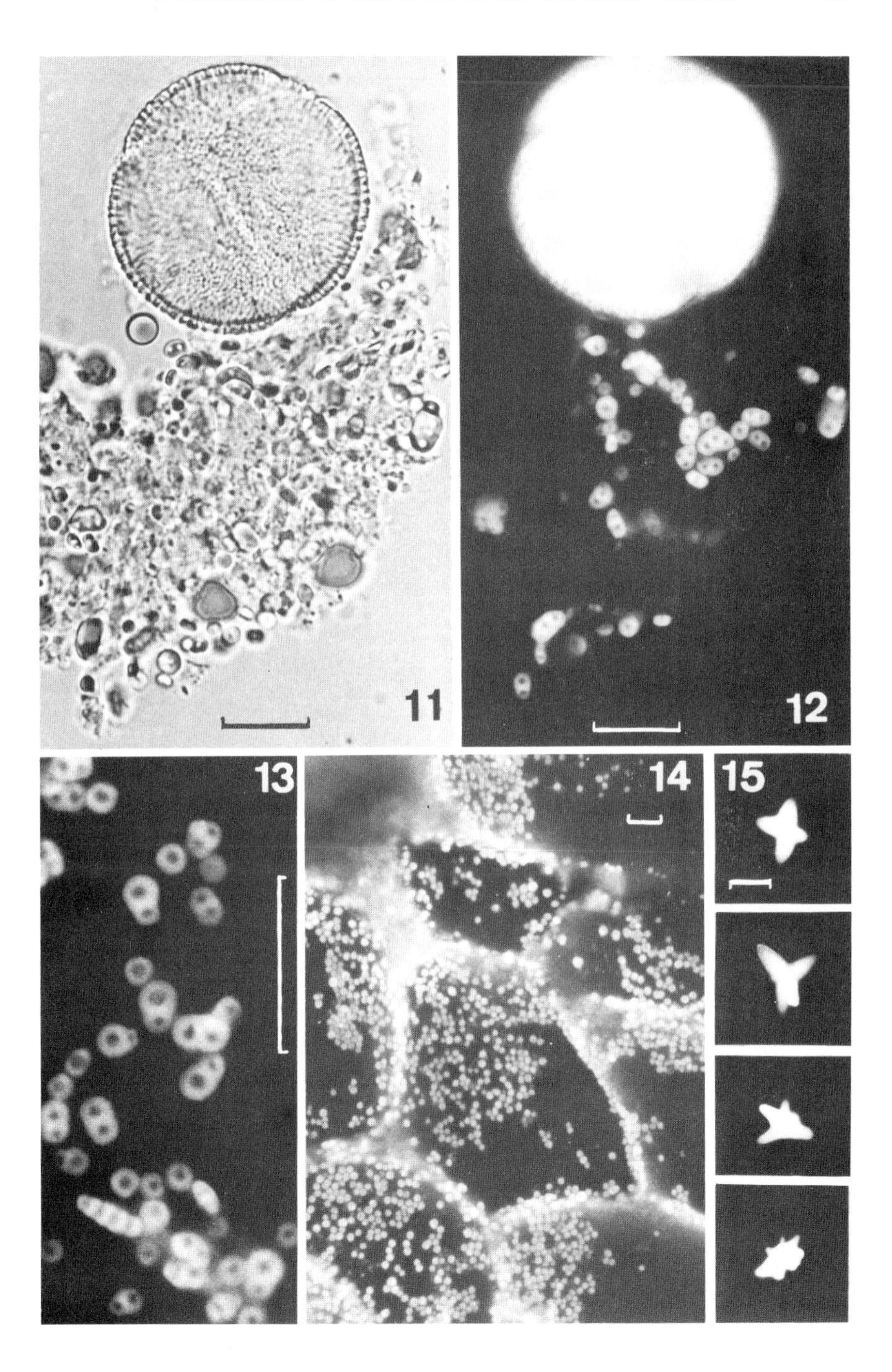
11
12
13
14
15

DISCUSSION ON DRS. WILLEMSE'S & DR. WATERKEYN'S PAPERS.

HESLOP-HARRISON: I would like to comment on Dr. Willemse's paper. One or two of his pictures showed a sporopollenin-like material inside the cytoplasm. Can he explain this?

WILLEMSE: I believe that most of the material is outside the cytoplasm, and none is really inside.

HESLOP-HARRISON: It may have been a sectioning artefact. I want to say something about the possible role of callose in pollen wall genesis. I cannot accept that callose plays any part in pattern formation. Sometimes electron microscope sections show invaginations of the exine which are reputedly filled with callose, but since callose is electron transparent, it cannot be identified.

LINSKENS: Could you not use enzymes such as callase to determine this?

HESLOP-HARRISON: Yes, but it has not been done.

LINSKENS: I would like to ask both authors what general conclusions they have arrived at from their studies of the fluorescence?

WATERKEYN: Fluorescence is altered by the chemical and physical states of the pollen grains.

WILLEMSE: Primary fluorescence is a rather general signal. It results from the various physical components, and is a mixed spectrum. Further work ought to try to separate the components of that spectrum out.

SIEVERS: Drs. Willemse showed a large number of Golgi vesicles in *Pinus* pollen, and he said that they may export the primexime or other cellulosic materials. I only know of cellulose extrusion by Golgi vesicles in the unicellular algae. What is your evidence in this case for saying that Golgi vesicles extrude cellulose?

WILLEMSE: Staining and examination of electron microscope sections. The tests were not specific, and perhaps I should have said 'polysaccharide like' material.

HESLOP-HARRISON: We know that Golgi vesicles export a great variety of materials, and we should hesitate to make any hard and fast decisions about what they do transport.

SIEVERS: Yes, but *not* cellulose. This is secreted on a molecular(not vesicular) basis by the plasmalemma itself.

HESLOP-HARRISON: That is a matter of opinion.

ROWLEY: With regard to the intine where this cellulose material is supposedly placed. Is there really evidence for the presence of cellulose in pollen grains?

SHAW: Yes. Zetzsche found good evidence thirty years ago. We have also isolated hydrolysed products, and most of the walls we have looked at contained about 10%

HESLOP-HARRISON: Why does Professor Rowley suspect the absence of cellulose?

ROWLEY: Well, there are a great many polysaccharides present, and the reason for my question was that Dr. D. Southworth examined three Composite species and was unable to find any cellulose, or at best, a very small amount.

HESLOP-HARRISON: At least there is a polysaccharide present in many places in pollen grains.

LINSKENS: Yes, and it is attacked by cellulase.

SHAW: Perhaps it is not cellulose but a chemically similar substance.

JONKER: I want to draw attention to the necessity of using a correct pollen morphological terminology. The term, foot layer has been used in different meanings. It is very important that pollen physiologists and those investigating pollen ontogeny are familiar with the pollen morphological terminology that they have to mention, e.g. the sporopollenin content of the different layers.

ULTRASTRUCTURAL AND CHEMICAL STUDIES OF POLLEN WALL DEVELOPMENT IN THE EPACRIDACEAE

JUDITH H. FORD.

SCHOOL OF BIOLOGICAL SCIENCES,
UNIVERSITY OF SYDNEY,
Sydney, New South Wales, Australia.

In the genus *Styphelia*, three members of each pollen tetrad degenerate; hence the separate roles of the pollen mother cell, the haploid microspore and the tapetum in pollen wall formation can be determined. Wall development was studied using both histochemical and electron microscopical techniques.

Primary ectexine and endexine layers form within the callose mother cell wall, directed by the diploid mother cell and influenced by a gradient present in the pollen mother cell cytoplasm. After callose dissolution sporopollenin is deposited on the primary ectexine by the tapetum and the lamellate endexine of the viable microspore is considerably thickened by the deposition on the lamellae of osmiophilic material synthesised in the haploid microspore cytoplasm. Subsequently tapetal spheres become attached to the mature ectexine.

Before microspore mitosis the cytoplasm expands and the wall, although not changed in volume, is greatly reduced in thickness, A third layer, the intine, is produced by the haploid microspore protoplast at this stage.

The mature pollen wall consists of three chemically and structurally distinct layers; the outermost ectexine in composed of sporopollenin, the endexine has a high lignin content while the intine is cellulosic.

Introduction

Controversy has ensued for many years from questions concerning the genetic control of spore morphology. Superficially the problem is very simple:- is the control primarily diploid: morphology being determined by the pollen mother cell (pre-meiosis); or is it primarily haploid: the microspore (post meiosis) establishing the pattern of development? Arguments have been put forward in support of both these alternatives.

1. Haploid Control.

Heslop-Harrison (1963-66) studied the chemistry and function of the callose wall which surrounds the P.M.C. during meiosis and separates the haploid microspores from one another after meiosis. Because of the effective isolation of haploid microspores by their individual walls, he suggested that wall morphology was genetically controlled by the haploid spore nucleus.

Mackenzie et al (1967) reported a reduction in the ribosomal population of pollen mother cells during mid-prophase of meiosis. This implied that the pollen mother cell may not be able to control the antigenic properties of the daughter pollen grains and Heslop-Harrison (1968a) postulated the sporogenously derived incompatibility was controlled by the diploid tapetal tissue. This was in accordance with his earlier view that the haploid microspore controlled wall development; the diploid tapetal tissue conferred its antigenic properties.

Godwin (1968) also supported "haploid control". He considered the segregation of different pollen types within species and found that such segregations only occurred either on separate plants or in heterostylous systems: there was no evidence for segregations within an anther. Godwin concluded that this information together with electron microscopical data strongly supported "haploid control".

Diploid Control

Considering Godwin's evidence, Heslop-Harrison (1968c) pointed out that this genetical evidence supported diploid rather than haploid control of wall development. He discussed ways in which his former observations could be brought into line with a theory of diploid control.

Other support for diploid genetical control has come from studies of wall development in aborted pollen grains (Rowley and Flynn 1969). Since the gross morphological and fine structure of pollen grains is similar in both functional and aborted pollen grains there is strong evidence that at least part of the wall structure is under diploid control.

The family Epacridaceae (Ord.Ericales) is subdivided into 2 tribes, the Epacrideae and the Styphelieae. In a series of papers on microsporogenesis in the Epacridaceae, Smith-White (1955-1963) observed that while in the Epacrideae pollen regularly developed in tetrads; the Styphelieae exhibited a diversity of pollen types. He termed this phenomenon the formation of "variable tetrad pollen".

In the genus *Styphelia* (tribe Styphelieae), following an apparently normal meiosis in the pollen mother cell, three nuclei regularly migrate to one end of the cell and degenerate. Only one microspore nucleus is functional. Because in each tetrad the walls of the functional cell and those of the aborting cells show different degrees of development, *Styphelia* provides valuable material for distinguishing between the roles of the microspore, the pollen mother cell and the tapetum in pollen wall formation.

Anthers of Styphelia viridis and S. triflora were studied using both light and electron microscopical techniques.

MATERIALS AND METHODS

i. Electron Microscopy

Anthers to be studied in the electron microscope were fixed in either (a) 0.1M phosphate buffered glutaraldehyde (1.5%) for 2 hours, then 6% for a further 2 hours, followed by post fixation in 2% osmium tetroxide in 0.1M phosphate buffer for 5 hours or (b) in 2% osmium tetroxide in 0.1M phosphate buffer for 2 hours followed by 2% uranyl acetate for 2 hours.

Specimens were then dehydrated through a graded series of ethyl alcohol and propylene oxide and finally were embedded in Araldite Resin M using a thermal vacuum embedding technique. Blocks were trimmed by hand and sectioned at 500-600 A° thickness on an LKB microtone 2, using glass knives.

Sections were collected on nitrocellulose/ carbon coated grids and stained with 2% potassium permanganate (Lawn, 1960) and lead citrate

(Venable and Coggeshall, 1965). They were examined on a Siemens Elmiskope 1 electron microscope at 60 KV and with a condenser aperture of 200μ. A 50μ objective aperture was used to examine most sections but a 30μ aperture was sometimes used to give increased contrast.

ii. Light Microscopy

The outer whorls of bracts were removed from the buds which were then fixed in either formalo-aceto-alcohol (F.A.A.: 5% formalin; 5% glacial acetic acid; 90%,70% ethyl alcohol) or in 1:3 acetic acid/ ethyl alcohol solution. Specimens were dehydrated through a tertiary butyl alcohol series and embedded in paraffin wax.

Sections of 10μ thickness were cut on a Jung rotary hand microtome and treated for histochemical studies. (Table 1.). Sections were mounted in polystyrene, Gurr's water soluble embedding medium, or glycerine jelly; whichever was appropriate to the technique.

RESULTS

1. Development Within the Callose Wall

Studies on pollen wall ontogeny have revealed that in most Angiosperms, a single wall layer, the "primexine", is formed at the completion of meiosis, within the mother cell wall. (Heslop-Harrison, 1963a,1963b,1964; Rowley, 1963,1964; Skvarla & Larson,196

Unlike these species, in the genus Styphelia, 2 distinct wall layers are formed within the callose mother cell wall. The outer and first formed of the two is similar to the primexine of other species while the inner has the morphological characteristics of the "secondary exine" described by Godwin *et al.* (1967). I shall refer to the outer layer as the primary ectexine and the inner as the primary endexine.

Primary Ectexine

At the time of completion of the cell plate between the functional and the aborting cells (there is usually an incomplete cell plate formed between the aborting cells), the pattern for the future ectexine is observed in the region of the callose wall adjoining the plasma membrane. Osmiophilic granular material is soon deposited on this template producing a well defined layer of rod-like structures (probacula), 0.43μ in height. A continuous basal portion of the same material is soon formed under this layer and the primary ectexine attains its maximum thickness of 1.23μ (Fig.1.).

The primary ectexine of the tetrad is not uniform in thickness but attains its maximum thickness at the pole of the tetrad distal to the aborting cells (Pole A) and its mimimum thickness at the pole of the

Fig.1. High power of primary ectexine within callose wall. OsO_4 fixation.

Fig.2a. Diagrammatic representation of the microspore tetrad showing the primary ectexine layer. Note the continuous increase in development of this layer from its minimum width at pole b (centre of the aborting cells) to its maximum width at pole a (centre of the functional cell).

Fig.2b. Tetrad, after the formation of the primary ectexine layer, is still enclosed in the callose mother cell wall. Note the reduced primary ectexine of the aborting cells compared with that of the functional cell. OsO_4 fixation.

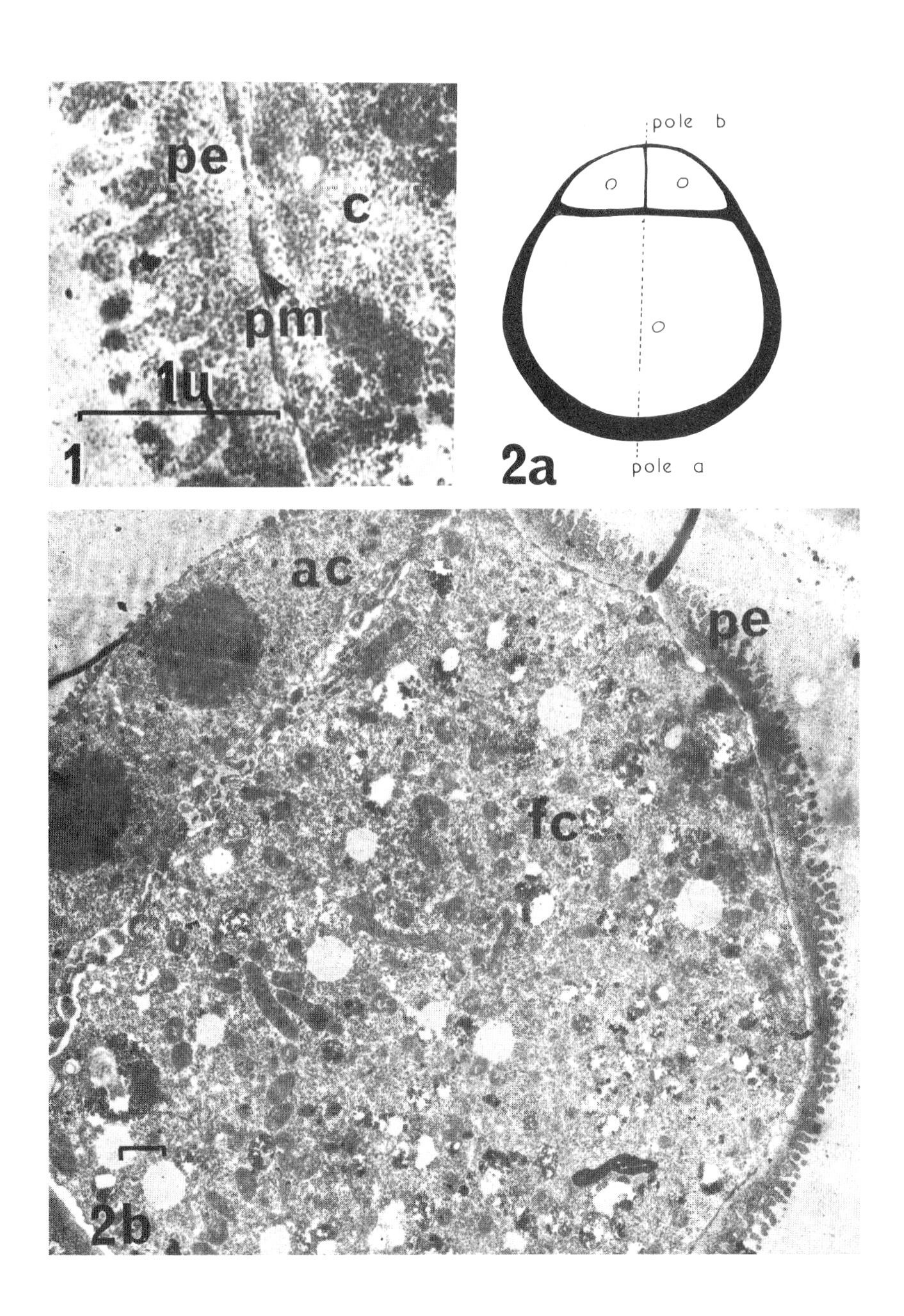
pe
c
pm
1μ
1
pole b
pole a
2a
ac
pe
fc
2b

aborting cells (Pole B), (Figs. 2a and 2b). This change in thickness is continuous from pole A to pole B and is exactly symmetrical about the pole axis.

Smith-White (1959) deduced that there must be a cytoplasmic gradient in the pollen mother cell of *Styphelia* to account for the migration and degeneration of 3 microspores in each tetrad. I have subsequently shown by electron microscopy, interference microscopy and microdensitometry (unpublished results) that the mechanism does involve a cytoplasmic gradient.

Since the wall pattern is polarized in exactly the same way as the cytoplasmic gradient controlling degeneration, it is very likely that exine patterning is influenced by this cytoplasmic gradient in the pollen mother cell.

There is, however, an alternative explanation which could account for exine polarization. If the haploid microspores normally determine exine patterning and the functional microspore in *Styphelia* is the only one which is capable of performing this function, then if the cell plate formed an incomplete diffusion barrier to the patterning substance, the tetrad would be polarized in the manner observed. This hypothesis can be discounted in 2 ways:-

1. It has not been possible in *Styphelia* to dissociate, in time, the processes of cell plate formation and primexine patterning. Adjacent cells in an anther locule may have neither or both of these structures; I have not observed any cells where one is present without the other. These two processes must both be very rapid and closely associated temporarily.

This observation has 2 implications:

a). the rapidness of the exine pattern elaboration suggests that "initiators" were present in the cytoplasm prior to the cell plate formation; probably before the end of meiosis and (b) a barrier (the cell plate) would have to restrict diffusion of the "initiators" greatly to induce the observed gradient in wall patterning, but (i) the cell plate is not a continuous layer and is unlikely to restrict diffusion to this extent and (ii) I predict that to produce a gradient of this magnitude, diffusion across the barrier would be very slow: too slow to account for the close temporal association of cell plate and pattern establishment.

Thus it is most unlikely that the "initiators" are synthesized by the haploid microspore after cell plate formation.

2. There is no evidence that the haploid microspore nuclei are differentially active at this stage. Staining with pyronin to detect RNA synthesis gives a low but equal incorporation of the stain in all nuclei at the time of primexine formation. At later stages there is an increase in RNA in the functional nucleus.

From these observations one can confidently say that the haploid microspore does not control primexine formation. The control lies in the pollen mother cell.

Soon after the cell plate and the pattern of the primary ectexine are established, osmiophilic material is deposited on them. Cytochemical and electron microscopical observations confirm that these walls have the same chemical composition, even though it has not been possible to establish their exact chemical nature.

Although they are chemically identical, the cell plate region (internal tetrad wall) fails to attain the elaborate pattern of the primary ectexine (Fig.3.). Since, in *Styphelia* callose does not normally penetrate into the cell plate region and in small regions where it does penetrate the ectexine pattern is established (Fig.4.) the usual absence of pattern between the microspores must be related to the absence of callose. A possible

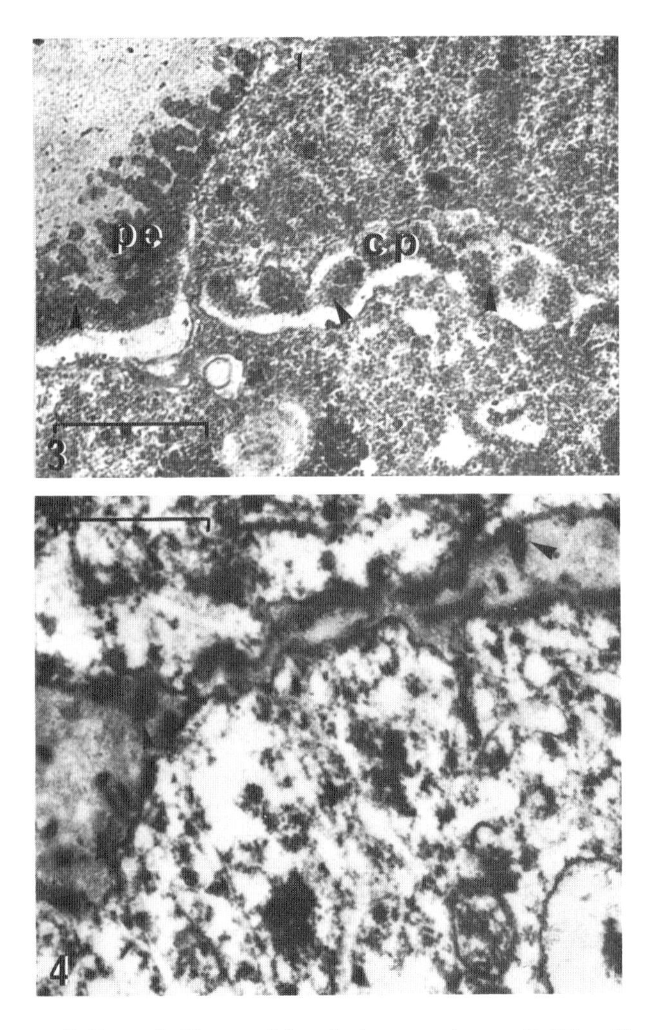

Fig.3. H.P. of the cell plate region of Fig.2b. Note that the cell plate is composed of the same dense, granular material as the primary ectexine (arrows). OsO_4 fixation.

Fig.4. Callose penetration into cell plate region in S.triflora. Note the initiation of probacula where the callose is present OsO_4 fixation.

explanation is that callose supplies some factor, either physical or chemical which is necessary for elaboration.

In speculations on the role of callose in ectexine formation, 2 alternatives are apparent: the first involves a chemical role, the second a physical role. I suggest that elaboration proceeds by either:

1. restricted enzyme degradation of the callose wall followed by deposition of cellulose. Degradation products of the callose may be used in cellulose synthesis, OR

2. deposition of cellulose in a specific pattern gradually compressing the callose and thus moving it away from the site of deposition.

In the first of these, elaboration would not proceed normally in the absence of callose because the substrate for degradation is not present. Without specific areas of degradation cellulose would be laid down evenly. Since the chemistry of a wall formed without callose is the same as one with callose - callose degradation products are obviously not essential. This does not eliminate the possibility that they are normally used in elaboration.

In the second alternative, the pattern would also be lost; but here the problem is a spatial one, perhaps related to stress factors, discussed by Wodehouse (1939) and haptophytic effects by Godwin (1968).

Skvarla and Larson (1966) described the initiation of exine patterning: "The plasma membranes become contorted and completely withdrawn from contact with the callose". High resolutions of sections at this time indicate that there is an electron translucent substance formed between the plasma membrane and the callose wall (Fig.5.).

Because there is not a space between the membrane and the callose, I suggest that the 2 former propositions are more valid than one of plasma membrane withdrawal. However, since there are too many unknown factors at this stage, I cannot differentiate between the 2 propositions.

In summary then, the pattern of the primary ectexine is controlled by the diploid mother cell cytoplasm which interacts with the callose wall to form the delicate organisation of this layer. The necessity for the presence of the callose wall is not understood.

Primary Endexine.

When the organisation of the primary ectexine is almost complete but when the callose wall is still intact, membranous structures can be seen immediately

Fig.5. H.P. plasma membrane region at telophase II. the pattern for the ectexine can be seen adjoining the plasma membrane. Osmiophilic material is starting to accumulate on the pattern. Glutaraldehyde fixation.

Fig.6. Membranous structures in a lamellar orientation are observed under the plasma membrane. OsO_4 fixation.

Fig.7. At stage I membranous structures in a lamellate orientation are observed under the plasma membrane. At stage 2 the plasma membrane breaks down and a new plasma membrane is formed inside the lamellar layer.

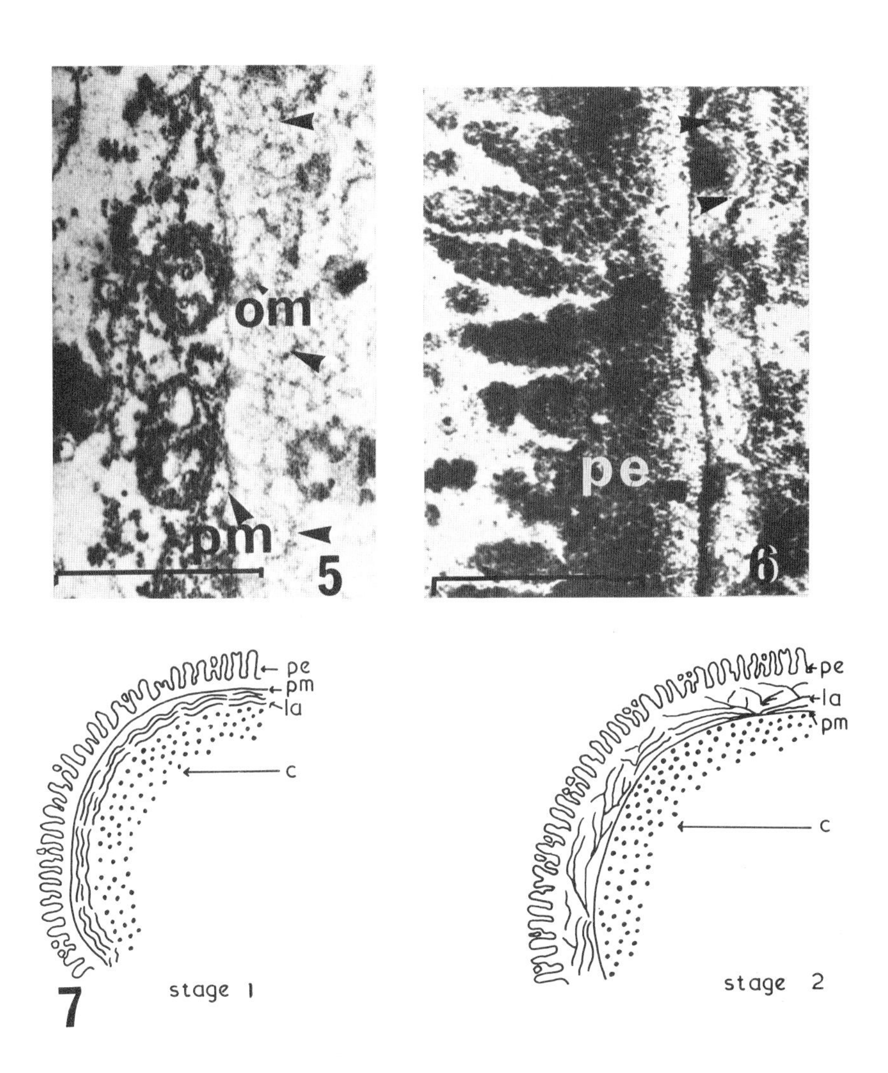
om
pm
5
pe
6
pe
pm
la
c
7
stage 1
pe
la
pm
c
stage 2

Fig.8. Lamellae are now observed outside the plasma membrane. They lie within an electron translucent matrix and a dense osmiophilic substance is deposited on the surface of those closest to the primary ectexine (arrow). The tetrad is still enclosed in the mother cell wall. OsO_4 fixation.

Fig.9. Deposition of granular material on the ectexine is complete and despite the changed chemical composition of the ectexine, the pattern of the primary ectexine is retained. Osmiophilic material is being produced in large organelles in the cytoplasm of the functional cell. I have referred to these organelles as lignin producing organelles (LO) glutaraldehyde fixation.

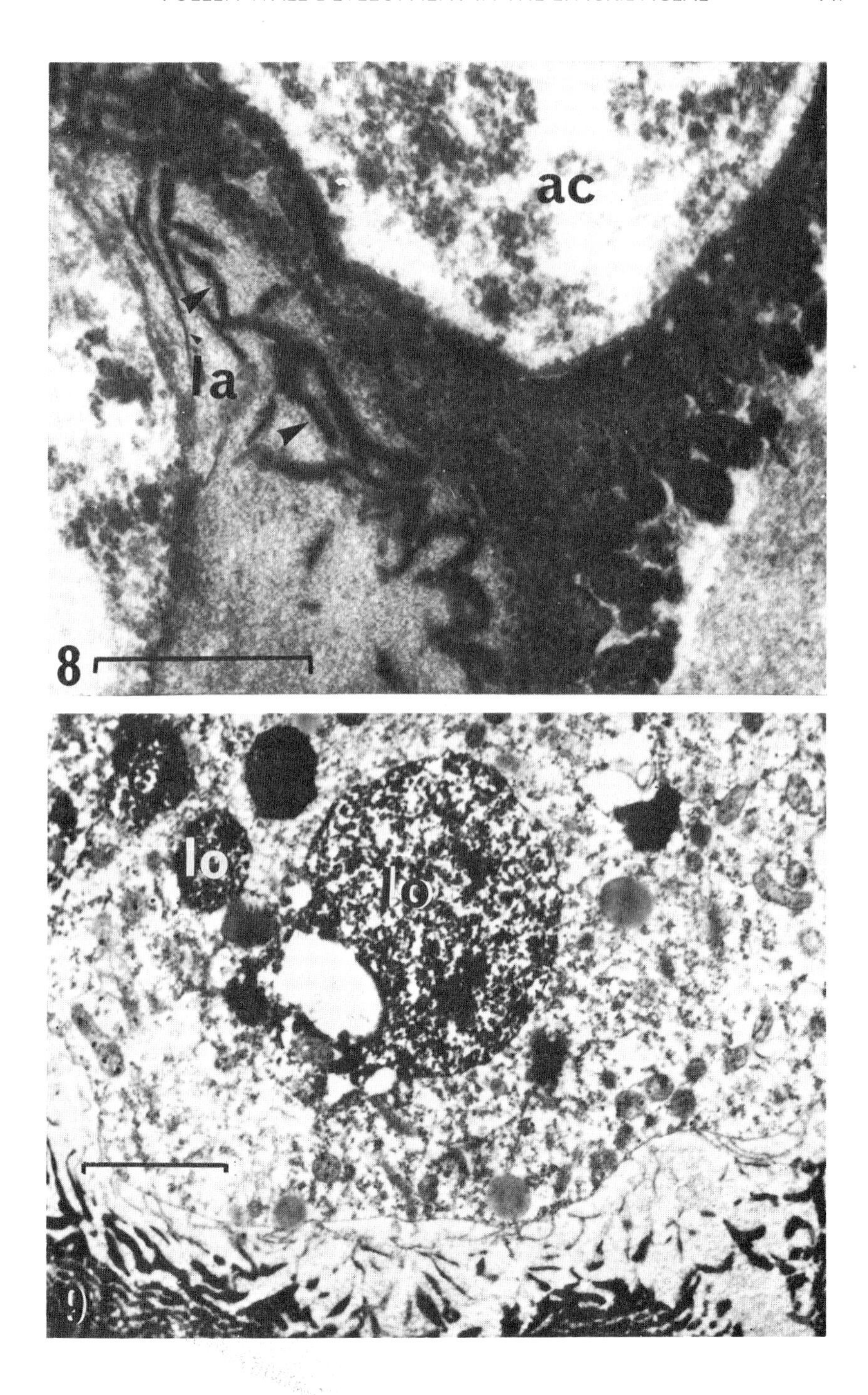
ac
la
8
lo
lo
9

under the plasma membrane of the functional cell (Fig.6.). These lamellae are produced in equal numbers under the ectexine of the outer and inner tetrad walls. Aborting cells rarely produce a significant number of these lamellae but those that are produced are affected, in the same way as the ectexine, by the gradient present in the tetrad. Thus lamellae production is at least influenced by the 2n pollen mother cell cytoplasm. Lamellae are first seen under the plasma membrane (Fig.6.). Shortly after their appearance the plasma membrane breaks down and reforms inside the lamellate layer. In some ways this process is similar to that of callose deposition early in meiosis (Skvarla and Larson 1966). As a result of lamellae formation the diameter of the cytoplasm is reduced and the thickness of the wall is increased proportionately (Figs. 7 and 8.).

Despite their early development, these lamellae closely resemble the strands of the "secondary exine" in _Ipomoea purpurea_ (Godwin _et al_., 1967) and those found in the pore region of some grasses (Rowley, 1964, Skvarla and Larson, 1966).

Further lamellae are produced only in the functional cell. The lamellae are produced in an electron-transculent, amorphous matrix, and as more are produced, those first formed are pushed out towards the ectexine where a dense osmophilic substance is laid down on the surface of each lamella. Since the deposition of this dense substance coincides with the production of an identical substance by the cytoplasm of the functional microspore (cytochemical and electron microscopical evidence), it is clear that this wall substance is synthesized in the cytoplasm of the functional haploid microspore (Fig.9.). This is in agreement with the findings of other authors who have attributed such depositions to an active microspore protoplast.

2. Development after dissolution of the Callose Wall.

a). Ectexine.

When the tetrad is released from the callose wall the primary ectexine, measured at the region of maximum thickness, is reduced from 1.23μ to 0.70μ. Concurrent with callose dissolution, the spore tetrads undergo rapid expansion and one might expect that this expansion would account for the observed thinning of the primary ectexine. Calculations on the volume of ectexine material before and after expansion indicate, however, that there is a significant reduction in the volume of material

Fig.10. As the tetrad is released from the mother cell wall, the primary ectexine is reduced from 1.23μ to 0.70μ. The tapetum now produces a granular material which is deposited on the primary ectexine, changing its chemical composition (Table II). Note the production of small vesicles between the ectexine and lamellate layers. glutaraldehyde fixation.

Fig.11. Vesicle production.

(a) Vesicles in the outer wall of the maturing tetrad. They are absent in the inner wall.

(b) Vesicles in the mature wall

(c) Production of lipoidal material in tapetal mitochondria.

(d) Lipoidal material at the surface of the tetrad.

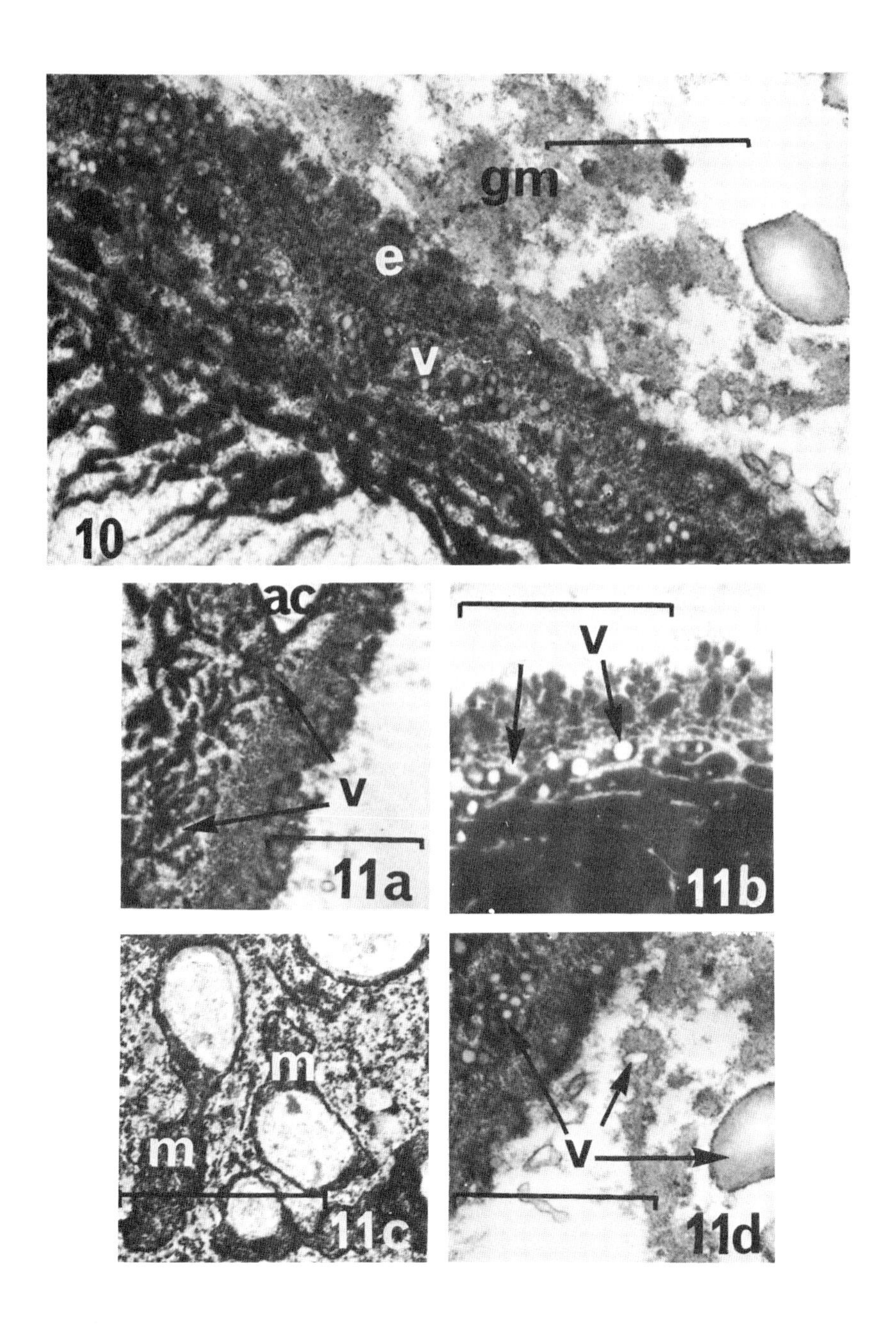
gm
e
v
10
ac
v
11a
v
11b
m
m
11c
v
11d

present. The cause of this reduction is not known but since the primary ectexine is not totally removed it is more likely that there is a structural breakdown than enzymatic degradation.

Immediately following dissolution of the callose, fine granular material is deposited on the primary ectexine of the tetrad (Fig.10.). The thickness of the ectexine of both the functional and of the aborting cells is increased by approximately 0.4μ. Since all parts of the tetrad wall are increased by the same amount, it is clear that the deposition of this material is not influenced by the microspores themselves. It is synthesized by cells surrounding the microspores, the diploid tapetum.

During deposition the detailed structure of the primary ectexine becomes slightly obscured although the initial pattern is recognisable. After deposition the pattern is resumed but its appearance is slightly changed. The rather coarse granulation has now become much finer; presumably its chemical composition has been changed.

Vesicle Production.

Not long after callose dissolution, small vesicles, approximately 750 A° in diameter, are seen between the ectexine and endexine layers of the tetrad on the walls surrounding both the functional and the aborting cells. They are rarely seen between the layers of the internal tetrad wall and when present in this wall, they are always located close to the outer wall (Fig.11a.). At first the vesicles are present in parts of the ectexine but they soon accumulate between the ectexine and endexine. A few vesicles, only a very small percentage of the total number, pass into the lamellate layer. The vesicles seem to be composed of a lipoid substance and since they are limited in distribution to the outer tetrad wall, they are presumably of tapetal origin.

Just prior to callose dissolution, globules of a lipoidal nature are formed within organelles (probably mitochondria) of the tapetal cells (Fig.11.c.). As the tapetal cell membranes break down these globules are released into the anther locule and are often seen near the surface of tetrads (Fig.11d.). It is likely that the tapetal globules are the source of the wall vesicles although intermediates between the two stages are rarely seen.

The vesicles persist throughout development and can be recognised in the mature wall (Fig.11b.).

I have seen no reference to these structures in the literature but since the tapetum is thought to be responsible for contributing many substances to the pollen grain wall; oils, pigments and perhaps incompatability substances (Heslop-Harrison 1968), it is not unreasonable to suppose that these vesicles represent one of the substances. Further research should indicate the function of these vesicles.

Ectexine Globules

As the endexine undergoes its rapid thickening, spheres, usually 4.0μ in diameter, are deposited on the ectexine. These spheres have the same electron density and amorphous appearance as the endexine. There are no differentiated regions in the ectexine for the attachment of these spheres which appear to be deposited randomly over the surface of the tetrad, (Fig.12c.) in equal numbers per unit area of the functional and the aborting cells.

The spheres are produced within the tapetal cells and are dispersed into the anther locule as the tapetal cells break down. Their initial development is very similar to that of Ubisch bodies in *Allium cepa* (Risueno *et al.*, 1969).

Fig.12. Sphere production.

(a) small spheres associated with elements of the endoplasmic reticulum.

(b) large sphere associated with a zone of radiating ribosomes.

(c) spheres seen near the tetrad.

(d) sphere "growing into" the ectexine.

Fig.13. Cytoplasmic diameter and wall diameter at a-o stages in development. Measurements were made on median sections only. The "wall diameter" is the width of wall in one cross section and is thus twice the actual wall thickness.

Stages represented by a-o.

a - telophase II

b - migration of nuclei; prior to cell plate formation

c - cell plate initiated but no development of the primary ectexine.

d - formation of probacula of primary ectexine

e - formation of foot layer of primary ectexine

f - lamellae formation and deposition of osmiophilic material (Fig.6.).

g-i - release of tetrad from callose wall and concurrent expansion of the protoplast j - further deposition of osmiophilic material on lamellae (Fig.9.)

k-l - further deposition on lamellae and some resynthesis of the cytoplasm

m-0 - the cytoplasm increases in diameter up to the stage of pollen grain mitosis (n) and although the wall diameter is decreased, a new wall layer, the intine, is formed.

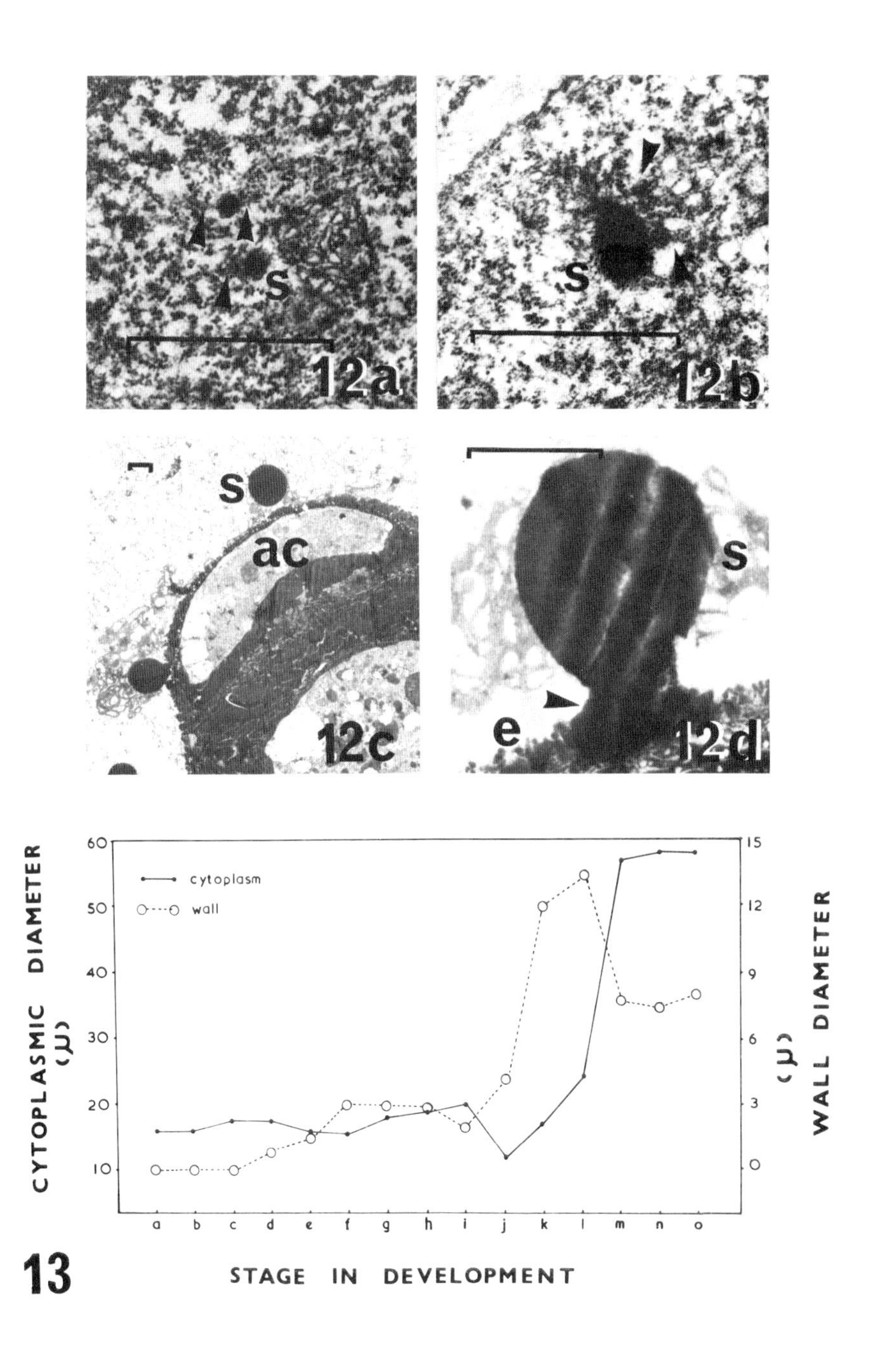
s
12a
s
12b
s
ac
12c
s
e
12d
cytoplasm
wall
CYTOPLASMIC DIAMETER (μ)
WALL DIAMETER (μ)
STAGE IN DEVELOPMENT
a b c d e f g h i j k l m n o
60
50
40
30
20
10
15
12
9
6
3
0
13

Before callose dissolution bodies are seen within membranes of the endoplasmic reticulum (Fig.12a.) and soon after as described in Helleborus (Echlin and Godwin 1968) they are associated with a zone of radiating ribosomes (Fig.12b.). The spheres increase in size and density and retain their association with the endoplasmic reticulum often even after the tapetal cells have broken down. Once they have reached the microspores they appear to "grow into" the ectexine (Fig.12d.), but even so they do not become strongly attached to the spores and they fall off easily during sectioning and after acetolysis without leaving obvious scars.

It is difficult to assess the relationship of these spheres to Ubisch bodies. Although their initial development is very similar, they do not appear to have a central "nucleus" (Risueno *et al* 1969) and they attain a size about ten (10) times that of most Ubisch bodies. They are also chemically dissimilar to Ubisch bodies (later section). At present there is only their similarity to initial development to suggest that they are homologous structures.

Cytoplasmic Expansion

Towards the completion of the deposition of osmiophilic substance on the lamellae, the cytoplasm begins to be re-synthesized and hence to expand. This expansion proceeds over a period of about 3 weeks. During cytoplasmic growth the wall is increased in diameter but is decreased in thickness (Fig.13.). There is no significant change in the volume of wall material present and the change in wall dimensions can be accounted for by a change in orientation of the wall components.

Late in the expansion period a new wall layer is formed inside the lamellate layer. This layer, the intine, is only formed in the functional microspore directed by the haploid microspore nucleus. At maturity, this intine attains a maximum thickness of 1μ.

Characteristics of the Mature Wall

The mature pollen wall of *Styphelia* is composed of 3 layers which differ from each other both structurally and chemically. The outermost of these layers, the extexine, is approximately 1μ thick and is organised into a series of bacula which arise from a continuous basal layer. It is composed of an

TABLE I

STAINING METHODS

Substances detected	Reagents	Reaction	Reference
Fats	Sudan 3 (saturated solution in 70% ethyl alcohol)	Fat: yellow to orange Fatty acids: unstained	Baker (1947)
	Sudan 4 (saturated solution in 70% ethyl alcohol)	Fat: orange to red Fatty acids: unstained	Gomori (1952)
	Sudan black B (saturated solution in 70% ethyl alcohol)	Fatty material is bluish-grey or blue-black	Burdon, Stokes & Kimbrough (1942)
Lignin	Safranin o (70% ethyl alcohol)	Chromosomes, nucleoli, cuticular and lignin substances stain red	Jensen (1962)
	Ammoniacal Basic Fuchsin	Lignified walls and cutin are stained red	Gurr (1965)

	Azure B	DNA & lignin stain clear blue green. RNA purple or dark blue	Flax & Himes (1952)
	Phloroglucinol	Lignin stains red	Johansen (1940) Siegel (1953)
	Monoethanolamine	Removes lignin	Bailey (1960) Siegel (1953)
"Sporopollenin"	90% acetic anhydride 10% conc. H_2SO_4	Sporopollenin remains after acetolysis	Erdtman (1943)
Cellulose	Aniline Blue (aq)	Cellulose walls and cytoplasm stain blue	Jensen (1962)
	Erythrosin	Cellulose tissues stain red	Chamberlain (1915)
	IKI/75% H_2SO_4	Cellulose is stained blue in IKI and is then hydrolysed by the acid to colloidal hydrocellulose	Johansen (1940)
Pectin	Ruthenium red	Pectin stains red	Gurr (1965)

TABLE II

Staining Reactions of the Layers of the Mature Pollen Wall in Styphelia

	Sudan 3,4 or black B	Safranin o	Ammoniacal Basic Fuchson	Cyanin	Azure B	Phloroglucinol	Remain after Acetylation	Remain after Ethanolamine	Erythrosin	Aniline Blue	IKI 75% H_2SO_4	Ruthenium red
Ectexine	+						+	?	Slightly +	+		+
Lamellate layer		+	+	+	+		+					+
Bobbles		+	+	+	+		+					+
Intine & Pore								+	+	+	+	
Cutin	+											
Lignin		+	+	+	+	+	+					
Cellulose								+	+	+	+	

extremely resistant lipid material, "sporopollenin". Sudan dyes stain this material and it is resistant to acetolysis and saponification by means of hot alkalis (Table II).

It has been established in the above study that the pattern of the primary ectexine is directed by the diploid pollen mother cell cytoplasm at the junction of the protoplast and the callose wall. It can now be added that the fine granular material deposited on this primary ectexine by the tapetum, is sporopollenin. It has not been possible to discover the exact chemical nature of the primary ectexine in *Styphelia* but chemical tests suggest that it has a cellulose component. Obviously the electron dense granular component observed in the electron micrographs (Figs. 1, 2b, 3, 4) is not cellulose thus it is most likely that the cellulose forms a matrix for this granular component as was suggested for *Cannabis sativa* (Heslop-Harrison, 1964). A lamellate layer 2μ thick, is present beneath the ectexine. These lamellae, because of the expansion of the protoplast, have lost much of their original rather random orientation and form fairly regularly spaced concentric lines.

In histochemical analysis the dense osmiophilic material of this layer and of the spheres does not give a positive reaction for lipids, but like sporopollenin is resistant to acetolysis. Because the material stains in ammoniacal fuchsin, azure B, safranin O, and cyanin; is autofluorescent and is removed by heating in ethanolamine at 97°C for 3 hours, it is concluded that this substance is a lignin. This "lignin" does not stain in phloroglucinol (Schubert, 1965) and it is not surprising that this extremely resistant lignin lacks the aldehyde groups characteristic of many lignins (Table II). In their chemical investigations, Shaw and Yeadon (1964) found a lignin component in spore membranes and suggested that the nonreactivity of this "lignin" was due to its deposition in a "sandwich" structure. If the "lignin" demonstrated in *Styphelia* is characteristic of spore membranes, it is unnecessary to postulate a sandwich structure since its staining properties can be attributed to a lack of aldehyde groups.

Heslop-Harrison (1968b) also suggested that the mature endexine had a lignin component which lacked aldehyde groups. He postulated, however, that this property of the lignin developed with maturation

that the lignin of the young spores possessed aldehyde groups. This is not true in *Styphelia*: the lignin does not change its properties with maturation.

During wall development lignin is synthesized in 2 regions of the anther locule - the lignin of the lamellar layer is synthesized by the protoplast of the functional microspore while the lignin of the spheres is synthesized by the tapetum. In both cases, when the lignin is removed by treatment with ethanolamine, a cellulose backbone remains. This cellulose backbone is synonymous with the electron translucent, amorphous matrix in which the lamellae are produced.

The innermost layer of the pollen wall, the intine, is approximately 1μ thick and is composed of cellulose (Table II.). This layer is synthesized by the functional microspore at a late stage in development.

SUMMARY

In the genus *Styphelia*, pollen wall development involves three main phases. It is possible in each of these phases to distinguish between the roles of the microspore and the tapetum.

1. Development within the callose wall

Two distinct layers are produced within the callose wall.

a. Primary ectexine

The synthesis of this layer is directed by the cytoplasm of the 2n pollen mother cell, but the presence of the callose wall is necessary (in some way, not yet understood), to produce its delicate organisation.

b. Primary endexine

The pollen mother cell protoplasm controls initial production of lamellae but later production and the synthesis of cellulose and lignin are largely controlled by the individual microspores. This layer is greatly reduced in the aborting cells.

2. Dissolution of the callose wall

a. Ectexine

When the tetrad is released from the callose wall, sporopollenin, which has been synthesized by the tapetum, is deposited on the primary ectexine.

b. Lamellate layer

There is no further production of lamellae after dissolution of the callose but this layer is greatly thickened by further synthesis of the cellulose matrix and by continued deposition of lignin on the pre-existing lamellae. The microspore controls the enlargement of this layer.

3. Cytoplasm Expansion

As the cytoplasm expands from its minimum diameter of 12μ to its maximum diameter of 58μ, the wall is correspondingly decreased in size. It is unusual for a lignified cell to expand but in this case the non-elastic lignin is passively stretched inside a cellulose matrix. As a result of this stretching the lignified lamellae of the endexine assume a new orientation and form fairly regularly spaced concentric lines.

A cellulose layer, the intine, is synthesized by the mature microspore.

Abbreviations

ac	aborting cell
c	cytoplasm
cp	cell plate
cw	cell wall
e	ectexine
en	endexine
fc	functional cell
gm	granular tapetal meterial
lo	lignin producing organelles
m	mitochondrion
ma	electron translucent matrix
om	osmiophilic material
pe	primary ectexine
pm	plasma membrane
po	pore
r	ribosomes
s	sphere
sp	sphere precursor
v	vesicles
la	lamellae

REFERENCES

Bailey, I.W. (1960) J.Arn.Arb.Vol.XLI (1) 141-151.

Baker, J.R., (1947) Quart.J.Microscop.Sci., 88: 463-465.

Burdon, K.L.,Stokes, J.C., and Kimbrough, C.E. (1942) J.Bact., 43, 717-724.

Chamberlain, C.J., (1915) Methods in Plant Histology (University of Chicago Press).

Echlin, P., and Godwin, H., (1969). J.Cell Sci., 5, 459-477.

Erdtman, G. (1943) An Introduction to Pollen Analysis. Waltham, Mass., U.S.A.

Flax, M.H., and Himes, M.H. (1952). Physiol. Zool., 25: 297-311.

Godwin, H., (1968) New Phytologist 67, 667-676.

Godwin, H., Echlin, P. and Chapman, B., (1967). Rev.Palaeobotan.Palynol., 3, 181-195.

Gomori, G., (1952) Microscopic Histochemistry. University of Chicago Press, U.S.A.

Gurr, E., (1965) The Rational Use of Dyes in Biology. Leonard Hill, London.

Heslop-Harrison,J., (1963a). Grana Palynologica 4: 7.

Heslop-Harrison, J. (1963b). Ultrastructural Aspects of Differentiation in Sporogenous Tissue. S.E.B. Symposia XVII 315.

Heslop-Harrison,J. (1964). Pollen Phys. & Fertilization. pp. 39-47. ed. H.F. Linskens, Amsterdam.

Heslop-Harrison, J. (1968a). Nature, 218:90.

Heslop-Harrison, J. (1968b) Canadian Journal of Botany, 46, 1185.

Heslop-Harrison, J. (1968c). Science, 161: 230.

Jensen, W.A., (1962) Botanical Histochemistry. W.H. Freeman and Co., San Francisco.

Johansen, D.A., (1940) Plant Microtechnique (McGraw-Hill, N.Y. and London).

Lawn, A.M. (1960). J.Biophys. Biochem. Cytol. 7: 197.

Mackenzie,A., Heslop-Harrison,J. and Dickinson, H.G. (1967) Nature 215: 999.

Risueno, M.C., Gimenez-Martin,G., Lopez-Saez,J.F. and Garcia, M.I.R. (1969). Protoplasma 67, 361-374.

Rowley, J.R. and Flynn, J.J. (1969) Pollen et Spores. XI, 2: 169

Schubert, W.H. (1965) Lignin Biochemistry. Academic Press. N.Y.

Shaw, G and Yeadon, A. (1964). Grana Palynol. 5, (ii) 247-252.

Siegel, S.M. (1953). Physiol. Plantarum. 6: 134-139.

Skvarla, J.J. & Larson, D.A. (1966). Amer.J.Bot. 53 (10) 1112-1125.

Smith-White, S.(1955). Heredity 9: 79-91.

Smith-White,S. (1959). Proc.Lin.Soc. N.S.W. vol. 84. Part 1. 8-35.

Smith-White, S. (1963). Proc.Lin.Soc. N.S.W. vol. 88. Part 2, 91-102.

Ubisch, G. (1927) Planta, Berlin 3, 490-5.

Venable, J.H. and Coggeshall, R. (1965). J.Cell Biol. 25: 407-413.

Wodehouse, R.P. (1939) Pollen Grains. McGraw-Hill N.Y.

ACKNOWLEDGEMENTS

I wish to thank Dr. A.R.H. Martin for his very helpful discussion of this work.

IMPLICATIONS ON THE NATURE OF SPOROPOLLENIN BASED UPON POLLEN DEVELOPMENT

John R. Rowley

Palynological Laboratory
Stockholm, Sweden

Abstract

The plasma membrane of tetrads consists of a unit membrane, amorphous layer (*ca*. 200 Å thick), and fibrils which bind cations; this free cell surface, like the intine, may be considered as a glycocalyx. The exine has a surface coating which binds thorium and other metals, indicating that the surface is anionic. Exine surface coatings are preserved by adding Alcian blue or ruthenium red to aldehyde fixatives indicating that these cationic dyes stabilize the coating substances. An argument is presented for the transfer of materials through the exine rather than its apertures, prior to intine formation, and it is shown that lanthanum and colloidal iron can pass through the exine. Lanthanum accumulates on lamellae in the nexine suggesting that the lamellae, so common during nexine formation in *Epilobium*, are not lost during development but are masked in the exine of mature pollen. The exine can be remarkably long lasting although reasons are set forth, based upon original observations and published evidence, for considering that the exine is a dynamic system with regard to form and permanence during development and germination.

Introduction

In connection with an international symposium on sporopollenin our concern is naturally enough more with the exine than the voluminous material needs of the pollen protoplast, unless a strong case can be made for an interrelationship between transport to the microspore cytoplasm and wall development and its final form. A role, even a very passive one, in the transport of nutrients is not one of the functions usually ascribed to the exine. The exine on mature pollen has been considered to serve as a possible vapor and UV barrier, and as a general protective and strengthening covering. Air sacs and surface configurations, or lack of them, have been linked with floating or flying with the wind or insects. None of these "functions" is at all relevant during pollen development within the anther.

The studies which Brooks presented at this symposium attest to the primeval production of sporopollenin as a coating for spores. Sporopollenin could be considered as a morphological or synthetic relic, but as a cytologist interested in pollen development I believe that gametophytes, which increase in volume as much as 1,200 X in a few days, are not phytogenetically neutral plants. From my point of view, sporopollenin as a substance, its precursors, or the manner of its production and deposition, must offer a highly successful means of transferring large volumes of material from the tapetum to the gametophyte, otherwise selection pressure would be negative for sporopollenin.

It is entirely rational and productive for us to concentrate on the pollen exine as it appears at maturity in each species in all families since the exine is most suitable for taxonomic characterization and most likely to be preserved in sediments, carried by insects, and so on. Pollen grains and spores often come to a characteristic state for a species at maturity, as chromosomes do at metaphase, although it is neither the beginning nor the end of a developmental sequence naturally leading to germination. Between species, mature pollen may not be representative of similar development, as Górska-Brylass (1970) so well states and defends, since anthesis takes place at various phases of pollen development and of generative cell and gamete differentiation. In this essay I wish to show cause for considering the microspore wall and underlying protoplast as an extremely dynamic system with regard to form, permanence and transported material. In support of this thesis I offer a synthesis of the results of my recent work and experiments done with collaborators. In doing the latter it is necessary, with apology, to refer to manuscripts and other informal documentation. Routine methodological data for illustrations will be included in figure descriptions. In my synthesis of literature on pollen development and germination pertaining to my essay I will not repeat citations or descriptions which are reviewed by Gullvåg 1966, Vasil 1967, Godwin 1968, Heslop-Harrison 1968, Rosen 1968, Echlin 1968, Heslop-Harrison and Dickinson 1969 and Risueño, Giménez-Martín, López-Sáez and Garciá 1969.

The expression of disbelief and concern from the geologists and chemists over the lack of priority for terms used by palynologists for exine parts is an aspect of the symposium which will probably not be included in the printed papers. I need, as I believe many palynologists do, parallel terms for exine subdivisions: one set must permit reference to exine parts without any compositional or ontogenetic overtones. This is basic and has to be strictly morphological while the second set should provide the freedom to express what is known about the composition etc. of "natural" subdivisions of the pollen wall. The latter set of terms must be open for definition by the investigator as to the specific level of the difference; in the future specialists will be able to specify chemical bonding, ratio of hydrogens, enzymes responsible for the synthesis etc. of exine subdivisions. I will use sexine and nexine (1 and 2) exclusively as I do not feel at liberty to superimpose my observations of resistance to degradation, staining, and ontogeny of exine subdivisions upon the existing terms for "natural" exine subdivisions.

I use tetrad for the 4-nucleate cytoplasmic mass resulting from meiosis until cytoplasmic division is completed, after which each uninucleate cell is a microspore until microspore mitosis when the two-nucleate (and two-celled) organism becomes a pollen grain. Both microspores and pollen grains are gametophytes, making that a useful, if somewhat unwieldy, term.

Some of the tests I have used are considered to be specific for mucosubstances or mucopolysaccharides or glycoproteins. I refer to them using the terms of the inventor of the stain procedure. Biochemistry reference works generally deplore the above terms, although I gather that mucopolysaccharides are mainly carbohydrates with a little protein tacked on and glycoproteins are proteins with some saccharides here and there. Kahn and Overton (1970) have sidestepped the problem by using the term acid polysaccharide protein.

The microspore as a free cell surface

The exine is at a free cell surface, and the free cell surface has been referred to as an ectobiological realm (Kalacker in preface to Davis and Warren 1967) which on red blood cells, amebas, pancreatic cells, and other intensively studied cells is highly specialized and selective with respect to the binding or passage of molecules and characteristic not only for the organism but the cells of organs and tissues as well.

The realm Kalacker referred to is the plasma-membrane including its surface specialization, which Bennett (1963) has termed the glycocalyx. According to Bennett (1969a) the glycocalyx may be attached or unattached to the cell membrane and always, so far as is known, contains sugars of some sort although these sugars are in the company of other substances such as amino acids, lipids, lignins, proteins, nucleotides, or other polymers.

In his synthesis of the function of glycocalyces Bennett says that, "In contrast to the cell membrane, glycocalyx components are readily pervaded by water, ions and small molecules and show low electrical resistance. From the position occupied by the glyco-calyx, these components can exert selective influences on substances in the extracellular medium and thus play a regulating role on materials which come into contact with the cell membrane itself." Glycocalyces show enormous variation in structure. The calcified matrix of bone and cell walls of woody plants are examples of robust and rigid glycocalyces. Bennett uses the plant cell wall as an example of a glycocalyx unattached to the cell membrane.

It is not my intention to suggest that the exine is a glycocalyx. In the first place it probably does not meet the requirement for saccharides and in the second place the glycocalyx is a specialization of the plasma membrane and while that implication might please me as a personal hypothesis, it cannot be expressed in public. The plasma membrane of microspores and pollen grains, however, is coated by a glycocalyx, and I will show evidence for that before illustrating several distinct exine surfaces and presenting evidence for the anionic nature of the exine surface.

To find out where, and if possible how, material entered the gametophyte we used colloidal iron for uptake experiments on *Populus tremula* and *Salix caprea* tetrads, microspores and pollen grains (Rowley and Dunbar, ms). Stamens were incubated in Imferon for 30 minutes, 2 hours, or 24 hours. The Imferon solution, obtained from Pharmacia Uppsala, Sweden, contained 50 mg. of dextran complexed with iron (Fe^{3+}) and 9 mg NaCl per ml of water. Stamen filaments were immersed in Imferon for the incubation period, then fixed before or following water chase in a fixative containing 0.2M glutaraldehyde, 0.1M cacodylate, 0.002M HCl and adjusted to a pH of 7.6 and 310 milliosmoles by the addition of 0.05M dextrose. Fixation in glutaraldehyde was for 2 hours at $0^{\circ}C$. Half of each anther was transferred to osmium tetroxide for 12 hours at room temperature. Dehydration was through an ethanol series starting with 30% alcohol and going to 100% alcohol within an hour. Following five changes of 100% acetone, anthers were infiltrated and embedded in Mollenhauer's epon-araldite mixture No. 1 (Mollenhauer 1964).

In young anthers containing tetrads, the callose special cell wall, the surface of the tetrad, and the tetrad contained ferric iron that was converted to Prussian blue by the reaction of Perl following a 30 minute incubation in Imferon. Glutaraldehyde-fixed and unstained sections showed that iron had accumulated on a foam outside the tetrad membrane. Sections of osmium-fixed and heavy metal-stained tetrads showed a well stained unit membrane at the tetrad surface covered by a weakly stained amorphous layer under the iron-coated foam (Fig. 1). Untreated and unstained tetrads had less density than the epon-araldite. Osmium-fixed and heavy metal-stained untreated tetrads showed a few fibrils of a weakly developed primexine between the plasma membrane and callose wall.

In colloidal iron-treated anthers the tetrad surface is similar to the cell membrane, amorphous layer and "fuzz" layer of the amoeba *Chaos chaos* (Revel and Ito 1967, Figs. 4 and 5). In both the tetrad and amoeba, positively charged colloidal metals are bound to the foam or fuzz but not to the tri-laminar cell membrane and amorphous layer. The plasma-membrane of *Populus* tetrads can be considered as having a glycocalyx.

Unstained pollen grains in Imferon-treated anthers had iron on the surface of the exine and within the intine (Fig. 2). Vesicles within protoplasts of treated grains contained iron on fibrils, similar in appearance to fibrils of the intine (Fig. 2). Iron in the intine was converted to Prussian blue by the Perl reaction, and after 24 hours incubation in Imferon, ferric iron was demonstrated within vacuoles in microspores and young pollen grains by this reaction.

Colloidal iron was translocated during our experiments from the vascular system of stamens

Figs. 1-15 Micrographs were taken with a Zeiss EM-9a using a 50μ objective aperture and Agfa-Gevaert 23D56 film developed for 5 minutes in Agfa-Gevaert G-150 diluted 1:5 or Dupont D-56 diluted 1:2.

Sections used for micrographs were picked up on 400-mesh copper grids without supporting films. Except for the material used for figures 9 and 10, section staining was done before the sections were mounted on grids, a wire loop was used for transfers, and sections were kept from clinging to the sides of disposable staining dishes by floating them within thin plastic rings cut from collecting vials.

Abbreviations used for wall subdivisions include:

E=exine
S=sexine
T=tectum
N=nexine
I=intine

Abbreviations used for fixations and stains are as follows:

GA=glutaraldehyde, see text for buffer
Os=osmium tetroxide
UAc=uranyl acetate, 1-2% aqueous, 5min., 50°C.
Pb=lead-citrate or lead-hydroxide
CI=colloidal iron
AB=Alcian blue
RR=ruthenium red
TH=Thorotrast
LA=lanthanum nitrate
SP=silver proteinate
T-SP=thiocarbohydrazide followed by SP
A-KOH-Cl=acetolysis followed by KOH, then chlorination.

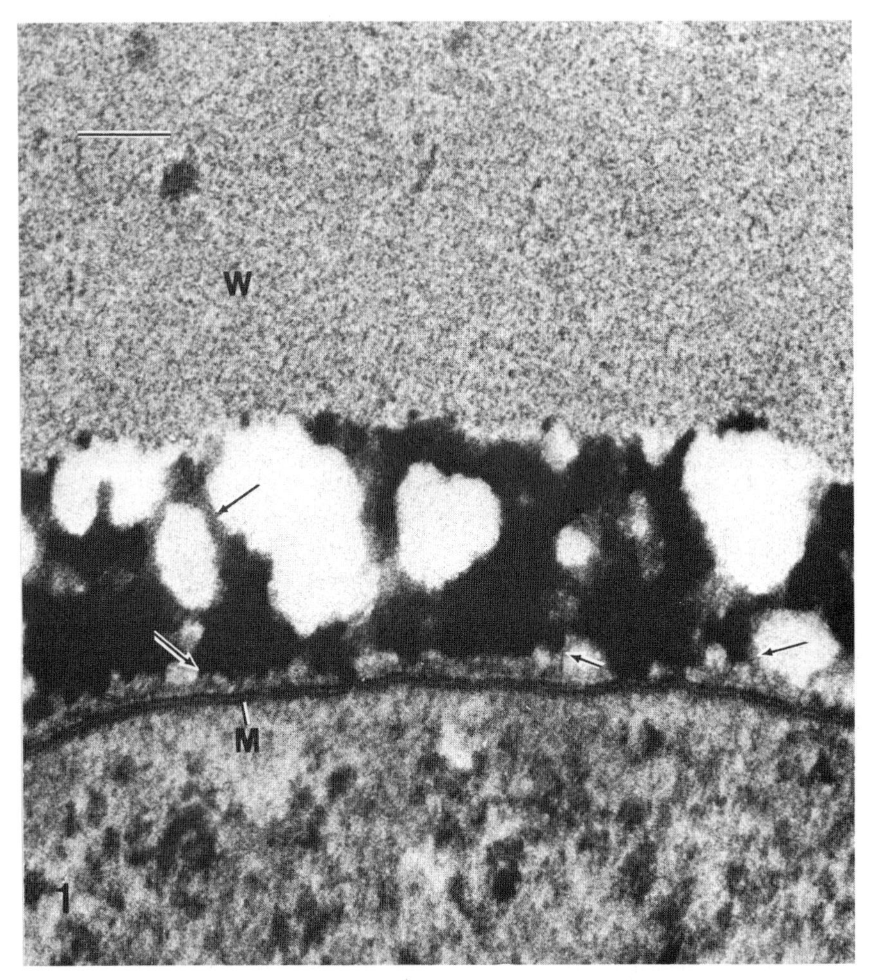

Fig. 1 The plasma membrane of P. tremula incubated in CI prior to GA fixation consisted of unit membrane (M), covered by an amorphous layer, and a coating (arrows) which binds CI. The callose special cell wall (W), with enclaves of iron, is at the top. Stained with Os, UAc and Pb. Line is ca. 0.1μm.

Fig. 2 Pollen of P. tremula incubated in CI before GA fixation. Iron in the intine (I) and in cytoplasmic vesicles (arrows) is emphasized by the absence of any other stain.

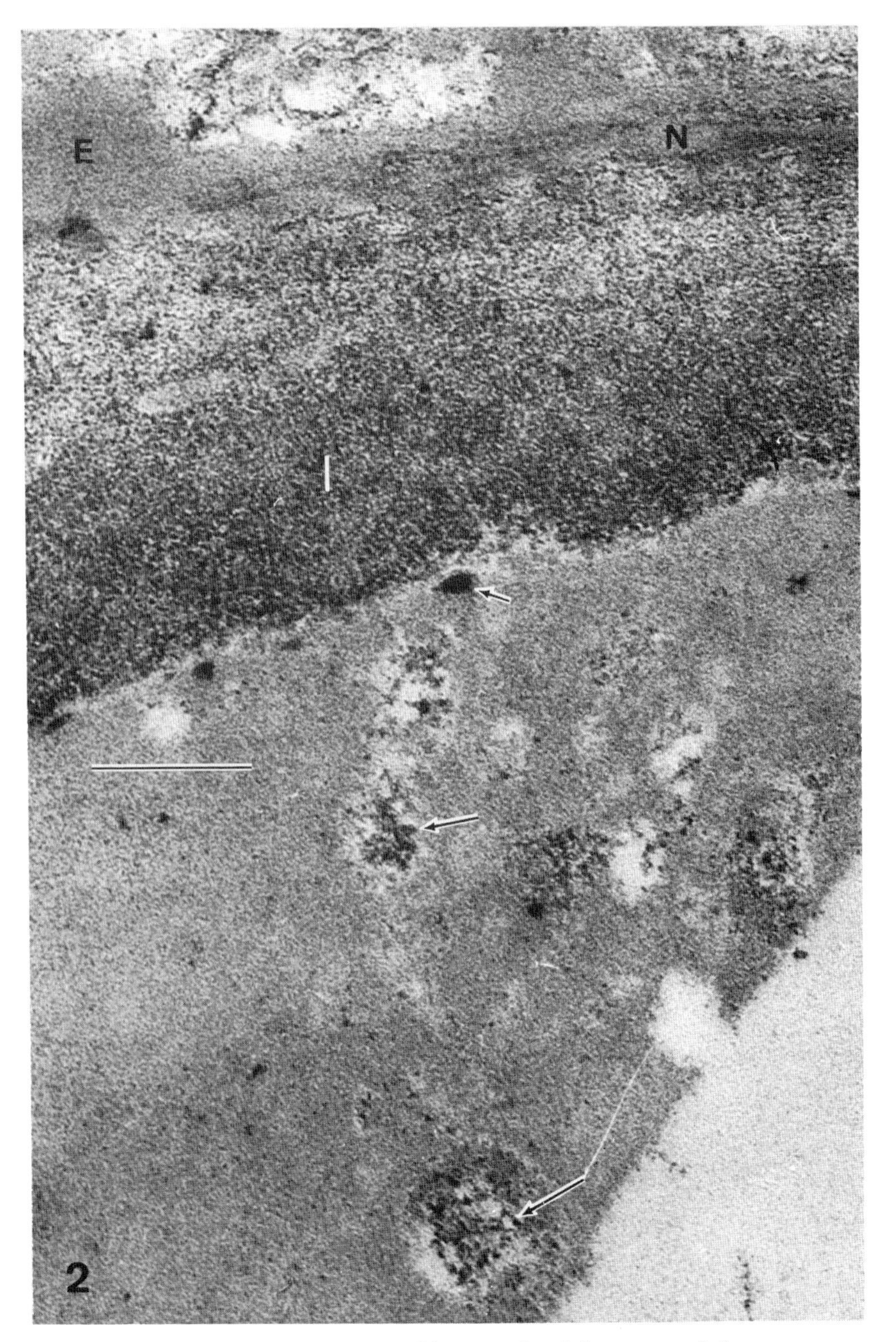

Fig.2 Description below figure 1. Line *ca*. 0.2μm.

through parietal and tapetal cells to the surface of microspore tetrads or intine of pollen grains. Iron accumulated on the tetrad surface or in the intine and was taken into protoplasts of microspores and pollen grains where after 24 hours it had become concentrated in vacuoles. Our results suggest that the intine may be considered as a glycocalyx, and that ferric iron bound to part of the glycocalyx may be taken into the protoplast by pinocytosis.

The exine surface

The surface of the exine has or is coated by: material organized on minor ridges (Sitte 1955, Fig. 27b and Rowley and Erdtman 1967, Figs. 5 and 10), a membrane-like skin (Dunbar 1968), or fibrils extending out perpendicularly from the exine (Dunbar and Erdtman 1969). The morphologically complex exine coat on pollen of *Aegiceras corniculatum* (Dunbar and Erdtman 1969, Fig. 2A) consists of a tubular rod (ca. 100Å in diameter) which joins a lamella located on surfaces of tapetal cells, Ubisch bodies and the exine proper (Rowley, *et al*. 1970, Figs. 4 and 8). Segments of *Aegiceras* anthers were fixed in glutaraldehyde, washed in 0.1M acetate buffer (pH 4.6), and incubated in a mixture of 1 ml Thorotrast in 0.9 ml acetate buffer and then washed in buffer at pH 4.6 before dehydration and embedding. Thorium was bound to exine and tapetal surfaces covered by the tubules (Rowley *et al*. 1970, Fig. 5). The wide use of Thorotrast to demonstrate the localization of mucosubstances at the fine structural level has confirmed Revel's (1964) claim for staining with colloidal thorium dioxide.

The exine surface coatings commented upon above were observed following a variety of routine methods used by electron microscopists and include no pretreatment at all in the case of surface replication of *Populus* pollen. Only a fraction of the micrographs published show exine coatings, implying that they are either rare or require special conditions for preservation and visualization in most species. A comparison of the *Populus tremula* exine surface in figure 3 with micrographs in Rowley and Erdtman (1967, Figs 5, 10) shows that visualization of the surface coat of *Populus* is greatly enhanced by appropriate preparation procedures. Fixation with glutaraldehyde in conjunction with Alcian blue resulted in a vast improvement in the visualization of the exine coating of *Populus*. The same Alcian blue method used with *Chamaenerion angustifolium* pollen resulted in the preservation of an exine coating we had never seen with routine fixatives (Fig. 4). The tectum of *Chamaenerion* pollen was covered by short tubules. According to Behnke and Zelander (1970) Alcian blue 8GX is a charged molecule with at least 2, probably 4, positively charged groups. The dye combines, probably by salt linkage, with polymers of high negative charge (Scott *et al*. 1964).

Sections of anthers containing the almost mature pollen shown in figures 3 and 4 were fixed in 0.1M glutaraldehyde in 0.1M cacodylate buffer at pH 6-6.5 for 18 hours at room temperature. Alcian blue was added to the fixative to produce a concentration of 0.5%. After a rinse in buffer, the anther wafers were fixed in 1% OsO_4 in 0.1M cacodylate buffer at pH 6-6.5 for 24 hours. Materials were dehydrated in an acetone series and embedded in epon-araldite. Sections were examined either unstained, or stained with uranyl acetate followed by lead citrate.

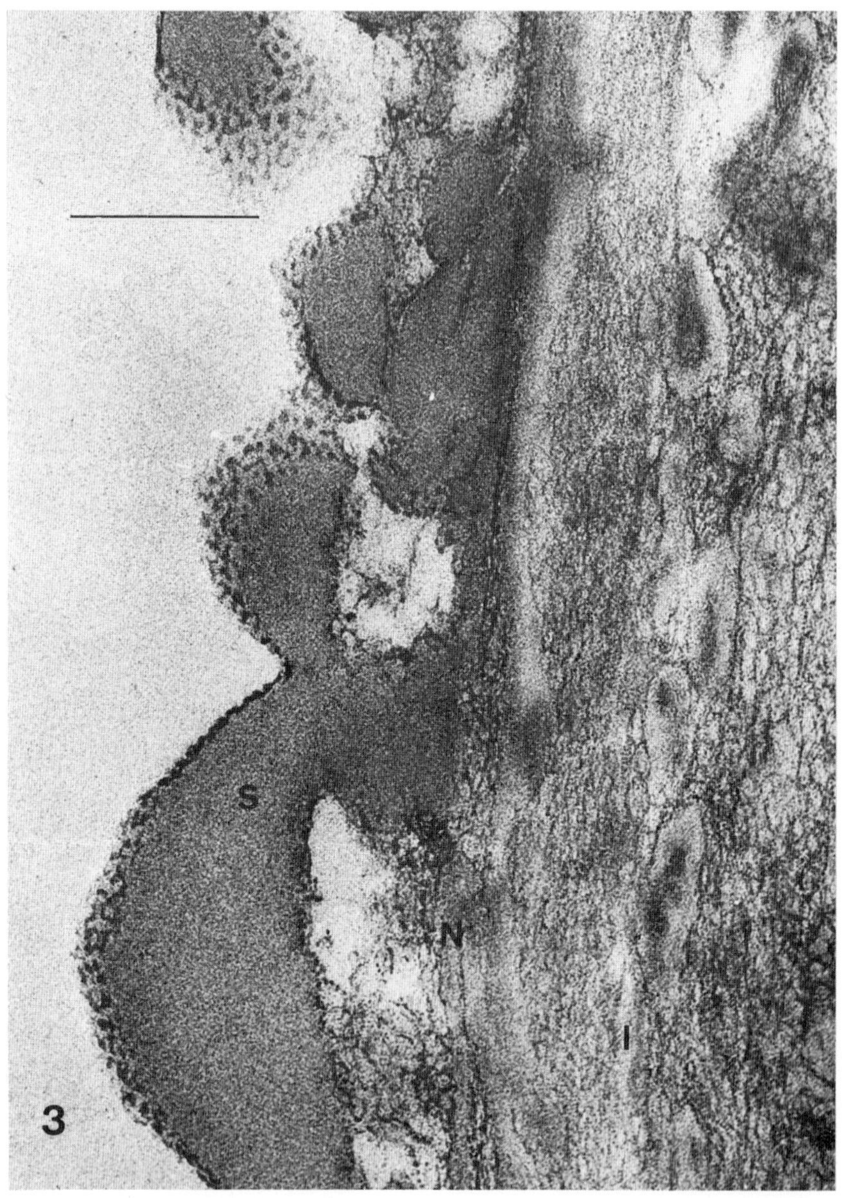

Fig.3 P. tremula pollen fixed in AB and GA to preserve exine surface coating. In glancing section the coating is seen, following Os, UAc and Pb staining as striations of light and dark material. Line ca. 0.2μm.

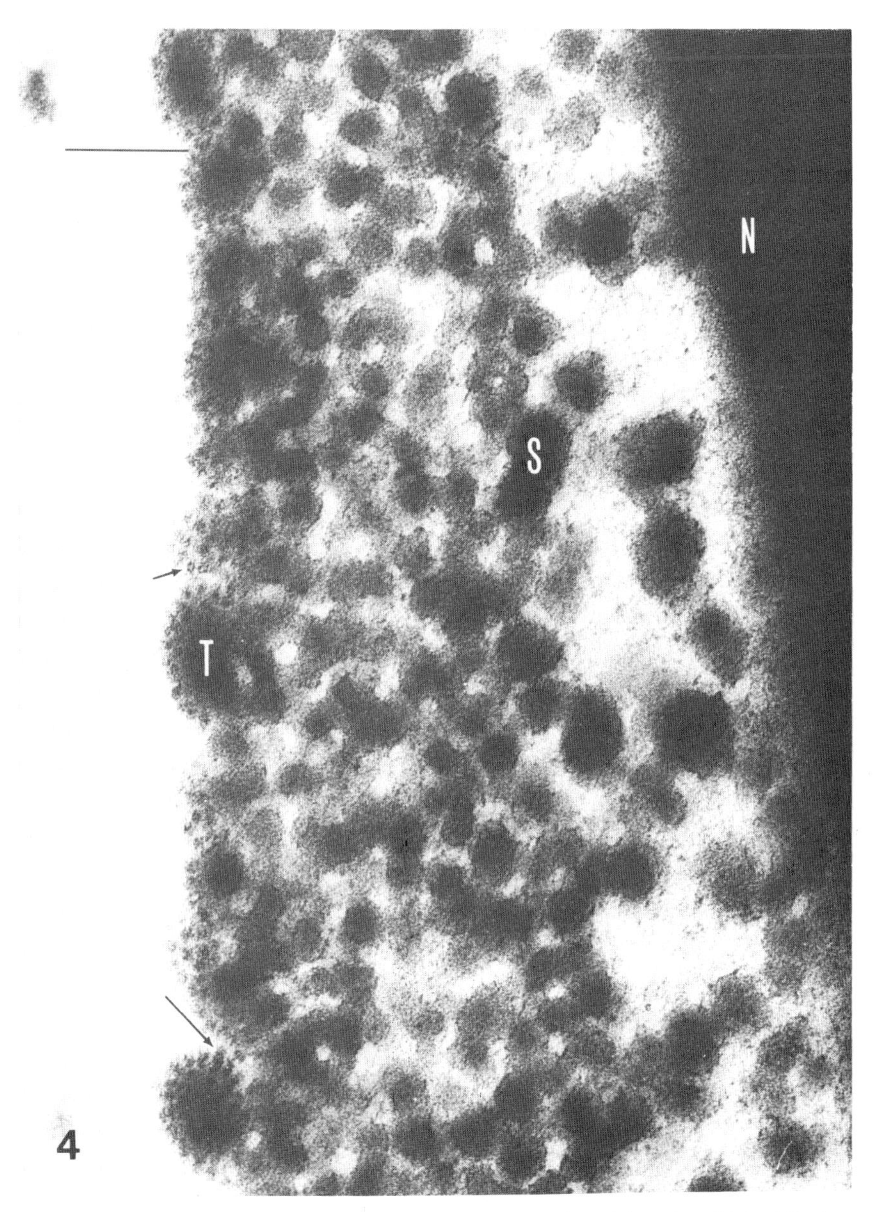

Fig.4 When C. *angustifolium* pollen is fixed with AB in conjunction with GA an exine surface coating (arrows) is preserved and readily visualized following osmium staining. Line *ca*. 0.2µm.

Use of Luft's ruthenium red procedure (1966) on Chamaenerion anther sections resulted in preservation of the exine coating shown in figure 5. For that material I fixed 0.1mm thick anther wafers in 1.2% glutaraldehyde in 0.1M cacodylate buffer at pH 7.4 for 1 hour at $4^{o}C$. Ruthenium red was added to the fixative to produce a concentration of 3000 ppm. After three rinses (10 minutes total) in buffer, the wafers were fixed in 2.7% OsO_4 in 0.1M cacodylate buffer at ph 7.4 for three hours at room temperature. The ruthenium red concentration was 3000 ppm. Wafers were rinsed in buffer for 10 minutes, dehydrated in an ethanol series and embedded in epon-araldite. Sections were stained with lead citrate. Commercial ruthenium red was used in staining the exine surface shown in fugure 5; however, Luft recommends a purification procedure to remove a violet complex. Ruthenium red obtained from K & K Laboratories Inc., Plainview, New York 11803, when purified following Luft's procedure (Brooks 1969) and used at a concentration of 1mg of ruthenium red/ml of fixative has preserved exine coatings as well as the Alcian blue treatment. Brooks reports that, "According to Luft (unpublished) a coupled reaction involving ruthenium red and OsO_4 occurs at exposed sites in tissues where acid mucopolysaccharides are located." Involvement in a full-scale discussion of the possibility of an acid polysaccharide protein exine surface coating is not warranted. We have not found either a polysaccharide (PAS positive) surface coating on pollen exines, using the methods of Thiéry (1967) for electron microscopy (Flynn and Rowley 1970, Fig. 5), or a protein coating, using the ninhydrin reaction on sections after a method devised by Flynn (1969) for electron microscopy. Southworth (personal communication ms.b) found that the exine is not stained by reactions for polysaccharides; however, she interprets the Azure B staining of the exine as indicating a high molecular weight polymer with anionic groups. She suggested that the anionic properties may be conferred by weakly acidic phenolic groups which are present in break-down products of sporopollenin according to Shaw and Yeadon (1966) and Brooks and Shaw (1968). I do not know if the anionic nature of sporopollenin can explain the binding of positively charged dyes and colloidal metals or the morphology of the exine coatings shown in figures 3, 4, and 5; however, the anionic property of the exine surface demonstrated by those micrographs is beyond question. The structural components of the exine coating are difficult to stabilize during fixation against subsequent dehydration and embedding. The affinity of aldehydes and osmium for these exine coatings must be poor, resulting in extraction during preparative procedures and failure to visualize them on most conventionally fixed pollen grains. In hundreds of micrographs of the *Nuphar* pollen fixed after many different treatments, Dr. James Flynn and I have not seen an exine surface coating. The *Nuphar* exine is not coated by periodic acid oxidizable material after aldehyde fixation, acetone dehydration, and epon-araldite embedding (Flynn & Rowley 1970, Fig. 5), but when incubated in thorium dioxide prior to aldehyde fixation, dehydration and embedding there was a thick and periodically irregular accumulation of thorium on the exine surface (Rowley *et al*. 1970, Fig. 6). Readers who check the appearance of figure 5 in the paper by Flynn and Rowley

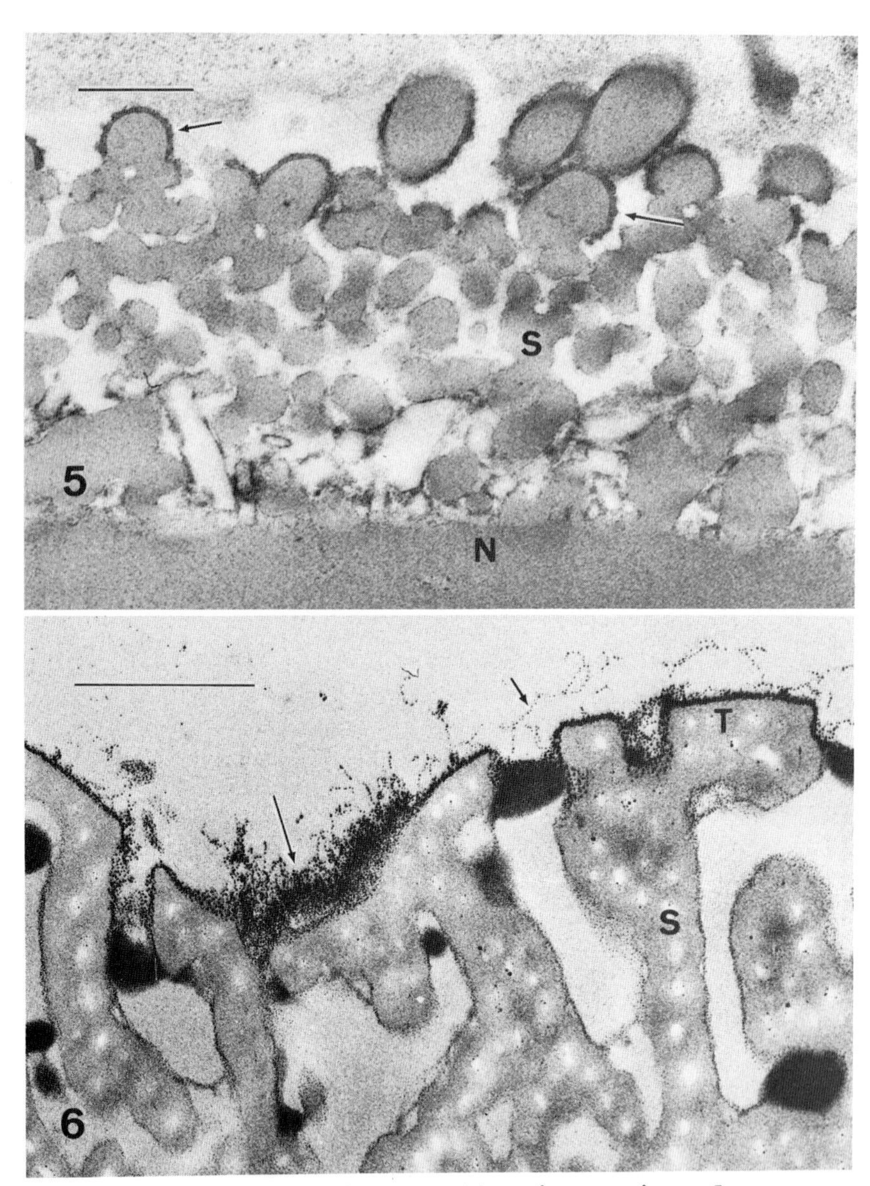

Fig. 5 The exine surface coating (arrows) on C. angustifolium fixed by RR+GA followed by RR+Os. Line ca. 0.2μm.

Fig.6 The exine surface coating (arrows) on S. canadensis pollen binds SP. Line ca. 0.6μm.

may note that the exine and its tubules are darkened following periodic acid oxidation, incubation in a ligand and staining with silver proteinate, the method of Thiéry (1967). Silver proteinate is bound to the untreated exine and exine surface as is shown in figure 6. The nearly mature *Solidago canadensis* pollen in figure 6 was fixed in glutaraldehyde, extracted in ethanol and acetone, hydrated, fixed in osmium and redehydrated before epon-araldite embedding as part of a series of extraction experiments done by Professor Walter Brown.

In figures 3, 4, 5, and 6 the emphasis is on exine coatings on the outer surface of the tectum. There is some contrast enhancement on the exine surface down to and on the nexine in those figures but no surface coating except on the tectum; all represent nearly mature pollen grains. I have not looked at young material treated with Alcian blue and ruthenium red; however, when lanthanum nitrate was added to the glutaraldehyde fixative (see below for methods), lanthanum was bound uniformly to all sexinous surfaces on microspores (Fig. 7) before the intine formed, but as the intine formed lanthanum accumulated largely on the nexine (Fig. 8). These are preliminary results, unsupported by experiments conducted elsewhere; however, the exine surface coating on the sexine of *Chamaenerion* may be altered around the time of intine formation.

Transfer of material across the exine

Tsinger and Petrovskaya-Baronova (1961) considered that the pollen wall, because of the presence of enzymes in both exine and intine, must be a physiologically active structure in the processes of interchange between the microspore protoplast and its environment (cf.

Rosen 1968). Their proposal implied that substances could pass through holes in the exine. I wish to present evidence that molecules can pass through the substance of the exine.

Transfer of colloidal iron through the nonaperturate exine of Populus, described in the foregoing section, is not adequate evidence for passage of iron by the substance of the exine since the laterally continuous part of the exine, the nexine, is not a sheet but a lattice composed of tapes. Microspores and young pollen grains of Epilobium montanum and Chamaenerion angustifolium, with thick exines and well separated pores, were selected to test for entry sites of an inorganic tracer. Lanthanum nitrate is considered neutral in charge and does not cross the cell membrane (Revel and Karnovsky 1967 and Overton 1968), presumably because it is in the form of a colloidal particle; hence, it was chosen as the tracer.

Anthers of Epilobium and Chamaenerion, cut in half to expose microspores and young pollen grains, were fixed in 0.1M glutaraldehyde, in 0.1M cacodylate-HCl buffer to which 1% lanthanum nitrate had been added. As suggested by Overton (1969) and Khan and Overton (1970) the mixture above was adjusted for a final pH of 6.9 and kept on the tissue for 24 hours, then rinsed for 30 minutes in buffer before dehydration. Dehydration was in an acetone series and embedding in Mollenhauer's epon-araldite mixture No. 1. Unstained sections showed that lanthanum had migrated through apertures and penetrated the nexine. It accumulated in the interbedded intine-nexine, between the cytoplasmic evaginations into the intine, and entered invaginations of the plasma membrane (Rowley and Flynn, in press, Figs. 7 and 8). Lanthanum also appeared to

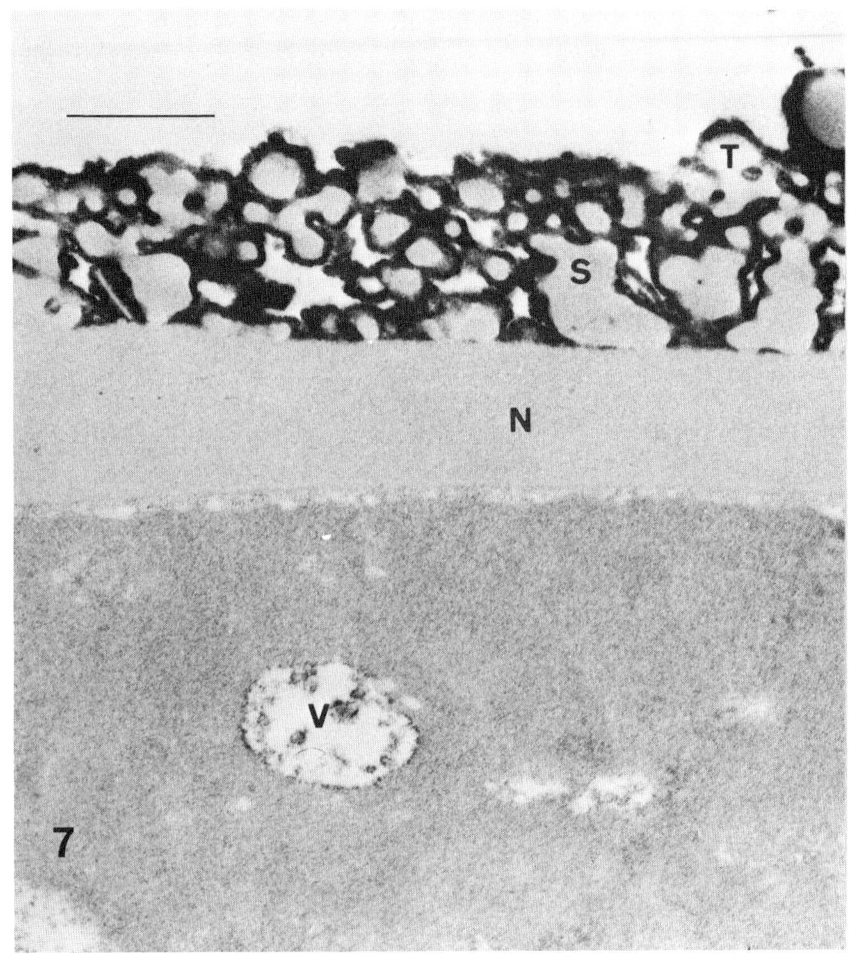

Fig.7 Sexine of C. angustifolium microspores prior to intine formation binds lanthanum nitrate. Following intine formation lanthanum accumulates on the nexine (cf. Fig.8). Stained only by LN. Line ca. 0.5μm.
Fig.8 LA introduced with GA passed through the nexine of C. angustifolium. Lanthanum accumulated within the arcade of the sexine, on lamellae within the nexine (arrows), in the intine, and within cytoplasmic channels (C). Stained only by LN. Line ca. 0.5μm.

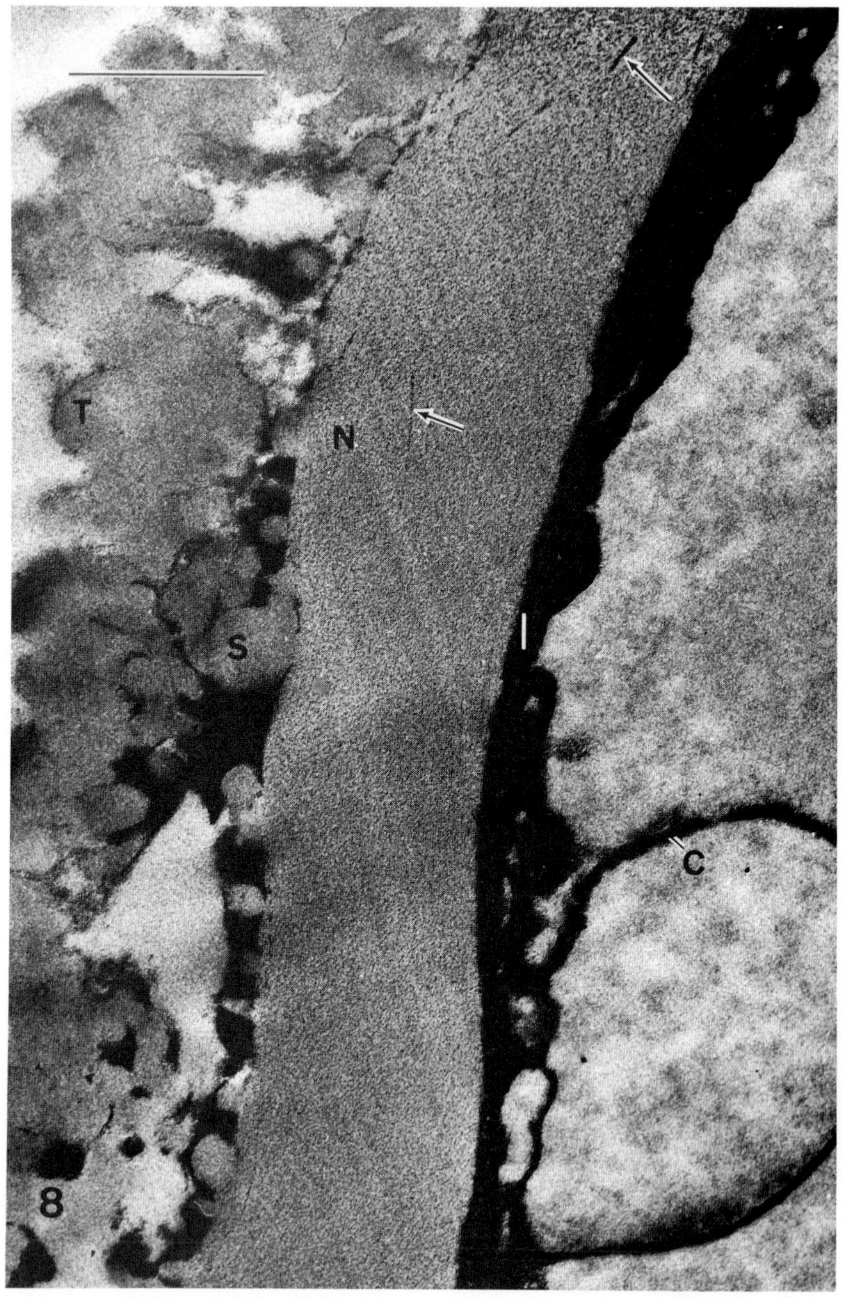

Fig.8 Description below Fig.7. Line ca. 0.5μm.

readily penetrate all parts of the nexine and to accumulate in passage at either side of lamellae (Fig. 8). Lanthanum accumulates at either surface of lamellae although these lamellae were not apparent in material stained with osmium, uranyl acetate and lead (Fig. 9). During microspore development in Epilobium and Chamaenerion the nexine contained many lamellae (Fig. 10). Results obtained with lanthanum suggest that the lamellae become masked during nexine maturation in these two genera rather than being eliminated from the nexine.

Lanthanum penetrates the pollen wall not only at apertures but through continuous parts of the exine as well. This cannot be considered as evidence that molecules pass through the exine (sexine and nexine) since the nexine in these onagraceous plants (which Lepousé and Romain (1967) refer to as endexine in Oenothera) may be more permeable than the sexine. There is lanthanum in both the sexine and nexine in the section shown in figure 8, but the open form of the sexine in Epilobium makes it unnecessary for lanthanum to migrate through the sexine going to or from the nexine. It is not necessary at this point to know whether molecules may migrate through the substance of sexine, as there are no examples, to my knowledge, of sexines without either interruptions or numerous holes through them. Any continuous exinous covering in angiosperm pollen is likely to be a nexine subdivision.

Apertures would seem to be the preferred routes for transfer of nutrients, and it is appropriate to question the value of exploiting the nonapertural exine regions in the transfer of substances. Nonaperturate grains should provide the answer to that question. I have looked at the nonaperturate pollen of Potamogeton natans, Matthiola incana and, of course, Populus and

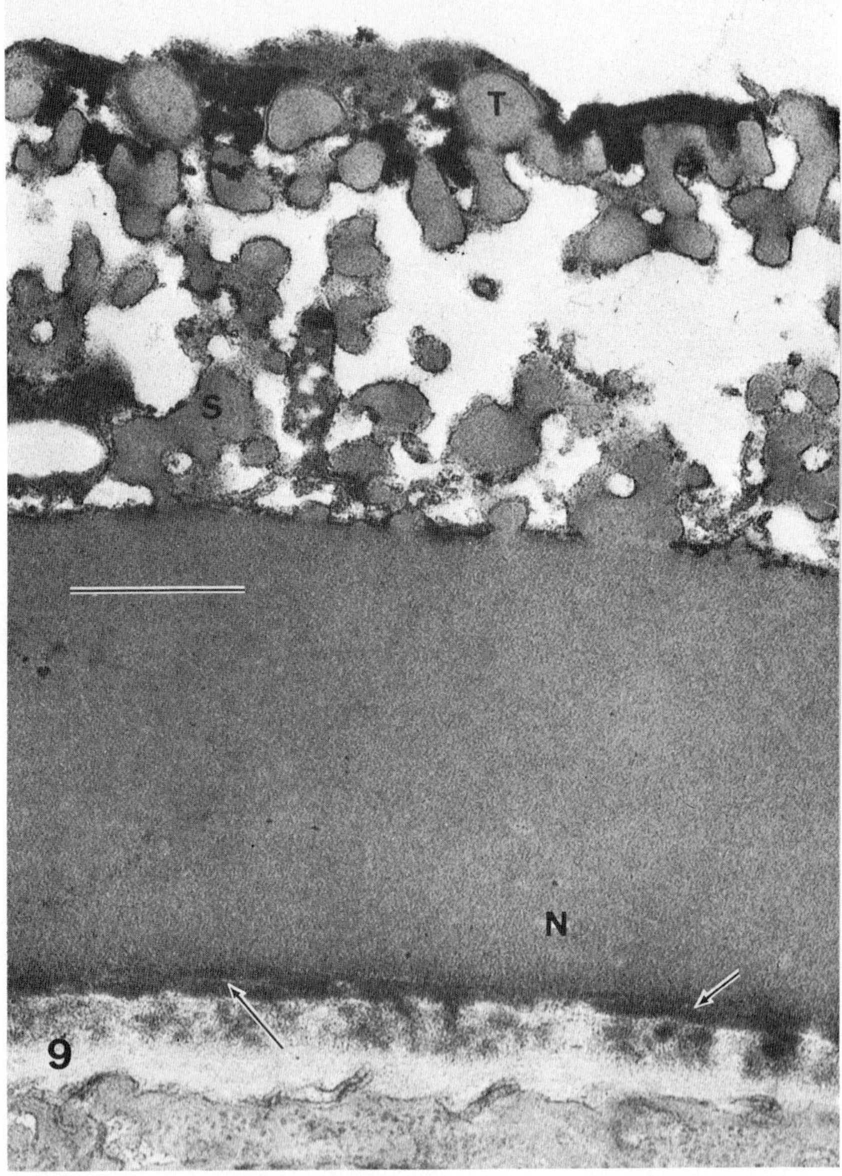

Fig.9 C. angustifolium pollen similar in stage to Fig. 8 but fixed in GA and stained with Os, UAc and Pb. The nexine-1 (N) appears uniform but lamellae are apparent in the nexine-2 (arrows). Line ca. 0.5μm.

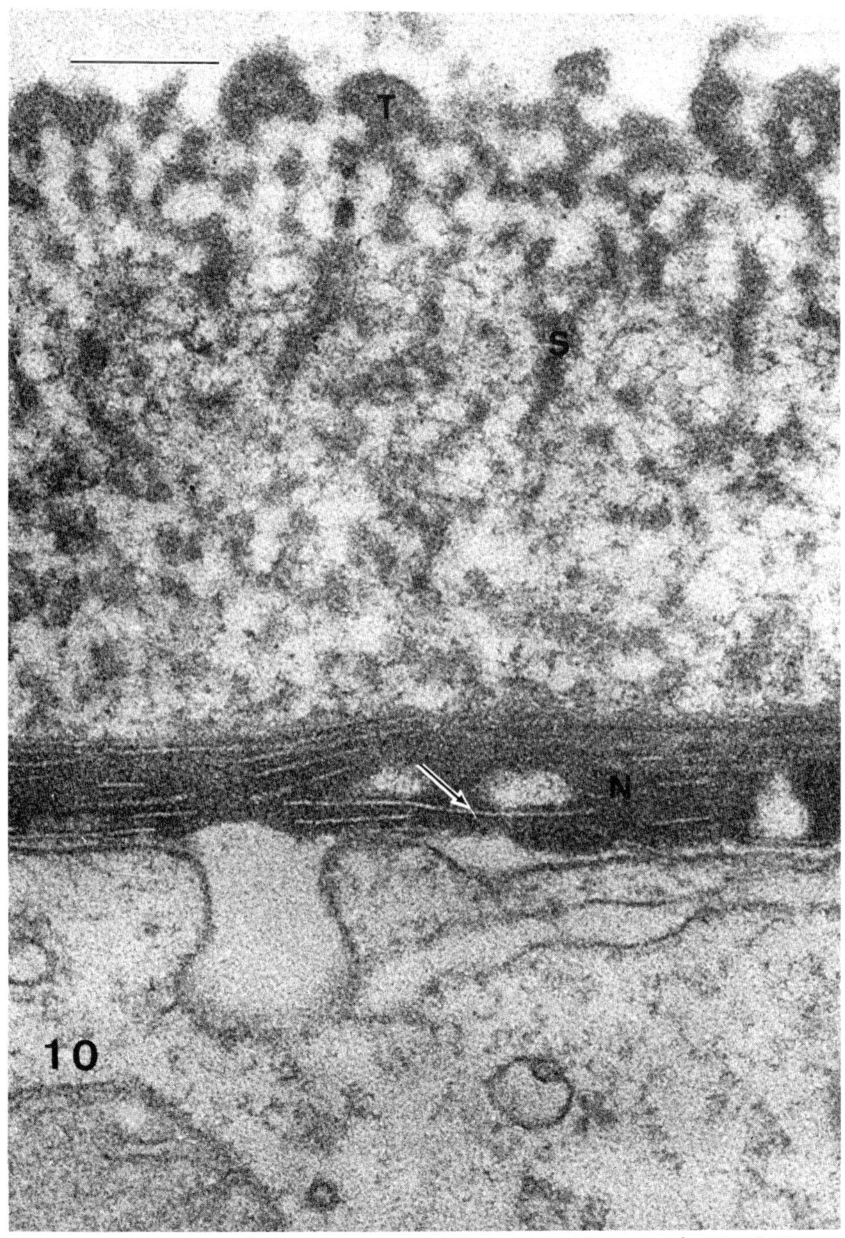

Fig.10 During microspore development the nexine of _E. montanum_ contains lamellae (arrow) and vesicles or channels. Stained with Os, UAc and Pb. Line _ca_. 0.2μm.

all are notable with respect to the discontinuity of the sexine and nexine-1. Dr. John Skvarla and I have studied pollen of *Zantedeschia* (Fig. 11) and *Canna* (Skvarla and Rowley 1970). In both of these the entire wall more closely resembles the germinal aperture of aperturate pollen grains than their nonapertural regions. The information from studies of nonapertural pollen types simplifies the problem somewhat as there is no doubt that nutrients do pass the wall of such grains. There is, anyway, no evidence that germinal apertures are important areas for the transfer of nutrients during pollen development.

Experiments on incorporation of radioactive compounds into RNA and proteins in microspores have not given information on the routes by which soluble molecules enter the gametophyte (cf. reviews by Vasil 1967, Heslop-Harrison 1968, Rodkeiwicz 1970, and Southworth, ms. a). In fact, labelled compounds that did not enter microspores in these incorporation experiments accumulated around the exine and not preferentially at apertures. However, when Young *et al.* (1966) germinated *Pyrus* pollen in a medium containing tritiated myo-inositol there was a localized incorporation of tritium into the pore regions. Using lanthanum nitrate as a label, Rowley and Flynn (in press) found lanthanum not only in cytoplasmic channels at apertural sites in *Nuphar luteum* and *Epilobium montanum* but also throughout nonapertural areas. Acid phosphatase reaction product was found throughout the intine under the nonapertural exine in *Tradescantia* (Horvat 1969) and Flynn and Rowley (1970, Fig. 10) show acid phosphatase reaction product within channels of the nonapertural exine of *Nuphar*. There are numerous examples of

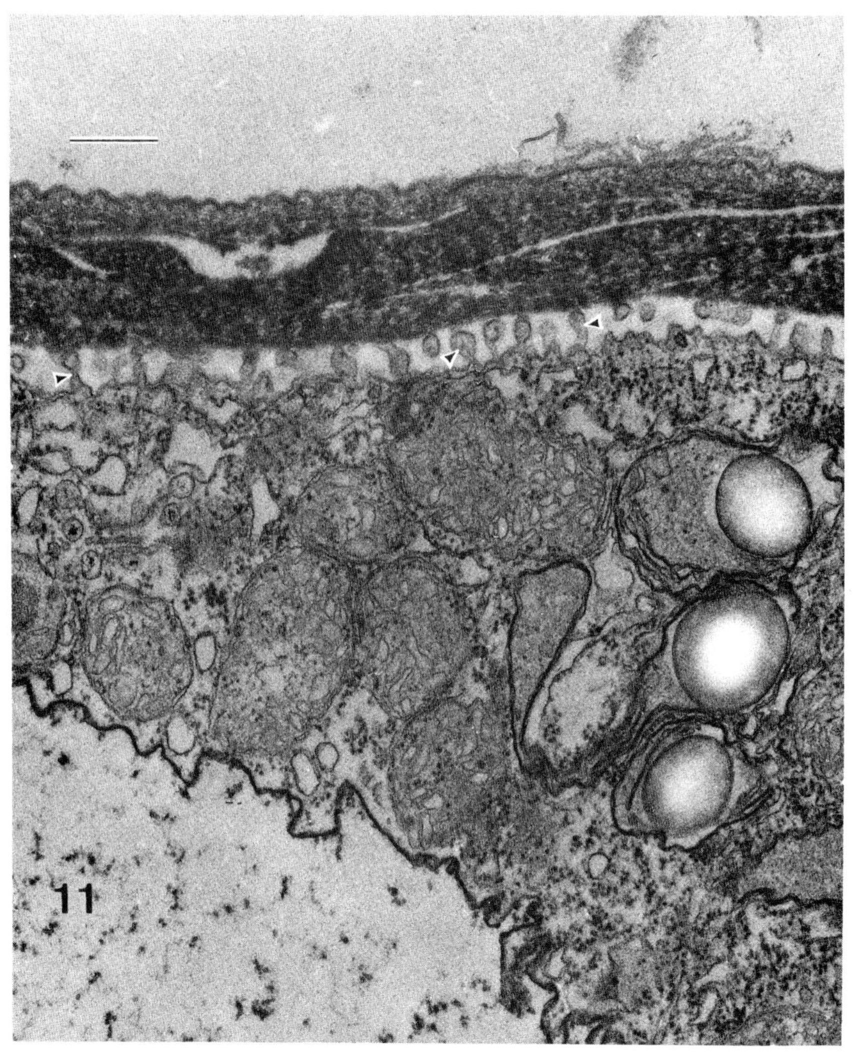

Fig. 11 Microvilli (arrow heads) cover the cytoplasmic surface of *Zantedeschia aethiopica* microspores. Material stained with Os, UAc and Pb. Line *ca*. 0.5μm.

vesicles and cytoplasmic evaginations in apertural regions (*e.g.* Flynn and Rowley 1970, Fig. 1) but the static morphological evidence for an active cytoplasmic surface is equally good for nonapertural regions (Figs. 8, 10-12) and Rowley (1964, Fig. 15). Another argument in favor of apertural involvement during pollen development is the rotation of grass microspores in the anther loculus so that the aperture faces the tapetum (cf. Vasil 1967). The specific rotation occurs, however, at about microspore mitosis, after the exine is well formed and as the intine is forming.

The thesis I am defending implies that exine forms at transfer sites for substances going to the gametophyte. If that concept is tenable at all, the sexine and nexine-1 should be asymmetrically deposited on grains which are physically prevented from rotating or prevented in some other way from having all parts of their surfaces uniformly near the tapetum or its secretion. The often cited observation of Gorczyński (1935, cf. Vasil 1967) on pollen of cleistogamous species of *Cardamine* indicated that exinous thickening occurred first adjacent to the tapetum. The exine became symmetrically thickened in Gorczyński's material as development progressed. In some pollen tetrads the sexine is incompletely formed on contact surfaces (Skvarla and Larson 1963) but this may be due to limited space. The best evidence for my point is in pollinia where space is available for sexine at the surface of inner tetrads but less and less exine forms from the outer surfaces toward the pollinia interior (Barth 1965, Chardard 1969, and Cocucci and Jensen 1969).

There are many examples of sexine asymmetry *i.e.*, aperture margins and distal and proximal poles, which in principle are derived from the primexine period and may perhaps also be supplemented by modification during volume increase of the microspore. Reduction in probacule height at apertures is well documented and reviewed by Vasil, Echlin, and Heslop-Harrison. In grains such as *Tradescantia* a considerable portion of the sexine forms within the callose wall (Horvat 1966, Mepham and Lane 1970), and the aperture margin becomes thin later in development, perhaps due to stretching (Rowley and Dahl 1962). The sexine does not become disproportionately thickened at aperture margins.

As Erdtman (1952) observed, the sexine of *Nelumbo nucifera* is not reduced in height at aperture margins or over apertures. This does not mean, according to my proposition, that substances were moved through the aperture since presumptive apertures are lacking in the primexine stage in *Nelumbo*. Uniformity in height of probacules and subsequent height uniformity of the sexine is strong support for Heslop-Harrisons probacule concept. Apertures in *Nelumbo* are inserted following the early microspore period after the sexine is formed (That point is given special consideration under the section *Exine dissolution during development and germination*).

Heslop-Harrison has shown that a unit of the endoplasmic reticulum closely parallels the presumptive aperture. That observation is especially well substantiated by the work of Echlin and Godwin and is well covered in the reviews cited. A layer of endoplasmic reticulum near the plasma membrane would, according to the cell membrane theory, result in an increase in electrical conductivity of that part of the plasma membrane (Stein 1967). I submit the hypothesis that during exine template formation, and for an indefinite period thereafter, apertures are not as likely to be sites for material transfer as are the nonapertural regions of the microspore surface. The developmental continuum between the primexine-probacule stage until the sexine and nexine-1 (when formed concurrently with sexine) are established is very inadequately known.

Two points, however, are fairly clear: the sexine, and nexine-1 when present, develop very rapidly (e.g. Rowley 1963, Heslop-Harrison and Dickinson 1969), and the number of sexinous processes is unchanged throughout pollen ontogeny (Angold 1967, also cf. review by Gullvåg). When the period of rapid sporopollenin deposition on the sexine and nexine-l is ended, then sporopollenin accumulation becomes far greater at apertures than elsewhere, but this occurs at the nexine--2 subdivision, not in the sexine and nexine-1. The sexine and nexine-1 change dramatically, although not to a greater extent at apertures than elsewhere, as is well recorded in the micrographs of Skvarla and Larson (1966, Figs. 18, 19, 22, 23).

The preferential formation of nexine-2 (nexine-1 if nexine-1 was not formed concurrently with sexine) about the time of intine formation and microspore mitosis is well documented and is summarized diagrammatically by Echlin (1968).

To test for migration of material through apertures I tried uptake experiments, using Thorotrast as a tracer. Following methods similar to those used by Padawer (1969) for mast cells, anthers cut at one end were incubated in Thorotrast at 20 and 25°C for various periods from 10 minutes to 24 hours before aldehyde fixation and epon-araldite embedding. Thorium did not enter microspores or pollen grains of _Epilobium_ or

Chamaenerion either before or after intine formation, although thorium did reach gametophytes and was bound to the sexine. (It was not removed from pollen isolated after incubation and washed in 0.1M acetate buffer at pH 4.5 for ten minutes.) In an effort to increase the possibility of microspores and pollen grains taking up Thorotrast I used a method invented by Chapman-Andresen (1962) in her studies on pinocytosis in amebae. She determined that 0.05% Alcian blue at 5°C would induce pinocytosis in amoebae but that the energy requirement for the process of pinocytosis was not met until the temperature was elevated. Amoebae transferred to media at 20°C without Alcian blue would begin to pinocytose and take up substances contained in the new media. *Epilobium obscurum* anther wafers were placed in 0.05% Alcian blue at 5°C for 10 minutes, rinsed in water at 5°C, incubated for 2 minutes to 24 hours at 20°C in 1% Thorotrast in 0.1M acetate buffer at pH 4.6. After a 10-minute rinse in acetate buffer the anther wafers were dehydrated in an acetone series and embedded in epon-araldite. Thorium did not enter microspores or pollen but Alcian blue did. I have not looked at the entire series but Alcian blue covered the sexine of pollen from wafers that had been in Alcian blue at 5°C for ten minutes; after 10 minutes at 20°C thorium was on the sexine and Alcian blue was not visualized; after 2 hours (Fig. 12) thorium was on the sexine and Alcian blue in vesicles at or near the plasma membrane and in the vacuole; and after 24 hours (Fig. 13) thorium still covered the sexine but Alcian blue was limited to the vacuole. Alcian blue appeared to enter the cytoplasm all around the grain and not only at

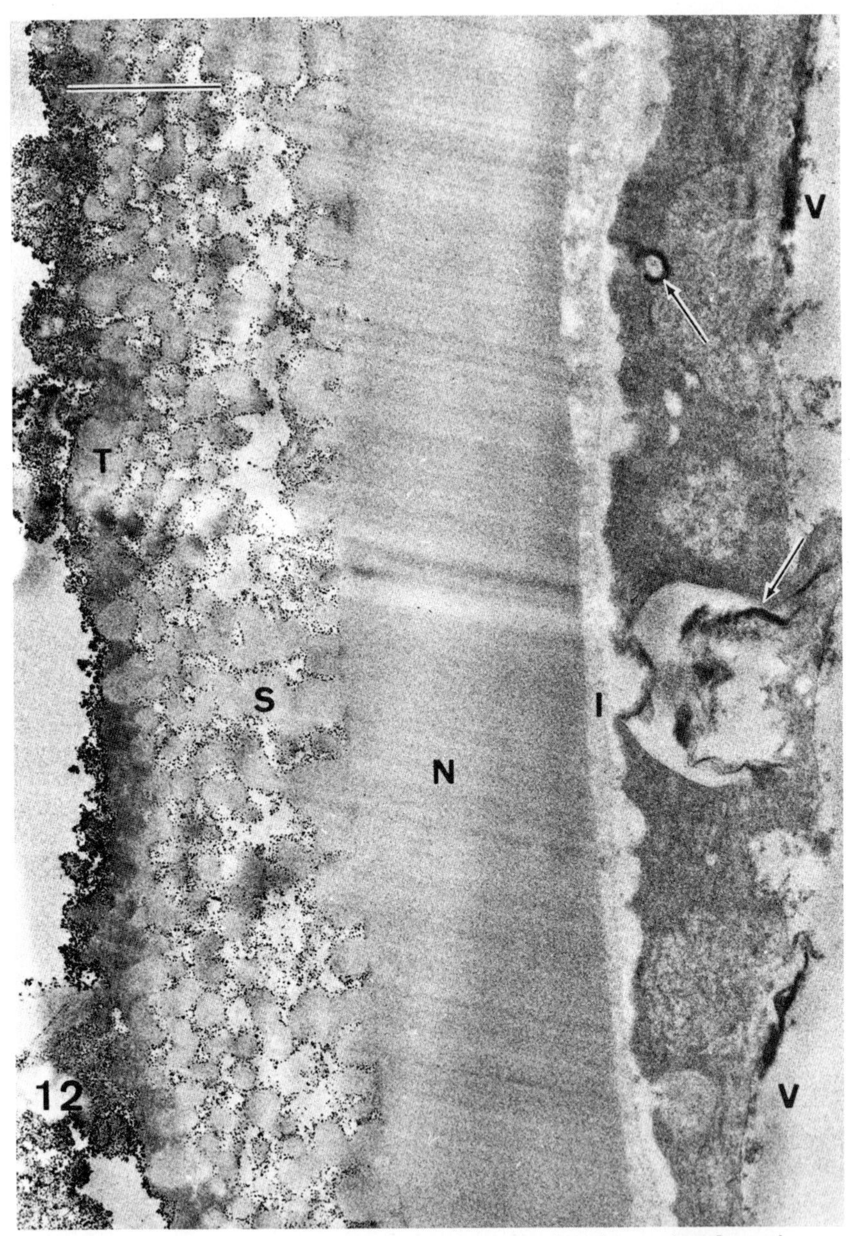

Fig.12 Thorium coats sexine and AB is in cytoplasmic vesicles (arrows) and vacuole (V) of _E. obscurum_ incubated in AB 10 min. then TH 2 hrs. Line _ca_. 0.5µm.

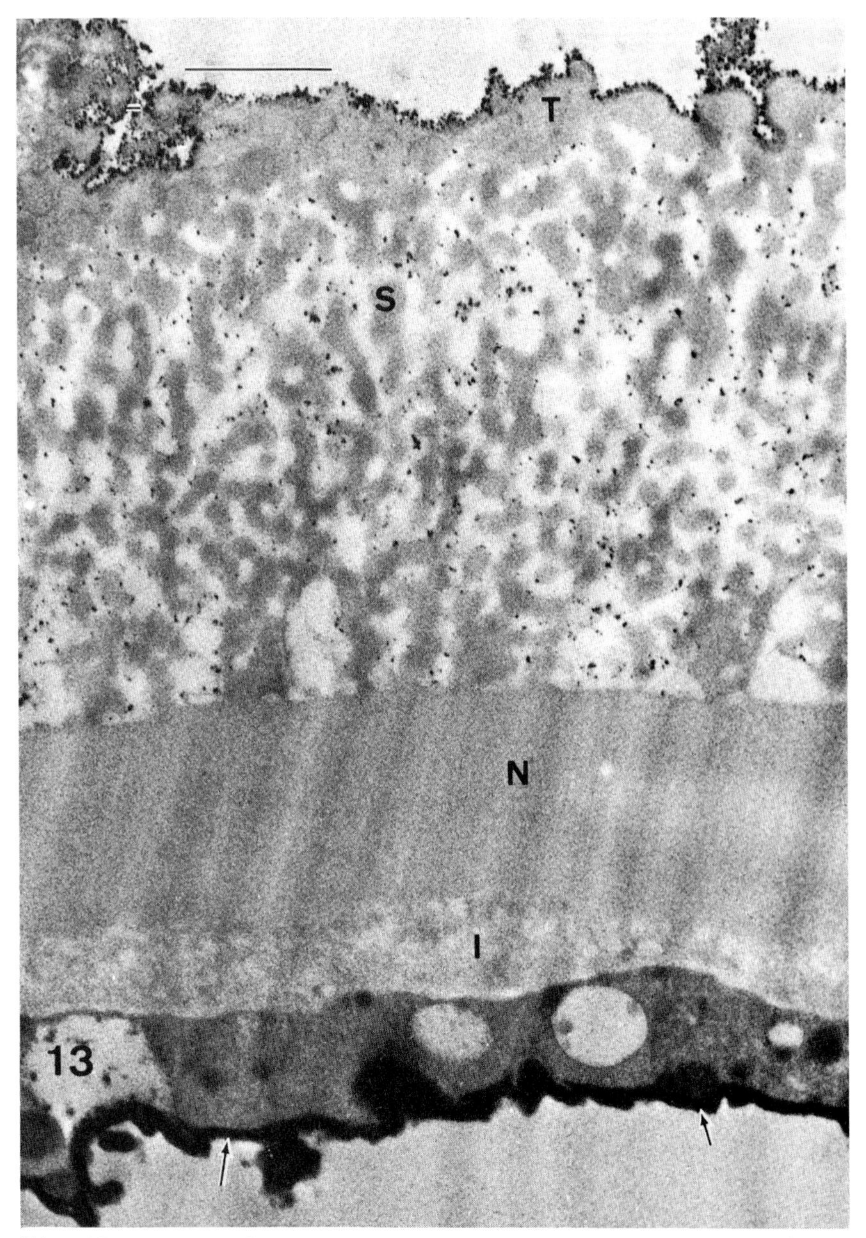

Fig.13 Incubations same as Fig.12 except 24hr. in TH. Thorium on sexine is mostly on tectum and AB lines the vacuole (arrows). Line ca. 5μm.

apertures. If Alcian blue did not pass through the nexine but only entered at apertures, then it must have migrated around through the intine in order to have been so widely available for uptake by the cytoplasm.

Exine dissolution during development and germination

Though pollen exine can be remarkably stable, Gherardini and Healey (1969) have demonstrated its dissolution during pollination. They not only were able to suggest from their experiments that the exine of *Pharbitis nil* pollen is enzymatically degraded when placed on a mature stigma but also that this degradation may actually be necessary for the successful completion of germination. Ehrlich (1960) published micrographs of exine during germination of *Saintpaulia* pollen in which the sexine was gone except for a lamella at the surface. Other examples of pollen wall dissolution occur which deal mostly with the intine (*e.g.*, Cocucci and Jensen 1969 and Flynn 1969).

Remobilization of the exine during development is more difficult to prove conclusively. In their summary of sporopollenin synthesis associated with the tapetum and microspores of *Lilium*, Heslop-Harrison and Dickinson conclude that there is no evidence that sporopollenin is ever remobilized once deposited on the exine. Sporopollenin may be remobilized during development of *Populus* pollen (Rowley and Erdtman 1967, Figs. 4 and 20). Acetolysed pollen of *Populus* at floral anthesis has no distinct nexine while earlier in development there were three or more lamellae. This apparent loss of nexine could be due to surface area increase although my calculations indicate that to be unlikely. The nexine in *Populus* consists of tapes, however, not continuous lamellae, and surface area calculations may be meaningless. There are two other phenomena concerning pollen wall formation that suggest autolysis of the exine: probacules and primexine are uniform and without interruption over the surface of *Nelumbo nucifera* microspores. Thus when the three long germinal apertures, characteristic of *Nelumbo* pollen, are introduced there must be precise local modification of the exine (Flynn and Rowley, in press). The apertures of *Nelumbo* may be introduced by a localized autolysis of sporopollenin. In work with *Chamaenerion* and *Epilobium* microspores we found that channels through the nexine were formed where there had been none before (Rowley and Nilsson, ms.). I would be willing to consider that the exine was in dynamic equilibrium between a sol and gel state during pollen development so that cytoplasmic evaginations might poke transitory channels through the exine. If not, then some kind of focal autolysis of the exine must be responsible for exine channel production. In *Zauschneria californica* the channels remain in the mature exine (Fig. 14) but in *Chamaenerion* and *Epilobium* they are filled by maturity (Fig. 9).

The sporopollenin added to the exine late in development to patch the nexine channels becomes as resistant as the "original" nexinous material. Dr. Butjo Prijanto and I chemically degraded the exine of a variety of pollen grains. The *Chamaenerion* pollen grain shown in figure 15 was acetolyzed for 1 min., boiled in 5% KOH for 10 minutes, and chlorinated for 10 seconds. This battery of treatments degrades the nexine in *Zauschneria* as well as *Chamaenerion* while only etching the sexine. Channels in the nexine of *Zauschneria* are enlarged and made more obvious but the degraded nexine of *Chamaenerion* showed no indication of

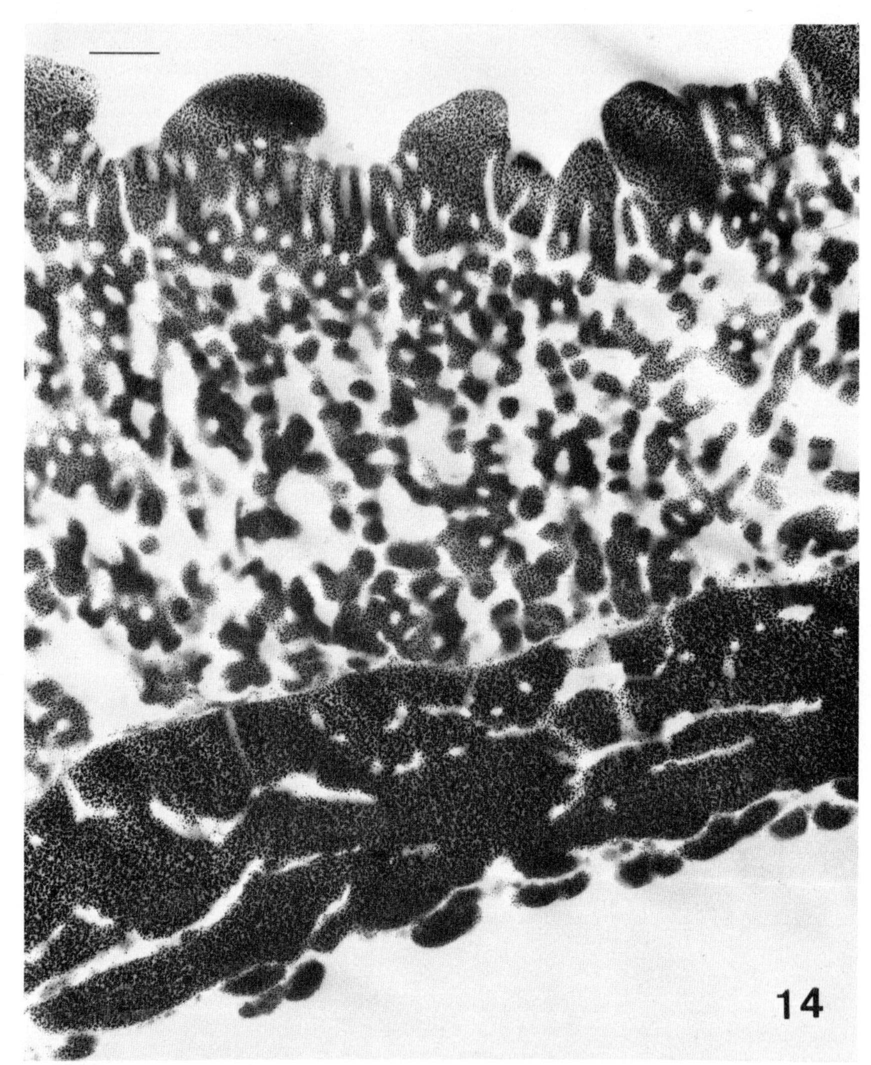

Fig. 14 Pollen of Z. californica acetolysed and Os stained before embedding and section staining by T-SP to amplify the Os. The osmophilia of the sexine differs from the nexine. Channels are permanent in both sexine and nexine in Zauschneria but only sexine in Chamaenerion (cf. Figs. 9 10, and 15) Line ca. 0.5μm.

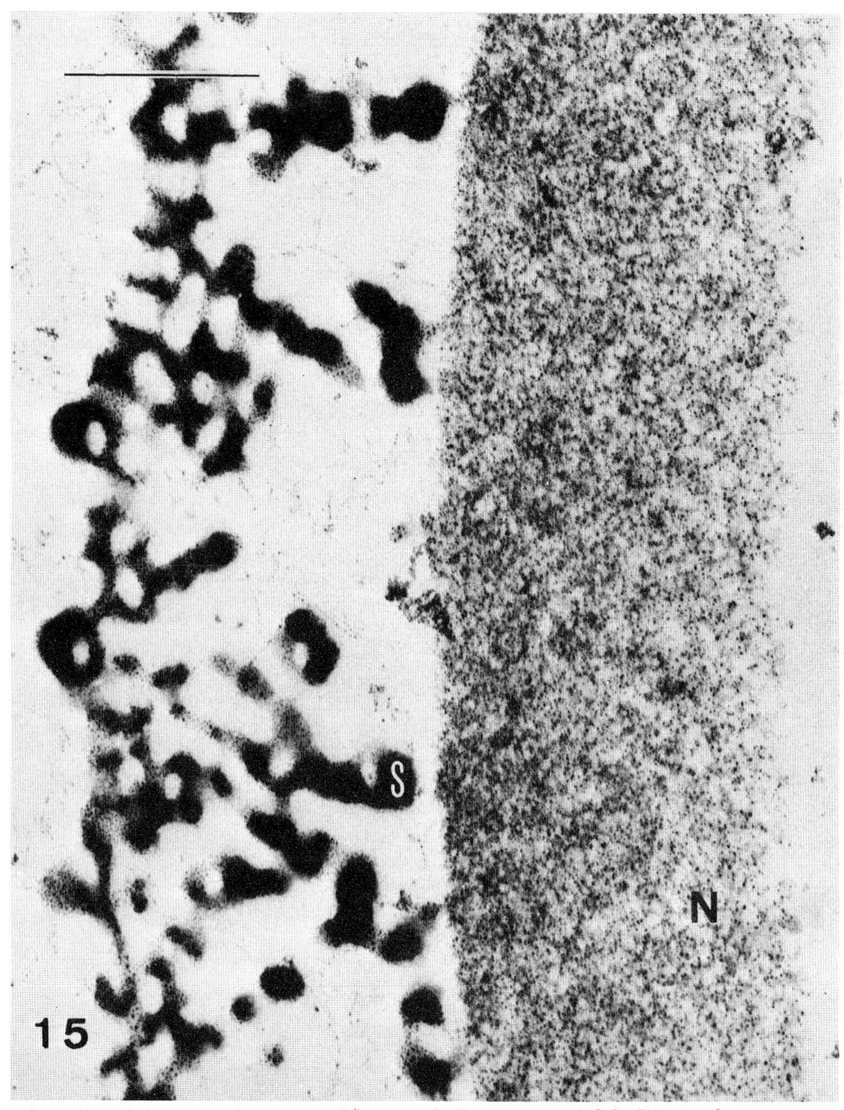

Fig.15 When mature pollen of _C_. _angustifolium_ is exposed to the battery of treatments A-KOH-Cl, then the nexine is degraded before the sexine. The channels of the nexine that were conspicuous during microspore development are not reexposed during degradation. Line _ca_. 0.5μm.

channeling (Fig. 15). Sporopollenin that fills nexinous channels of Chamaenerion late in development when the exine stains as it does in figure 9 must have similar properties, with respect to our treatment series, to the rest of the nexine. I have prepared a report (Prijanto and Rowley, ms) covering some of our results on the selective destruction of the exine. Dr. Butjo Prajanto was killed in an automobile accident; the death of this fine man is a great loss to the world.

General discussion

The mediation of genetic control over the exine with all of its elaborate processes and fenestrations is by no means clear. Techniques and results described by Heslop-Harrison for the symposium will add considerably to the eventual proposal of a model for genetic control for the interactions which govern wall establishment on pollen. Rogers and Harris (1969) made observations all palynologists interested in wall formation will want to consider in great detail. They made crosses of Linum tetraploids and diploids which resulted in sterile triploid plants. As a result of the irregular chromosome pairing at meiosis in such plants some chromosomes did not move to spindle poles but were left scattered in the cytoplasm. A number of small cells formed after cleavage of the pollen mother cells, and these Rogers and Harris considered on statistical grounds to be formed around the scattered chromosomes. The correctness of that statistical deduction seems assured but does not greatly alter the significance of the observation that the diminutive cells form germ pores and a "typical" exine, in spite of a greatly reduced and variable genome or complete lack of chromosomes. In their extremely lucid discussion they say that wall patterning cannot be directly related to the genome of the grain, and that consideration must be given to either the cytoplasm or to information filtered through the callose special cell wall from the nuclei of the large spores which would probably contain the necessary quota of genes. They recognize the evidence against the diminutive grains being fed information through the callose wall, as I do also, but I would amend their cautiously worded alternative to cytoplasmic control to add that, if information could come from neighboring spores, it might also come from the tapetum.

The experiments of Harris and Rogers (1969) have given dramatic cause for reassessment of the role of the nucleus in exine formation. The work of Joshi (*e.g.* 1968) is no less spectacular although much more difficult to sort out. Her chemically induced autoploids and translocation heterozygotes resulted in a great variety of pollen grain shapes and sizes because of the chromosome change induced by the chemicals, colchicine-gammexine. An exine similar to that of controls formed on treated sporogenous cells which did not divide at all after meiosis when cytokinesis failed, and on microspores with a variety of ploidy levels and chromosome aberrations.

It took a long series of experiments to find a method of showing pinocytosis in plant cells. Mc Laren and Bradfute (1966) summed up the problem in their statement that pinocytosis is a dynamic process, not a static end result and that no one has visually observed the dynamic events with plants although Jensen and McLaren (1960), Bhide and Brachet (1960), Cocking (1965) and others (cf. review by Clowes and Juniper 1968) had suggested that proteins and viruses may enter plant cells by pinocytosis. The experiments of Mayo and Cocking (1969) with polystyrene latex particles have shown the initial phases of the pinocytotic uptake of that large tracer and removed doubt of the capacity of plant cells for pinocytosis. The specialized conditions of their experiments on tomato fruit protoplasts isolated by enzymatic treatment to remove the cell wall leaves uncertain, as Mayo and Cocking indicate, the extent to which pinocytosis is a normal activity of plant tissues. Pinocytosis is defined (Bennett 1969a) as a selective, energy consuming, physiological process which involves uptake of material by the cell from the surrounding medium, utilizing the capacities of the cell surface for specific binding, membrane flow, configurational change, caveola formation, membrane fusion, recombination and vesiculation. Pinocytosis can be divided into three steps, namely: attachment of substances to the cell surface, formation of pits or plications by membrane movements, and pinching off of vesicles into the cytoplasm from the base of pits. In the plant cell it is further required that the material must first pass through the cell wall. Even if it assumed that pinocytosis is common in plant cells and that the removal of the cell wall in the experiment of Mayo and Cocking only enabled a relatively large marker to reach the protoplast surface, pinocytosis in pollen might seem of doubtful utility due to the position of the protoplast within an often thick and seemingly impervious exine as well as inside an intine with ultra-filtering capacity similar, in all probability, to cellulosic cell walls. Since microspores are nonphotosynthetic organisms, nutrients for growth of the gametophyte must come from the external environment and the amount of nutrients taken up has to be great since both mass and volume increase (Dahl *et al*. 1957).

It is not my claim that results with Imferon and Alcian blue clearly show pinocytosis in developing pollen grains. However, I do claim that gametophytes are excellent cells for experiments on pinocytosis in angiosperms. They offer a free cell surface and make possible the selection of stages either with an exposed plasma membrane (tetrad) or a plasma membrane covered by an exine or by both intine and exine. The plasma membrane includes a glycocalyx, and the exine surface and/or a coating on the exine binds colloidal iron, thorium, silver, ruthenium red, Alcian blue and lanthanum under the same experimental conditions which have been proved so productive in the study of free cell surfaces on animal cells (Fawcett 1965, Marshall and Nachmias 1965, Bennett 1969a and b). Hall's (1970) work on pinocytosis and ATPase activity at the cell surface in root tip cells makes a better case for the importance of the process than does uptake of cationic inorganic tracers. Cation-stimulated ATPases have been found in cell wall fractions from several plant species (*e.g*. Dodds and Ellis 1969, Hall and Butt 1969) and cytochemical studies have shown that this uptake of cations is probably associated with the plasma membrane and plasmodesmata of root cells (Hall 1969). Hall (1970) cautiously suggests from recent work on cells of *Zea mays* root tips that the mechanism of ion transport in these root cells involves invagination of the plasma membrane and the uptake of ions in small vesicles which later release their contents to the cytoplasm. Hall's cytochemical observations of ATPase activity in root cells are interpreted, by him, as supporting a pinocytotic concept of active ion transport which utilizes ATP as the energy source.

The results presented give ample cause to doubt that the exine is an impervious wall. Ferric iron and lanthanum pass through the substance of the wall itself as well as through its interstices. In the test materials, substances can be considered as available for uptake by the pollen cytoplasm everywhere over its surface.

Acknowledgments

The work was supported by the Swedish Natural Science Research Council and National Science Foundation of the USA (grant GB 7077). Space and equipment for electron microscopy were made available by Professor F. Fagerlind, Morphological Botany Department, University of Stockholm.

References

Angold, R. E. (1967). Rev. Palaebot. Palynol. 3, 205.

Barth, O. M. (1965). Pollen et Spores 7, 429.

Behnke, O. and Zelander, T. (1970). J. Ultrastruct. Res. 31, 424.

Bennett, H. S. (1963). J. Histochem. Cytochem. 11, 14.

Bennett, H. S. (1969a). in Handbook of molecular cytology (A. Lima-de-Faria, ed.), p. 1261. North Holland Pub. Co. Amsterdam.

Bennett, H. S. (1969b). ibid., p. 1294.

Bhide, S. V. and Brachet, J. (1960). Exptl. Cell Res. 21, 303.

Brooks, J. and Shaw, G. (1968). Grana palynol. 8, 227.

Brooks, R. E. (1969). Stain Technol. 44, 173.

Chapman-Andresen, C. (1962). Compt. rend. Lab Carlsberg 33, 73.

Chardard, R. (1969). Rev. Cytol. et Biol. vég.32, 67.

Clowes, F. A. L. and Juniper, B. E. (1968). Plant Cells. Blackwell Sci. Pub., Oxford.

Cocking, E. C. (1965). Biochem. J. 95, 28.

Cocucci, A. and Jensen, W. A. (1969). Planta 84, 215.

Dahl, A. O., Rowley, J. R., Stein, O. L., Wegstedt, L. (1957). Exptl. Cell Res. 13, 31.

Davis, B. D. and Warren, L., eds. (1967). The specificity of cell surfaces. Prentice-Hall, New Jersey.

Dodds, J. J. A. and Ellis, R. J. (1966). Biochem. J. 101, 31 p.

Dunbar, A. (1968). Grana Palynol. 8, 14.

Dunbar, A. and Erdtman, G. (1969). Grana Palynol. 9, 63.

Echlin, P. (1968). Scientific American 218, 80.

Ehrlich, H. G. (1960). J. Biophys. Biochem. Cytol. 7, 199.

Erdtman, G. (1952). Pollen morphology and plant taxonomy. Angiosperms. Almqvist and Wiksell, Stockholm.

Fawcett, D. W. (1965). J. Histochem. Cytochem. 13, 75.

Flynn, J. J. Jr. (1969). Ph. D. Thesis. Univ. of Massachusetts, Amherst.

Flynn, J. J. and Rowley, J. R. (1970). Zeiss Information 76, 40.

Flynn, J. J. and Rowley, J. R. (in press). Experientia.

Gherardini, G. L. and Healey, P. L. (1969). Nature 224, 718.

Godwin, H. (1968). New Phytol. 67, 667.

Gorczyński, T. (1935). Pers. Acta Soc. Bot. Polon. 12, 257.

Górska-Brylass, A. (1970). Grana 10, 21.
Gullvåg, B. M. (1966). Phytomorph. 16, 211.
Hall, J. L. (1969). Planta 85, 105.
Hall, J. L. (1970). Nature 226, 1253.
Hall, J. L. and Butt, V. S. (1969). J. Exp. Bot. 20, 751.
Heslop-Harrison, J. (1968). Science 161, 230.
Heslop-Harrison, J. and Dickinson, H. G. (1969). Planta 84, 199
Horvat, F. (1966). Grana Palynol. 6, 416.
Horvat, F. (1969). Pollen et Spores 11, 181,
Jensen, W. A. and McLaren, A. D. (1960). Exptl. Cell Res. 19,414.
Joshi, S. (1968). Cytologia 33, 345.
Kahn, T. A. and Overton, J. (1970). J. Cell Biol. 44, 433.
Lepousé, J. and Romain, M. F. (1967). Pollen et Spores 9, 403.
Luft, J. H. (1966). Federation Proc. 25, 1773.
Mayo, M. A. and Cocking, E. C. (1969). Protoplasma 68, 223.
Marshall, J. M. and Nachmias, V. T. (1965). J. Histochem. Cytochem. 13, 92.
McLaren, A. D. and Bradfute, O. E. (1966). Physiol. Plant. 19, 1094.
Mepham, R. H. and Lane, G. R. (1970). Protoplasma 70, 1.
Mollenhauer, H. H. (1964). Stain Technol. 39, 111.
Overton, J. (1968). J. Cell. Biol. 38, 447.
Overton, J. (1969). J. Cell Biol. 40, 136.
Padawer, J. (1969). J. Cell Biol. 40, 747.
Prijanto, B. and Rowley, J. R. (ms). Selective destruction of the exine.
Revel, J.-P. (1964). J. Microscopie 3, 534.
Revel, J.-P. and Ito, S. (1967). in The specificity of cell surfaces. (B. D. Davis and L. Warren,

eds.). Academic Press, New York.

Revel, J.-P. and Karnovsky, M. J. (1967). J. Cell Biol. 33,C 7.

Risueño, M. C., Giménez-Martín, G., López-Sáez, J. F., and -García, M. I. R. (1969). Protoplasma 67, 361.

Rodkiewicz, B. (1970). Planta 93, 39.

Rogers, C. M. and Harris, B. D. (1969). Amer. J. Bot. 56, 1209.

Rosen, W. G. (1968). Ann. Rev. Plant Physiol. 19, 435.

Rowley, J. R. (1963). Grana Palynol. 3, 3.

Rowley, J. R. (1964). in Pollen Physiology and Fertilization (H.F. Linskens, ed.). North-Holland Pub. Co., Amsterdam.

Rowley, J. R. and Dahl, A. O. (1962). Pollen et Spores 4, 221.

Rowley, J. R. and Erdtman, G. (1967). Grana Palynol. 7, 517.

Rowley, J. R., Flynn, J. J., Dunbar, A., and Nilsson, S. (1970). Grana 10, 3.

Rowley, J. R. and Flynn, J. J. (in press). Cytobiologie.

Rowley, J. R., and Dunbar, A. (ms). Transfer of colloidal iron from sporophyte to gametophyte.

Rowley, J. R. and Nilsson, S. (ms). Pollen grain ontogeny in Epilobium and Chamaenerion.

Scott, J. E., Quintarelli, G., and Dellovo, M. C. (1964). Histochemie. 4, 73.

Shaw, G. and Yeadon, A. (1966). J.Chem. Soc. (C), 16.

Sitte, P. (1955). in die Chemie der Pflanzenzellwand. (E. Treiber, ed.). Springer-Verlag, Berlin.

Skvarla, J. J. and Larson, D. A. (1963). Science 140, 173.

Skvarla, J. J. and Larson, D. A. (1966). Amer. J. Bot. 53, 1112.

Skvarla, J. J. and Rowley, J. R. (1970). Amer. J. Bot. 57, 519.

Southworth, D. (ms. a.). Incorporation of radioactive precursors into developing pollen walls. Ph. D. Thesis, Univ. Calif., Berkeley.

Southworth, D. (ms. b.). Cytochemistry of pollen walls. ibid.

Stein, W. D. (1967). The movement of molecules across cell membranes. Academic Press, New York.

Thiéry, J.-P. (1967). J. Microscopie 6, 987.

Tsinger, N. V. and Petrovskaya-Baronova, T. P. (1961). Dokl. Akad. Nauk. USSR 138, 466.

Vasil, I. K. (1967). Biol. Rev. 42, 327.

Young, L. C. T., Stanley, R. G., and Loewus, F. A. (1966). Nature 209, 530.

DISCUSSION ON PROFESSOR J. ROWLEY'S PAPER

HESLOP-HARRISON: In the interests of accuracy, the permeability of the exine should be defined rather carefully at each developmental stage. I believe that the mature exine, when impregnated with pollenkitt, tryphine, etc., is a remarkably impermeable membrane as far as water, water vapour and material in aqueous solution is concerned. All of the structural apertures etc. are plugged with matrix material.

ROWLEY: When you put pollen grains into buffer, then about 15 antigens come out in 30 seconds at maturity.

HESLOP-HARRISON: Now this is exactly the point I want to make. We have shown that except in certain cases where the exine is porous over a good deal of the surface, the enzymes and antigens are mostly lost through the apertural regions (Knox and Heslop-Harrison, _J. Cell Science_, 6, 1 - 27, particularly Fig.20., see also Heslop-Harrison, this volume - Ed.)

ROWLEY: Yes, Young, Stanley, and Lewis used labile material to demonstrate where material enters and leaves pollen grains. In this work, the aperture was the first area to be labelled, but this was at germination and I think that during development, it would be unlikely for the pollen grain to use only the aperture.

HESLOP-HARRISON: That's right. In early developmental stages, the wall might well be porous enough to allow a fairly general passage of materials. But we should be cautious about supposing this is true throughout. Recently we have been doing some experimental work with grass pollen, following the course of vacuolation after release from the tetrad. Firstly there is a period of expansion without vacuolation when the protoplast becomes extremely hydrated. Then vacuolation begins and the pollen grain enlarges very considerably. When this happens, polarisation related to the aperture is imposed on the system. The nucleus moves across away from the aperture and stays within an arc of about 30° opposite the aperture, and the first vacuole forms in the apertural region. Now, I can see no reason for this happening unless the aperture is involved in polarisation and allows the ingress of material at this stage. By using fluorescent tracers, it is possible to show that fluorescence develops first in the apertural region and gradually expands away from there until the whole grain is fluorescent. This indicates that the main ingress is through the apertural region.

PRODUCTION OF SPOROPOLLENIN BY THE TAPETUM

Patrick Echlin, Botany School, Cambridge, England

Abstract

It has long been known that the developing pollen grains and tapetal cells within the maturing anther pass through what appear to be opposing phases of metabolism. The tapetal cells, which are initially in a stage of active synthesis, gradually pass into a phase of decline and senescence, while the pollen grains continue to develop and differentiate. There are, however, important differences in the timing and extent of these events, particularly when a comparison is made between the secretory and periplasmodial types of tapetum. During the tapetal senescence, it is tacitly assumed that some of the degradative products of the tapetal tissue are utilized by the maturing pollen grains.

The bulk of the sporopollenin in the anther is found in the pollen grain walls. During the early phases of development, when the pollen grains are enveloped by an impermeable callose wall, the sporopollenin is most certainly formed from precursors located within the pollen grain cytoplasm. Following the dissolution of the callose layer and the release of the microspores into the anther cavity, there is a bulk deposition of sporopollenin on the pollen grain wall. Small pores in the furrow region would appear to provide a route through which soluble precursors could pass into the pollen grain cytoplasm. The tapetal cells of some plant species form Ubisch bodies (orbicles) which are released into the anther cavity. The precursors of these bodies are formed in embayments of the endoplasmic reticulum which are a characteristic feature of the tapetal cytoplasm at this stage of development. Similar profiles of endoplasmic reticulum are also seen in the tapetal cytoplasm of those species which do not form Ubisch bodies. There is a close correlation between the time of bulk sporopollenin deposition in the pollen grain wall and the appearance of Ubisch bodies at the surface of the senescing tapetal cells. Polymerized sporopollenin is only rarely found within the cell cytoplasm and is characteristically deposited outside the cell membrane. This deposition appears to be centred on thin electron transparent lamellae, which may represent either a synthetic site for the polymerization of soluble precursors or a passive framework on which the sporopollenin is deposited. Although both developing microspores and senescing tapetal cells have the ability to synthesize sporopollenin, it is only in the wall of the former that it is deposited in a regular and reproducible manner.

Introduction

In the past decade, much fruitful research has been engendered when a multi-disciplinary approach to a problem has been adopted. This conference on sporopollenin represents such an interface. In this context, the substance of my contribution is deliberately broadened, and will present little that is not already known to workers closely concerned with the ultrastructure and ontogeny of pollen. The paper is directed more to the chemists and geologists who are contributing so much to the better understanding of the nature and biological role of sporopollenin. It is hoped that each of us, specialists in our own field, may more readily comprehend the problems and results obtained by others, and in turn use these findings to further our own work.

Development of the anther

The tapetum is the innermost wall layer of the microsporangium of the anther, and is the tissue in closest contact with the developing pollen grains (Fig. 1). Typically, the tapetum is composed of a single layer of cells characterized by densely staining protoplasts and prominent nuclei. The tapetal cells in the developing anthers of angiosperms are often enlarged and may be multi-nucleate or polyploid. The tissue plays an essential nutritive role in the formation of the pollen grains, as all the food materials for the developing pollen grains must either pass through the tapetum or be metabolized by it.

Shortly after its differentiation in a stamen primordium, a young anther as seen in cross section consists of a mass of cells, each with the potential for further division. Although most of these cells may possess the ability to form pollen grains and tapetal cells, sporangium initiation is restricted to separate areas which come to correspond to the corners of the anther. At each of these regions, a discrete number of hypodermal cells

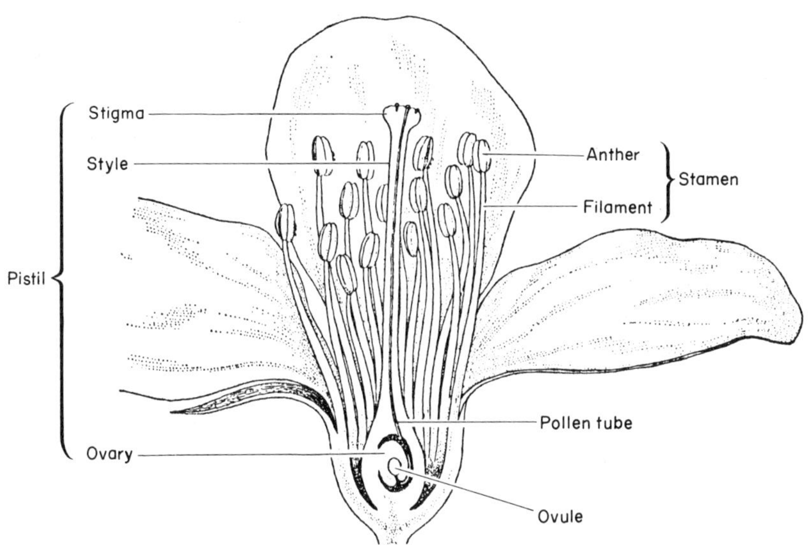

FLOWER CROSS SECTION shows a cluster of stamens, each with a pollen filled anther at its tip, surrounding a central pistil with a sticky stigma at the tip and an ovary in the base. One pollen grain adhering to the stigma has grown a long tube reaching the ovary.

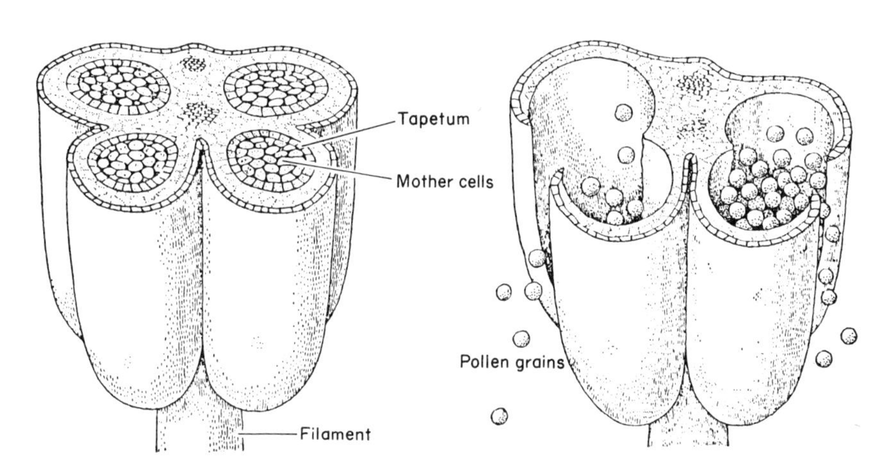

ANTHER CROSS SECTIONS show the two kinds of tissue (left) that respectively give rise to mature grains of pollen and to the tapetum, or inner wall of the anther cavity. When mature (right), the anther splits open and the loose pollen in its cavities is dispersed. From "Pollen" by P. Echlin.

undergo periclinal divisions, and this archesporial tissue gives rise to primary parietal cells to the outside and primary sporogenous cells to the inside. The primary sporogenous cells in turn give rise to the sporogenous tissue which matures to form the pollen grains, and the primary parietal cells undergo repeated periclinal and anticlinal divisions to form the several layers of the anther wall and the tapetum. A simplified developmental sequence to illustrate the origins of the various regions of the anther is shown in Fig. 2. For a more detailed consideration of the early embryonic events in the formation of the tapetum and the pollen grains, reference should be made to the work of Maheshwari (1950).

The significant point in this present discussion is that both the tapetal cells and the pollen grains have developed from the same archesporial tissue. The mature pollen grains are bi- or even tri-nucleate haploid cells which have arisen from diploid pollen mother cells, and the tapetal cells are diploid, frequently polyploid cells which have also arisen from diploid cells.

The Tapetum

Two major types of tapetum are recognized, based on their behaviour during pollen grain development. The glandular or secretory type of tapetum, where the cells remain in their original position, but progressively become more disorganized

Figure 2. The Ontogeny of the Anther

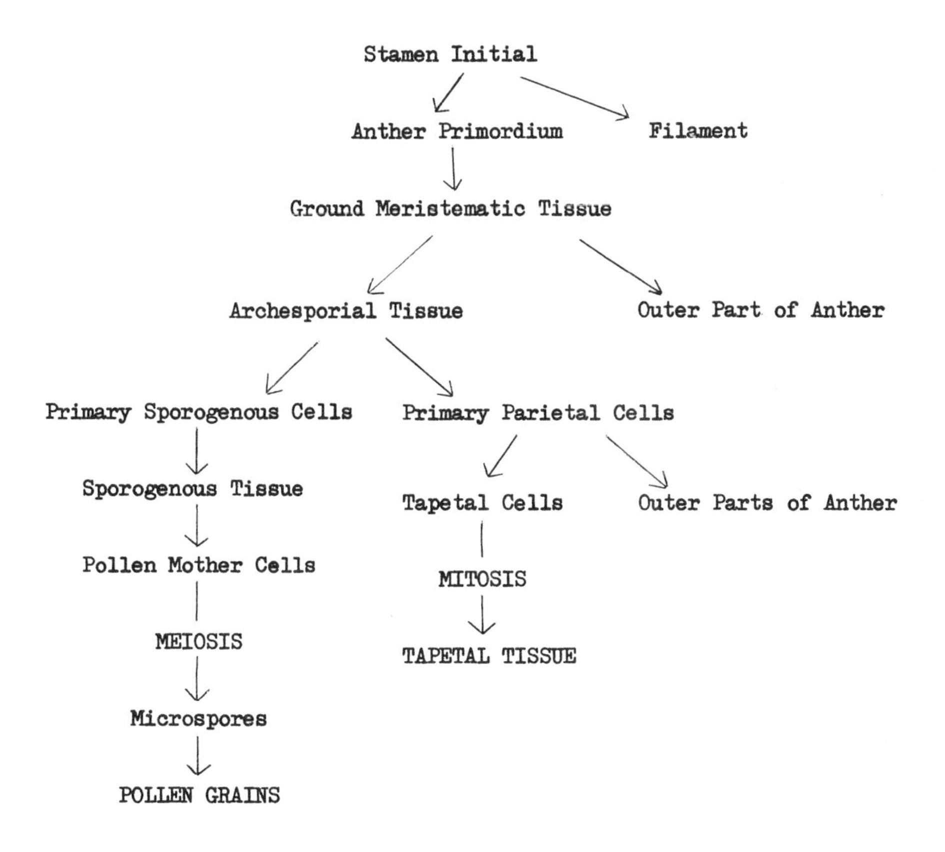

and finally undergo complete autolysis. The amoeboid or periplasmodial type of tapetum which is characterized by an early loss of tapetal cell walls and the intrusion of the tapetal protoplasts among the developing pollen grains, followed by the fusion of these protoplasts to form a tapetal periplasmodium.

The development of both types of tapetum has been followed in a considerable number of plants, using the light microscope, and the details are well documented (see for example Maheshwari (1950) and Davis (1966)). A number of studies have also been made using the electron microscope, although these have mostly been on the secretory type of tapetum. A brief description will be given of the ultrastructural details of the development of both kinds of tapetum, as it provides considerable information as to how the tissue may be involved in the production of sporopollenin. However, it is difficult, if not impossible, to separate the development of the tapetum from the development of the pollen grains, as the two are closely inter-related. In the descriptions of the development of the two types of tapetum, the timing and stage of development will be related to the developmental stage of the maturing pollen grain. A more detailed and comparative consideration of these events is given in the paper by Echlin (in press).

Development of the Amoeboid or Periplasmodial Tapetum

The only descriptions concerning the development of this type of tissue are to be found in the brief report by Heslop-Harrison (1969) and the more extensive study by Mepham and Lane (1969b) on *Tradescantia*. In *Tradescantia* the tapetum possesses an organized and apparently functional ultrastructure, and as development proceeds the cells undergo reorganization rather than the degeneration which is characteristic of the secretory type of tapetum.

During the pre-meiotic stage of sporogenous cell development, the tapetal cells become progressively more vacuolate, and there is evidence of small vesicles, derived possibly from dictyosomes, being discharged from the cell. These vesicles are thought to contain enzymes which would hasten the breakdown of cell walls for the next stage in development is a fairly rapid dissolution of the tapetal cell wall. At about the time the pollen mother cells are undergoing meiotic divisions to form the microspores, a wave of lysis extends throughout the anther loculus, stripping the tapetal cells of their carbohydrate (pectic?) cell walls. The net effect of these activities is an anther cavity containing microspore tetrads infiltrated by tapetal protoplasts which now contain significantly more polysaccharide material.

Towards the end of pollen mother cell meiosis the new periplasmodial tapetal cytoplasm undergoes a further reorganization. Callases which Mepham and Lane believe to be derived from the tapetal protoplasts, progressively divest the microspore tetrads of the callose coat and the long amoeboid-like tapetal cells penetrate between the newly released immature pollen grains.

The tapetal cell metabolism then appears to change from one primarily associated with carbohydrate formation and degradation to the synthesis of lipid material. It is significant that lipid globules appear within the plastids and are progressively extruded from them into the tapetal cytoplasm. Parallel cytochemical studies confirm the presence of unsaturated lipids and phospholipids within the tapetal protoplasts. Towards the end of pollen grain development, at about the stage of the final mitosis, the tapetal protoplasts appear more vacuolate and a second phase of polysaccharide synthesis appears to be in train. Finally, changes in the anther cuticle and connective cause extensive dehydration of the tapetal cytoplasm, and at anthesis it is deposited as a thin tryphine layer on the surface of the pollen grains.

The significance of the close relationship which develops between the tapetum and the pollen grain wall has already been discussed in some detail by Mepham and Lane (1968, 1969a, 1969b), Godwin (1968) and Echlin (1968).

Aside from the ultrastructural details, the main feature of the development of this type of tapetum is its close and intimate association with the pollen grains, and the maintenance by the tissue of a high degree of structural and presumably functional organization until just prior to anthesis. Clearly the tapetum must be having a considerable influence over the transport of material to the pollen grains, and while Mepham and Lane do not consider that nutrients are derived directly from the tapetum they do concede that such materials must pass through it. In light of the recent work on the biosynthesis of sporopollenin, it is worth noting that Mepham and Lane (1970) have recently shown that the bulk of exine deposition on the pollen grain of *Tradescantia* occurs whilst the microspores are still invested with the callose layer. Once released from the tetrad, there is no further ektexine sporopollenin deposition, but synthesis of endexine sporopollenin continues over most of the inner surface of the pollen grain.

Mepham and Lane (1970) find narrow channels through the endexine as the young microspore expands after release from the tetrad. The microspore cell membrane may protrude into these channels, and could make contact with the membranes of the tapetal protoplasts which infiltrate the inter bacular spaces during the later stages of development. Such membranous contacts could provide a route for the passage of sporopollenin precursors into the developing pollen grain.

Flynn (in press) has also demonstrated the presence of micro-tubules in the pollen wall of *Nuphar* and tubules between the exine and tapetal surface in *Aegiceras*. These micro-tubules may thus provide a route for the transfer of material from the tapetum to the pollen grain. The intra-exinous micro-tubules seen in *Nuphar* are similar in many respects to the membranous profiles seen in the lipid exudate of mature *Tradescantia* pollen.

It is unfortunately not entirely clear whether the stage in tapetal development characterized from the work of Mepham and Lane by the appearance of lipid globules initially in the plastids and subsequently in the cytoplasm, corresponds to the stage in development of the pollen grain where endexine channels appear and maximal deposition of endexine sporopollenin takes place.

Development of the Secretory Tapetum

Considerably more work has been carried out on this type of tissue, and for a summary of the significance of these studies reference should be made to the papers by Echlin (in press) and Echlin and Godwin (1968a, 1968b, 1969). For the sake of completeness in this present record, a brief description will be given of the events which occur in *Helleborus foetidus*.

During the earliest stages of development, the tapetal cells are very similar to the amoeboid type of tapetum previously discussed. The cells, although having less readily recognizable organelles, have numerous dictyosome derived vesicles, and evidence of polysaccharide synthesis.

By the pre-meiotic stage of pollen grain development, the tapetal cytoplasm contains more ribosomes, polyvesicular bodies, and pro-Ubisch bodies (see later). The tapetal cells reach the peak of their maturity at about the tetrad stage, but then rapidly go into senescence terminating in complete dissolution.

At the critical phase of pollen grain wall initiation, the common wall between the tapetal cells and the anther cavity appears less electron dense and there is a space between it and the cell membrane. By the phase of ektexine sporopollenin deposition, the tapetal cell walls have virtually disappeared, although the remainder of the cytoplasm appears reasonably intact. As pollen wall development proceeds through to endexine sporopollenin deposition and cellulosic intine formation, the tapetum becomes progressively more disorganized. Finally, the limiting membrane is ruptured, and the remainder of the cytoplasm covers the pollen grains and frequently fills the inter-bacular cavities. It is worth noting that in most secretory tapetums which have been examined the stage of maximum sporopollenin deposition coincides with the stage of tapetal degeneration.

It can thus be seen that the development of the two types of tapetum are quite different. The principal difference being the apparent functional longevity of the periplasmodial type compared to the short lived secretory tapetum. As might be expected, the timing of the various changes in the secretory tapetum shows some small species variation, but the general trend is that whereas the development of the tapetum goes through an initial phase of synthesis followed by degeneration, the developing pollen grains continue along a synthetic path.

Having given a brief review of the ultrastructure and ontogeny of the tapetum, it may now help to consider some of the synthetic abilities of this tissue, with particular reference to sporopollenin metabolism.

Development of Ubisch Bodies

One of the distinctive features of the secretory tapetum of many angiosperms is that during the tetrad stage they often acquire particles of sporopollenin at the cell surface. Although most of these particles are on the inner locular face, they may also be found along the transverse walls and occasionally on the distal face.

Ubisch bodies, or orbicles, are spheroidal structures found in the anthers of many higher plants. They generally occur in large numbers, and are usually only a few micrometres in diameter, although some larger aggregation of particles may occur during later stages of development.

The walls of the coating of Ubisch bodies consists of sporopollenin, and is usually thick in proportion to the total size of the Ubisch body. It may be significant that the overall shape and in some instances the surface architecture of the Ubisch body exhibits a similarity with the pollen grain with which it is associated. It would thus appear that Ubisch bodies may represent a capacity by the anther for organizing similar but less complex sporopollenin structures by smaller cytoplasmic units. With the final autolysis of the tapetal cells, the Ubisch bodies tend to lie irregularly upon the remnants of the tapetum amongst the maturing pollen grains.

The following description of the development of Ubisch bodies is based largely on the work carried out in our own laboratory, and for details reference should be made to the appropriate entries in the literature.

The first sign of Ubisch body formation is the appearance in the tapetal cells of a number of medium electron dense bodies surrounded by a limiting membrane. The structures are referred to as pro-Ubisch bodies and in many respects resemble the spherosomes described by a number of workers. The variability between the electron density of the pro-Ubisch bodies in *Helleborus*, and similar structures and spherosomes described by other workers, is probably due to variation in fixation procedures (Mishra and Colvin, 1970). An examination of the developing microspores at the same stage in development reveal similar structures in the cytoplasm underlying the wall region.

As anther development proceeds, the pro-Ubisch bodies increase in number, particularly in that sector of the cell nearest the anther cavity. At the time of late tetrad formation, the pro-Ubisch bodies are surrounded by a zone of ribosomes which radiate from the bodies like the spokes of a wheel. By the time recognizable exinous elements of the pollen grain wall are evident, profiles of endoplasmic reticulum are in close association with the pro-Ubisch bodies. There appears to be an open continuity between the lumen of the pro-Ubisch bodies and the cisternae of the endoplasmic reticulum, and the material in both is of similar electron density and structure. This strongly suggests that both ribosomes and the endoplasmic reticulum are involved in the synthesis and metabolism of sporopollenin precursors.

The pro-Ubisch bodies are extruded from the tapetal cytoplasm into the anther loculus where they rapidly become invested with a layer of electron dense material which in some instances appears to be deposited on both sides of an electron transparent layer of unit membrane dimensions. Chemical tests indicate that this material is sporopollenin, although initially it is probably in a relatively low state of polymerization.

This electron dense material is only deposited after the pro-Ubisch body has been released from the tapetal cytoplasm and has never been observed to occur while the body is still within the cell. This observation, coupled with the observations on the formation of exine in the maturing pollen grain wall, indicate that sporopollenin deposition is an extracellular event as there are as yet no well documented instances of sporopollenin being deposited within the cytoplasm. However, there is now some evidence (based on UV fluorescence studies) on *Pinus* (Willemse: this symposium) which indicate that sporopollenin may in fact be deposited within cells. It is not presently understood the degree to which this sporopollenin may or may not be polymerized. Risueno et al. (1969) have earlier reported that sporopollenin deposition may occur within the cisternum of the endoplasmic reticulum of *Allium*. However, it is a moot point whether, topologically, the cisternum of the endoplasmic reticulum is inside the cell cytoplasm.

The Ubisch bodies, as they are now called, become progressively more electron dense due, it is thought, to increased polymerization of sporopollenin. The density and apparent rate of sporopollenin deposition continues to parallel the deposition of exinous material on the pollen grains. In the later stages of development, this deposition is associated with thin lamellae of unit membrane dimensions, similar to those seen associated with the deposition of endexine in the pollen grain wall of the same species. In some instances multiple Ubisch bodies may form, due to the coalescence of a number of smaller units. At no stage in development has it been possible to find Ubisch bodies free within the anther cavity: they invariably remain associated with the surface of the tapetal cells. However, at the time of the final dissolution of the tapetum, the Ubisch bodies and remaining pro-Ubisch bodies become intermingled with the pollen grains in the anther sac. There is a close morphological resemblance in electron density and structure between the pro-Ubisch bodies released from the tapetum and the lipo-proteinaceous tryphine which is deposited on the pollen grains during the final phases of development.

Function of Ubisch Bodies

If we adopt a purely mechanistic approach we may consider the Ubisch bodies simply as particles released into the anther loculus as a consequence initially of tapetal metabolism and subsequently of tapetal autolysis. These particles which are resistant to autolysis may themselves be considered to retain a limited synthetic ability (ability to polymerize sporopollenin precursors).

Taking this approach one step further, it may also be possible to consider that Ubisch bodies represent one of the many synthetic abilities of the anther. In the microspores, the sporopollenin is deposited in an orderly fashion to form the delicate and reproducible patterns of the pollen grain wall, and although this synthetic capacity to produce sporopollenin is also present in the tapetum of some plants, the controlling mechanisms are either absent or non-functional.

However, some workers have sought to ascribe a definitive function to Ubisch bodies, and a list of current ideas is given below. A critical appraisal of the validity of these various functions, together with the appropriate references, is to be found in the paper by Echlin (in press).

1. Ubisch bodies are a transport mechanism for the conveyance of polymerized sporopollenin from the tapetum to the maturing pollen grain wall. (This is now thought unlikely by many workers.)
2. They represent in part a conveyance mechanism for sporopollenin precursors from the tapetum to the pollen grain wall, and that the electron dense bodies to which they correspond in electron micrographs is an unfortunate consequence of fixation.
3. They have, an as yet not fully understood, function in the dispersal of pollen.
4. They provide a non-wettable layer lining the anther sac from which pollen grains are readily detached.

Tapetal Membranes

This refers to the acetolysis resistant membrane, made of sporopollenin, surrounding the whole tapetum and enclosing the pollen grains like a sac.

Banerjee (1967) demonstrated the presence of these membranes in a large number of grasses, all of which are characterized by the secretory type of tapetum. The membranes which can form over the inner and outer faces of the parietal tapetum consist jointly of a region of Ubisch bodies; a stranded layer on a fenestrated back layer, and a layer of micro-rods. Because his studies were made on single-stage replicas of acetolysed material, it is difficult to separate out which portion of this complex tapetal membrane was derived from the inner face and which part from the outer face of the tapetum.

What is clear is that these membranes originated as a result of secretions from the tapetum, which reach the peak of their activity at the time of maximum deposition of exine on the pollen grain walls.

Heslop-Harrison (1969) has described a similar structure in certain Compositae confined to the outer face of the parietal tapetum of the periplasmodial type. The membranes persist for some time and it is considered that one possible function is to act as a culture sac enclosing the developing pollen grains and the labile tapetal protoplasts. Dickinson (this present symposium) describes a similar peritapetal wall in *Pinus banksiana*.

Pollenkitt Formation

Although the term "Pollenkitt" was originally used to describe the whole content of the mature anther, there is a tendency to restrict the term to a description of an oily layer found on the outside of mature pollen grains from many insect pollinated plants. Heslop-Harrison (1968a, 1968b) and Heslop-Harrison and Dickinson (1969) have been able to show, in *Lilium* at least, that this material is mainly lipid and contains carotenoids. These carotenoids first appear as highly osmiophillic globules in the tapetal cytoplasm, but are only released into the thecal cavity at the time of the dissolution of the tapetum. It is still uncertain whether this and similar material is unpolymerized sporopollenin precursors involved in Ubisch body formation, or whether it does really represent another quite separate synthetic ability of the anther.

The biological function of this material is uncertain. It may represent only the remains of pro-Ubisch bodies which were not secreted in the normal way. Other workers suggest it may act as an insect attractant, afford protection against the damaging effects of UV radiation, or as a sticky material to ensure attachment to insect bodies. Further suggestions include involvement in the genetic incompatibility mechanisms of some plant species, or because of its hydrophobic nature even aid in pollen dispersal.

Further work is obviously necessary before a function(s), if any, may be ascribed to this curious material.

Tryphine Formation

To avoid undue confusion in the literature, the term Tryphine is considered to comprise a complex mixture of substances derived from the breakdown of the tapetal cells during the final stages of anther maturation. Thus, although it is possible to recognize cytoplasmic elements such as strands of endoplasmic reticulum and pro-Ubisch bodies in the tryphine, much of the material consists of morphologically unrecognizable structures. The deposits of tryphine on the surface of pollen grains is never as extensive as the pollenkitt deposits which occur on some pollen grains.

It would be interesting to chemically identify tryphine, for although it clearly contains materials which are lipidophillic in nature, it also contains many other substances as well.

Tryphine appears to be present in wind and animal pollinated plants, and there is no certainty concerning any role it may play in the life of a plant.

Contribution by the Tapetum to the Pollen Grain Wall

It is unlikely that the tapetum makes any significant contribution to either the cellulosic primexine or to the sporopollenin based primary exine (early ektexine). This is because in most plant species which have been examined, the callose [β 1-3 glucan] which surrounds the young microspores, is a most effective barrier to all but the smallest molecules. Also, both the secretory and periplasmodial type of tapetum have not entered the phase of decline generally associated with the release of metabolites. It is thus most likely that during the early stages of pollen wall formation the sporopollenin is formed from precursors located within the pollen grain cytoplasm. The medium electron dense spherosome-like structures reported by Echlin and Godwin (1968b) probably represent deposits or synthetic sites of such precursors.

In all plants which have been examined so far during the phase of secondary (endexine) deposition,this is accompanied by either a) the progressive degeneration of the secretory tapetum, b) a change towards the metabolism of lipid material in the periplasmodial tapetum, or c) the production of Ubisch bodies where these are seen to occur. It is difficult to concede that the tapetum is not making some contribution to the sporopollenin of the developing pollen grain wall, although the definitive proof is still lacking.

It is presumed that any such material passing from the tapetum to the microspores would be a precursor of unpolymerized sporopollenin (protosporopollenin?), and one may postulate three ways by which this material could be laid down in the anther cavity.

1. Solubilized precursors could be deposited on pre-existing sites, such as the thin lamellae or tapes on and within the maturing exine.
2. The precursors could be deposited or polymerized in the form of Ubisch bodies.
3. The precursors could be in the form of small granules or accretions, which in turn are laid down onto pre-existing structures.

Much has already been written elsewhere and in this present symposium on the role carotenoids are thought to play in the production of sporopollenin to obviate the necessity of reiterating the details. Although there is now general agreement that carotenoids probably are involved in sporopollenin synthesis, the mechanism(s) by which its deposition is controlled or organized is far from certain.

In any event, the developing pollen grain must obtain precursors from one external source or another as in some species there is up to thirty-fold increase in volume without an apparent increase in vacuolation. An examination of the few plant species which have been studied fails to reveal substantial reserves in the pollen grain cytoplasm which might account for this significant increase in growth. Should the tapetum make no contribution to the maturing pollen grain, the only remaining source of metabolites is the callose layer surrounding the microspores, but this is only conjecture as there is no evidence regarding the ultimate fate of the breakdown products of this substance.

The opposite view is held by Mepham and Lane (1969b) who consider that in Tradescantia at least the exine is secreted, controlled and polymerized by the microspore itself, and that the tapetum makes no direct contribution to it.

It is difficult to differentiate satisfactorily between the lipid material thought to be extruded by the pollen grain cytoplasm and the periplasmodial lipid deposited at the pollen grain surface. Much of the evidence of Mepham and Lane is based on the presence of lipid in the inter-baculoid cavities, and it is difficult to see how this material passes out from the pollen grain cytoplasm. However, in a later paper, Mepham and Lane (1970) show that the endexine becomes discontinuous in the apertural region, as the pollen grain expands.

There is thus conflicting evidence whether the tapetum makes any contribution to the sporopollenin of the pollen grain wall. While the evidence is in favour of such a contribution being made by plants with a secretory tapetum, this is apparently not corroborated in the single plant with a periplasmodial tapetum which has been studied in any detail.

Control of Sporopollenin Deposition

An important question still remains unanswered concerning the mechanism(s) controlling the deposition of sporopollenin which results in the characteristic sculpturing or patterning of the pollen wall of a particular species or of a closely related group of plants.

Although this is beyond the scope of this present paper, brief mention must be made of the current thoughts on the subject. It is generally agreed that the cellulosic primexine provides a template on which the sporopollenin is deposited. What is not presently understood is how this template is controlled, and there are four systems by which pattern determination may be initiated.

1. Sporophytic control either through interaction between the developing microspores or the surrounding tapetal tissue.
2. Sporophytic control from relict information derived from the engendering diploid mother cells.

3. Gametophytic control through information inherited and segregated at meiosis.
4. Haplophytic control as postulated by Wodehouse (1935).

A detailed consideration of this subject is given in the papers by Echlin (1969), Heslop-Harrison (1968c, 1968d) and Heslop-Harrison and Dickinson (1969). The consensus of opinion was until quite recently that the template is under gametophytic control. An exception to this is in a paper by Heslop-Harrison (1968d) where, while not entirely excluding gametophytic control, he considers that the genetic evidence favours sporophytic control. A recent paper by Rogers and Harris (1969) strengthens the case for diploid control of spore wall patterning. These workers found that during the development of Linum microspores miniature pollen grains with an apparently normal exine developed around incomplete chromosome complements. They found a close correlation between the amount of exine laid down with the amount of cytoplasm in spores.

Thus while the evidence now favours sporophytic control of wall patterning, the mechanism by which it is brought about is far from certain, and this remains one of several problems concerning the metabolism,deposition and organization of sporopollenin in biological systems. While it is becoming more evident that the tapetum does contribute material to the developing microspore, there is little biochemical data indicating the true identity of the material and how this contribution is effected and controlled.

Acknowledgement

The author is grateful to Miss Ruth Braverman for her patient secretarial assistance.

References

Banerjee, U.C. 1967. Grana Palynologica 7, 365-377.

Davis, G.L. 1966. Systematic Embryology of the Angiosperms. John Wiley. New York.

Echlin, P. 1969. Ber. Deut. Bot. Ges. 81, 461-470.

Echlin, P. (in press). in Pollen and Pollen Physiology. Butterworth and Co. (London).

Echlin, P. and Godwin, H. 1968a. J. Cell. Sci. 3, 161-174.

Echlin, P. and Godwin, H. 1968b. J. Cell. Sci. 3, 175-186.

Echlin, P. and Godwin, H. 1969. J. Cell. Sci. 5, 459-477.

Flynn, J. in Pollen and Pollen Physiology. Butterworth and Co. (London).

Godwin, H. 1968. Nature 220, 389.

Heslop-Harrison, J. 1968a. New Phytol. 67, 779-786.

Heslop-Harrison, J. 1968b. Nature 220, 605.

Heslop-Harrison, J. 1968c. Developmental Biology Supplement 2, 118-150.

Heslop-Harrison, J. 1968d. Science 161, 230-237.

Heslop-Harrison, J. 1969. Can. J. Bot. 47, 541-542.

Heslop-Harrison, J. and Dickinson, H.G. 1969. Planta 84, 199-214.

Maheshwari, P. 1950. McGraw Hill Book Company. London : New York.

Mepham, R.H. and Lane, G.R. 1968. Nature 219, 961-962.

Mepham, R.H. and Lane, G.R. 1969a. Nature 221, 282-284.

Mepham, R.H. and Lane, G.R. 1969b. Protoplasma 68, 175-192.

Mepham, R.H. and Lane, G.R. 1970. Protoplasma 70 (1), 1-20.

Mishra, A.K. and Colvin, J.R. 1970. Can. J. Bot. 48, 1477-1480.

Risueno, M.C., Gimenez-Martin, G., Lopez-Saez, J.F. and Garcia, M.I.R. 1969. Protoplasma 67, 361-374.

Rogers, C.M. and Harris, B.D. 1969. Amer. J. Bot. 56 (10), 1209-1211.

Wodehouse, R.P. 1935. McGraw Hill. New York.

NOTES ON THE RESISTANCE AND STRATIFICATION OF THE EXINE

G. Erdtman, Palynological Laboratory, Stockholm

Abstract

Partial decay of the exine in pollen grains of _Epilobium_ and _Geranium_ in mineral soils is described. An example is provided of exine corrosion, probably due to microbial activity, in birch pollen. Suggestions are made concerning "interbedded zones" in the nexine of _Cobaea_ and _Passiflora_. Bacular conditions at or near the transition from the sexine to the nexine need more study. Cf. "stalactites", "ordinary (infratectal) bacules", and "quasi-ordinary bacules" provided (at least temporarily) with a distinct, root-like extension into the nexine etc. In conclusion it may be asked, if we - by altering, in one way or another, the "natural physiology of pollen development" - are able to get further insight into the mechanisms of still enigmatic morphogenetic etc. processes.

Resistance to decay

The outermost film or part of the exine in certain plants seems to be more resistant to decay and/or certain chemicals than the rest of the exine. Examples: pollen grains of *Epilobium angustifolium* and *Geranium silvaticum* in mineral soils underlying raw-humus in Swedish Lapland. Of the former as a rule only the outermost film of the tectum remains. For the very fine structure of this part in fresh pollen grains see, e.g., Erdtman 1969, Plate 98B:f-g and Plate 102. Of *Geranium silvaticum* only fragments of the characteristic reticulum remain. The exine remains of *Epilobium* are considerably larger than the corresponding parts of the same pollen type occasionally found in nearby peat bogs.

Corrosion possibly due to microbial activities

More than ten years ago we made a simple experiment with acetolysed birch pollen grains in Solna. Nothing has been published about it and I mention it here, in passing only, since it may be of some interest in connection with exine corrosion resulting - I assume - from bacterial activity. Pollen slides were made and sealed in the usual way. The mounting medium was distilled water, not glycerol jelly. The status of the pollen grains was checked from time to time. After four or five weeks, if I remember correctly, a marked change was observed: the exine, i.e., the tectum, began to exhibit a delicately scabrous surface, crammed with details more or less similar to hieroglyphics, Arabic letters, or fine details similar to those later so carefully photographed and described in Grana palynologica (Praglowski 1962). I recollect getting the impression - correct or wrong - that a certain corrosion was going on along special lines since more resistant micro-areas (cf. the nano--spinules in Erdtman 1969, Plates 46:1, p. 305, and 49:6,7, p. 34, and in Praglowski 1962, Plates 6-9; cf. also *Alnus incana*, Erdtman 1969, Plate 13, p. 77) could be seen more distinctly after attacks upon less resistant interspinular etc. microareas. (For the occurrence of bacteria - *Cryptococcus luteolus* - in connection with birch pollen, see Colldahl and Carlsson 1968.)

Radial perforations and/or channels in the exine of mature pollen grains

The presence of small tectal perforations and/or "hanoperforations" or "nanochannels" and/or the presence or more or less radial channels ("tubuli") through nexine-1 seems to indicate a tendency restricted, as it seems, to the very solid and resistant outer, morphogenetically defined exine-1. In contradistinction the possibly slightly weaker and younger exine-2 (corresponding to nexine-2) seems to be devoid of radial channels and perforations. In this context attention should be drawn to Dr. Erika Stix's observations regarding the orientation of leptones in tectum and nexine-1 on one hand and in nexine-2 on the other (Stix, 1964). The title of her paper is "Polarisationsmikroskopische Untersuchungen am Sporoderm von Echinops banaticus" (cf. p. 291-292: "Bei zwei untersuchten Arten, Echinops banaticus und Epilobium angustifolium löscht die Exine nicht einheitlich aus, vielmehr ist der innere Teil sowohl der nativen als auch der azetolysierten Exine negativ, der äussere Teil positiv". (On the alleged nexine-2 character of the nexine in an onagraceous plant - Oenothera biennis - cf. also Lepouse & Romain 1967). "Die Grenze liegt bei Epilobium zwischen Sexine und Nexine und bei Echinops in der Nexine. Nexine-1, Bacula und Tegillum sind positiv, Nexine-2 und Intine sind negativ sphäritisch. Eine ähnliche Aufteilung kommt auch bei Ipomaea vor: dort ist die Sexine isotrop und die Nexine negativ sphäritisch").

Tectal perforations and/or nanochannels are fairly common. Radial perforations and/or nanochannels through nexine-1 have been encountered, e.g., in Zea mays (Rowley, 1960; Larson et al., 1962) and in some large nyctaginaceous pollen grains with thick exine (Erdtman 1952).

"Interbedded Zones" (Larson 1966)

Here only two cases - Cobaea penduliflora and Passiflora caerulea - will be discussed. The mature pollen grains are reticulate (eureticulate; "mature" might here be defined as the stage when normal pollen tubes can be produced or have just been produced). The muri consist of an upper solid tectal layer or part and a lower, infratectal baculate zone. In addition to the infratectal bacules there are smaller (thinner) luminal bacules with free ends.

In acetolysed pollen grains from a herbarium specimen of Cobaea penduliflora investigated about 20 years ago the presence of more or less fragmented remains of the pollen walls revealed that each of the solid, glistening bacules, infratectal as well as luminal,had a centripetal, rootlike extension into the lower, nonsculptured, seemingly less solid part of the exine, viz. the nexine (Erdtman 1952, frontispiece Fig. 2 and Fig. 193 a-f, p. 330).

Electron micrographs have shown that the conspicuous bacular roots in Cobaea penduliflora are surrounded by an irregular "network" of thin sporopollinin threads, strands, or "lamellae". There are also a number of more or less horizontal (transverse) sporopollinin "rootlets" fastened to or, perhaps more likely, branched off from the vertical and solid main roots. This "network" has apparently the same electron opacity as the bacules and their roots. It is noteworthy that there are no signs of "nexinous bacular roots", i.e., of "interbedding", in slides of other species of Cobaea or in slides made from other specimens of Cobaea penduliflora. Can the "interbedding" be explained by the existence of a particularly "plastic stage", perhaps of short duration, during which the distal parts of the larger bacules increase in size and amalgamate (Erdtman, 1952, Figs. 193 a-c,e) forming the tectal part of the conspicuous muri. In contradistinction the small (luminal) bacules do not increase in width and they never amalgamate.

At the same time the "interbedded" condition gradually disappears. When the non-solid part of the nexine becomes as solid and electron opaque as the roots the nexine - which seems to be of nexine-1 character - appears to become uniform throughout, i.e., the roots disappear as morphologically recognizable details. It is still not known, at least as far as I am aware, if there is a nexine-2 in the pollen grains of Cobaea. It should also be mentioned that the "root stage" once observed in Cobaea penduliflora, may - or may not - be some sort of a monstrosity.

The "interbedded" layer described in Passiflora caerulea (Larson (1966)) can hardly be a monstrosity. We were not, however, able to trace it in UV micrographs (made in the late fifties) of thin sections (about 0.5 μm in thickness) through fresh, non-acetolysed pollen grains of the same species. Pending more detailed investigations, it may not be out of place to point at the logical possibility that the interbedded condition here, as maybe the case also in Cobaea penduliflora, is nothing but a transient stage of development.

References

Colldahl, H. and G. Carlsson 1968. Acta allergol. 23: 387-395.

Erdtman, G. 1952 (reprinted 1966). Pollen Morphology and Plant Taxonomy. Angiosperms (An Introduction to Palynology. I). Stockholm and Waltham, Mass. (1952), New York and London (1966).

Erdtman, G. 1969. Handbook of Palynology. An Introduction to the Study of Pollen Grains and Spores. Copenhagen and New York.

Erdtman, G. Further observations on the roots of bacules and/or pila-like elements in the pollen walls of

References (cont.)

Cobaea penduliflora (manuscript; in print 1970).

Larson, D. 1966. Bot. Gaz. 127:40-48.

Larson, D., J. Skvarla & W. Lewis 1962. Pollen Spores 4:233-246.

Lepouse, J. and M.-F. Romain 1967. Pollen Spores 9:403-413.

Praglowski, J. 1962. Grana palynol. 3(2):45-65.

Rowley, J. 1960. Grana palynol. 2(2):9-15.

Stix, E. 1964. Grana palynol. 5:289-297.

THE PRESERVATION OF SPOROPOLLENIN MEMBRANES UNDER NATURAL CONDITIONS

K. Faegri

Botanical Museum, University of Bergen, BERGEN, Norway.

When I accepted the invitation to speak at this Symposium on the fossilization of sporopollenin, it was with some misgivings because I am afraid there is very little to be said on the subject from the point of view of a practising pollen analyst. The problems related to the fossilization of sporopollenin membranes have been known for many years. We have no lack of primary data, but primary experiences have been mainly concerned with other problems, so our solution to this has been to avoid all so-called unsuitable deposits where difficulties could arise and to concentrate on deposits where these difficulties were absent or minimal. Unfortunately it is not possible to examine 'good' deposits every time, and even practising pollen analysts have occasionally to study altered deposits. In these latter cases we are under constant fear that differential destruction of the walls might distort our pollen diagrams (ie.pollen species A may be preserved, and pollen species B destroyed. Consequently 100% of species A may be registered when in reality the original pollen rain contained 50% A and 50% B). This kind of problem is constantly with pollen analysts, because we are not dealing with the difficulties inherent in the pollen deposits examined so our approach and results are completely empirical. This means that I am not able to give any answers, but I can discuss a few of the problems that pollen analysts encounter in their studies. Some of the more detailed studies on degradation are known and have been presented by Dr Havinga (this Volume) and Dr Elsik has also produced evidence of microbiological decay of sporopollenin (1966 and this volume).

My discussion will therefore be in more general terms. I have already used the words fossilized and fossil and Russian workers usually use the word 'buried' to mean fossil. This leads to the popular misconception that everything that is buried sufficiently deeply and for long periods of time is a fossil. This is not the case, and although we can not go into detail of the theory of fossilization, we are able in general terms to state that if the actual substance is still present in its original composition it is not a fossil, but a sub-fossil. I am not aware of any true fossil pollen grains, which would mean either the grains were preserved as imprints or casts or alternatively that the original pollen wall material had been replaced by some mineral (silica, pyrite, calcium carbonate).

Usually there are various quantities of sub-fossil sporopollenin from the different geological ages and very often it is stated that this material has undergone no change and is present in the deposit virtually in the state it was released from the anther. Sometimes colour changes and size alteration of the grains are observed, but the differences in reaction of the different pollen grains to chemical reagents show that some alterations must have taken place. We are ignorant of these processes taking place, but they have not been considered sufficient to worry the pollen analyst and have therefore been continually ignored.

We know that by treating recent pollen grains in a suitable manner (usually by treatment with boiling 10% potassium hydroxide solution for varying periods of time) it is possible to make the grains behave more or less like fossil material with regard to size alteration. I have always considered that this KOH treatment did not alter the chemical structure in any appreciable way, but that it might probably dehydrate the exine material or some similar type of reaction. The dehydration of the wall can also be imagined to take place during the process of sub-fossilization (the correct term that should be used), but this is a suggestion or rather a problem which our chemical colleagues may assist in answering. There must be some slight chemical or structural changes taking place in the sporopollenin molecule during sub-fossilization.

Next I would like to consider another very enigmatic property of pollen grains. It is usual in making pollen preparations to treat the material with acetolysis mixture and then mount the sample in glycerol jelly. When these slide preparations are examined after about two years, various things may have happened to the samples. The grains may be preserved in the original preparation and found to have retained their original size and surface details. Alternatively, examination of another sample treated in exactly the same way which may even be of the same species may show only big blobs from which all features have disappeared. Cushing (1961) showed that part of this alteration was simply a compression effect due to the weight of the cover-slip. This Cushing effect is probably not the full reason for the alteration, although it explains why some pollen grains become flattened it does not explain why other grains lose all their structural features.

This phenomena of the variation in sample after treatment with acetolysis mixture and preparation of the slide is a complete enigma to pollen analysts. The reasons why some preparations remain constant whilst other preparations show large alteration remains a mystery. We have all kinds of hypotheses, sometimes we have suggested the alteration was caused by the remains of the acetolysis reagents and so we removed these very carefully from the preparations, but some preparations were still degraded, whilst other samples which were not washed especially carefully to remove the reagents retained their structure. Similarly with any other treatments that were carried out, some samples retained their shape whilst other samples were degraded showing that there is something unusual about the acetolysis reaction.

However, some facts appear to emerge from these sub-fossilization studies: Firstly, time is a relatively minor factor. Given stable conditions, the changes in the sporopollenin structure after the period of incorporation in the sediment and the period immediately following are very slow processes and any alteration will, in most cases, have taken place at an infinitely slow rate. I would like to suggest there are two definite phases in these deposition processes. Phase one starts the moment the pollen grain is ripe, and continues until it is deposited and the next layer of deposit has settled on top. During this early stage, before the pollen grain is blanketed under the next layer of sediment, the pollen grain is exposed to various external conditions that may cause changes to occur. After the grain has become incorporated in the sediment the changes are much more regular, and usually fewer in number resulting in the alteration processes being much slower.

Secondly, enzymatic breakdown of pollen walls in sediments is usually very slight. Enzymes normally have difficulty digesting material of long-chain molecules, since they can not usually attack the middle of the chain and so must start their digestion at the end of the molecule. This kind of attack is rather difficult and apparently, most enzymes have great difficulty in finding suitable points of attack on long-chained molecules. Although the sporopollenin molecule is rather vulnerable to oxidative chemical attack, apparently only very few enzymes are able to attack the material. Usually, all natural substances have some known enzyme that will break it down, otherwise we would quite simply be submerged in these materials that have been repeatedly produced during geological ages. Sporopollenin may be no exception, to enzymic degradation, but at the moment we do not know very much about these enzymes which are able to digest this material. I was charmed by the idea presented at the Symposium by Heslop-Harrison that the final stages in the formation of sporopollenin may be a non-enzymatic reaction, which suggests that there are no enzymes able to reverse the process if the chemical equilibrium is disturbed.

Nevertheless, sporopollenin does break-down in a number of ways:-

i) Although stigmatic break-down does not usually concern geologists, it should not be forgotten. There are indications that there is some break-down of the exine on the stigma, but the exines remaining on the stigma are not usually considered by biologists. Once the pollen grain has germinated everybody studies the pollen tube produced and the exines remaining on the stigma are neglected in fertilization studies as well as by pollen analysts. Thus, there is hardly any reference to this aspect of exine decay.

ii) In the studies of the pollen analyst the break-down of the pollen grain does not end on the stigma, especially in studies of the grains of the anemophiles which are some of the more interesting examples. These pollen grains are sifted all over the landscape, and may be subjected to physical disintegration (eg by being ground up in moving sand, etc.). These physical disintegration phenomena leads one to wonder whether the degradation of pollen grains in sediments is always due to chemical oxidation or could be caused by a mechanical grinding effect. Similar mechanical grinding can be seen from the fate of diatom frustules which are resistant to oxidation.

iii) It is important to mention at this stage that temperature effects on pollen grains during diagenesis of rocks is included under physical effects and it is known that sporopollenin can withstand high temperatures.

iv) On the surface of pollen grains there are often a lot of fungi, chytrids and other microbes growing. The presence of these organisms on the grain has been wrongly interpreted as attacking the pollen wall. This is not the case, most of these organisms on the pollen grain surface are living on the cytoplasm and do not degrade the exines. The exines are left intact; the organisms pass into the interior of the pollen grain and digest the cytoplasm and may often also destroy the intine. This explanation pertains to many fungi that are found on the surface of pollen grains and it is important to note that it also pertains to the insects that habitually live on pollen grains. There is no report of these insects having developed an enzyme system to digest the pollen exines. Thus pollen grain exines are preserved undamaged in their excreta (see Professor Harris's contribution in the discussion section to this chapter). This preservation again shows that enzymatic break-down of exines is a very difficult job. Having said this, we should not forget that traces of fungal and bacterial breakdown have been published (Elsik 1966) and probably these are not as scarce in reality as they are in the literature. Otherwise, I think we should have found more 'fimmenite' deposits in the world than have been reported.

In the degradation of pollen grain exines there must be some kind of large-scale chemical break-down taking place, which may be non-enzymic oxidation in nature. This conclusion is verified, as any pollen analyst knows, by the preservation of exines being very dependant on the pH of the deposit. In neutral or even alkaline deposits degradation of the exines is very rapid, and the pollen grain remains in the deposit are present as very poor fragments.

Different species of pollen show differing preservation in sediments, which would indicate that either different types of sporopollenin are present in the pollen grain wall or alternatively, that sub-fossil exines consist of sporopollenin plus some other component. The variation in the percentage of sporopollenin in the wall may explain this difference in preservation. This variable behaviour is not directly related to the thickness of the exine, since rather thick-walled, impressive-looking exines can be degraded rapidly whilst other walls containing a thinner exine can be much more resistant to degradation. This suggests that there must be a chemical explanation for the different relative resistance of exines to degradation in addition to physical factors.

When the pollen grain and spore exines are listed in order of their increasing resistance to degradation various inconsistencies are seen. There is a group of pollen grains which always comes out near the top of the list as being very resistant and there is another group that always comes out at the bottom of the list as being very susceptible to decay. In between these two extremes, however, there is apparently no reproducibility of the order of resistance to decay of the various pollen grains. This middle sequence is rather random and difficult to explain. A possible explanation for this is that the primary and secondary exines (to use the 'neutral' term coined in this country) may behave differently to various chemical reagents. The observations of Dr Havinga, that pollen grains that have been initially abraded or in any way scarified appear to be more susceptible to later degradation is interesting. But, in general, by the exclusion of oxygen and in low pH deposit conditions (acidic) there is apparently no degradation of pollen grain walls and these along with spores can be recovered from deposits of any age showing that sporopollenin membranes are remarkably resistant to break-down.

In some deposits, we can imagine cases where all the original plant constituents will have been destroyed and only the pollen grains or their exines, due to their resistivity would be preserved. During deposition proteins, low molecular carbohydrates, pectin-like substances, cellulose and lignin are degraded and what is left in the deposit at the end of diagenesis will be cutins from the plant cuticles and sporopollenin from the pollen grain and spore walls. There is also material present in some fungus hyphae - which is of unknown structure - which is also very resistant to degradation. By this kind of plant debris break-down a residual resistant organic concentrate , what one might call a secondary fimminite,is produced. Asecond type of deposit consisting practically only of pollen grains is known. This type of deposit, unlike a primary fimmenite deposit originally contained various plant constituents which have been completely destroyed during diagenesis to leave an organic rich residue consisting solely of pollen grains.

In addition to pollen grains, there is one other soil constituent - sand - which is equally resistant to decay. In decay processes where the deposit constituents are gradually destroyed it is sometimes the case that the final composition of the deposit is a mixture of sand and pollen grains. This kind of deposit brings me to some ideas that I have been considering for some time of the rather extraordinary possibilities of the podsolization of soils.

The podsol profile consists of:

i) an A_o-layer of more or less living things;

ii) an A_1-layer of raw humus (this term is now considered unfashionable);

iii) an A_2-layer, the podsol or bleach sand layer.

There are also layer B, C, D etc. which are not important in this context.

The theory postulated for the podsol profile, and it is applicable in many cases especially in the majority of the cases of podsolization, is that a continuing decending water current from excess precipitation brings down acidic substances, unsaturated humic-like colloids and other mineral soil components (eg sesquioxides) that are leached out of the upper parts of the soil. These various soil components are deposited further down the profile. This results in the A_2-layer finally consisting of nothing but the non-leachable bleached sand. This model gives the usual explanation to account for the layer of bleached sand, but it has not explained to the pollen analyst the observation that sometimes there are also a great deal of pollen grains in the layer (Beijerinck 1934). The idea of pollen grains travelling down through a deposit is not normally accepted by pollen analysts, but a valid explanation for the presence of pollen grains in the bleached sand layer was not forthcoming. One interesting theory put forward is that if the raw humus containing pollen grains etc.

undergoes very slow oxidation, then the amount of organic matter (which may constitute 98-99% in the upper parts) will gradually decrease further down the profile, because the majority of the organic constituents will be oxidatively degraded, and result in only a layer of sand remaining.

According to the podsol profile theory a raw humus profile contains very few rain worms. When these worms are present in the profile they are very small in size and usually move in the horizontal plane whilst the rain moves vertically. Alternatively, in certain types of temperate soil (the brown earth) there is very little vertical water movement, because the water that penetrates during wet periods evaporates during the dry periods. In these temperate soils there are big rain worms which can move vertically 4-5 meters giving vertical rain worm movement and horizontal water movement. In this latter soil, sand from below will be brought up to higher layers by the rain worms causing the distribution of mineral particles throughout the soil profile. A typical raw humus contains very little mineral particles, but some may be deposited in the humus accidentally by the rain worms or may become included by aerial deposition. In sub-recent layer pollution will cause incorporation of many more mineral particles than would occur in underlying deposits, but these are not specially relevant to this discussion. Such a modification of/alteration tothe ordinary theory of formation of the podsol profile would explain/demand that with continually proceeding slow oxydation in/of the organic parts of the raw humus deposit the layer immediately above the bleached sand should show a high concentration of oxidation-resistant material, of which in the next stage mainly sand and pollen grains would be left. This does in no way preclude that the lower layers of the bleached sand may have been formed as demanded by the classical theory.

In a profile of the type indicated above, the number of pollen grains per unit volume is about 100 in the fresh raw humus (this figure is uncalibrated, so the absolute number of grains is unknown), but is increased to 500 grains per unit volume immediately above the bleached sand. In the same layers it is interesting that the percentage of mineral particles rose from 2 to 12% between the raw humus layer and the layer immediately above the bleached sand. This indicates that in the lower layer (adjacent to the bleached layer) the more resistant components have been concentrated by oxidation. This means that the bleached sand layer, or rather the upper parts of the bleached sand layer may not be autochthonous, or leached in situ, but that the individual grains of sand constituting the bleached sand layer were once accidentally incorporated into the original raw humus. During the period whilst these mineral grains were present in the humus, they were individually leached and only formed the A_2-layer after being concentrated from the rest of the soil (except for pollen grains in these cases) by oxidation.

The time span involved in the formation of this podsol profile is about 700, or perhaps the bottom of the profile may be 1000 years old. This refers to the podsol profile from Øvre Grove,

Voss, W. Norway, recently described elsewhere (Faegri,1970). In these studies the time of formation of the profile does not mean very much, since other examples are known which are considerably older. The time of formation of these profiles is dependent on climate, topography, etc. (Mamakova 1968).

There is no reason to think that the process described above is the way in which a podsol profile is generally formed, but it could explain the unusual occurrences of pollen grains in places where they are not expected to be found.

Finally there is one important observation to be made. So far the above discussion has been applied to what may be called primary podsols (ie. soils in which the podsolization process has been continuous), but many of the podsols that exist today were at one time brown earths. In this situation pollen grains may have been transported by the activities of rainworms. This means that the occurrence of pollen grains in the sand layers of these profiles may be explained in a different manner. On the other hand the oxidation rest hypothesis is equally applicable to such soils during their podsolization phase. It is usual that when profiles have been formed from the brown earth stage, the pollen grain exines are usually rather badly degraded.

In my introductory remarks I stated there were many problems in pollen analysis and there are very few explanations. The pollen analyst is faced with these problems every day and we have not got the tools to solve the difficulties. We very much look for assistance to the chemists now they have the tools for study thanks to the investigations of Shaw and Brooks.

References

Beijerinck, W. 1934 Proc. kon. aka wetensch. Amsterdam, 37, 93.

Cushing, E.J. 1961 Pollen et Spores, 3, 265.

Elsik, W.C. 1966 Micropaleontology, 12, 515.

Faegri, K. 1970 Colloquium geographicum, 12, 125.

Mamakova, K. 1968 Arb. univ. Bergen Mat.-naturv. ser., 4.

DISCUSSION ON PROFESSORS G. ERDTMAN & K. FAEGRI'S PAPERS.

HARRIS: I suggest the eating of humus by small animals may contribute to pollen destruction. I once had reason to look at pollen digestion: bees, ducks and rabbits cause no obvious destruction. But the goat does. Cedar pollen passes through the goat apparently unchanged, but then it continues to be excreted for many days and at the same time the exines become more delicate and sketchy. The peculiar digestion of a ruminant is responsible; only half the rumen contents is cudded and swallowed to the next part of the stomach, while half remains fermenting in the rumen , along with the next meal. Here there is intense anaerobic acid bacterial action. I also persuaded the goat to eat a cake of mixed pollen (unused store from a hive). These grains were in great variety. I didn't identify them but I noted that after digestion many kinds had vanished, so that the composition of the mixture changed.

FAEGRI: I quite agree with Professor Harris's interesting observations, but feel that there is very little real information yet about the animals which can digest exines.

HAVINGA: Professor Erdtman described results whereby the inner parts of the exine vanished before the outer parts. Most of my work shows that the outer parts vanish first, and so it seems that a great variety of microorganisms may act on spores and that it is unsafe to generalise.

ERDTMAN: The tectums appear to be physico-chemically different from the rest of the exine, as shown by Professor Rowley's work, and this may axplain the observations that I made long ago.

HESLOP-HARRISON: May there not be an observational artefact here, in that the sexine is readily identified because it is patterned?

FAEGRI: If you stain preparations, then you have a better chance of detecting fragments, and usually, the percantage of fragments not identified by a trained pollen analyst is comparatively small (say 25% - not serious).

ROWLEY: In our experiments, it was not always the tectum that went first - sometimes, e.g. in Malva, the tectum went after the spines. If, as Professor Heslop-Harrison suggested, and you were only left with the endexine of Malva or Epilobium, it would be a mysterious object indeed. In Betula, the labral parts of the aperture were most resistant as well as areas near the channels through the sexine.

VAN GIJZEL: In practice, it is very difficult to distinguish the layers of badly corroded pollen. However, U.V. fluorescence can be a great help in detecting more or less destroyed grains.

JONKER: Professor Faegri mentioned that microorganisms which can attack sporopollenin are not known, but he also mentioned the discovery of microorganisms within the grains. How can they find their way in, if they do not degrade sporopollenin?

FAEGRI: Through the apertures.

JONKER: This is covered with a nexine.

FAEGRI: This is usually ruptured.

HESLOP-HARRISON: This is an important point. The sporopollenin is nearly always fragmented at the apertures; it is very rarely a complete plate. In most species it forms a series of loosely linked granules or platelets, set in a matrix probably polysaccharide in nature, which is presumably attacked by enzymes from deeper in the intine, or from the cytoplasm,during germination 'See Roland, Pollen et Spores 8, 409 - 419, 1968; see also Heslop-Harrison - this volume).

VAN GIJZEL: It seems from these two lectures that pollen diagrams may be unreliable, because it will very rarely be possible to take into account all the factors controlling the assemblage which will be fossilised during subsequent burial. Furthermore, we perhaps ought to try to define exactly what we mean by a fossil; i.e. when is a pollen grain actually a fossil?

JONKER: It is very difficult. I know one spot where the same species of diatoms has been living since the Eocene. Now there are living diatoms and dead ones, and fossil ones, so where must we draw the line between merely dead, and fossil?

A PALAEOBIOLOGICAL DEFINITION OF SPOROPOLLENIN

W.G. Chaloner & G.Orbell, Dept. of Botany & Microbiology,

University College London.

Abstract

The term "spore" has been applied to single-celled and small multi-cellular propagules from a wide range of organisms. When the cell walls of these spores are encountered as fossils, there is no secure criterion of morphology for separating all algal spores from those of all vascular plants; nor of separating all pollen grains from micro- and isospores of vascular plants. The application of the term sporopollenin not merely to all these spores and pollen grains but also to kerogen and organic material in meteorites is questionable because of the biological origin implicit in the term. We need to know whether fossil organic matter other than spore and pollen exines has similar - or distinct - physico-chemical characteristics. Fossil spore wall material from diverse sources shows minor differences in physical properties such as specific gravity. Flotation on a density gradient may be one means of segregating such spores when differences of morphology alone are not sufficiently characteristic.

Introduction

The term sporopollenin, coined by Zetzsche and his students has quite a long history. The German chemist John (1814) referred to pollenin, as the inert material left after extracting tulip pollen with alcohol and dilute alkali; Braconnot (1829) used the term in a similar sense, applying it to *Typha* pollen, after extracting various soluble components. In terms of modern usage, both John and Braconnot were using the word pollenin not only for the exine, but also for the underlying intine of cellulose - in fact the whole of the sporoderm or spore coat. Zetzsche and his students were the first to use sporonin as the corresponding constituent of pteridophyte spore walls (Zetzsche and Huggler 1928) and they also coined sporopollenin, emphasising the similarity of composition of spore and pollen walls but using a plural (generic) term when describing both collectively. It was these workers who restricted the concept to the chemically most inert outer wall layer of pollen grains and spores - that is, the exine of current usage, excluding the cellulosic intine which they removed with phosphoric or sulphuric acids.

It is important to acknowledge that there were two aspects involved in the coining of the term sporopollenins - the biological source, and the actual chemical composition. The source of Zetzsche and Huggler's material was the pollen of several seed plants, and the spores of *Lycopodium clavatum*. They also looked at several sources of fossil sporopollenins - "pollenin" from the Tertiary Geiseltal lignite, "sporonin" from a Carboniferous spore-coal, and "tasmanin" from Tasmanite, a rock largely of algal origin.

Spores, Living and Fossil

Before considering sporopollenin as a term applicable to fossils, it is necessary to review the meaning of "spore" and "pollen" as these words are applied to living plants.

The term spore has been used so widely in biology that its meaning has really become very broad indeed. A spore is any kind of single-celled or few-celled body produced by an organism and destined to function as a propagule - that is, to give rise to (directly or as part of a sexual process) a new individual. Spores in this broad sense are produced by all the major groups of plants including fungi and algae as well as the higher plants, but also of course by bacteria and protozoa. So that bodies, properly to be called spores, range from the free-swimming flagellated zoospores of algae to angiosperm pollen grains with an elaborately ornamented outer layer of sporopollenin.

There is nothing necessarily resistant about spores - some may in fact be very ephemeral bodies, being viable for perhaps only a few hours, or even less if conditions are unfavourable. Others (including some fungal and bacterial spores, and under exceptional circumstances, the spores of higher plants) may survive in a viable state for several years. Generally speaking, the more long-lived spores have a fairly thick wall, resistant to microbial attack and relatively impervious to moisture.

The early palynologists - and especially palaeopalynologists dealing with fossil material - realised that the resistant bodies encountered in maceration residues of rocks were mainly vascular plant spores (including pollen grains), and it appeared that generally fungal, algal and bacterial spores were rarely preserved. So that when palynologists referred to "fossil spores" it was generally assumed that these represented land plant spores or pollen grains. But it has become apparent in recent years that a number of the resistant organic fossil bodies called spores by the early workers (for example hystrichospheres, *Tasmanites*) are not land plant spores at all, but the resting cysts of unicellular algae. For example, some of the bodies within this group show the characteristic equatorial and longitudinal flagellar grooves or plate symmetry of the living group of algae called Dinoflagellates. It now appears that quite a large proportion of "hystrichospheres" are in fact the fossil resting cysts of dinoflagellates; but this leaves a considerable residue of "hystrichospheres" without such distinctive morphology, referred to as the Acritarchs, of which the biological affinity is enigmatic and almost certainly very diverse. (For a recent treatment of the status and nomenclature of these bodies see Schopf 1969, Evitt 1969). In addition to the Dinoflagellates which produce spores capable of surviving as fossils, we know that at least one other group of quite unrelated algae produces a non-motile stage with a tough, inert wall which readily fossilizes. This group, the Prasinophyceae, includes the living alga *Pachysphaera* and the fossil genus *Tasmanites* (Fig.1F) which has been shown by Wall (1962) to be indistinguishable from the cystose phase of *Pachysphaera*. So that we have among the acid-insoluble fossils referred to as spores, occurring along with the exines of land plant spores, at least two widely separated groups of algae.

So long as we are dealing with living plants, of which

something is known of the biochemical composition and character of the life cycle, we generally know at least whether we are dealing with a bacterium, a fungus, an alga or an archegoniate plant (the group comprising the bryophytes - mosses and liverworts - and the vascular plants or tracheophytes). But in the case of fossils of which we have only the resistant spore coat, dispersed from the parent plant, often altered by diagenesis, it is much more difficult to resolve the botanical affinity of the spore. For example, it is by no means clear which Acritarchs represent unicellular algal cysts, and which may be propagules of multicellular plants. We know that some of the plants of Devonian time, probably adapted to life on land, formed spores which are hard to distinguish from some of the smooth-walled supposed planktonic algal cysts. An example of this is in the spores produced by the Lower Devonian plant *Parka decipiens* (Fig. 1A). This was a flat, multicellular structure which must in life have been rather like a large thallose liverwort. *Parka* was very thoroughly investigated by Don and Hickling (1917), and little has been added to our knowledge of the plant since that time. Although it is somewhat problematical in character, most palaeobotanists are agreed that *Parka* was not a vascular plant; indeed, it lacks any evidence of either vascular tissue or stomata, although it probably lived on a partially exposed land surface. It may have had a mode of life comparable to some modern thallose liverworts, although from what we know of its structure it can only have had a most superficial resemblance to that group. Within cavities in its thallus, *Parka*

FIG. 1. Various types of fossil spore, and their parent plants.

A. *Parka decipiens*, from the Lower Devonian of Scotland. The dark circular areas represent spore masses within the plant.

B, C. Spores extracted from *Parka* by maceration; the curved markings represent the impressions of adjoining spores, resulting from compaction in a sporangial mass. They appear to be random in distribution, rather than representing tetrad markings.

D. *Protosalvinia furcata*, from the Upper Devonian of Ohio, U.S.A.; a bifurcating thallus with a spore tetrad (lower centre).

E. A tetrad of spores from *Protosalvinia furcata*, showing the clear triradiate mark.

F. *Tasmanites* sp. A marine planktonic alga, from the Rhaetic (Triassic) Bone Bed of southern England.

G. A spore of *Horneophyton*, an early vascular plant from the Lower Devonian Rhynie chert, Scotland, showing a clear triradiate mark.

H. A smooth-walled acritarch from the Pre-Cambrian Torridonian Sandstone of Scotland. (Preparation by Dr C. Downie).

Magnification: Each single line is 25 u in length; the double lines are 2.5 mm in length. Figs. B and C are at the same magnification.

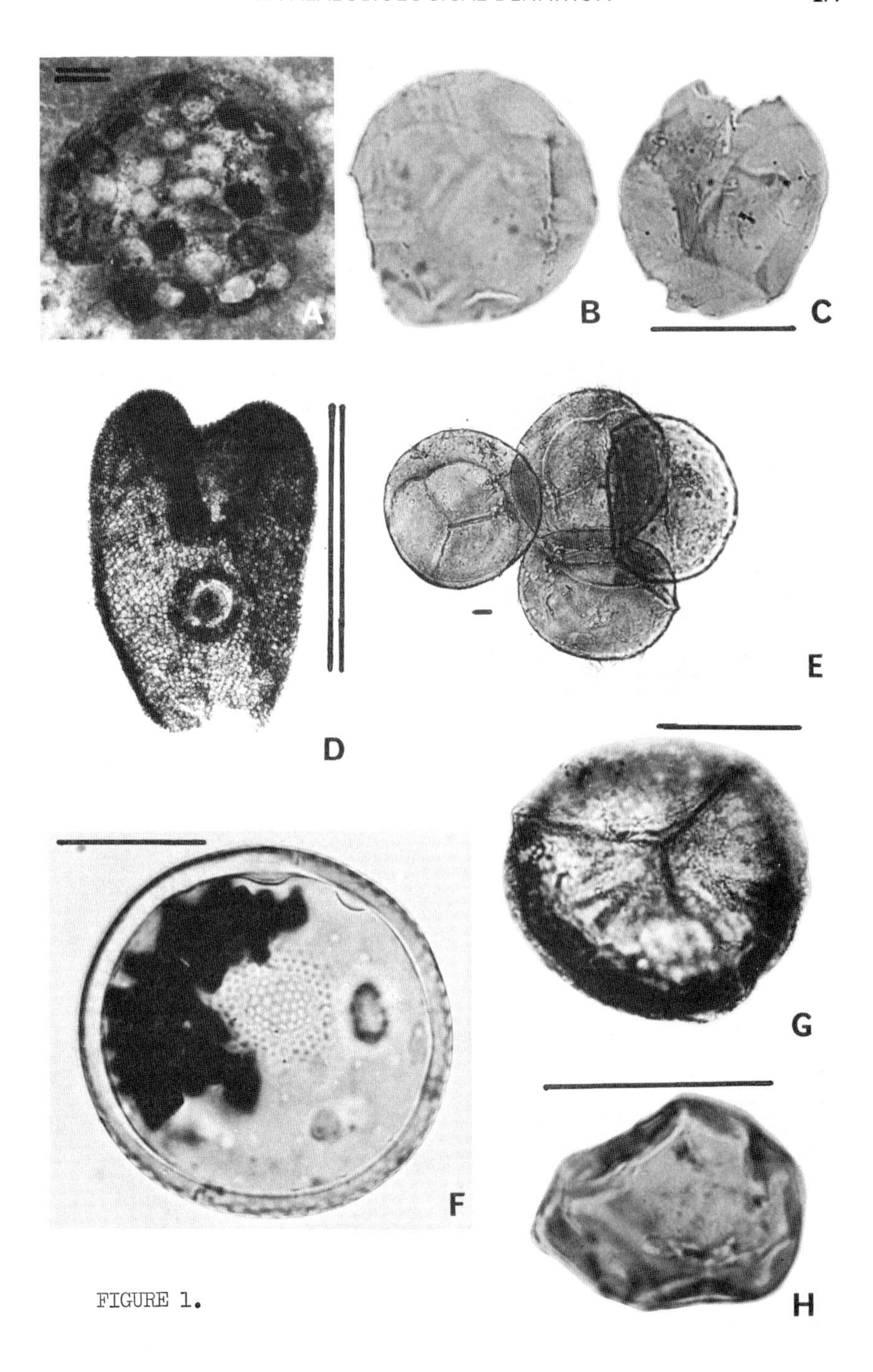

FIGURE 1.

produced masses of spores (Fig. 1B,C) which were presumably liberated as reproductive bodies. These spores, seen individually, appear to have been simple smooth-walled spheres, lacking a tetrad mark, so that they resemble the spherical planktonic cysts called *Leiospheridia*, some of which have been compared by Wall (1962) to the green alga *Halosphaera*. So that where we are dealing with dispersed fossil spores in the form of a single-layered simple collapsed spherical structure, they may represent either single-celled planktonic bodies (such as the living *Pachysphaera* or *Halosphaera*) or the spores produced inside the body of a large multicellular plant such as *Parka*. On present evidence, there is no basis for distinguishing between the diverse sources of such simple fossil spores. When, for example, we find in Pre-Cambrian rocks such smooth-walled spherical spores (Fig. 1H) we do not know whether these represent planktonic cysts or the dispersed spores of a more elaborate plant of the *Parka* type.

Even the possession of a triradiate mark, indicating formation within a tetrad, is not peculiar to the vascular plants. Although spores showing a triradiate suture are found within the sporangia of the early vascular plants (e.g. *Horneophyton*, Fig. 1G), triradiate spores are also formed by living bryophytes; and perhaps more significantly, at least one Devonian plant, believed to be of algal rather than vascular-plant affinity, produced spores with a clear triradiate mark. This is the genus *Protosalvinia* (*Foerstia* of some authors) generally regarded by palaeobotanists (see Arnold 1952, Andrews 1961) as being a non-vascular plant. *Protosalvinia*, in common with some living red algae, produced resistant spores in tetrads within cavities in the surface of its dichotomising thallus (Fig. 1D). The spores show a clear triradiate mark which coincides with a suture, which presumably functioned as in living pteridophyte spores as the site of germinal exit (Fig. 1E).

In summary, it may be said that many of the spores formed by lower plants which are resistant enough to become fossilized show regrettably little unambiguous evidence of their origin. If we look at Devonian plants alone, it can be seen (Table 1) that smooth, spherical spores of the type included for the most part in the acritarchs, may represent either individual cysts formed from unicellular algae, or spores formed within a sporangium by a multi-cellular plant. Equally, triradiate spores may represent either true vascular plant spores, or they may have been formed by plants which are on present evidence non-vascular (a situation seen of course also in the living bryophytes, to which *Protosalvinia* may or may not be related).

The distinction between spores and pollen

The distinction between the microspores of living hetero-sporous pteridophytes and the pollen of seed plants is perfectly clear-cut. Microspores are generally believed to have been the evolutionary antecedents to pollen, in which sense pollen and microspores may be said to be homologous. In their behaviour in

Table 1: Types of spores formed by various Devonian plants

Spore Morphology	Origin
Spore with a triradiate suture	1. Spores formed in tetrads in a sporangium, as in modern pteridophytes. e.g. Horneophyton (Fig.1G) 2. Spores formed in tetrads by a non-vascular plant. e.g. Protosalvinia (Fig.1D)
Spore spherical, more or less unornamented, without tetrad mark	3. Spores produced in masses by a multicellular non-vascular plant. e.g. Parka (Figs.1,B,C) 4. Cysts formed by unicellular algae. e.g. Tasmanites (Fig.1F)

living plants, they are none the less very different; microspores germinate on their own, normally in soil moisture or in water, to release free-swimming gametes; while pollen grains only fulfil their biological function to the plant when they germinate within the micropyle of a seed, or on a stigma, eventually to fertilize an egg within the ovule. These differences in behaviour are accompanied by considerable differences of structure, evident in the exine of mature pollen grains and spores. One of the most significant of these differences is in the polarity of the germination exit with respect to the tetrad in which the spore or pollen formation took place. In all those living homosporous and heterosporous vascular plants in which there is evidence in the mature spore of the arrangement in the tetrad, germination is proximal, usually through a suture on the site of the monolete or triradiate mark corresponding to the side of the spore facing into the tetrad (Fig. 2A, B', B). In contrast to this, in the pollen of gymnosperms and angiosperms, the germination is either distal (on the face originally directed outwards from the tetrad) or equatorial (commonly with three equatorial apertures) or else it may occur through one of a number of globally distributed apertures (Fig. 2D, E).

There is now rather well documented evidence in the fossil record of the evolution of seed plants from heterosporous ancestors (Pettitt 1970). In this process, the stage at which the megaspore was retained within an ovule was the critical point in the origin of a seed plant, and that at which by definition the microspores became pollen. Now the pollen of those few Carboniferous seed ferns for which it is known (e.g. *Crossotheca*, see Potonié 1967) still retained proximal germination, like their free-sporing ancestors (Fig. 2C). This early type of pollen, which retained proximal germination, has been referred to as

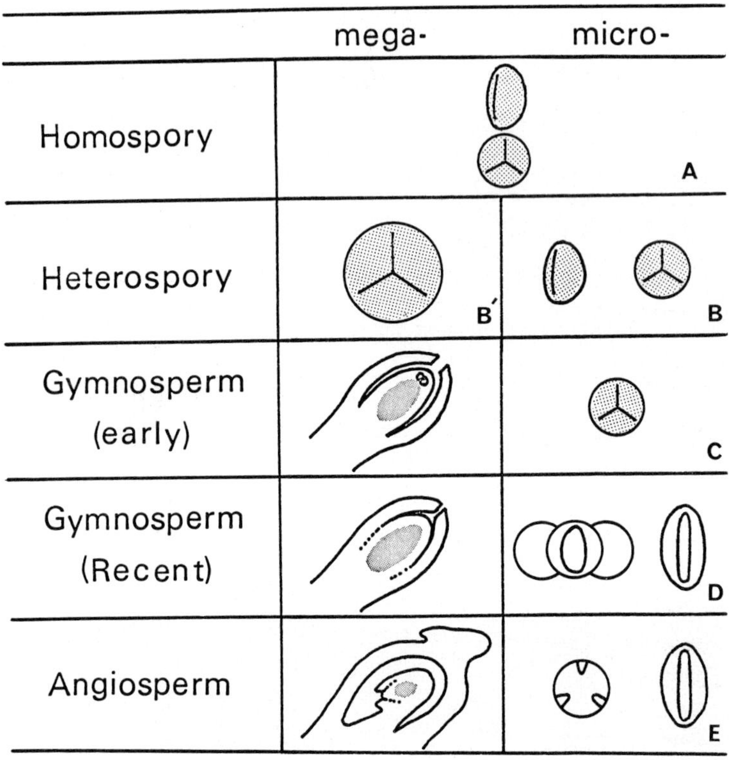

	mega-	micro-
Homospory		A
Heterospory	B′	B
Gymnosperm (early)		C
Gymnosperm (Recent)		D
Angiosperm		E

FIG. 2. A diagram showing the evolutionary progression (A-E) from the homosporous condition to the angiosperm seed. Spores oriented with the proximal pole towards the observer are shown with open stippling; those with the distal pole towards the observer are unstippled. Megaspores retained in the seed are shown with close stippling. Note that germination at a proximal tri-radiate mark was retained (C) after enclosure of the megaspore in a seed; this condition (prepollen) is seen in Carboniferous pteridosperms.

prepollen (Schopf 1938, 1949, Chaloner *in press*).

The nature of the distinction between pollen and spores is commented on here, in order to emphasise that although it is clearly defined in living plants, it is harder to recognise with confidence in dealing with dispersed fossil spores. In this respect we have a problem analagous to the algal/vascular plant spore distinction dealt with above; we can only make the distinction between pollen and spores *sensu stricto* by a process of supposition, based on similarity to spores of which the parent plant is known; and we must accept that there is a category of pollen (prepollen) some members of which are to all practical purposes indistinguishable from microspores.

The recognition of sporopollenin in fossils

Shaw and his students (Shaw 1970) have already demonstrated that there are significant features of agreement in physical and chemical attributes of spore wall material from sources ranging from fossil algal cells (*Tasmanites*) to angiosperm pollen exines. They compare the infra-red absorbtion spectra and the pyrolysis products of wall material from these several sources, and primarily on this basis extend the use of the term sporopollenin from the exines of vascular plant spores to the wall substance of algal cysts. Other workers have taken exception to this usage: Cooper and Murchison (1970) point out that sporopollenin was coined with reference to the exine material of vascular plants, and suggest that it "should certainly not include the polymerised cell contents of aquatic algae and bacteria". But an even more controversial extension lies in the use of sporopollenin for organic matter of extraterrestrial origin. Shaw (1970) argues that very similar infra-red absorbtion spectra and pyrolysis products are shared by sporopollenin and the inert amorphous inorganic matter in ancient sediments (kerogen) and in carbonaceous meteorites. In terms of these and other attributes, kerogen and meteoric inert matter are said by Shaw to "have a common identity" with sporopollenin derived from the pollen of living plants.

The critical question in the controversy is what constitutes an acceptable criterion by which sporopollenins can be positively identified. If sporopollenins were to be regarded simply as a group of compounds defined in terms of certain characteristic physical and chemical attributes alone, this use of the term could not be called to question. But most workers, including Shaw, regard sporopollenin as having a connotation automatically linking it with a biogenenic source. Shaw, for example, regards the presence of material "unquestionably of the sporopollenin type" occurring in the oldest sedimentary rocks as suggesting "that advanced life was present at a very early stage of the earth's formation" (Shaw 1970). But of course this is only true so far as we can be confident that the characteristics used to identify sporopollenin are peculiar to the products of "advanced life", and cannot possibly represent material of some other derivation. It is important that we should confirm that other inert organic matter - for example, fossil cuticles, and the supposedly abiogenic organic matter occurring in association with some igneous rocks (Sylvester Bradley and King 1963) show a different and clearly distinguishable version of these same properties. Until this can be done, it seems premature to extend the term sporopollenin to cover such a wide range of substances.

The resolution of different types of fossil spore wall material by density gradient centrifugation

The high degree of similarity between types of sporopollenin from widely different sources, both fossil and living, demonstrated by Shaw and his students, has tended to draw emphasis away from the minor differences in composition which are of considerable biological interest. As a possible clue in elucidating the sources of various types of plant microfossil, these differences are quite as significant as the relative sameness of sporopollenins as a class. It has been noted that differences in exine composition of pollen and spores are reflected in their differential resistance to microbial attack (Sangster and Dale, 1964, and Havinga in this volume). Such differences in exine composition may still be perceptible even after diagenesis accompanying fossilization, and are presumably responsible for minor colour differences between species, or variation in staining properties both between different species and between layers of the exine within a single spore (Leffingwell *et al*. 1970). Even in highly coalified material, such differences may be evident both in colour, observed in thin section, or in reflectance of a polished surface (Cooper & Murchison, this volume). A possibility arising from these observations is that of using minor differences in specific gravity to physically separate fossil spores of varied biological origin; such differences might reflect merely the circumstances of their fossilization history (e.g. re-working, etc.) or where the process of incorporation of a fossil assemblage has been relatively uniform, it may reflect original diversity of composition, and hence of biological provenance of the spores.

It has become a common practice in extracting fossil spores from clastic sediments to use centrifugation in a heavy liquid to separate fossil spores from the disaggregated matrix. In this procedure, specific gravities from about 1.8 to 2.2 have generally been advocated (e.g. Hughes et al 1964; Staplin et al 1960).

We have followed this procedure by subjecting the "float" containing spores to further centrifugation on a density gradient, and examining the resulting density fractions. Ultra-centrifugation on density gradients is commonly used to separate fractions of different density from living cell extracts (Britten and Roberts 1960), and we have adopted a simple modification of this procedure (B. Spratt, _pers. comm._ 1970) for fossil spore assemblages.

A liquid with a density gradient of S.G. 1.35 to 1.55 is made, using zinc bromide solutions of these values, running them in progressively changing proportions into a centrifuge tube (Fig. 3). Starting with 25 ml. of each, the heavier liquid is run, via the tube containing the lighter, which is subjected to constant stirring by a magnetic stirrer, and through a hypodermic

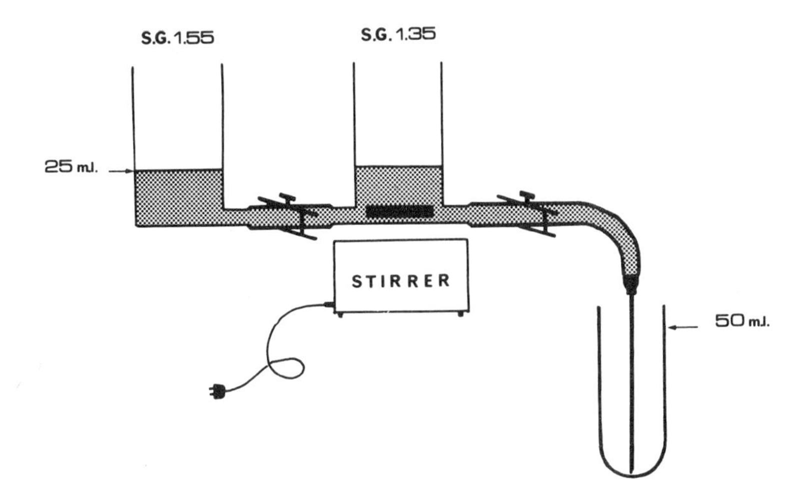

FIG. 3. Apparatus for the preparation of a density gradient of specific gravity between 1.55 and 1.35. The zinc bromide solution of these two strengths is contained in the tubes at left, from which it passes into the centrifuge tube at right. (After Britten & Roberts, 1960).

needle into a centrifuge tube. The progressively heavier fractions displace the lighter fractions upwards as they enter the bottom of the tube, to produce a more or less linear density gradient. The fossil assemblage, already centrifuged from the rock residue by flotation in a zinc bromide solution at S.G. 1.8 is then introduced at the top of the density gradient prepared in this way, and is then centrifuged for half an hour at ca. 4000 revs. per minute. This procedure has been applied by one of us (G.O.) to a spore assemblage from the British Rhaetic containing (among other constituents) many triradiate spores, pollen (including *Classopollis* sp.) and various planktonic acritarchs and dinoflagellates. A perceptible banding occurs in the tube, and the individual bands of microfossil material may then be drawn off with a hypodermic needle and counted separately. Figure 4 shows the results of one such separation. It should be noted that the histograms are constructed so that the percentages sum to 100 vertically and not horizontally across the diagram. This means that the percentages represent the fraction of any one constituent of the assemblage with respect to its total occurrence in all the fractions counted. It can be seen that the greatest contrast in distribution behaviour is between triradiate spores and the acritarch/dinoflagellate population, the latter having a higher modal specific gravity value than the former.

In an experiment with a more extended specific gravity range (S.G. 1.35 - 2.10) dispersed particulate organic matter, resembling kerogen, and fragments of black fusainous material

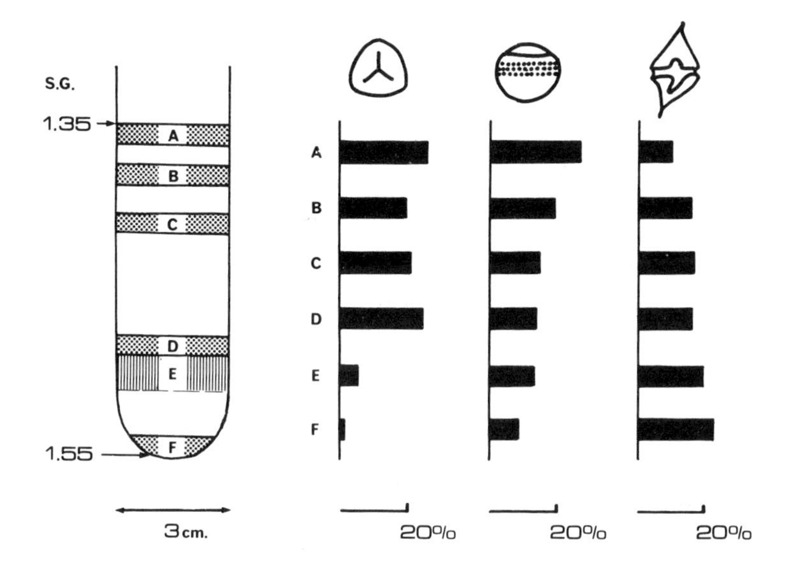

FIG. 4. Result of centrifugation of a British Rhaetic fossil spore assemblage on a density gradient of zinc bromide solution. The bands (A-F) represent concentrations of the fossil material; the distribution of three of the constituents of the assemblage in these bands is shown in the histograms at right. These represent tri-radiate spores, <u>Classopollis</u> plus <u>Circulina</u> and Acritarch/dinoflagellates respectively.

(? detrital charcoal) form two separate bands below those containing the microfossils. The amorphous organic matter (? kerogen) band lies below the fusainous one.

The significance of the actual banding of the fossil spore assemblage is not clear, since each band does not represent a single obviously uniform biological entity; the banding may represent an actual discontinuity in specific gravity of the fossil spore walls, or possibly a non-linearity of the density gradient, and further work in progress should clarify this. But the changing spore concentrations up the gradient presumably reflect actual differences in composition of the spore wall material, or at least their differences after alteration by diagenesis. The fact that such differences persist and appear to have a biological basis seems to be of potential value in palaeopalynology. For example, the behaviour of one particularly enigmatic microfossil, *Rhaetipollis germanicus* Schulz, present in the fossil assemblage only in rather small numbers, follows that of the triradiate spores and pollen rather than the acritarch/dinoflagellate fraction. This suggests that in its composition at least, this microfossil of unknown affinity is closer to the land plant spores than to the plankton.

BIBLIOGRAPHY

ANDREWS, H.N. 1961. Studies in Paleobotany, New York, 487 pp.

ARNOLD, C.A. 1952. Svensk. Bot. Tidskr. 48, 292-300.

BRITTEN, R.J. & ROBERTS, R.B. 1960. Science, 131, 32-3.

BRACONNOT, H. 1829. Ann. Chim. Phys., 42, 91-105.

CHALONER, W.G. in press. Geoscience and Man.

COOPER, B.S. & MURCHISON, D.G. 1970. Nature, 227, 194-5.

DON, A.W.R. & HICKLING, G. 1917. Quart. J. Geol. Soc. Lond. 71 (4), 648-66.

EVITT, W.R. 1969. in Aspects of Palynology, New York, 439-79.

HUGHES, N.F., JEKHOWSKY, B. de, & SMITH, A.H.V. 1964. 5th Congr. Int. Strat. Geol. Carb., 1095-1109.

JOHN, 1814. J. Chem. Phys. 12 (3), 244-52.

LEFFINGWELL, H.A., LARSON, D.A. & VALENCIA, M.J. 1970. Bull. Can. Petr. Geol. 18 (2), 238-62.

PETTITT, J. 1970. Biol. Rev. 45 (3), 401-15.

POTONIÉ, R. 1967. Forschungsber. Land. Nordrhein-Westf. 1761, 1-310.

SANGSTER, A.G. & DALE, H.M. 1961. Can. J. Bot. 39, 35-43.

SCHOPF, J.M. 1938. Ill. State Geol. Surv. Rept. Inv. 50, 1-55.

1949. Ibid. 142, 681-724.

1969. in Aspects of Palynology, New York, . 163-92.

SHAW, G. 1970. in Phytochemical Phylogeny. London, 31-58.

STAPLIN, F.L., POCOCK, S.J., JANSONIUS, J. & OLIPHANT, E.M. 1960. Micropaleo. 6 (3), 329-31.

SYLVESTER BRADLEY, P. & KING, R.J. 1963. Nature, 198, 728-31.

WALL, D. 1962. Geol. Mag. 99, 353-62.

ZETZSCHE, F. & HUGGLER, K. 1928. Annalen, 461, 89-108.

ASPECTS OF SPORIN

ON THE AROMATISATION OF SPORIN AND THE HYDROGEN DENSITY OF THE SPORIN OF CARBONIFEROUS LYCOPSIDS

R. POTONIE & K. REHNELT

Geologisches Landesamt, 415 Krefeld, De Greiff-Strasse, 195, Germany

Abstract

The material of Palaeozoic Pteridophyte spore exines is often fairly uniform. Layering can be distinguished neither through variation of structure nor through colour differences. In some cases, however, e.g. in the Filices, we see the beginnings of a differentiation; in such cases, the outer layer of the fossil material usually appears to have been more strongly aromatised, and therefore was initially less stable. A reversal of this situation is to be found in the taxonomically higher Whittleseyoideae of the Upper Carboniferous. From its darker colouration in transmitted light, the inner layer, in this case appears to have altered its chemical structure more than the lighter outer layer (the exoexine). The chemical cause of these relationships is discussed. The degree of aromatisation and the hydrogen density,H4, may be of particular significance.

Introduction

Sporopollenin is a fairly general term. It covers not only the substance of the exines of spores and pollen grains, whether these be Recent or fossil, but also the cell wall material of the alga Tasmanites.

Past work

SHAW & YEADON (1964) studied the composition of Recent sporopollenin of higher plants. They found that it consisted largely of an aliphatic chain. Fossil sporopollenin differed from this through the gradual increasing abundance of cyclic nuclei; in the course of fossilisation, a slow aromatisation takes place. We call this fossil aromatised sporopollenin, sporin; namely the substance without mineral inclusions (POTONIE & REHNELT, 1969). SOUTHWORTH (1969) reported on the UV absorption of the spore wall, and thereby contributed to our knowledge of sporopollenin.

Present work

Remarks on the aromatisation of sporin as compared with sporopollenin.

During such fossilisation processes as coalification, or saprofication, the spore exines of Pteridophytes behave in different ways depending on whether the layers developed during the life of the spore consist of chemically similar or chemically variable substances. Those layers which were most resistant chemically have been less aromatised than the more labile layers of the same spore exine. They therefore appear lighter in transmitted light and in fluorescence, but darker in reflected light, and they become more intensely coloured by stains (but only after they have been macerated).

Even with the help of the above methods, and also with

the use of staining techniques, no difference can be observed between the material of the intexine and exoexine of the spores of many Palaeozoic Pteridophytes. However, when, in transmitted light, the spore exines of Palaeozoic Lycopodiales or Filices do show colour differences in the various layers, such differences usually reveal a formerly greater resistance of the inner layer (intexine). This layer appears lighter in transmitted light and during fluorescence, but darker in reflected light and also shows a better uptake of stains, the latter only after maceration with HNO_3 + $KClO_3$, whereas the former differences can be seen in the raw material.

Exceptions do exist, however, and such exceptions are to be found mainly in the Carboniferous Whittleseyoidea, which belong to the Cycadofilicales of the Pteridospermopsida. One is tempted to interpret this as a phylogenetic advance, since, in the Whittleseyoidea, this change occurs at the same time as the disappearance of the raised λ-monoletum (Lambda-monoletum) and the development of an endolambda. The lambda of the monolete mark is now composed of a material that appears darker in transmitted light, and therefore consists of the more strongly aromatised material of the intexine which is surrounded by less aromatised exoexine substance in such a way that the monoletum is no longer raised above the upper surface of the exine; it may be described as an endomonoletum.

The changes of colour and reflectivity of fossil spore exines in the course of coalification has been known for a very long time, but it is only recently that K. Rehnelt has shown that the exine material of Lycopsids in a relatively low rank coal is correspondingly less aromatised than the exine material of a higher rank Westphalian coal. The exine composition in Recent spores is largely aliphatic (SHAW & YEADON, 1964), but in the course of coalification, this aliphatic part becomes more and more aromatised.

In the same way as I had with F. ZETZSCHE when he visited my laboratory in Berlin, so I was able to provide K. REHNELT with material to which no palaeontological or stratigraphic objection could be raised. In what follows, I provide a further note to complete the results already given in the literature quoted below.

2) On the hydrogen density of Carboniferous Lycopsid sporin.

As a result of the ozonolysis of sporopollenin, SHAW & YEADON (1964) were able to isolate a 6,11-dioxy-n-hexadecane-diacid and other mono- and dicarboxylic acids. Among other units for sporopollenin, they were therefore able to erect the structural unit shown in Fig.1.

As a result of our studies (POTONIE & REHNELT, 1969, 1970) on fossil spore exines, we are of the opinion that in SHAW & YEADON's structural unit (Fig.1) both the keto groups separated in the chain by four methyl radicals already represent precursors of a cyclohexene ring. However, in view of the experimental conditions, a partial ozonisation cannot be excluded. The possibility of a selective ozonisation (BECKER, et al. 1967, p.253) for example in sporin has also been taken account of by POTONIE, REHNELT, STACH & WOLF (1970) (see the structural formulae given in pages 491 - 492 of that paper).

For details of the ozonolysis of cyclohexene derivatives we would refer the reader to HARRIES et al.(1906), GRIEGEE et al. (1954), and SUTHERLAND (1961). Cyclohexene structures are preferentially formed during the dimerisation of carotenoids (MOUSSERON-CANET et al., 1968); and they also appear in the chain of C_{50}-carotenoids (SCHWIETER, et al., 1969). These findings do not contradict BROOKS & SHAW's (1968) hypothesis on the formation of sporopollenin from carotenoid substances.

One of the possible structural elements of sporopollenin could therefore be represented by the structural formula (Fig.2). When such material is converted to sporin during fossilisation, aromatisation inevitably leads to an H4 electron density (structural formula , Fig.3), i.e. four adjacent hydrogen atoms attached to a benzene ring. We have already demonstrated such an electron density in Lycopsid sporin (POTONIE & REHNELT, 1969, 1970) (see plate 1, I.R. spectrum from 500 cm^{-1} to 1000 cm^{-1}).

NOWAK (1969) has independently proposed a high H4 electron density for exinite.

The sporin structural element (Fig.3) shows the same relative abundance of elements as the Sporin-mixture-C isolated by us from the exines of Carboniferous Lycopsids (See POTONIE & REHNELT, 1969).

From the 6,11-dioxy-n-hexadecane-diacid we can also derive the structural element in Fig.4, which, by aromatisation of both cyclohexene rings will also lead to a composition similar to that of the exines of Carboniferous Lycopsids.

When they have been completely aromatised, these structural elements will contain no methylene chains more than four units long. In the spectra we measured (see plate 1) no chain vibrations with n >4 could be seen, but the possibility of overlap cannot be completely excluded. According to THOMPSON & TORKINGTON, 1945 (quoted in BELLAMY (1955)) polyethylene makes a doublet at 721 -732 cm^{-1}.

COLTHUP (cited in BELLAMY, 1955) suggests that the $\vartheta(R_3C\text{-}S\text{-})$ band lies between 570-600cm^{-1}, $\vartheta(R_2CH\text{-}S\text{-})$ between 600-630cm^{-1} and $\vartheta(R\text{-}CH_2\text{-}S\text{-})$ between 630-660cm^{-1}. Since the spectrum of the sporin of the genera <u>Laevigatisporites</u>, <u>Tuberculatisporites</u>, <u>Setosisporites</u>, and <u>Triangulatisporites</u> (see plate 1) shows no sharply defined bands in these regions, organic sulphur bonds are probably not present to any great extent. According to OELERT (1965,1967) the in-phase γ-oscillations of the hydrogen electron densities H1 to H5 only just fall within the region of the valency vibration of $\vartheta(\text{-}C\text{-}S\text{-})$, i.e. between 600 - 700 cm^{-1}.

Recent measurements show that the maximum of the broad absorption band No. 9 of sporin-mixture-B (see POTONIE & REHNELT, 1969, p. 266, Table 1) lies at 1612 cm^{-1}. Several authors (see OELERT)ascribe hydrogen bonds linking quinoid groups (R=C=O=........H-R) to band 9.

To us it appears significant that the wave numbers of the various hydrogen electron densities shift as a result of the maceration of the sporin of sub-bituminous coals. These shifts which may attain valuse up to 10cm^{-1} are closely related to substitution in the exine material (see table 1).

As early as 1965 (BELLAMY, p. 65) described the influence of nitrosubstitutions on the position of the (CH)-wagging vibrations. According to him, the bands shifted towards higher frequencies, and he found that the amount of shift paralleled the degree of substitution.

When nitrosubstitution is far advanced, the bands finally reach a position where further substitution of nitro groups has no further effect on them.

As table 1 shows, the individual absorption bands behave differently with respect to wave number shift when an increasing number of nitro groups are substituted in a ZETZSCHE-formula (90 C atoms).

References

BECKER, H., et al. Organikum, Organisch-Chemisches Grundpraktikum. 6th. Edition, Berlin, 1967, 253.
BELLAMY, L.J., Ultrarot-Spektrum und chemische Konsitution. Darmstadt, 1955.
BELLAMY, L.J., The infra red spectra of complex molecules. London, 1962.
BROOKS, J., & SHAW, G., Grana Palynologica, 8, 1968, 227-234.
BRÜGEL, W., Einführung in die Ultrarotspektroskopie,4th edn,1969.
GRIEGEE, R., BLUST, G., & ZINKE, H., Chem.Ber. 87, 1954,766-768.
HARRIES, C., & NERESHEIMER, H., Ber.dtsch.chem.Ges., 39,1906, 2846 - 2850.
LUONGO, J.P., Appl.Polymer Symp.,1969,10, 1969, 121-129
MOUSSERON-CANET,M., LERNER,D., & MANI, J.-C., Bull.Soc.chim. Fr. 11,1968, 4639-4645: ref. from Chem.Zbl.,1969,57,1158.
NOWAK, D., Zum chemischen Bau von Exiniten, Diss.TH Aachen,1969.

OELERT,H.H., Erdöl & Kohle, Erdgas-Petrochemie, 18,1965,876-880.
OELERT,H.H., Brennstoff Chem.,48, 1967, 331-339.
OELERT,H.H., Z. analyt.Chem.,231, 1967, 81-105.
OELERT,H.H., Z.analyt. Chem.231, 1967, 105-121.
POTONIE,R., & REHNELT,K., Bull.Soc.Roy.Sc.Liege,38,1969 259-273.
POTONIE,R., & REHNELT,K., Bull.Soc.Roy.Sc.Liege,(in press,1970).
POTONIE,R., REHNELT,K., STACH,E., & WOLF,M., Fortschr.Geol. Rheinld. &Westfal., 17, 1970, 461-498.
SCHWIETER,U., & LIAAEN-JENSEN,S.,Acta chem.Scand.,23, 1969 1057-1058: Ref. from Chem.Zbl.,1969, 54---0869 & Chem.Zbl.1969, 63---0272.
SHAW,G., & YEADON,A., Grana Palynologica,5, 1964, 247-252.
SHAW,G., & YEADON,A., J. Chem.Soc.(C),1966, 16-22.
SOUTHWORTH, D., Grana Palynologica, 9, 1969, 5-15.
SUHADOLC,T., & HADZI,D., Liebig's Ann.Chem.,730, 1969,191-193.
SUTHERLAND,J.K., Chem. & Ind., 1961, 1607 - 1609.

Table 1.

Material		Sporin-Mixture-B	Sporin-A (artefact)	Nitro-sporin-A
Number of nitro groups in Zetzsche formula		0	2	9-10
Band number & hydrogen electron density.		(cm^{-1})	(cm^{-1})	(cm^{-1})
18	H1	910	915	915
19	H2	795	798	805
20	H4	775	778	785
21	H4, H5	745	750	-
22	H4, H5	690	695	700
23	ar	530	532	540
24	ar	465	468	472

Table 1. ar = aromatic structure (after OELERT).
Sporin-A and nitro-sporin-A are from the exines of the genus *Laevigatisporites*.
Sporin-mixture-B is from the genera *Laevigatisporites*, *Tuberculatisporites*, *Setosisporites* and *Triangulatisporites*.

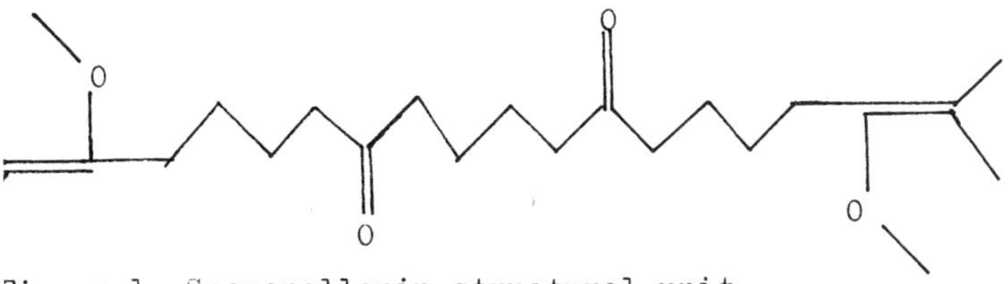

Figure 1. Sporopollenin structural unit (after SHAW & YEADON, 1964)

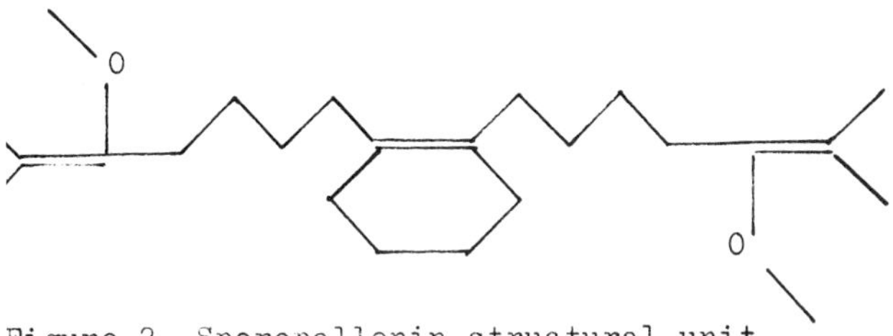

Figure 2. Sporopollenin structural unit (our interpretation)

Aromatisation

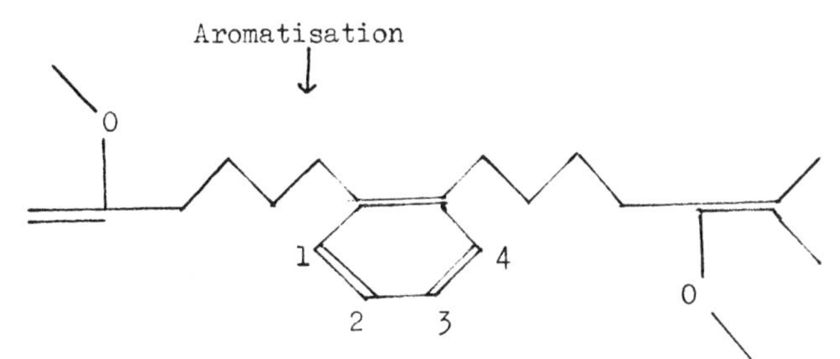

Figure 3. Sporin structural unit with hydrogen electron density H4

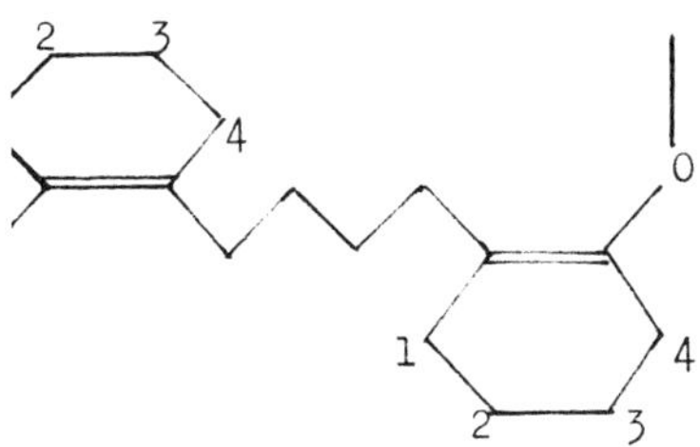

Figure 4. C_{16}-structural unit

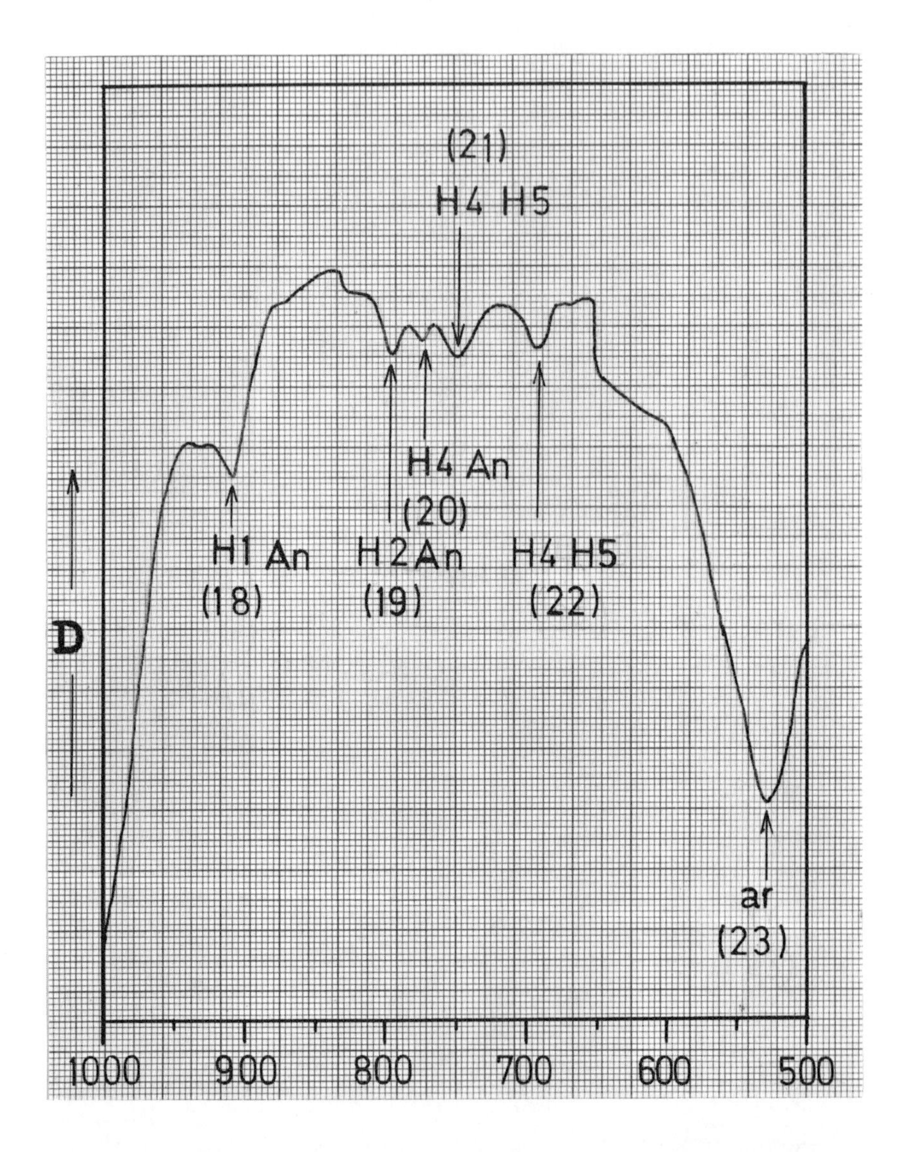

Plate 1: Section of the I.R. spectrum of sporin-mixture-B (Exine material from Laevigatisporites, Tuberculatisporites Setosisporites and Triangulatisporites) between 500 & 1000 cm^{-1}. The arrows indicate which absorption peaks are due to the various hydrogen electron densities H1 - H5 of the aromatic part of the molecule (after OELERT). An = superimposed inorganic peak (ash fraction); ar = aromatic bands (see POTONIE & REHNELT, 1969,1970).

THE CHEMISTRY OF SPOROPOLLENIN

G. Shaw, School of Chemistry, University of Bradford, Bradford, 7.

Abstract

The remarkable resistance of pollen and spore exines to biological decay and non-oxidative chemical attack is a major contributor to the existence of the science of Palynology. The particular type of chemical material which has these properties, called sporopollenin, has been isolated from a wide variety of pollen grains and spores of both higher and lower plants. Early chemical work by Zetzsche and co-workers produced a method for isolation of a standard sporopollenin preparation and established that the polymer was highly unsaturated, contained hydroxyl and C-methyl groups, and that these properties were similar for all the types of sporopollenin examined whether from a recent or fossil source. Later work by Shaw and co-workers has indicated that sporopollenins are oxidative polymers of carotenoids and/or carotenoid esters. Synthetic analogues may be prepared by ionic catalysed oxidative polymerisation of various types of carotenoids; and the products are very similar to the natural materials. The relationship between carotenoids and sporopollenin has been further confirmed by comparison of ozonisation and potash fusion products and by labelling experiments. The biosynthesis of sporopollenin in the sporogenous tissue is seen to resemble a typical suspension polymerisation process in which the globules undergoing polymerisation are known as orbicules. The polymerising globules which will contain carotenoids are laid down on the microspore sac in a manner which is governed by the pre-formed architecture of the sac which in this way operates as a template.

Contents

Introduction

Palynology is one of the newest of the earth sciences. The term, derived from the Greek *palynin* ("to scatter"), normally covers specific areas in the study of pollen grains and spores, and those which have perhaps so far excited most interest have been related to the general fields of palaeobotany and palaeontology. The existence of the subject is centred around the characteristic, specific and quite unusual properties of pollen grains and many spores and spore-like bodies (e.g. hystrichospherids) which, after dispersal by various environmental forces, may, given the right conditions, survive for many millions of years as recognisable structures. Indeed pollen and spore grains are the most ubiquitous of fossils and more widely distributed in time and space than any other representative of living matter. During the processes which occur after deposition, many pollen or spore grains uniquely preserve their characteristic morphological structure by virtue of the great resistance of their exines (outer part of the pollen grain wall) to both biological decay and chemical (especially non-oxidative) attack; and undoubtedly much of the original organic molecular structure of the exine material is also preserved, according to the severity of the particular metamorphic charges to which it has been exposed.

Vast quantities of pollen are produced at flowering, especially from wind pollinated trees and grasses. Estimates suggest that the spruce forests of Sweden alone may produce up to 75,000 lbs of pollen per annum; but, little is known about the amounts of algal, fungal or bacterial spores produced in nature. (Faegri and Iversen, 1964). Pollen grains possess shapes which arc sufficiently characteristic to enable histological diagnosis of many species to be made with considerable accuracy. This morphology is, of course, frequently retained during the fossilisation or preservation processes so that diagnostic tests can generally be applied with almost equal accuracy even after the passage of many millions of years.

The resistance of sedimentary pollen grains to decay

parallels their resistance to many organic and inorganic chemicals. When pollen grains are treated successively with organic solvents (hot aqueous sodium hydroxide, and hydrochloric acid) the cytoplasmic contents are removed, but, generally, the wall remains intact. When this material is further treated with either 72% sulphuric acid or better, 85% phosphoric acid for about four days, polysaccharides are removed and an exine remains which generally retains the shape of the original grain. The exine material, so prepared, appears to be reasonably homogeneous, generally nitrogen-free (small, less than 1% amounts, may occasionally be present) but may contain small amounts of inorganic matter which can prove difficult to remove. The exine material so derived from the spore or pollen grains is called sporopollenin (Zetzsche and Huggler, 1928) and it is the resistance of this material to chemicals (e.g. those used during the isolation procedure) and to biological decay, that underpins the whole science of palynology. It is only in recent years, however, (Shaw, 1970) that it has been possible to gain an insight into the chemical nature of this unique and novel bio-polymer. This article will attempt to outline some of the findings in this field.

Early Experiments with Sporopollenin

The earliest recorded observations bearing on the chemistry of the exine of pollen grains appear to be those by John (1814) and Braconnot (1829). These authors recognised the toughness of the outermost exine material compared with the remainder of the wall substance and were able to give rough estimates of the amounts of exine present in the pollen grain of a typical plant. Braconnot introduced the term "pollenin" to describe this material but with minor exceptions the only other work of any substance to be carried out on materials of this type, until our own recent work, was that of Fritz Zetzsche, who with collaborators published, in the period 1928-1937, a series of papers directed in large measure to a study of the walls of pollen grains and spores of both living and fossil material. Zetzsche and Huggler (1928) coined the phrase "Sporonin" to describe the exine material from *Lycopodium clavatum* and observed that these spores had exines very similar indeed to those derived from pollen grains. Accordingly they finally used the collective noun sporopollenin as a general term to describe the resistant exine material prepared from either pollen grains or spores by specific chemical processes. It must be realised, of course, that the use of the sporopollenin in this context is not intended to imply that all pollen grains or spores necessarily contain this material. It will also be understood that during the particular chemical processes used in the isolation of sporopollenin, some modification of the chemical structure may occur (e.g. hydrolysis of ester groups) but the fact that the morphology of the pollen or spore grains generally remains unaltered during these processes would suggest that the basic skeletal structure is essentially unaltered. It may be noted, that all pollen investigated so far has been found to contain the same sporopollenin-like substances although in widely varying amounts (Tobler) but in addition pollen from certain aquatic plants has been reported to contain an exine not of the normal type (Faegri and Iversen, 1964). Very few spore walls however have been examined and our own admittedly limited results, discussed later in this article, suggest that there may be marked differences between asexually and sexually derived spores. Only the latter apparently contain sporopollenin-like material.

The pioneer work of Zetzsche and co-workers (1928-1937) established a number of facts about the basic chemistry of sporopollenin. These may be summarised as follows:

1) A sporopollenin-like material is a constant and characteristic component of all the pollen grain and spore walls examined. The material is composed essentially of carbon, hydrogen and oxygen in ratios which varied somewhat from species to species but with a stoichiometry quite unlike that possessed by any other naturally occurring polymer. Following elemental analyses, Zetzsche described his materials in terms of empirical formulae based arbitrarily on a C_{90} unit and we have generally adopted this same standard for comparative purposes. It must be realised, however, that adoption of this unit does not imply any specific type of molecular fragment or structure. Typical analyses for various sporopollenins derived from different species are recorded in Table 1.

2) Zetzsche, and for that matter other early workers, noted the extreme resistance of sporopollenin to chemical reagents and at the same time produced evidence which indicated that certain fossil spore wall materials were similarly composed of a sporo-

pollenin-like substance; the other cellular components having presumably been decomposed with time. Zetzsche was especially interested in tasmanite, the so called "white spore coal" from the Mersey River, Tasmania, the similar boghead coal which is composed almost entirely of spore exines, and related materials from various brown coals.

3) Zetzsche *et al* (1928-1937) showed that although the cytoplasmic contents of the pollen grain could be removed by solvents and hot alkali and acid treatment, the resulting wall required further vigorous treatment over several days with either cuprammonium hydroxide (only partly successful), 40% hydrochloric acid (chlorine introduced into the molecule), 72% sulphuric acid (sulphur introduced into the molecule) or best, warm 85% phosphoric acid for about a week, in order to produce a relatively "pure" and apparently homogeneous sporopollenin which did not contain phosphorus. They were also able to suggest that the phosphoric acid or related treatments removed the inner cellulose layer (intine); evidence (Zetzsche *et al*, 1937) for this included:

a) histological examination of the stained (by iodine and concentrated sulphuric acid) intine layer showed a deep blue colour which is a characteristic of cellulose when treated in this manner.

b) Following the vigorous acid hydrolysis, cellobiose (identified as the octa-acetate) a characteristic hydrolysis product of cellulose, was isolated. The intine content of *Lycopodium clavatum*, the main species examined, was about 10%

of the total wall, and its partial cellulose nature was confirmed by examination of the insoluble material which remained after treatment of the walls with diacetyl nitrate in acetic acid. This removed the outer layer of sporopollenin, and the remaining insoluble residue showed the staining properties characteristic of the intine and hydrolysis with acid and acetylation of the products gave crystalline octa-acetyl cellobiose. The cellulose content of various exines are recorded in Table 1.

In retrospect, Zetzsche and his co-workers clearly spent a great deal of time in an attempt to obtain, by classical methods, some idea of the chemical nature of the sporopollenin molecular structure. However, the results obtained were meagre although subsequently to prove suggestive of a specific type of monomer precursor. Almost all the detailed chemical work was carried out on either exine material derived from *L. clavatum* or *Pinus sylvestris* because of their ready availability and the fossil material tasmanin, subsequently shown to consist of the planktonic algal spore, *Tasmanites punctatus* (Muir and Sarjeant, 1970). The more important results may be summarised as follows:

a) Sporopollenin contains hydroxyl groups. Zetzsche was able to suggest the number of hydroxyl groups per C_{90} unit from quantitative hydrolysis of fully acetylated material. A few typical figures are outlined in Table 1.

b) Sporopollenin contains C-Me groups. These were determined by the Kuhn-Roth oxidation with chromic acid. This method, of course, does not identify gen-dimethyl (CMe_2) groups and the results (see Table 1) on polymeric material can only be regarded as affording minimal values.

c) Sporopollenin contains substantial unsaturation. In particular it was found to react extremely readily with substances like bromine in carbon tetrachloride to give a bromosporopollenin, with about 50% bromine content, although some of the introduced halogen could, of course, have resulted from substitution rather than addition reactions, and, in addition, the structure was readily and rapidly cleaved to soluble materials by oxidising agents, including nitric acid, potassium permanganate and chromic acid but especially ozone.

d) Zetzsche found that the only satisfactory way of degrading the sporopollenin structure without over vigorous treatment was to use ozone followed by hydrogen peroxide (Zetzsche *et al* 1937), when the polymer was degraded to simple organic compounds. These were examined and shown to consist in the main of simple (C_3-C_6) dicarboxylic acids (in ratio of 1:2:1:1) which accounted for about 40% of the total molecular structure. The same mixture of acids was obtained from all the sporopollenins examined, including fossil material. With hindsight the formation of such large amounts of these specific yet simple dicarboxylic acids was highly suggestive of an equally specific, single type of monomer precursor of the sporopollenin polymer structure. A less specific monomer mixture would normally be expected to give rise to a far wider variety of organic compounds and acids after the ozone/hydrogen peroxide reaction.

TABLE 1

Composition of the wall of some micro- and mega-spores including pollen grains and synthetic alalogues

Material	% Wall[2]	% Cellulose[2]	% Sporopollenin[2]	Empirical Formula[3]	C–Me/C_{90}unit	OH/C_{90}unit	Ref[4]
Aspergillus Niger	-	-	-	$C_{90}H_{115}O_{10}$			c
Vitamin A palmitate polymer	-	-	-	$C_{90}H_{150}O_{13}$			b
Selaginella kraussiana wegaspore	-	-	27.8	$C_{90}H_{124}O_{18}$			b
Corylus avellann	-	2.7	8.5	$C_{90}H_{138}O_{22}$		11	a
Phoenix dactylifera	-	-	-	$C_{90}H_{150}O_{23}$	3.45		a
Picon orientalis	-	-	-	$C_{90}H_{144}O_{25}$		14	a
Chamaenerion angustifolium	7.9	2.8	5.1	$C_{90}H_{145}O_{25}$			b
Taxus baccata	-	-	-	$C_{90}H_{138}O_{26}$	2.74		a
Lycopodium clavatum	26.1	2.7	23.4	$C_{90}H_{144}O_{27}$	2.04	15	b
Chara corallina	7.6	4.4	3.2	$C_{90}H_{117}O_{28}$			b
Pediastrum duplex	8.7	5.4	3.3	$C_{90}H_{121}O_{28}$			b
Festuca rubra	18.0	3.8	14.2	$C_{90}H_{138}O_{29}$			b
Alnus glutinosa	16.8	5.1	10.7	$C_{90}H_{122}O_{30}$			b
β-Carotene polymer	-	-	-	$C_{90}H_{130}O_{30}$			b
Festuca pratensis	13.0	3.3	9.7	$C_{90}H_{141}O_{30}$			b
Rumex acetocella	13.0	6.7	6.3	$C_{90}H_{142}O_{30}$			b,c
Narcissus pseudonarcissus	2.4	0.5	1.9	$C_{90}H_{143}O_{30}$			b
Equisetum arvense	-	14.1	1.8	$C_{90}H_{144}O_{31}$		13	a
Polulus balsaminifera	1.8	0.4	1.4	$C_{90}H_{146}O_{31}$			b
Ceratozamia mexicana	-	-	20.1	$C_{90}H_{148}O_{31}$			b
Rumex thyrsiflorus	9.9	4.2	5.7	$C_{90}H_{139}O_{32}$			c
Lilium henryii carotenoids polymer	-	-	-	$C_{90}H_{110}O_{33}$			b
Mucor mucedo (±) zygospore	9.5	5.25	4.25	$C_{90}H_{130}O_{33}$			b
Quercus robur	7.1	1.3	5.8	$C_{90}H_{144}O_{33}$			b

Chenopodium alba	15.6	4.4	11.2	$C_{90}H_{144}O_{35}$		
Fagus silvatica	9.4	2.6	6.8	$C_{90}H_{144}O_{35}$		
Lilium henryii	8.4	3.1	5.3	$C_{90}H_{142}O_{36}$		
Gladiolus X pandion	13.2	8.9	4.3	$C_{90}H_{127}O_{36}$		
Phleum pratense	6.2	2.7	3.5	$C_{90}H_{146}O_{36}$		
Lilium longiflorum	8.4	3.3	5.1	$C_{90}H_{144}O_{37}$		
Rumex acetosa	11.2	7.0	4.2	$C_{90}H_{144}O_{37}$		
Pinus canadensis	23.9	7.0	16.9	$C_{90}H_{150}O_{37}$		
Pinus contorta	26.2	5.5	20.7	$C_{90}H_{150}O_{37}$		
Lilium henryii carotenoids/ Carotenoid ester polymer	-	-	-	$C_{90}H_{148}O_{38}$		
Picea excelsa	21.6	5.8	15.8	$C_{90}H_{148}O_{38}$		
Pinus radiata	30.0	6.6	24.4	$C_{90}H_{149}O_{44}$		
Pinus silvestris	29.8	6.0	23.8	$C_{90}H_{158}O_{44}$	1.7	13

1 Except where otherwise stated the material refers to pollen or micro-spores of the particular plant, algal or fungal species.

2 The figures refer to percentages by weight of the original spore or pollen grain.

3 The formulae are recorded on an arbitrary C_{90} basis to facilitate comparison with other published formulae (Zetzsche et al, 1937)

4 (a) Zetzsche et al, 1937.

(b) Brooks and Shaw, 1970, Shaw 1970.

(c) Green Holleyhead and Shaw, 1970.

Zetzsche's results, although somewhat scant, nevertheless pointed to the existence of a new type of biopolymer, highly unsaturated, containing hydroxyl and other oxygen functions and C-Methyl groups and broken by ozone and hydrogen peroxide largely into a specific mixture of simple dicarboxylic acids. Such a combination of groups would not perhaps normally be associated with high stability to chemical or biochemical attack, but stability could be explained by a "ladder type structure", whereby it is necessary to break the ladder at opposite parts of the same rung simultaneously in order to degrade the polymer and this would, of course, be difficult.

Later Chemistry

In the early 1960's Shaw and co-workers (Shaw *et al* 1964-1970) commenced a re-investigation of the chemistry of sporopollenin in the hope that application of the many and various new separatory and analytical techniques of chromatography, spectroscopy etc. which had become available during the post-war period would enable new information about the sporopollenin structure to be unearthed. Most of the new studies were carried out on sporopollenin from *L. clavatum* and *P. sylvestris*. Initially, the general conclusions reached by Zetzsche and outlined above were confirmed. It soon became clear that oxidation was the only degradative technique of value and of the numerous reagents tried, ozone was found to be the most useful. In addition to the formation of simple dicarboxylic acids by this last method, traces of longer chain dicarboxylic acids were obtained together with evidence for the formation of substantial amounts of simple mono-carboxylic acids with both straight and branched chains (Shaw and Yeadon, 1964).

A puzzling feature about the oxidative degradation products was the sharp division into short chain dicarboxylic acids and long chain monocarboxylic acids; the two classes accounting quantitatively for almost the whole carbon skeleton of the sporopollenin structure. At the time this peculiar spectrum of organic acids defied interpretation in terms of any known simple type of naturally occurring monomer unit.

Other oxidative techniques including sodium hypochlorite, chromic acid and strong nitric acid were also investigated but were found to be inferior to ozone. The only other degradative technique which gave new specific information was potash fusion at about 250°C when, in addition to formation of the characteristic mixture of dicarboxylic acids (C_2-C_6), a well defined pattern of phenolic acids was obtained which was similar for all examples studied but reasonably characteristic of the particular pollen source e.g. *L. clavatum* sporopollenin gave p-hydroxybenzoic, m-hydroxybenzoic acids and protocatechuic acid, whereas *P. silvestris* gave mainly p-hydroxybenzoic and m-hydroxybenzoic acids. These results (Shaw and Yeadon, 1964, 1966) suggested that sporopollenin might contain a lignin component and indeed the quantities of phenolic acids produced (assayed colorimetrically) allowing for the loss of phenols of the amount known to occur during a typical lignin decomposition, a figure of about 10-15% lignin content could be tentatively assumed for the sporopollenin structure of

L. clavatum. However, methylation of the polymer with diazomethane followed by potash fusion of the methylated material gave no methoxy-aromatic acids as would be expected if phenolic materials were present in the original polymer molecule. Further work subsequently led to rejection of the lignin concept and the formation of phenolic acids of the particular type noted is now readily explained in another manner.

The major problem concerning the chemistry of sporopollenin was to establish generally if the polymer was derived from a specific type of monomer unit. That this concept might be true followed from the spectrum of organic compounds which were constantly obtained from different types of sporopollenin, and in addition since all known naturally occurring polymers of any consequence have a simple monomer origin there seemed to be no reason why sporopollenin should be an exception. It was clear, however, that further progress in this field could only be achieved by temporarily abandoning the direct approach to the structure and adopting some alternative procedure. We ultimately decided to reverse the order of investigation and use a biochemical method in which a search would be made for likely monomer precursors of sporopollenin. This was achieved by examining the simple organic compounds present in the anthers of a plant at different intervals of time. These results will be discussed in section IV.

Biochemistry of Sporopollenin Production in Lilium Henryii

a) Carotenoids and Sporopollenin

The primary objective in all the work on sporopollenin structure was to obtain information about the nature of any specific monomer unit from which the polymer may have been derived. It had become increasingly clear from both our own and Zetzsche's studies that chemical degradative methods were of limited value and were not readily going to produce the sort of information required. Accordingly we decided to change the direction of the research and investigate the chemical products formed during the development of the anther tissue in a particular plant. *Lilium henryii* was chosen on the advice of Professor Heslop-Harrison since:

1) A good deal of information was available about the development of pollen in the anther tissue of this plant.

2) The large size of the anthers made measurements and isolation techniques easier than is the case in many plants.

In preliminary experiments (Brooks and Shaw, 1968a), about 100 bulbs of *L. henryii* were grown under greenhouse conditions and at various stages of bud growth the anthers were removed, weighed and their dimensions recorded. At the same time small parts of the removed anthers were examined microscopically and the various morphological stages of anther and pollen grain development assigned to a time and bud and/or anther size. The anthers were further extracted with a variety of organic solvents and with aqueous media; the various extracts being examined for the presence of organic chemicals. A proportion of the plants (10-15%) was allowed to mature, the pollen collected and the sporopollenin isolated using the general methods which had been used for *Lycopodium* and *Pinus* pollen. Elemental and other analyses of the

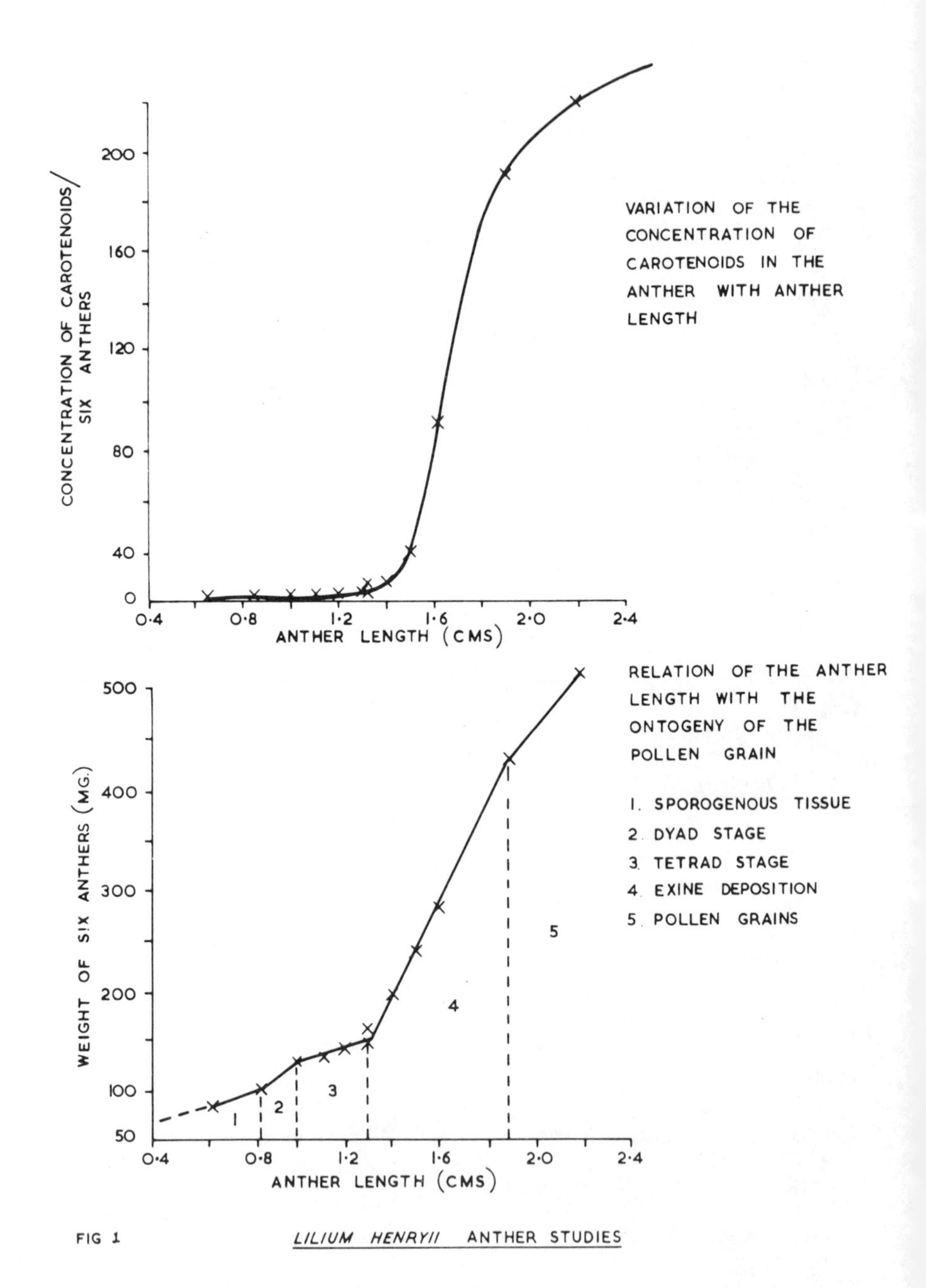

FIG 1 *LILIUM HENRYII* ANTHER STUDIES

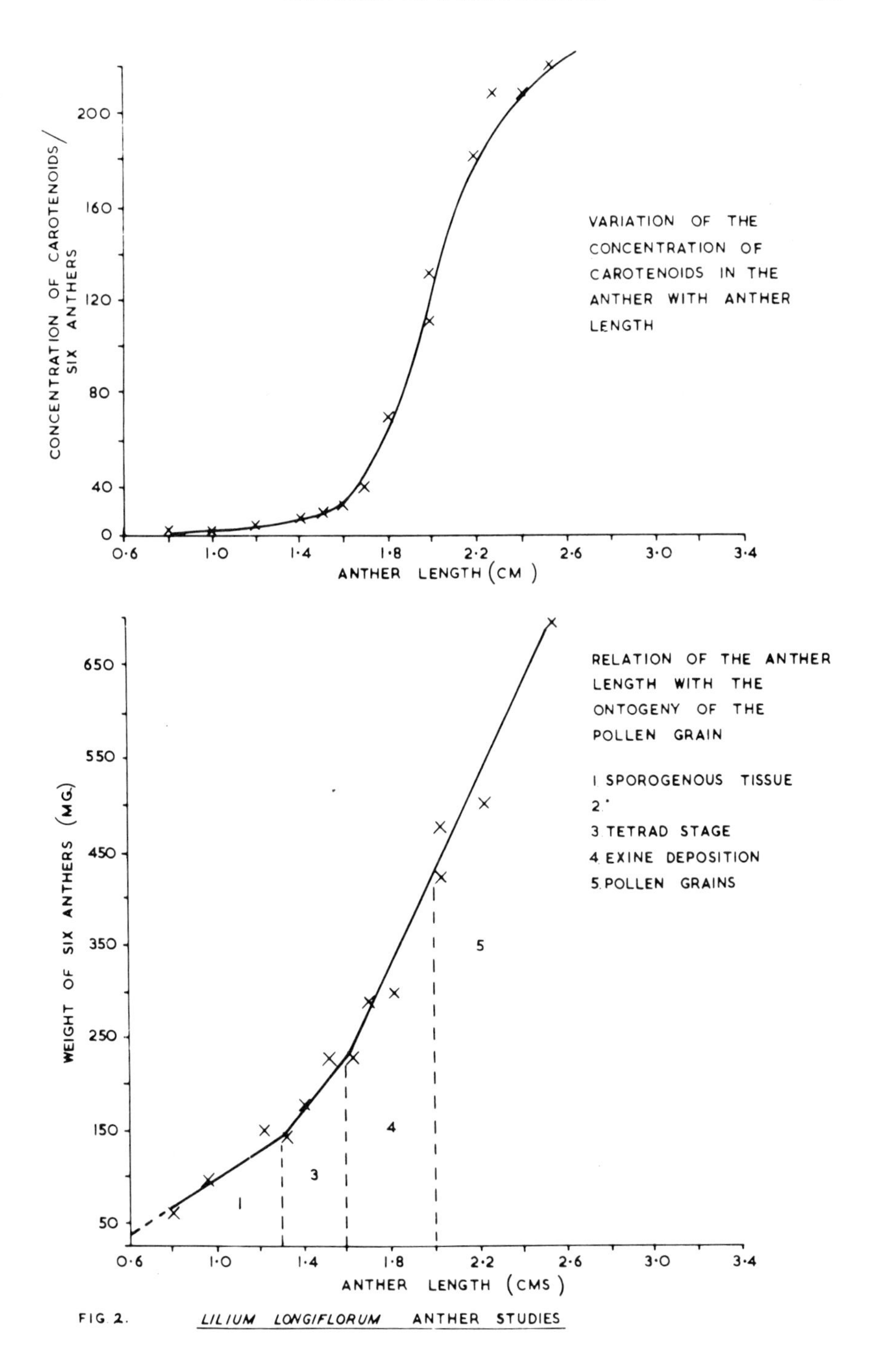

FIG. 2. *LILIUM LONGIFLORUM* ANTHER STUDIES

Lilium sporopollenin showed that it was very similar in all respects to similar sporopollenins derived from other plants (See Table 1).

It soon became apparent from an examination of the extracts that the only material being formed in any major amount and in turn which accompanied pollen grain development in a quite dramatic manner was the carotenoids. The term is necessarily used here and throughout the text to include the whole range of carotenoid-like compounds and their simple derivatives. Thus it will include all hydroxylated and oxygenated materials in addition to hydrocarbons although these have frequently been termed xanthophylls. It will especially include derivatives such as esters and also possible colourless partly reduced precursors of the Phytoene type. Figures 1 and 2 show the relationship between anther length and carotenoid concentration in both *L. henryii* and *L. longiflorum*. In addition, the variation in the ultra-violet absorption of the carotenoid extracts with anther length (Fig. 3) implies, possibly, some small variation of carotenoid type during anther development. Subsequent work with the gourd *Cucurbita pepo* (var. hundredweight) has suggested however that the carotenoid composition in the anther tissue remains remarkably constant during its development. The carotenoids were by far the major class of compound present in the extracts which contained, in addition, minor amounts of hydrocarbons and simple fats. The particular carotenoid mixture from *L. henryii* consisted of a mixture of free carotenoids and carotenoid esters in the approximate ratio of 2.2:1 respectively. Hydrolysis of the chromatographically purified carotenoid mixture with alkali and examination of the products using g.l.c. showed them to consist of a mixture of straight chain and branched chain (10%) fatty acids with palmitic acid as the major components; (Table 2). Examination of the free carotenoids from the saponified mixture indicated that the major component (estimated as approximately 80% of the total) was antheraxanthin. It immediately became clear from an examination of the basic skeletal structure of the carotenoid and especially the carotenoid ester molecules that they provide the very features required of a monomeric unit which by a, presumably enzymic, oxidative process would produce a macromolecular structure analagous to that of sporopollenin. In particular such a unit would be expected to give an insoluble lipid-like molecule which would retain a substantial amount of unsaturation, would contain hydroxyl and C-Me groups and would, after ozonisation, give rise to the hitherto puzzling mixture of simple dicarboxylic acids (derived in the main from the carotenoid stem) and longer chain saturated mono-and di-carboxylic acids derived largely from the ester part of the carotenoid ester structure. It will be realised that a small percentage of the carotenoid esters will contain unsaturated acids and these would undoubtedly be a source of both longer mono-and di-carboxylic acids, some hydroxylated and oxy-fatty acids and medium chain length fatty acids and in addition would provide a focus for subsidiary polymerisation reactions. The presence of carotenoids in plant anthers has been known for many

TABLE 2

The Fatty Acids from the saponified Carotenoid - carotenoid ester extract from *Lilium henryii* anthers - as methyl esters

Acid	Carbon number											
	C_7	C_8	C_9	C_{10}	C_{11}	C_{12}	C_{13}	C_{14}	C_{15}	C_{16}	C_{17}	C_{18}
straight-chain	0.33	-	0.40	0.30	0.32	0.16	1.62	-	-	80.20	-	6.66
branched-chain	trace	0.25	trace	0.20	1.20	0.80 0.25	-	2.72	-	4.64	-	-

Percentage of fatty acids in the mixture

straight-chain acids = 89.99 per cent

branched-chain acids = 10.06 per cent

The percentage of each component in the mixture was calculated from the area under the component peak of the chromatograms using a disc integrator (Disc Instruments Inc.USA) fitted to the Honeywell-Brown recorder and also by manual geometric measurements.

years and they are equally well known to occur in many of the lower plants, including algae and fungi, which produce spores. The occurrence of carotenoids in plant material, especially the floral parts, has generally been a puzzling feature since they have had little apparent function and this has given rise to suggestions about the inefficiency of natural selection in this context (Burnett, 1965).

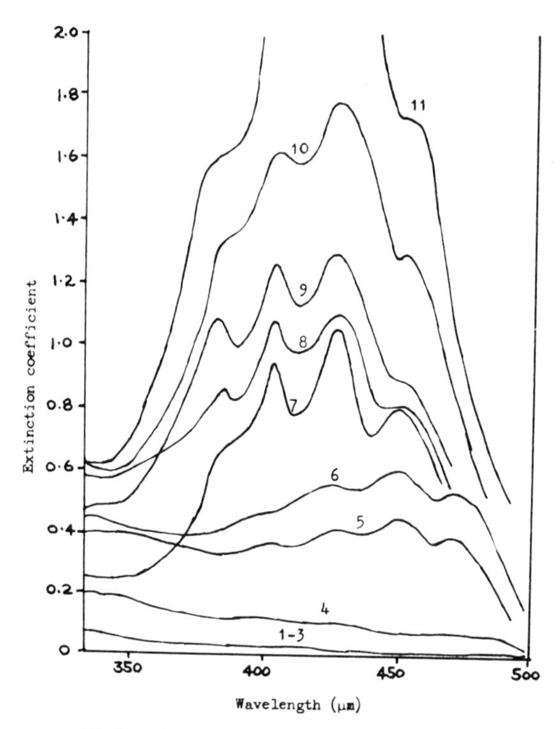

Fig 3. Extinction coefficients for the total pigment extracts of Lilium henryii anthers plotted against wavelength. (Sample numbers 1 to 11 see Table 1 for other details.)

b) Carotenoid Esters

Carotenoid esters are of wide occurrence in nature and several (especially palmitates) (Fig. 5) have been crystallized directly from plant extracts, but in general their chemistry has been little studied.

Examples of reported carotenoid esters include:

1. Dipalmitate of Lycophyll (Lycophyll $C_{40}H_{56}O_2$ MW=568) crystallized from benzene-methanol as purple needles (M Pt 76°) when extracted from *Lycopodium esculentum*, *Solanum dulcamara* and *Solanaceae*. (Zechmeister and Cholnoky, 1936).
2. Helenien ($C_{72}H_{116}O_4$) a dipalmitate of adaptinol crystallized as red needles (M Pt 92°) from alcohol and occurs in *Helenium autumnale L.*, *Compositae* and other flowers. (Kuhn and Winterstein, 1930).
3. Physalien ($C_{72}H_{116}O_4$) a dipalmitate of zeaxanthin occurring naturally in *Physalis* flowers. Crystallizes as fine yellow or red crystals (M Pt 97°) from benzene. One kilogram of air dried flowers yields 4g of zeaxanthin (after saponification of the dipalmitate). (Kuhn and Winterstein, 1930).
4. Astacein ($C_{72}H_{108}O_6$) a dipalmitate of astacin (3, 4, 3', 4' tetra-keto-β-carotene). This pigment is especially found in crustacea, but also in algae, protoza, sponges, fish and reptiles. It crystallizes as almost square red leaflets (M Pt 121°) from petroleum ether. (Kuhn and Lederer, 1933), (Karrer *et al*, 1934-1936).
5. Dipalmitate of Astaxanthin ($C_{72}H_{112}O_6$) found in plants and

animals, crystallizes as purple needles (M Pt 72°) from benzene. (Tischer, 1941).

6. Carotenoid esters in Sunflower oil. Sunflower oil contains esters of carotenoids (0.05 to 0.35 per cent). (Zolochevskii and Sterlin, 1967).

7. Fatty acid esters of carotenoids from flowers of *Forsythia intermedia*, *Taraxacum officinale*, *Tussilago farfar* and *Impatiens noli-tangere*. These all contain C_{12}, C_{14} (main components), C_{16} and C_{18} long-chain saturated mono- and di-esters of:-

   cryptoxanthin; cryptoxanthin epoxide; zeaxanthin; lutein; lutein epoxide; antheraxanthin; and violaxanthin.

   The pattern of the esterification in the four plants is similar and "seems to be of widespread occurrence". (Kleinig, Nietsche, 1968).

8. Carotenoid ester pigments in *Mytilus edulis* and *M. californianus*. Both contain the same range of pigments. Alloxanthin is the most concentrated pigment with esters of alloxanthin and mytiloxanthin. (Campbell, 1970)

9. Carotenoid esters in *Lilium longiflorum*. Twelve carotenoid pigments were isolated and identified:

   two cis-mutatoxanthin esters (mainly palmitates);
   four cis-antheraxanthin esters;
   three cis-zeaxanthin esters;
   one cis-antheraxanthin and two unidentified carotenoids.

   Antheraxanthin esters were 91.7 to 94 per cent of the total

pigments. (K. Tsukida and K. Ileuchi, 1965)

10. Esters of siphonaxanthin and siphonone have been identified in Siphonales (Green algae) group. (Goodwin, 1970).

11. Fatty acid esters of carotenoids. A whole series of carotenoid esters with both saturated and unsaturated ester groups occur in nature. (Jensen, 1970)

c) Oxidative Polymerisation of Carotenoids

To test whether in fact a relationship existed between carotenoids and sporopollenin structure we decided to examine the oxidative polymerisation of various carotenoids under different conditions and with different catalysts. In methyene chloride solution using an ionic catalyst (boron trifluoride or similar compounds) and in the presence of oxygen, the purified carotenoid ester extract of *L. henryii* gave a high yield of an insoluble oxygen-containing polymer which possessed a high degree of unsaturation and had properties virtually identical with those of the natural sporopollenin obtained from the *L. henryii* pollen grain. The properties of the synthetic and natural materials were correlated by comparison of various properties including: insolubility in solvents; resistance to similar chemical reagents including acetolysis with acetic anhydride and sulphuric acid; elemental analysis. (Table 1); infra-red spectra (Shaw, 1970); the products of ozone degradation (Tables 3, 4, 5); products of potash fusion; and comparison of polymers directly, using pyrolysis-gas chromatography (Shaw, 1970). A number of other simple carotenoids including β-carotene, and carotenoid esters

TABLE 3

The Empirical Formulae of some Natural and Synthetic Sporopollenins and the amounts of Acids produced by Ozonization

Material	Empirical formula	percentage acids produced by ozonization		
		Branched-chain mono-acids	Straight-chain mono-acids	Di-carboxylic acids
Lilium henryii pollen exine	$C_{90}H_{142}O_{36}$	29.3	38.4	32.9
Lilium henryii carotenoid-carotenoid ester co-polymer	$C_{90}H_{148}O_{38}$	28.3	33.6	36.4
Lilium henryii carotenoid co-polymer	$C_{90}H_{110}O_{33}$	64.1	3.4	33.4
β-carotene co-polymer	$C_{90}H_{130}O_{30}$	61.3	1.9	35.0
Vitamin A-palmitate co-polymer	$C_{90}H_{150}O_{13}$	28.3	47.6	25.8
Lycopodium clavatum spore exine	$C_{90}H_{144}O_{27}$	16.7	45.6	37.7

Analyses of the monocarboxylic acid fractions from the ozonolysis of L.henryii, L.clavatum sporopollenin and the material from L.henryii carotenoids-carotenoid esters, using more efficient packed columns and coated capillary column gas chromatography has given more accurate estimates of the ratios of straight-chain to branched-chain monocarboxylic acids. Larger quantities of the straight-chain acids are present than first reported (Brooks and Shaw, 1968a) and this gives further weight to our suggestions of sporopollenin genesis from carotenoid esters, the acids from which in nature are mostly straight-chain. The dicarboxylic acid assays are much the same as previously recorded.

TABLE 4

Comparison of the mono-carboxylic fatty acids produced by ozonization of some natural and synthetic sporopollenins

Material	Acids	Carbon number										
		C_7	C_8	C_9	C_{10}	C_{11}	C_{12}	C_{13}	C_{14}	C_{15}	C_{16}	C_{17}
Sporopollenin of *L.henryii*	Straight-chain	-	0.05	0.57	1.14	trace	4.02	-	8.71	-	14.60	-
	Branched-chain	-	0.05	0.10	7.96	9.82	6.90	1.25	0.71	0.59	1.92	-
Synthetic oxidative polymer from the carotenoids-carotenoid esters of *L.henryii*	Straight-chain	-	1.68	0.32	0.25	0.91	3.91	-	4.32	-	17.47	-
	Branched-chain	0.10	0.82	-	6.40	1.73	0.82	5.10	2.06	-	3.26	-
Sporopollenin from *L. Clavatum*	Straight-chain	-	3.20	-	1.74	-	4.65	-	4.99	-	16.97	-
	Branched-chain	0.36	2.30	0.30	0.27	-	2.04	2.00	5.33	-	4.17	-
Oxidative polymer of the anther carotenoids from *L.henryii*	Straight-chain	0.27	-	-	0.59	trace	0.45	0.50	1.58	-	-	-
	Branched-chain	-	0.40	7.92	7.35	1.84	20.4	23.7	2.54	-	-	-
Oxidative polymer of β-carotene	Straight-chain	-	-	-	1.47	0.37	-	-	-	-	-	-
	Branched-chain	trace	0.88	5.15	4.06	7.36	2.95	4.42	-	-	-	-
Oxidative polymer of Vitamin A palmitate	Straight-chain	-	-	-	-	-	0.08	-	trace	-	47.5	-
	Branched-chain	0.32	2.40	0.26	8.80	6.80	-	8.73	-	trace	0.95	-

TABLE 5

Comparison of the di-carboxylic acids produced by ozonisation of some natural and synthetic sporopollenins

Material	Dicarboxylic acids - Carbon number							
	C_3	C_4	C_5	C_6	C_7	C_8	C_9	C_{10}
Sporopollenin from L. *henryii*	6.00	7.32	4.07	8.37	1.28	-	2.11	2.75
Oxidative polymer from the carotenoids and carotenoid esters of *L. henryii*	7.38	11.75	5.85	9.60	2.98	-	2.22	3.55
Sporopollenin from *L. clavatum*	7.73	11.60	-	6.15	5.35	1.50	3.90	1.46
Oxidative polymer of the anther carotenoids from *L. henryii*	3.20	9.20	6.30	9.02	2.45	-	3.30	-
Oxidative polymer of β-carotene	2.80	14.70	6.42	9.62	-	1.47	-	-
Oxidative polymer of Vitamin A palmitate	3.05	8.34	4.66	6.26	2.93	-	-	-

TABLE 6

Oxidative polymers of some carotenoids and carotenoid esters

Molecular Formula	Polymer (C_{40}) (C_{72})	Polymer (C_{90})	Sporopollenins	
$C_{40}H_{56}O_2$ Zeaxanthin	$C_{40}H_{69}O_{18}$	$C_{90}H_{156}O_{40}$	P. silvestris	$C_{90}H_{158}O_{44}$
			P. contorta	$C_{90}H_{150}O_{37}$
			L. longiflorum	$C_{90}H_{144}O_{57}$
$C_{72}H_{116}O_4$ Zeaxanthin di-palmitate	$C_{72}H_{108}O_{19}$	$C_{90}H_{135}O_{24}$	L. clavatum	$C_{90}H_{144}O_{27}$
			Taxus baccata	$C_{90}H_{138}O_{26}$
			Corylus avellana	$C_{90}H_{138}O_{22}$
$C_{40}H_{56}$ -carotene	$C_{40}H_{59}O_{11}$	$C_{90}H_{132}O_{30}$	M. mucedo	$C_{90}H_{130}O_{33}$
			P. duplex	$C_{90}H_{121}O_{28}$
$C_{40}H_{56}O_2$ iso-zeaxanthin	$C_{40}H_{48}O_6$	$C_{90}H_{111}O_{13}$	A. niger	$C_{90}H_{115}O_{10}$
$C_{72}H_{116}O_4$ iso-zeaxanthin di-palmitate	$C_{72}H_{94}O_{17}$	$C_{90}H_{118}O_{23}$	Chara corallina	$C_{90}H_{117}O_{28}$
			Alnus glutinosa	$C_{90}H_{122}O_{30}$
$C_{40}H_{54}O$ Echinone	$C_{40}H_{51}O_5$	$C_{90}H_{115}O_{11}$	A. niger	$C_{90}H_{115}O_{10}$
$C_{40}H_{52}O_2$ Canthaxanthin	No polymer	No polymer		

have been similarly polymerised, and in most cases polymers similar to natural sporopollenins were obtained (See Table 6). It will be noticed that it is possible to ascribe to the synthetic materials, naturally-occurring sporopollenin analogues from both plant, fungal and algal materials (Table 4) although it must not of course be assumed that the relationship of the molecular formulae implies any specific carotenoid monomer. The figures at least do show however that the range of naturally occurring carotenoid types may be echoed in a similar range of sporopollenins. It is of particular interest to note (Table 6) that the terminal allylic positions are of special importance (Green, Holleyhead and Shaw, 1970). When both positions are fully oxygenated as with canthaxanthin no polymer was obtained, whereas when one of the positions is fully oxygenated (echinenone) the resulting polymer tends to have a low oxygen content. On the other hand, free allylic positions combined with esterification of hydroxyl groups generally results in a high oxygen contend in the polymer. This last type of material appears to predominate in plant pollen grains although it is important to examine a large number of species before coming to any definite conclusion on this point.

The naturally occurring sporopollenin from _L. henryii_ and the synthetic oxidative polymer derived from its extracted carotenoids give after potash fusion, an identical mixture of phenolic acids, namely p-hydroxybenzoic acid and similar amounts of _m_-hydroxybenzoic acid and protocatechuic acid (Brooks and Shaw 1968a, Shaw 1970). The synthetic polymer derived from β -carotene similarly with potash, gave p-hydroxybenzoic acid as the main product. It is clear that the earlier suggestion of a lignin type component in sporopollenin could now be abandoned since aromatisation of the carotenoid skeletal units are clearly capable of producing aromatic materials under various forcing conditions including potash fusion. β-Carotene of course is well known to give aromatic compounds when heated at relatively low (160°) temperatures (Kuhn and Winterstein 1932). The property of ready aromatisation is of special importance in considerations of the relationship of sporopollenin to the various types of kerogen of the so-called "aliphatic" and "aromatic" types (Brooks and Shaw, 1968b, 1969a, 1969b, 1970). The results confirm that sporopollenins are oxidative polymers of carotenoids and carotenoid esters; and in the anthers of *L. henryii* the polymerisation process commences at an early stage of pollen development at about the time when the microspores are about to leave the tetrads.

Further confirmatory evidence for the role of carotenoids in sporopollenin production came from a detailed quantitative study of the acids produced by ozonisation of both natural and synthetic materials. The relative amounts of both mono- and di- carboxylic acids produced by ozonisation of *L. henryii* sporopollenin were almost identical with those produced from the synthetic oxidative polymer prepared from the unhydrolysed *L. henryii* carotenoids-carotenoid esters (Tables 3, 4, 5). Similar "spectra" of carboxylic acids were obtained from β-carotene and *L. henryii* free carotenoid polymers (prepared by hydrolysis of the esters with alkali). Many of the ozone degradation products have been further identified by mass spectroscopy (Brooks and Shaw 1970).

d) Sporopollenins of Algal and Fungal origin

Although most of our experiments have been carried out with sporopollenins derived from pollen grain exines or micro-spores of *Lycopodium* we have naturally been especially interested in examining mega-spores of the higher plants and numerous types of spores from algae and fungi many of which have been only vaguely described and where there seems to be some confusion about the use of the word spore (which we assume to have an essentially sexual connotation) and cyst (normally a thickened cell). We have now examined the megaspore of *Selaginella kraussiana* which contains a high percentage of sporopollenin (Table 1) and the spores of *Aspergillus niger*, asexual and (±) spores of *Mucor mucedo*, spores of *Chara corallina* and *Pediastrum duplex*, a cyst of *Prasinocladus marinus* (algal "swarmer"), and asexual spores of *Pithophora oedogonia*. The results of preliminary examinations of these materials are recorded in Table 7. Of special interest are the spores of the *Mucor* (Brooks and Shaw 1969b). During formation of the (±) zygospore which contains (unlike the asexual spore) sporopollenin, mating of the (+) and (-) strains is accompanied by a marked enhancement (100 fold) of carotenoid production (Gooday, 1968) which is readily seen on culture plates. This appears to be under a type of hormonal control by the substance trisporic acid (Gooday, 1968). The *Mucor* produces β-carotene almost exclusively

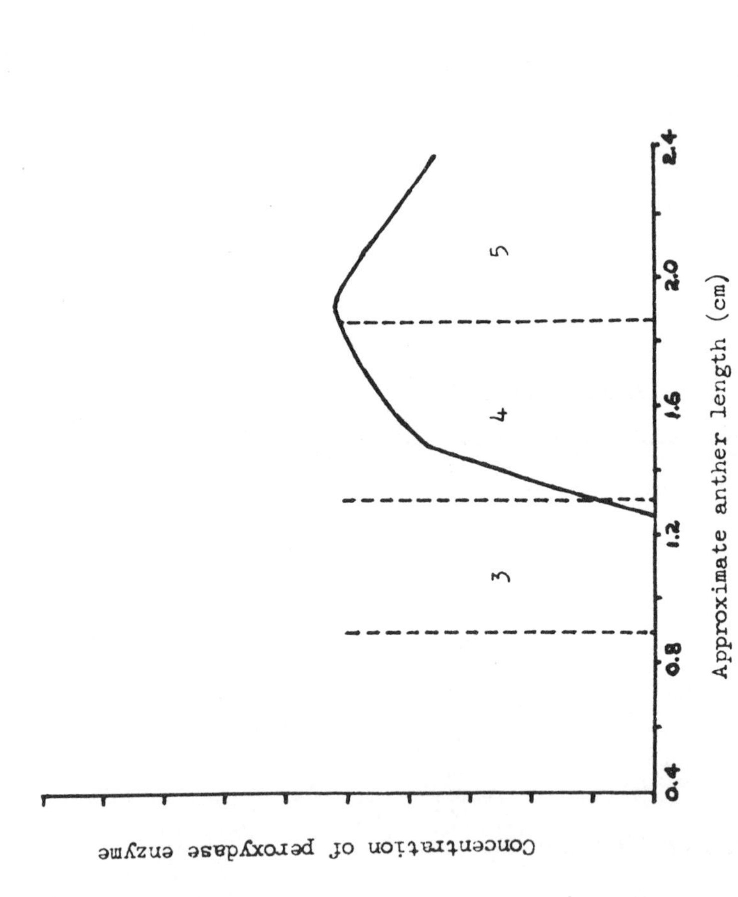

Fig 4. The relationship of anther length of Lilium henryii with concentration of peroxydase enzyme (Linskens[23]).
(3 tetrad stage; 4 exine deposition; 5 mature pollen grains)

and hence one would expect its sporopollenin to be similar to the synthetic oxidative polymer of β-carotene. Comparison of the elemental analyses (Table 1) and other properties - infra-red spectra, pyrolysis g.l.c., potash fusion products etc. (Brooks and Shaw, 1969b) - showed that the *Mucor* sporopollenin was virtually indistinguishable from the synthetic material. In contrast the *Mucor* asexual spores, which are quite different in appearance (small colourless envelopes as distinct from the large sculptured dark brown-yellow spheroidal zygospores) quite certainly do not contain sporopollenin and readily dissolve following the normal chemical isolation treatment, or after acetolysis.

Radiochemical Studies

The use of labelled chemicals offered a potentially valuable aid, both to the determination of structure and to information concerning the mechanism of the biological reactions which lead to the formation of the sporopollenins from carotenoid precursors. Our work in this field (Green, Holleyhead and Shaw 1970) is currently only at a preliminary stage but some of the applications and uses include the following:

a) *Determination of specific functional groups in the sporopollenin molecule*. The natural sporopollenin from *L. clavatum* and the synthetic material derived from β-carotene were treated with ^{14}C labelled acetic anhydride for varying lengths of time, to give information about the minimum number of hydroxyl groups present in the structure.

b) *Incorporation Studies*. We have also been able to examine

TABLE 7

The Composition of the Wall of the Spores of some Lower Plants

Material	Class	% Wall	% Sporopollenin	% Cellulose	Empirical Formula
Mucor mucedo	Fungal				
(±) spore		9.50	4.25	5.25	$C_{90}H_{130}O_{33}$
asexual "spores"	Fungal	20.80	-	20.80	-
Aspergillus niger (van Tieghen)	Fungal	10.9	4.8	6.1	$C_{90}H_{115}O_{10}$
Chara corallina	Algal	7.60	3.20	4.40	$C_{90}H_{117}O_{28}N_{1.5}$
Prasinocladus marinus (cienk) Waern	Algal Swarmer Cyst	14.20	-	14.20	-
Pithophora oedogonia	Algal	approx 5.00	-	approx 5.00	-

TABLE 8

Specific Activity Determinations of Sporopollenin from Cucurbita pepo (var. hundredweight)[a]

Source of Label[b]	Specific Activity of Label Source	Specific Activity of Sporopollenin[c]	Isotope dilution Factor
Sodium Acetate	6.8×10^6	209.8	3.2×10^4
Sodium Palmitate	6.2×10^7	278.0	2.2×10^5
Labelled carotenoids[d]	3.2×10^3	1,874.0	1.70

a Activities expressed as disintegrations min^{-1} mg^{-1}

b 5uC

c Represents the mean of 3 determinations

d Derived from ^{14}C labelled sodium palmitate.

in more detail the incorporation of ^{14}C labelled carotenoids into sporopollenins of various plant and fungal species. We have in particular examined the incorporation of carotenoids into the pumpkin (*Cucurbita pepo* var. hundredweight) which is known for its rapid growth, generous production of male flowers, large sized pollen grains, (pumpkin pollen is reputed to have the largest known grains up to 0.25 mm diameter), and not least its ability to flower under greenhouse conditions during the late autumn when the experiments were carried out. In preliminary experiments 5 μC each of uniformly labelled ^{14}C sodium acetate and ^{14}C sodium palmitate in aqueous solution at pH 7.0 were introduced simultaneously using a specially constructed capilliary feed into plants which had first been debudded. After about three weeks the male flowers were collected and the pollen isolated from the anthers. The carotenoids were extracted from the anther tissue with organic solvents and purified by the usual methods, including: washing with 6% aqueous potassium hydroxide-methanol (1:1), water, and purification by chromatography on activated alumina with monitoring of eluted fractions. Carotenoids were eluted as a single discrete band and the constant specific activity is recorded in Table 8. The mature pumpkin pollen grains were treated by our earlier methods to obtain the sporopollenin fraction and in this way the cellulose-free material was obtained thoroughly washed to remove all extractable radioactivity in a yield of about 10% of the original pollen grain weight. Specific activities are recorded in Table 8. Difficulties in grinding and cutting the material to a uniform particle size resulted in indeterminate self absorption effects when attempts were made to determine the level of radioactivity directly. Consequently in all determinations we have used a modified Schoniger combustion technique (Kely *et al*, 1961), with absorption of ^{14}C labelled carbon dioxide in ethanolamine, or in a solution of hyamine hydroxide in methanol followed by incorporation into a xylene based liquid scintillator. Quenching was made by the channels ratio method (Bush, 1963). The labelled carotenoids were dissolved in a small amount of Mazola Corn Oil and the solution emulsified with aqueous $^{M}/_{100}$ sodium palmitate. The emulsion was injected directly into pumpkin plants in which bud formation was just evident and the injection was made as close to the last bud as possible. After flowering the pollen was collected and the sporopollenin isolated therefrom. The specific activity of the material was determined using the combustion technique and the results are recorded in Table 8. They clearly show a high degree of incorporation of anther carotenoids into the sporopollenin.

Post-tetrad Ontogeny of Sporopollenin Derived from Pollen Grains of Lilium Henryii

During the last decade there have been rapid extensions in knowledge of the fine details of the stages of development during formation of the pollen grain wall in the anther sporogenous tissue (Heslop-Harrison, 1968a,b, Rowley, 1962 1963, Godwin and Echlin, 1968). Nevertheless, this work, which has been largely of a histological nature, has not, with minor exceptions, had any biochemical basis. By combining the botanical data with the information obtained, by us, on the role of carotenoids and carotenoid esters in sporopollenin genesis, it is possible to outline, in an admittedly rough manner, a possible sequence of chemical events which lead alternatively, in the anther of a typical plant, to the deposition of exine material.

1. The microspores are released from the tetrads at an early stage in anther development. They have a transparent wall which appears to consist essentially of polysaccharides, mainly cellulose and a hemi-cellulose or pectin-like material. Even at this stage the microspore possesses most of the morphology of the final pollen grain and it may therefore be regarded as a genetically produced template of the pollen grain.

2. At about the time, or perhaps slightly before, the microspores emerge from the tetrads, transparent globules with a roughly spherical nature seep from the tapetal cells. These globules are known as the pro-orbicules and in _L. henryii_ they are about 3 to 5 μ in diameter. After leaving the tapetum the pro-orbicules become increasingly opaque in electron micrographs During this period they develop a thin skin of sporopollenin on their surfaces and at this or a slightly later stage are known as orbicules. The pre-sporopollenin which has formed in this manner is later transferred to the microspore sac, and the biochemical changes taking place during this period may be interpreted in chemical and biochemical terms as follows:

a) The pro-orbicules consist of a solution of carotenoids and carotenoid esters in a fat and/or hydrocarbon solvent, stabilised presumably by an emulsifying agent, perhaps of a protein origin. The average size of the globules is typical of the size normally characteristic of a standard suspension or emulsion polymerisation process. We find that carotenoid esters are much more soluble in fat solvents than free carotenoids, and this may be an important factor in the subsequent polymerisation processes.

b) When the pro-orbicules leave the tapetal environment, they come into contact with an aqueous phase which contains a specific enzyme system and a source of oxygen, which may, of course, be molecular or chemical. Linskens has recently (1966) reported the variation in anther protein patterns, and shown that during pollen development in _L. henryii_ the protein pattern indicates a phase-specific variation wherein each cytological stage shows patterns which control and co-ordinate the sequence of biochemical events essential for the development of the pollen. Of special interest to our hypothesis is the reported evidence for a peroxidase enzyme system that enters into pollen development at the tetrad stage and reaches a maximum concentration in the early post-tetrad period during which sporopollenin deposition is occurring (See Fig. 4).

The enzyme will presumably operate as an ionic catalyst since an alternative free radical oxidation seems unlikely. We have only been able to induce oxidative polymerisation of carotenoids with ionic catalysts (such as BF_3, $AlCl_3$, etc), and attempts to use free radical catalysts with and without photochemical illumination have always proved unsuccessful.

c) Co-polymerisation of the carotenoid mixture occurs in what appears to be a fine suspension polymerisation process in <u>L. henryii</u>; at least, the average orbicule diameter (1-5 μ) here is substantially larger than that of a swollen micelle (0.01 μ) or average latex particle (0.04-0.08 μ) of the ideal emulsion polymerisation system (Duck, 1966). In addition, most observers agree that during development of the orbicules a skin of sporopollenin forms on their surface corresponding to the typical behaviour observed in the suspension process where a polymer skin forms on the surface of the droplets followed by diffusion of further monomer particles to the surface and repetition of the polymerisation process.

d) The part-polymerised pro-orbicules/orbicules ultimately contact the microspores where a transfer of sporopollenin occurs. Presumably, at this stage the globules are sticky and the polymerisation process is incomplete. The globules arrange themselves on the surface of the microspore template wall so as to maintain the shape. There may be, of course, further variation in the final external architecture, either by expansion of the pollen grain which would result in invasion of the template into the sticky layer of deposited sporopollenin or by contraction, which might be expected to cause additional surface features to arise including elongation of the orbicules with formation of baculae type exine structures, examples of which may be found in many pollen grains.

e) The anther tissue invariably possesses excess polymerising orbicules and these, after completion of the polymerisation process ultimately give rise to hard globules of sporopollenin ("Ubisch bodies") which have an independent existence within the anther tissue. A certain amount of excess orbicular material coats tissues other than the microspores (Pettitt, 1966, Banerjee, 1967, Heslop-Harrison, 1969). There is no evidence that sporopollenin is ever remobilised metabolically once deposited in the anther (Heslop-Harrison, 1968) and this suggests that flowering plants do not contain enzymes which are capable of degrading the polymer. Recent observations have suggested that the independent orbicules, formerly assumed to have no specific function, may, by forming a carpet over the entire surface of the anther wall create an environment to which the final pollen grains do not readily adhere, and can therefore be more readily dispersed (Dickinson and Heslop-Harrison, 1969).

The ontogeny of sporopollenin formation in algal and the (±) *Mucor* spores will clearly differ in detail from that of pollen since the "orbicular" matter will presumably arise from within the spore by way of algal filaments or fungal hyphae. Certainly in *Mucor* during zygaspore formation tiny yellow globules (presumably composed of a lipid solution of β-carotene) are readily visible in the hyphae. Whether the sporopollenin in these species is laid down as an exine or intine is uncertain, but this is clearly a field worthy of further study.

References

BANERJEE, U.C. (1967). Grana Palynologica 7, 365.

BRACONNOT, H. (1829). Ann. Chim. Phys. 2, 42.

BROOKS, J. and SHAW, G. (1968a). Nature, 219, 532.

BROOKS, J. and SHAW, G. (1968b). Grana Palynologica, 8 (2-3), 227.

BROOKS, J. and SHAW, G. (1968c). Nature, 220, 678.

BROOKS, J. and SHAW, G. (1969a). Nature, 223, 754.

BROOKS, J. and SHAW, G. (1969b). In "Pullman International Conference on Pollen" (Heslop-Harrison, J. ed.) Butterworth, London (in press).

BROOKS, J. and SHAW, G. (1970). Nature, 227, 195.

BURNETT, J.H. (1965). In "Chemistry and Biochemistry of Plant Pigments" (T.W. Goodwin ed.). Academic Press, London and New York.

BUSH, E.T. (1963). Anal. Chem., 35, 1024.

CAMPBELL, S. (1970). Comp. Biochem. Physiol., 32, 97.

DICKINSON, H.G. and HESLOP-HARRISON, J. (1969). Planta, 84, 199.

DUCK, E.W. (1966). In "Emulsion Polymerisation" in "Encyclopedia of Polymer Science and Technology", 5, 801, Interscience, New York, London and Sydney.

FAEGRI, I. and IVERSEN, J. (1964). In "Textbook of Pollen Analysis", Blackwell, Oxford.

GODWIN, H. and ECHLIN, P. (1968). J. Cell. Sci. 3, 161.

GOODAY, G.W. (1968). Phytochem. 7, 2103.

GOODWIN, T.W. (1970). In "Algal Carotenoids", Phytochemical

Society Meeting, University of Liverpool, April, 1970.

GREEN, D., HOLLEYHEAD, R. and SHAW, G. (unpublished results).

HESLOP-HARRISON, J. and MACKENZIE, A. (1967). J. Cell. Sci. 2, 387.

HESLOP-HARRISON, J. (1968a). Canad. J. Bot. 46, 1185.

HESLOP-HARRISON, J. (1968b). Science, 161, 230.

HESLOP-HARRISON, J. (1969). Canad. J. Bot. 47, 541.

JENSEN, S.L. (1970). In "Recent Chemistry of the Carotenoids", Phytochemical Society meeting, University of Liverpool, April, 1970.

JOHN (1814). Journal Fur Chemie und Physik, 12, 244. (cf. Zetzsche and Huggler, 1928).

KARRER, P. et al (1934). Helv. Chim. Acta. 17, 412, 745; idem (1935) 18, 96; idem (1936) 19, 479.

KELY, R.G., PEETS, E.A., GORDON, S. and BUYSKE, D.A. (1961), Ann. Biochem. 2, 267.

KLEINIG, H. and NIETSCHE, H. (1968). Phytochemistry, 7, 1171.

KUHN, R. and LEDERER, E. (1933). Ber. 66, 488.

KUHN, R. and WINTERSTEIN, A. (1932). Ber. 65, 1873.

KUHN, R. and WINTERSTEIN, A. (1930). Ber., 63, 1489; idem (1930) Naturwiss., 18, 754.

LINSKENS, H.F. (1966). Planta, 69, 79.

MUIR, M.D. and SARJEANT, W.S. (in press).

PETTITT, J.M. (1966). J. Linn. Soc. (Bot.), 59, 253.

ROWLEY, J.W. (1962). Grana Palynologica, 3 (3), 3.; Science, 137, 526.

ROWLEY, J.W. (1963). Grana Palynologica, 4 (1), 25.

ROWLEY, J.W. and SOUTHWORTH, D. (1968). Nature, 213, 703.

SHAW, G. and YEADON, A. (1964). Grana Palynologica, 5 (2), 247.

SHAW, G. and YEADON, A. (1966). J. Chem. Soc., (C), 16.

SHAW, G. (1970). In "Phytochemical Phylogeny", p. 31, (Harborne, J. ed.), Academic Press, London and New York.

TISCHER, J. (1941). Z. Physiol. Chem. 267, 281.

TSUKIDA, K. and ILEUCHI, K. (1965). Bitamin, 32, 222.

ZECHMEISTER, L. and CHOLNOKY, L. (1936). Ber., 69, 422.

ZETZSCHE, F. and HUGGLER, K. (1928). Annalen, 461, 89.

ZETZSCHE, F. and KALIN, O. (1931). Helv. Chim. Acta, 14, 517.

ZETZSCHE, F., KALT, P., LIECHTI, J. and ZIEGLER, E. (1937). J. Prakt. Chem., 148, 267.

ZOLOCHEVSKII, V.T. and STERLIN, B. Ya. (1967). Maslozhir Prom., 33, 7.

DISCUSSION ON DRS. CHALONER & SHAW'S PAPERS.

FAEGRI: I disagree with Dr. Chaloner and Professor Heslop-Harrison that the general term sporopollenin can or should be restricted morphologically, taxonomically, or phylogenetically. On the other hand, the general term most probably comprises a number of special sporopollenins. It is *a priori* rather improbable that for example the wall substance of *Lycopodium* spores is identical with that of an angiosperm, but they are both in the general chemical class of the sporopollenins, however badly this may be defined chemically (cf. e.g. the alkaloids). The difficulty of defining the taxonomic relationships of *sporae dispersae* (where I am in complete agreement with Dr. Chaloner) would suggest that the general terms should be used in a general sense. If Dr. Chaloner is prepared to qualify sporopollenin to e.g. vascular plant sporopollenin, then I can quite agree with him. Sporopollenin unqualified is a chemical term and contains no implication of origin . I don't like to see it coming out of meteorites either, but if that is what the chemists call the substance that they find, then we have to accept their findings.

SHAW: Perhaps I could comment on the meteorite occurrence. We have only made suggestions about the sporopollenin like nature of the meteoric material, but have never had sufficient material to do a complete chemical study.

MUIR: Dr. Brooks and I have been looking at some of these meteorite residues under the Stereoscan, and comparing them with the chemically similar organic substances from the Onverwacht Chert)Early Precambrian), and we have found distinct morphological differences between the two. The Onverwacht material had, generally, amorphous, filamentous, and sphaeroidal particles present. But the meteorite material contained curious hexagonal box-like structures which would be difficult to relate to any living organism. Morphological studies should be carried out in parallel with chemical work.

POTONIÉ: Yes, morphological information is important. In Carboniferous studies, we found that coal petrologists never used the term sporopollenin, but only exinite. It is necessary to try to reach agreement on such matters.

HARRIS: I welcome Dr. Shaw's vigorous definition of sporopollenin as a chemical residue: This can be applied logically and results in the wide recognition of sporopollenin in bodies of varied origin. We should also look at cutin which emphatically is not sporopollenin. Now, though the cuticles I know are dissolved in alkali they are intensely resistant to acid maceration (as resistant as many spores) and apparently last indefinitely in the presence of water (without oxygen. I strongly suggest the before we extend the term "sporopollenin" beyond vascular plants we should look in the same ways at recent and fossil cuticles. If they prove to have the properties of "sporopollenin" then we should reconsider the use of that term.

FAEGRI: I wish to draw your attention the work of Mme Jentys-Szaferova, who oxidised _Betula_ and _Corylus_ exines with chromic acid, with the result that in the pollen wall, there appeared 2 - 3 more resistant lamellae. The lead given here should be followed.

HESLOP-HARRISON: What does Dr. Shaw think of the reported experiments of dissolving sporopollenin in ethanolamine?

SHAW: Dr. Brooks tried for two years to dissolve _Lycopodium_ spores in this solvent, but to no effect. The experiment probably did not replicate the previous results where the esperiment was carried out on only a few pollen grains on a microscope slide on the presence of oxygen.

HESLOP-HARRISON: These is no doubt that on a microscale surface features disappear, so this may be the answer.

FAEGRI: When I tried this experiment, in a test tube, the entire exine disappeared, and only the intine and contents were left. This was performed at just below boiling point and all the exines disappeared within ten seconds.

ROWLEY: I have dissolved considerable quantities of pollen, but I think that _Lycopodium_ spores are very resistant.

SHAW: Perhaps we should try ethanolamine again.

BROOKS: I believe that the degradation products from cuticle are different from those of sporopollenin.

CALDECOTT: Yes, also, I think it is true to say that people who have worked with cuticles have chosen those that were easily hydrolysed.

SOME CHEMICAL AND GEOCHEMICAL STUDIES ON SPOROPOLLENIN

J. Brooks

School of Chemistry, University of Bradford, Bradford 7, Yorkshire

(present address British Petroleum Co. Ltd., BP Research Centre, Exploration and Production Division, Sunbury-on-Thames).

Abstract

The chemical study of various modern and fossil spore walls of gymnosperms, angiosperms, pteridosperms, fungi and algae show a majority to be composed of sporopollenin. The composition of spore walls with respect to exine and intine content, and differences in the chemical structure of their sporopollenin are discussed as possible phytochemical parameters in plant classification. Chemical studies on microfossil walls and associated insoluble organic matter present in sediments upto 3,400 to 3,700 million years old suggest these may have a common identity with sporopollenin present in modern spore exines.

When spores are deposited in sediments, in addition to micro-biological and chemical alterations, there are three main physico-chemical factors that cause changes to occur; temperature, time and pressure. An examination of the thermal alteration of spore walls is presented and the potential of sporopollenin as a possible progenitor of oil and as an indicator of sedimentary temperature are discussed.

Introduction

The earliest reported chemical observations on the exine of pollen grains seems to be those by John (1814) and Braconnot (1829). John (1814) investigated various plant pollens and analysed more than six varieties which he claimed "include a wealth of botanical and particular physiological data". He examined the extractable components and also the residual material, which he claimed was particular to pollen and formed its predominant constituent, which he called pollenin. John examined mainly tulip pollen along with various other species and extracted their contents with organic and inorganic reagents and was able to identify malate of ammonia, sulphate and phosphate compounds, gummy parts, sulphur, resin, oil, much sugary non-crystallisable material, a little waxy material (cerin), volatile parts, pigments, a "cheese-type" protein present in the extract and an insoluble pollenin residue. John in announcing his new substance (pollenin) pointed out the differences between protein and pollenin:- "The latter reacts to acid differently from protein, not withstanding the fact that it produces a bitter substance with nitric acid; it is quite insoluble in alcohol, ether, water, oil of terpentine, naphtha or carbonate and caustic alkalis, and produces, in dry distillation, apart from ammonia, an acid phlegm. It is, however, noteworthy that the elements of pollenin in different types of pollen are present in different proportions........". "Now pollenin may be treated with all these solvents, whilst still retaining its outstanding property, that, when thrown in the fire it will, in the manner of resinous bodies, make the sparks fly".

Braconnot (1829) examined:- "...des envelopes speratiques du pollen du typha epuisee par l'eau bouillante (pollenine) ...." and reported that:- "... pollenine d'une nature particular et materiere colorante jaune = 25.96 per cent ....", of the total pollen.

Thus it was John (1814) and Braconnot (1829) who introduced the term pollenin (pollenine) to describe the tough resistant material which makes up the outer wall of pollen grains. The next chemical studies on these wall materials did not take place until 1928, when Zetzsche (1928) published the first of a series of papers devoted to the study of the wall of pollen grains and spores. Zetzsche (1928) introduced the term sporonin to describe the wall material from the spores of *Lycopodium clavatum*, which because of their ready availability have become a major source for chemical studies of spore walls. The resistant material which Zetzsche obtained from either pollen grain or spore walls showed the same general chemical character, which led Zetzsche (1931) to introduce the collective name sporopollenin to describe this chemical substance present in both pollen grain and spore walls.

Recent studies (Shaw and Yeadon 1964, 1966; Brooks and Shaw 1968, 1969 and 1970) have extended the chemical studies of sporopollenin and show that resistant walls have much wider occurrence

OCCURRENCE OF SPOROPOLLENIN

POLLEN GRAINS	SPORES	FOSSIL 'SPORES'
ANGIOSPERMS	SELAGINELLALES	TASMANITES
GYMNOSPERMS	LYCOPODIALES	DINOFLAGELLATES (?)
	ALGAE	CARBONIFEROUS MEGASPORES
	FUNGI	

TABLE I

than previously realised (Table 1). Sporopollenin has been identified in the spore wall of higher plants, some algae and fungi, in the fossil wall of *Tasmanites* and Carboniferous megaspores. Recent chemical examination on the fossil wall of a dinoflagellate concentrate (from the Kingak Formation, North Slope Alaska, 150 million years old) suggest it to be composed of a material with similar chemical properties to *Tasmanites* walls.

Since sporopollenin has such wide occurrence, attempts have been made to calculate the amount produced by nature. The amount of pollen produced by various common forest trees (Table 2) during a period of fifty years gives an indication of the vast quantities

AMOUNT OF POLLEN PRODUCED BY VARIOUS FOREST TREES.

SPECIES		AMOUNT PRODUCED BY ONE TREE IN 50 YEARS.
SPRUCE	(PICEA ABIES)	20,000 grams. (0.40 cwt.)
BEECH	(FAGUS SYLVATICA)	7,600 grams. (0.15 cwt.)
PINE	(PINUS SYLVESTRIS)	6,000 grams. (0.12 cwt.)
HAZEL	(CORYLUS AVELLANA)	2,800 grams. (0.06 cwt.)
ALDER	(ALNUS SP.)	2,500 grams. (0.05 cwt.)
SILVER BIRCH	(BETULA VERRUCOSA)	1,700 grams. (0.04 cwt)

TABLE 2

produced. Vast amounts of tough walled resistant spore walls are also produced by lower non-flowering plants (algae, fungi and related species) in marine surroundings, but it is impossible to estimate this production of spores from these lower plants, although concentrated deposits of these thick-walled spores (eg *Tasmanites*) have a world-wide occurrence.

Application of Sporopollenin Studies to Phytochemistry

In any phytochemical study by using chemical examination and data of plant constituents it is hoped that a classification of plants and species may develop. It is important in these studies that the material chosen is characteristic of the plant and independent of seasonal and environmental changes. Probably the most characteristic plant material for study is the wall of the pollen grain and spore.

This material is botanically important as the covering for the carrier of genetic material in the plant and may therefore be expected to have characteristic properties and structure and be independent of external changes. Also by studying fossil pollen and spore preservation it may be possible to achieve some historical classification.

For phytochemical studies to be meaningful many species must be examined, but from the already reported pollen and spore wall analyses (Zetzsche 1928; Kwaiatkowski and Lubliner-Miahowska 1957; Shaw and Yeadon 1966; Brooks and Shaw 1969; and Brooks 1970) it is possible to observe certain phytochemical trends developing:-

1. the wall content of selective pollen grains and spores vary from *Selaginella kraussiana* (31.8 per cent) to *Populus balsaminifers* (1.8 per cent);
2. the wall of some pollen grains retain their original shape after treatment to remove their contents (eg *Pinus, Fagus, Quercus, Phleum pratense, Narcissus pseudonarcissus*), but other pollen grains such as *Lilium* and *Rumex* lose their shape and become an amorphous mass of sporopollenin bearing very little histological relationship with the original exine. It is thought that a possible explanation for this break-up of the wall is due to the thinness of the exine. This does not seem the reason, because pollen grains with small amounts of sporpollenin in their wall (eg *Narcissus pseudonarcissus*) retain their

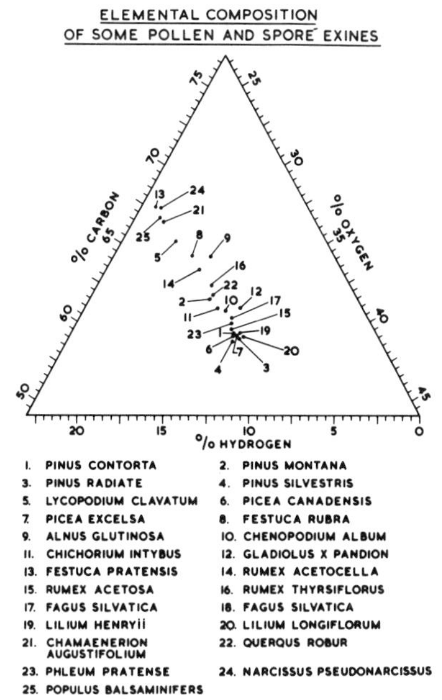

TABLE 3

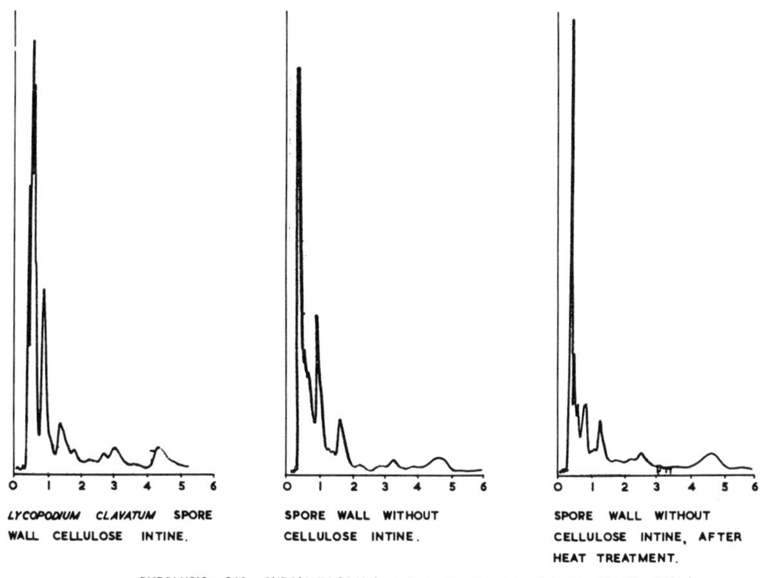

PYROLYSIS – GAS CHROMATOGRAMS OF *LYCOPODIUM CLAVATUM* SPORE WALLS.
(PYROLYSIS TEMPERATURE 710°C)

FIGURE 1

morphology after removal of their contents and intine, whilst other species (eg *Lilium*) containing larger amounts of sporpollenin in their wall become amorphous.

3. in spite of minor variations in stoichimetry of the molecular species and architectural variation of the exines, sporopollenin seems to represent a specific coherent group of bio-polymers with related chemical composition considering the wide range of species examined (Table 3).

Examination of the chemical composition of four *Pinus* species gives analyses fitting into the general pattern for sporopollenin, but also shows significant variation between the species. Groups of other plant genera, *Picea*, *Rumex*, *Festuca* and *Lilium* show (Table 3) collective similarities within the species, but significant differences can also be observed (eg variation in the oxygen content of the *Rumex* species) which after further clarification may give an objective chemical analysis to assist in phytochemical classification.

An interesting relationship is observed when the amount of sporopollenin exine in the wall of different species is compared with the amount of cellulosic intine content (Table 4); *Selaginella kraussiana* wall contains 27.8 per cent exine and 3.0 per cent intine by weight of the megaspore (ratio exine: intine = 9.27:1), whilst *Rumex acetosa* wall (ratio 0.60) contains very different amounts. Table 4 shows a few examples of exine to intine ratios for different species: *Selaginella kraussiana* megaspore (ratio 9.27) which is considered to be an early type of plant in the terrestrial evolutionary sequence has a high percentage of sporopollenin to cellulose in its walls; *Lycopodium clavatum* spores (ratio 8.67), has a slightly reduced amount of exine to intine wall content and examination of species of more advanced evolutionary development, gymnosperms (eg *Pinus*) show an intermediate ratio (3.70 to 3.97) whilst the angiosperms, considered to have a still more recent evolutionary history show much smaller (0.61 to 1.71) ratio of exine to intine. In fact some angiosperms (eg *Rumex acestosella* and

RELATIONSHIP BETWEEN SPOROPOLLENIN AND CELLULOSE CONTENT OF EXINES FROM DIFFERENT PLANT SPECIES

PLANT GROUP	EXAMPLE	% SPOROPOLLENIN	% CELLULOSE	% SPOROPOLLENIN / % CELLULOSE
SELAGINELLALES	SELAGINELLA KRAUSSINA MEGASPORE	27.8	3.0	9.27
LYCOPODIALES	LYCOPODIUM CLAVATUM SPORES	23.4	2.7	8.67
GYMNOSPERMS	PINUS SILVESTRIS	23.8	6.0	3.97
	PINUS MONTANA	27.7	7.1	3.90
	PINUS CONTORTA	20.7	5.5	3.76
	PINUS RADIATE	24.4	6.6	3.70
ANGIOSPERMS	LILIUM HENRYii	5.3	3.1	1.71
	LILIUM LONGIFLORUM	5.1	3.3	1.55
	RUMEX THYRSIFLORUS	5.7	4.2	1.26
	RUMEX ACETOCELLA	6.3	6.7	0.94
	RUMEX ACETOSA	4.2	7.0	0.60

TABLE 4

VARIATION IN MOLECULAR FORMULA OF SPOROPOLLENIN FROM DIFFERENT PLANT SPECIES.

PLANT GROUP	EXAMPLE	% SPOROPOLLENIN	MOLECULAR FORMULA
SELAGINELLALES	SELAGINELLA KRAUSSINA MEGASPORE	27.8	$C_{90}H_{124}O_{18}$
LYCOPODIALES	LYCOPODIUM CLAVATUM SPORES	23.8	$C_{90}H_{144}O_{27}$
ANGIOSPERM	LILIUM HENRYii POLLEN GRAIN	5.3	$C_{90}H_{142}O_{36}$
GYMNOSPERM	PINUS SILVESTRIS POLLEN GRAIN	23.8	$C_{90}H_{158}O_{44}$

TABLE 5

acetosa) posses a wall composed of a larger amount of intine than exine. These chemical results and observations suggest that there has been evolutionary development reducing the thickness of the sporopollenin exine combined with a possible thickening of the cellulosic intine. Larson and Lewis (1961) reported that *Lycopodium* has a dominant outer lamellated layer (exine), whilst in gymnosperms the thick inner layer is lamellated (eg *Pinus*) and in the angiosperms the lamellated layer (exine) is considerably reduced. If this relationship holds true for other species; lower plants, gymnosperms and angiosperms, the ratio of amount of exine to intine in the wall could offer some useful information about evolutionary relationship in plants.

Fossil Sporopollenin

Fossil pollen and spores present in sedimentary deposits have to be considered in relation to a number of factors that have exerted an influence on them during the time between their dispersion and their extraction from the sediment for examination. It is important that the history of fossil pollen and spores is considered because this will give better understanding to palynology.

During the time that spores are transported from their origin there are various agencies that can act and change their distribution, concentration and state of preservation. The original distribution and concentration of spores in a sediment can be influenced by:-

flotation, sorting, mixing, resettling and aquatic and aerial reworking and several other factors that may alter the occurrence of spores during sedimentation. These alterations result in quantitative and qualitative change in the analysis of the recovered pollen and spores from a deposit or sediment (Tschudy 1969).

When a pollen or spore fails to reach its intended destination, it soon perishes and the cytoplasm, genetic material and cellulose intine are rapidly destroyed leaving the outer most layer(s) of wall (exine). These exines, because of their extra-ordinary resistance to chemical and microbiological attack survive for long periods and often remain in rocks where all other biological material has been distorted or destroyed making them the most widely occurring plant fossil. There are several different groups of organic-walled microfossils which made significant contributions to the total organic matter in sediments. Many are related to algae and fungi, others are of unknown affinity, but all are characterised by having resistant walls. The earliest known organic walled microfossils are the single-celled organisms and are considered related to algae and fungi (Schopf 1967). The majority are resting spores of algae and are simple rounded-bodies (sporomorphs) sometimes with spined forms and mainly occur in marine sediments suggesting a phytoplanktonic origin.

Probably one of the best known groups of microfossils is the tasmanites which are fossil organisms with a very resistant tough organic wall and can be readily isolated from sediments by standard extraction techniques. Their morphology is relatively simple and constant (Muir and Sarjeant 1970); they are spheroidal to bean-shaped, but are often found to be disc-shaped due to compression during sedimentation. It is thought that these 'spores' were originally colourless to yellowish, but the fossils range in colour from yellow to red-brown to almost black, depending on their diagenetic history within the sediment. Their size varies from about 100μ to over 600μ and the outer wall is thick and possesses a system of pores (Muir and Sarjeant 1970).

The first record of *Tasmanites* was made by Hooker (1852) who found them in Silurian deposits of England. Later, Dawson (1871) described them as sporangia ("spore cases") from various scattered locations in northern United States and Canada. *Tasmanites* occur in high concentration in the locality of La Trobe on the Mersey River, Tasmania in a Permian deposit, and are also known as Tasmanite, Tasmanin and "White Coal" deposits. Ralph (1865) identified these spherical bodies as algae and Newton (1875) named them *Tasmanites punctatus*. In 1939, Eisenack (1938) described tasmanites from the Silurian of the Baltic region and after much terminological disagreement classified them as *Tasmanites erraticus*. Recent work (Burlingame, Wszolek and Simoneit 1968; Collett 1968) has shown that tasmanites from Alaskan tasmanite present in the late Jurassic of northern Canada and Alaska belong to this class of unicellular organisms.

Wall (1962) identified *Tasmanites* and suggested a close affinity to the extant spherical green alga *Pachysphaera pelagica* both of which he placed in the Leiosphaeridae. Cane (1968) has drawn attention to the relationship between *Pachysphaera* and *Tasmanites* and its close parallel with variants of *Botryococcus braunii* which is often found in vast quantities in Southern Australia and in certain lakes in Siberia. Cane (1968) comments on Wall's opinion that "*Pachyspaera* is regarded as a living representative of the fossil genus *Tasmanites*", would equally well apply to *Botryococcus* which contributes both to the extant Coorongite and to fossil fuel Torbanite.

The *Tasmanites* are little known to geologists, which is suprising when one considers their wide geological distribution and considerable stratigraphical age range (Table 6) and that they were first

FORMULAE OF SOME FOSSIL SPORE EXINE WALLS

MATERIAL	SOURCE	AGE	FORMULA
TASMANITES (ALASKA)	ALASKA NORTH SLOPE	230 MILLION	$C_{90}H_{121}O_{15}$
TASMANITES PUNCTATUS	MERSEY RIVER, TASMANIA	250 MILLION	$C_{90}H_{132}O_{16}$
TASMANITES HURONENSIS	LAKE HURON, N. AMERICA	350 MILLION	$C_{90}H_{134}O_{17}$
TASMANITES ERRATIUS	BALTIC REGION, EUROPE	450 MILLION	$C_{90}H_{133}O_{11}$
GLEOCAPSAMORPHA PRISCA	UNKNOWN	500 MILLION	$C_{90}H_{131}O_{17}$
GEISELTALPOLLENIN	GEISELTAL, NR. HALLE IN EAST GERMANY	250 MILLION	$C_{90}H_{129}O_{19}S_7N$
LANGE-SPORONIN	SAAR BROWN COAL, EUROPE	250 MILLION	$C_{90}H_{82}O_{17}N$
VALVISISPORITES AURITUS	KANSAS LONE STAR LAKE, U.S.A.	250 MILLION	$C_{90}H_{102}O_{16}$

TABLE 6

recognised over 120 years ago and first classified over 90 years ago. The species have been described under a variety of names and misidentified as spores of land plants, pollen and diatoms, but they are now considered to be related to living planktonic algae and have been placed in the class *Prasinophyceae*. As long ago as 1899, microchemical tests were performed by Ostenfeld (1899), these have been repeated and additional tests carried out by Wall (1962), who found the membrane of *Tasmanites* (and *Leiophaeridia*) to be extremely resistant to maceration in "Schultze's reagent" and showed no change after 13 hours treatment. Oxidation with sulphuric acid (this is the reported reagent, but the acid is probably nitric acid) turned the wall bright yellow and later brown. Heating with concentrated hydrochloric acid produced no effect. There was no reaction with cellulose reagents, but slight colouration of the cell contents was shown after treatment with Ruthenium red (a stain used for identification of pectic acids). It was concluded from these tests that the wall consisted of a complex lipoid substance with little or no cellulose, whilst the inner fraction of the wall could contain pectic substances. Neither Ostenfeld nor Wall were able to find silica in the wall:they continued to "suspect" its presence.

Although most of the tests on fossil exines proved negative, this has not stopped ideas and suppositions being made about the exine being composed of cutin, waxes and resins, chitin and even silica and these often wrong conclusions are now repeatedly quoted in botanical literature. Since chemical analysis of spore exine show very small or zero nitrogen content, the presence of chitinous substances (poly-glycosamines) is incorrect, with additional evidence being provided by their general resistance to hot mineral acids and strong alkali during the extraction from sediments. These extraction techniques would readily hydrolyse and degrade cutin (Mazliak 1968), resins and waxes (Brooks 1967) and chitinous material (Jeuniaux 1965).

Zetzsche (1937) attempted a systematic chemical study of fossil exines and compared their oxidative degradation products with those obtained from modern pollen and spore exines. The results showed that the fossil walls of Tasmanin (*Tasmanites punctatus*), Lange-sporonin and Geilseltalpollenin ("fimmenite") were composed of sporopollenin. Eisenack (1966) analysed the spore walls of *Tasmanites erraticus* (also called a "leiosphere") and *Gleocapsamorpha prisca* (a colonial algae) and showed that these materials had very similar chemical composition and were closely related, not only to each other, but also to the probable composition of the walls of dinoflagellates (Eisenack 1963). These observations of Zetzsche and Eisenack on the thick walled organic wall of fossil spores indicated that they are composed of a similar chemical substance - sporopollenin.

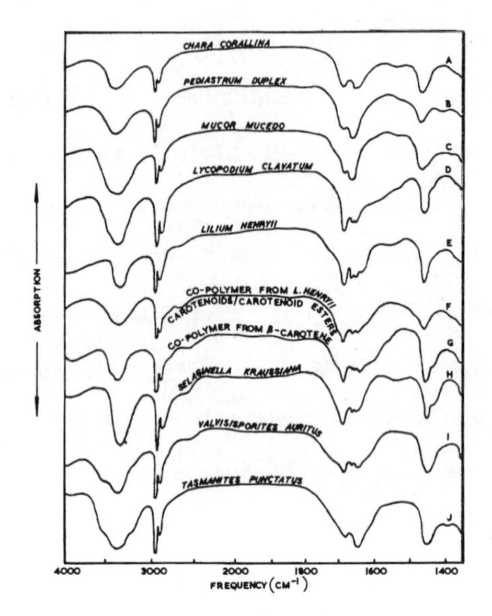

THE INFRA-RED SPECTRA OF SOME NATURAL AND SYNTHETIC SPOROPOLLENINS.

FIGURE 2

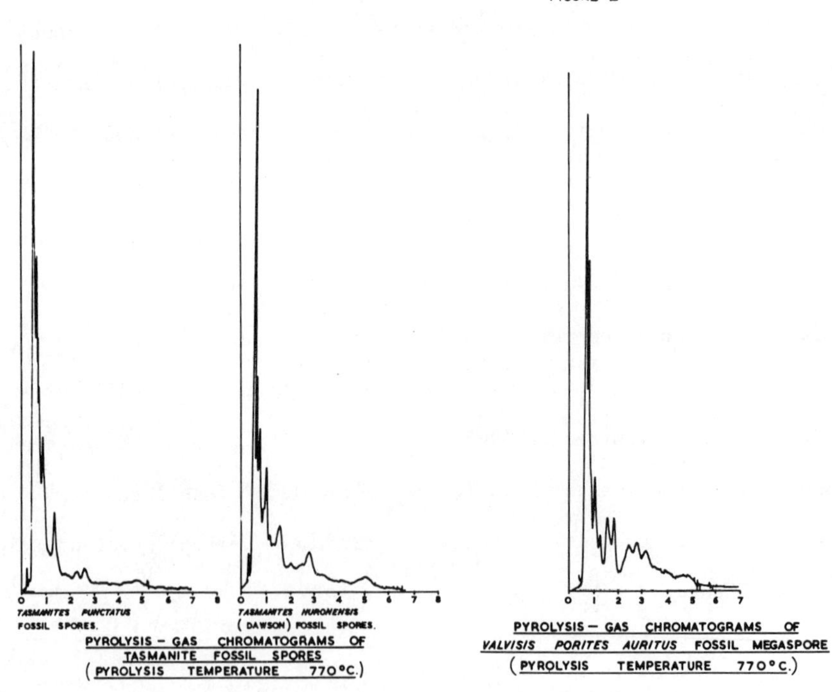

PYROLYSIS – GAS CHROMATOGRAMS OF TASMANITE FOSSIL SPORES (PYROLYSIS TEMPERATURE 770°C.)

PYROLYSIS – GAS CHROMATOGRAMS OF *VALVISIS PORITES AURITUS* FOSSIL MEGASPORE (PYROLYSIS TEMPERATURE 770°C.)

FIGURE 4

Recent Studies on Fossil Walls

It was assumed (Brooks and Shaw 1969) that if nature provided a resistant organic exine for spores and pollen grains of higher plants, then sporopollenin may be a common component in all spores that have resisted chemical and microbiological degradation and become "fossilised". The chemical nature and composition of various vascular plant pollen exines, *Lycopodium clavatum* spore exine and *Selaginella kraussiana* megaspore exine were examined (Brooks and Shaw 1969) and shown to be composed of sporopollenin. Similar comparative chemical tests for sporopollenin were applied to *Tasmanites* species and *Valvisisporites auritus* fossil megaspore (from the Lawrence Shale of the Lone Starr Lake Region, Kansas) and these walls proved to have the characteristic properties of sporopollenin (Table 6; Figures 2 and 4), except that the oxygen content of the fossil species was significantly lower than is found for sporopollenin present in modern vascular plants and that small amounts of sulphur and nitrogen are occasionally detected in fossil spores. The presence of sulphur and nitrogen can be readily duplicated in the laboratory by mild treatment of sporopollenin from modern spores with nitrogen- and sulphur- containing compounds and with elemental sulphur (Table 7) suggesting the presence of these elements in sporopollenin from fossil walls to have been added during deposition or fossilisation. The lower oxygen

PRESENCE OF SULPHUR AND NITROGEN IN FOSSIL SPOROPOLLENIN.

GEILSELTALPOLLENIN ("FIMMENITE")	$C_{90}H_{129}O_{19}S_7N$
LANGE — SPORONIN	$C_{90}H_{82}O_{17}N$
LYCOPODIUM CLAVATUM SPORES	$C_{90}H_{144}O_{27}$
TREATED WITH SULPHUR COMPOUNDS	$C_{90}H_{140}O_{24}S_5$
TREATED WITH NITROGEN COMPOUNDS	$C_{90}H_{144}O_{25}N_{11}$

TABLE 7

content of fossil spore exines was thought to be caused by chemical and/or physical processes occurring during diagenesis, when chemical dehydration of sporopollenin would significantly alter the hydrogen and oxygen content of the material. Dehydration of sporopollenin is still an important process affecting fossil exines in the sediments (see later), but examination of the oxygen content of sporopollenin from lower plants showed these also contained lower oxygen content similar in value to that of fossil sporopollenin. This similarity in chemical composition (especially oxygen content) of sporopollenin in lower plants (eg *Selaginella kraussiana* $C_{90}H_{124}O_{18}$) suggest that lower oxygen content may be associated with changes in biochemical processes during evolution of the spore and pollen wall. This association of sporopollenin from lower plants with lower oxygen content is consistent with analyses of fossil spore walls from lower plants.

Oxidative degradation (using chromic acid) of Alaskan tasmanite and *Tasmanites punctatus* and analysis of the products by gas chromatography-mass spectrometry showed the products to be a mixture of straight-chain (maximum component C_{16}), branched-chain mono-carboxylic acids and di-carboxylic acids as well as smaller amounts of keto-acids and aromatic di-carboxylic acids (Burlingame, Wszolek and Simoneit 1968).

The oxidative degradation (using ozone and acetic acid solvent) of *Tasmanites punctatus* wall and analysis of the degradation products by gas chromatography showed a similar mixture of mono-carboxylic acids to that obtained from other sporopollenin degradative experiments.

Stable Carbon Isotope Studies on Tasmanites Fossil Walls

Organic carbon exhibits a range of $^{13}C/^{12}C$ ratios, since during photosynthesis plants selectively absorb $^{12}CO_2$ rather than $^{13}CO_2$ (Wickman 1952; Bowen 1960 and Park and Epstein 1960). In effect the plant behaves as an "isotope separator" concentrating the ^{12}C isotope in the plant and in this manner photosynthesis exerts a major control either directly or indirectly on the distribution of the stable carbon isotopes in plants. Marine, non-marine and terrestrial organisms show different stable carbon isotope values ($\delta^{13}C$). These differences can be largely attributed to the fact that during photosynthesis marine organisms utilize carbonate and bicarbonate ions from the ocean, the non-marine organisms use the carbonate and bicarbonate from fresh water, whereas land plants use the isotopically lighter carbon dioxide in the atmosphere (Figure 16) and this is mirrored in the the different $\delta^{13}C$ value for plants from these various environments. Mass spectral analysis of the ratio of ^{12}C to ^{13}C in carbon dioxide produced from a complete combustion of plant material gives an indication of the environment in which they originated (Silverman 1962).

Although tasmanites are abundant from the Silurian to Cretaceous (Table 6), it has never been fully established whether or not these planktonic algae are marine or non-marine in origin (Muir and Sarjeant 1970). The stable carbon isotope value ($\delta^{13}C$) for Alaskan tasmanite and *Tasmanites punctatus* (Table 8) suggest that the organisms making

STABLE CARBON δC^{13} RATIO OF TASMANITE SPORE WALLS.

SAMPLE	δC^{13} VALUE
TASMANITES (ALASKA)	−23·2
TASMANITES PUNCTATUS	−9·0

TABLE 8

up Alaskan tasmanite ($\delta^{13}C$ = -23.2) had their origin in brackish water (near-shore water) since the $\delta^{13}C$ value is intermediate between the average value for marine algae ($\delta^{13}C$ = -17) and non-marine plants ($\delta^{13}C$ = -25). Whilst the value for *Tasmanites punctatus* (δ^{13} = -9.0) suggests this species originated in marine environments.

Palaeotological studies (Collett 1968) on the geological origin of *Tasmanites* indicate they are found in high concentration in some marine clays or muds (suggesting a shallow water deposition) and the present palynological indications are that *Tasmanites* had a marine and/or brackish origin and were collectively wind-blown together on shallow or near shore marine water to give high concentration of material that subsequently sedimented. This theory of a marine and/or brackish origin for *Tasmanites* is supported by the stable carbon isotope measurements.

These preliminary $\delta^{13}C$ studies on *Tasmanites* suggest this analytical technique may provide valuble additional information to the origin and environment of palynomorphs and become important as a new geochemical parameter to assist in palynological interpretations.

Occurrence of Fossil Spore Concentrates in sediments

Spores are amongst the most conspicuous constituents in thin sections of coal, but in many coals they also constitute bulk to the mass. The later applies especially to many cannel and splint coals and to certain layers or benches of bright coals (Sprunk,

Selvig and Ode 1938). There are various spore coal deposits (Cane 1968) and the *Tasmanites* present in enormous concentration in the Mersey River area, South of La Trobe, Tasmania has been examined by various workers (Zetzsche 1937; Cane 1968; Burlingame, Wszolek and Simoneit 1968; Brooks and Shaw 1969; 1970). We have also examined a sample of spore coal, which was hand selected from a band of dull coal in the Beeston Seam at Peckfield Colliery, Yorkshire. The maceral composition of the sample (Table 9) contained a high proportion

THE MACERAL COMPOSITION OF THE SPORE-COAL MATERIAL.

MATERIAL	PER CENT BY VOLUME
VITRINITE	2·0
EXINITE[a]	87·0
INERTINITE	10·7
PYRITE	0·3

a THE EXINES ARE MAINLY OF THE SAME TYPE & BELONG TO THE GENUS DENSOSPORITES ($C_{90}H_{120}O_{19}$)

TABLE 9

of exinite mainly of the same type and belong to the genus *Densosporite* (Smith 1968). This spore-coal after treatment with mineral acids and warm caustic solution, left a dark residue, and examination showed yellow-brown spore bodies (composition $C_{90}H_{120}O_{19}$), which had all the characteristic properties of sporopollenin.

Thermal Alteration of Sporopollenin

When organic matter is deposited in sediments, in addition to microbiological and chemical alterations there are three main physicochemical factors, temperature, time and pressure, that cause changes to occur. Spore exines found in coals from different ranks show gradual changes in colour from yellow to light brown to dark brown-blackish due to the varying degree of heating. Spores from other sedimentary deposits, such as shales show similar colour changes with temperature.

A knowledge of the degree of heating of the contained organic matter of a sediment is important for several reasons. Ecological studies have used pollen and spore morphology to estimate the extent and nature of vegetation in different areas and in different eras; if exines have been too strongly heated they become opaque and show no morphological structure which makes classification difficult.

Secondly, the degree of heating of exines in a sediment can be regarded as an index of the level of temperature to which other organic matter has been subjected.

Thirdly, there may be a correlation between sedimentary properties and the changes occurring in the chemical structure of sporopollenin exines when heated. In studies on sediments it is possible to estimate temperature variation of sediments from their inorganic crystal structures and mineral composition. These methods are, however, most effective for relatively high sedimentary temperatures (about 400°C), and estimates of low temperature history are more difficult to correlate. There appears to be potential in using alteration in the chemical structure of sporopollenin of recovered fossil spores to estimate low temperature sedimentary history.

Rogers (1865) first demonstrated that a correlation existed between fossil pollen and spore composition, oil and gas occurrences and the degree to which associated sediments had been heated. Later, White (1920) described a relationship between the limits of known commercial oil and gas occurrences and the variation in intensity to which the sediment had been headed from studies on the carbon contents of the organic matter in the surrounding coals. White (1925) also suggested that organic matter in sediments with carbon analyses of 66 to 77 per cent was the highest value which would allow occurrence of any oil. Many of the correlations between carbon value of the organic matter present in sediments and the degree of heating has been associated with studies on coalfields. But since in many areas of the world, coals are either absent or rare, this method of studying thermal history of sediments is of limited use. Extensive palynological studies have shown that in all areas of the world where sedimentary formations occur there is a relatively large accumulation of pollen and/or spores often in association with other plant material.

Organic matter in sediments occurs in three different forms, which may or may not be combined (Brooks and Shaw 1968) as

1) morphologically intact organic matter of known biological origin;
2) amorphous insoluble organic matter (often called "kerogen"), and
3) soluble organic components.

Geochemical methods enable these combinations of organic matter to be examined and analysed. The various techniques of extraction concentrate the insoluble organic matter from the sediment and microscopic examination identifies the morphological material (spores, pollen, dinoflagellates etc.) and the amorphous organic matter. These microscopic observations, along with other physico-chemical analyses (Correia 1969) and biological data when statistically analysed by computer gives an evaluation of the degree of evolution of hydrocarbons and the presence of other organic matter in sediments (Correia 1969).

Two physico-chemical methods have been developed to examine the level individual spores have been heated. The simplest and fastest method, estimates the degree of heating from an evaluation of the colour changes of pollen and spore exines. The second and more successful method has been a correlation of the change in translucency of pollen grains when heated (Gutjahr 1966). When *Quercus robur* pollen grains were heated their translucency varied with the degree of heating and from these studies various interesting factors emerged:-

1) slight changes in colour translucency of the pollen occurred when they were heated at 100°C for one week;
2) more measurable changes occurred when pollen was heated at 150° to 200°C for one week, and
3) a graph of translucency of the pollen against time (Figure 5) shows that at each temperature (100°, 150° and 200°), the rate of change (measured by pollen translucency) decreased rapidly and after 50 hours a level is reached when no further changes occur (Gutjahr 1966).

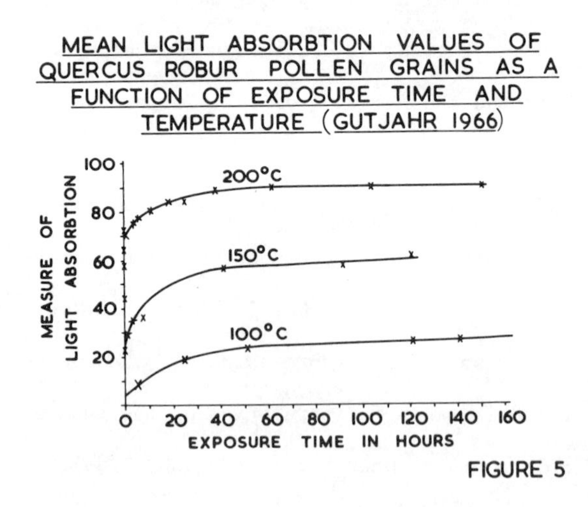

FIGURE 5

Results show that temperature causes predominant changes to occur in sedimentary organic matter (Staplin 1969). Although the effects of low temperature on pollen grains for a short time is not very important, the effect of higher temperature on pollen is much more significant (Gutjahr 1966). Slight changes in the appearance of pollen grains occur at temperature of 100° for one week, but more drastic changes occur at higher temperatures (150° to 200°) in a much shorter time and when pollen is heated to 200°C the grains become opaque and measurement of translucency becomes difficult.

A new method (van Gijzel 1967) which studies the autofluoresence phenomena shown by pollen and spores when irradiated with ultra-violet waves is now being developed. Fluorescence phenomena of pollen and spores is closely related to the physical and chemical nature of the wall constituents and can be applied to various palynological and geochemical problems including relationship in fluorescence colour differences in pollen and spore types of different age and also to changes in fluorescence occurring with pollen and spores of different thermal histories (see van Gijzel, this volume).

Our experiments have attempted to obtain conditions which are similar to those in a sediment, the duplication of sedimentary rock pressures (which can be as high as 2000 to 10000 lbs per square inch) could not be attempted. We examined the effect of heat on the

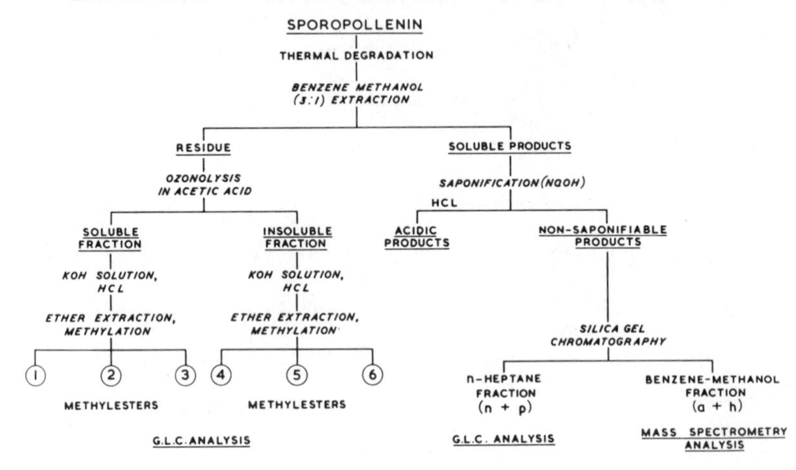

FIGURE 6

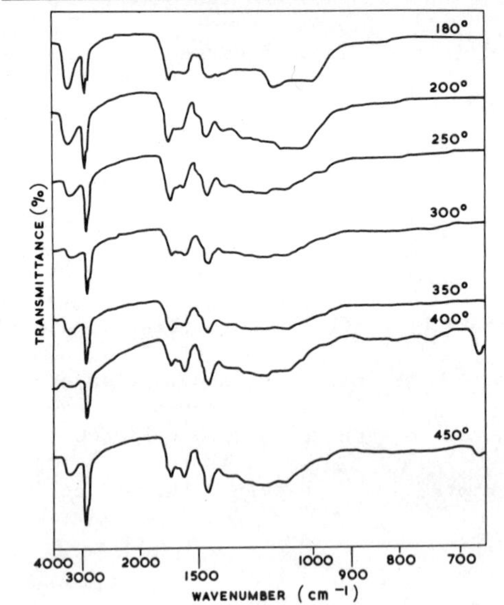

FIGURE 7

cellulose-free wall of *Lycopodium clavatum* spores at temperatures 180° to 450°C and attempted to correlate changes in chemical structure of the sporopollenin wall at the different temperatures. We chose an arbitrary 60 hours time scale as a convenient period of study, but as can be seen from Figure 5, this period was ideal and appears to allow changes occurring in the exine to proceed almost to completion.

A weighed amount of *Lycopodium clavatum* spore exines were placed in a glass sleeve and gently tapped to ensure good packing and placed in the central portion of a stainless steel bomb. The bomb was evacuated at 0.2mm for 30 minutes to remove all the air (especially the oxygen) and put into a furnace, which was heated to the required temperature (200° to 450°). After 60 hours the bomb and contents were cooled to room temperature and the valve opened to release the gaseous products. At this stage no attempt was made to analysis the more volatile gaseous products. The residue and soluble products were separated and examined using the techniques illustrated in Figure 6.

Infra-red examination of the extracted residue (Figure 7) shows when sporopollenin is heated there is gradual reduction in absorption at ~3,500 cm^{-1}, corresponding to hydroxyl groups; a gradual increase in carbon-carbon double bond absorption (~1,600 cm^{-1}) and after heating at 400° and 450°C there is aromatic carbon absorption

SPOROPOLLENIN

—AN OXIDATIVE CO-POLYMER OF CAROTENOID—CAROTENOID ESTERS

CAROTENOID

β-CAROTENE

ANTHERAXANTHIN

CAROTENOID ESTER

ANTHERAXANTHIN DIPALMITATE

SPOROPOLLENIN CONTAINS :—

CH_3 CH_3 —OH

$CH_3(CH_2)_nCOO$—

$CH_3(CH_2)_nCH_2$

FIGURE 10

ALTERATION IN ELEMENTAL COMPOSITION (C,H,O) OF SPOROPOLLENIN WHEN HEATED

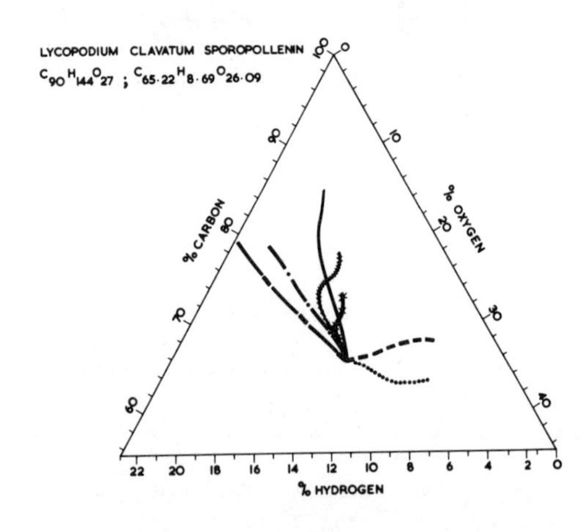

AFTER HEAT TREATMENT — 0°C → 400°C
—LOSS OF HYDROXYL GROUPS
—LOSS OF CO_2
—LOSS OF $-CH_2$ GROUPS
—LOSS OF -H ATOMS
LOSS OF HYDROXYL GROUPS AND-H ATOMS

FIGURE 11

(600 ~ 700 cm^{-1}). This shows that when sporopollenin is heated there is loss of hydroxyl groups (probably by dehydration) and increased aromaticity of the material.

Recent chemical studies (Brooks and Shaw 1968 and 1969) have shown that sporopollenin is an oxidative co-polymer of carotenoids and carotenoid esters which are chemically bound together into a matrix. In the polymer, there may be specific groups (Figure 10) that are not directly bound into the structure resulting in hydroxyl, C-methyl, and ester groups and ether linkages being preferentially degraded in addition to the polymer matrix becoming more aromatised by heating. It has been known (Winterstein and Kuhn 1933) for many years that carotenoids are aromatised when heated at quite low temperatures. In order to test this hypothesis of the processes taking place during the thermal alteration of sporopollenin, a series of theoretical curves have been determined of variation in carbon, hydrogen and oxygen content of the polymer when the various functional groups are degraded (Figure 11). Comparison of the theoretical curves with the curve from the experimental results suggested that during thermal alteration of sporopollenin there are two main processes taking place; chemical dehydration of the polymer by loss of hydroxyl groups and a gradual increase in aromatic nature. These results are in good agreement with the infra-red studies (Figure 7) and with preliminary X-ray examination of the thermally altered sporopollenin which indicates an increase in aromatic nature with increased temperature (Brooks and Wood 1970). Although the chemical alterations may be complex, these observations of thermal alteration in the structure of sporopollenin, based on partial knowledge of the chemical structure show an empirical relationship between C, H and O composition, various physical properties and temperature which will prove useful in future studies of sedimentary temperature and in estimation of maximum and minimum temperature history of organic deposits.

Examination of the Soluble Products from Thermally degraded Spropollenin

Various attempts (Sprunk, Selvig and Ode 1938) have been made to examine the chemical and physical properties of spores, especially thermal decomposition of their exine wall. Spores of coal forming plants (*Calamites, Lepidodendra, Sigillaria* and *Sphenophylls*) were extracted from coals by crushing, washing and floatation and the extracted spores ($C_{90}H_{101}O_{7.9}N_{0.6}S_{0.9}$ and $C_{90}H_{105}O_{9.6}N_{0.6}S_{0.9}$) when thermally decomposed at 500°C gave 51.2 per cent volatile matter (major constituents hydrogen, methane, ethane, carbon monoxide and carbon dioxide) and 48.8 per cent carbonised residue.

Later work (Macrae 1943) on the thermal decomposition of two samples of solvent extracted spore exine concentrates from the Beeston Seam, Micklefield Colliery, containing not less than 90 per

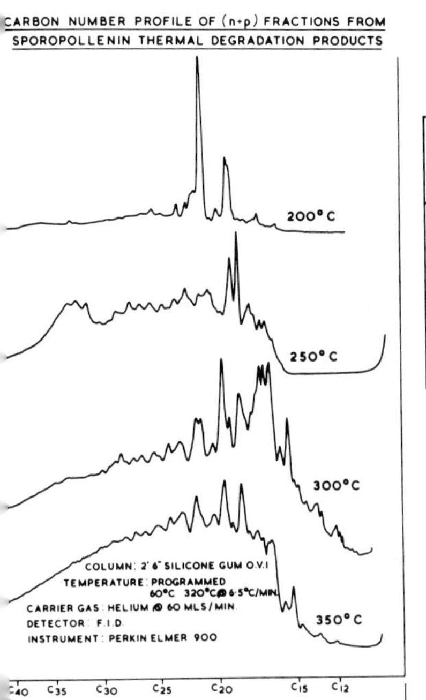

FIGURE 8

PRODUCTS FROM THERMAL DEGRADATION OF SPOROPOLLENIN

(5g OF LYCOPODIUM SPOROPOLLENIN)

TEMPERATURE	INSOLUBLE RESIDUE	% VOLATILES	% SOLUBLE	% WT. ACIDIC COMPDS. OF SOLUBLES	% WT. OF NON-SAP COMPDS. OF SOLUBLES
200°	4·8g	3·6	1·4	56	44
250°	4·7g	5·4	1·6	39	61
300°	4·6g	4·8	3·2	24	76
350°	4·0g	12·4	7·6	10	90
400°	3·6g	15·2	12·8	2	98
450°	2·9g	17·4	24·6	—	100

TABLE 10

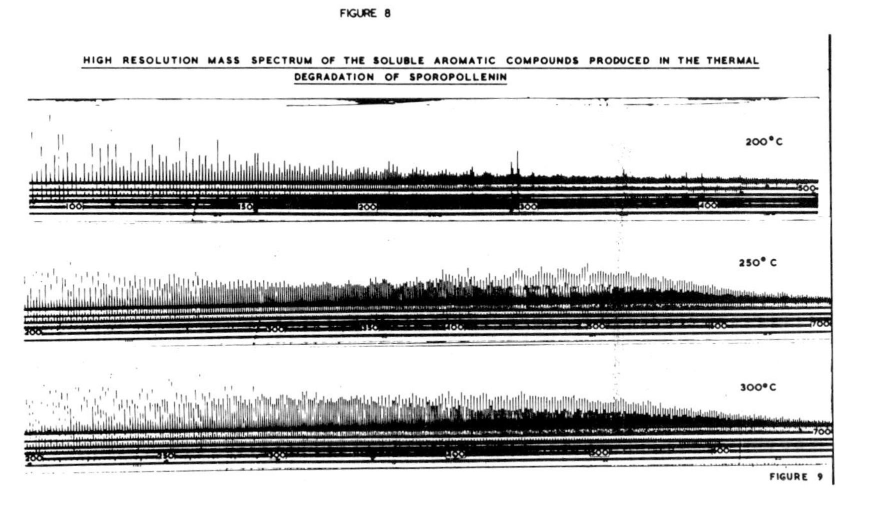

FIGURE 9

cent of exines were heated in *vacuo* at temperatures upto 350°C for periods of 135 days and 180 days. These studies showed that the evolution of volatile products (oxides of carbon, hydrogen sulphide, saturated and unsaturated hydrocarbons and hydrogen) was temperature dependent. The exines also yielded oils, especially heavy oil which showed a very sudden increase in production about 275°C. The yield of heavy oil was almost 30 per cent by weight of the exine and this mixture contained some 90 per cent of neutral substance (80 per cent of which were soluble in light petroleum). These neutral oils contained, according to the temperature of degradation, 35 to 45 per cent oxygenated compounds and other material removable by ferric chloride, 35 to 50 per cent of unsaturated and aromatic hydrocarbons and 15 to 25 per cent of saturated paraffin and naphthalene compounds (Macrae 1943).

Our studies on the thermal degradation of *Lycopodium clavatum* extines gave increased amounts of volatile components from 3.6 to 17.4 per cent at decomposition temperatures of 200° to 450° (Table 10). Similarly amount of soluble products increased from 1.4 to 24.6 per cent for similar temperature increases. These soluble materials were fractioned (Figure 6) and examined. The carbon number profile of the (n+p) fractions (Figure 8) showed an increasingly complex mixture (carbon number range C_{12} to C_{40}) with increased decomposition temperature. High resolution mass spectral hydrocarbon-type analysis (Figure 9) of the (a+h) chromatographic fraction showed a very complex mixture of aromatic (including a major amounts of phenolic components) and heterocyclic compounds were present. The aromatic products from thermal decomposition at 200° contain components with molecular weight up to 500, whilst the aromatic products produced at 300° had components with molecular weight over 700 (Figure 9).

Steuart (1912) suggested that kerogen (the insoluble amorphous organic matter present in sediments) was produced from various kinds of organic matter by the action of microbes under special circumstances, or on the other hand it could have been produced from the remains of certain kinds of vegetable matter like *Pinus* or lycopod spores. In order to test this hypothesis, Steuart prepared an artifical oil-shale from Florida fuller's earth and *Lycopodium* spores (which represented the supposed lycopodiaceous *Lepidodendron* spores from a forest near to a lagoon) and studied its thermal degradation. The degradation products have very similar properties to the soluble matter from the Torbanehill mineral and Higher Shales, which result in Steuart ascribing the organic matter from spores as contributing in some degree to the kerogen of oil shales (Steuart 1912).

Brooks and Shaw (1969 and 1970) have produced chemical evidence suggesting that sporopollenin derived from land and marine plants may contribute to the insoluble organic matter (kerogen) found in sediments

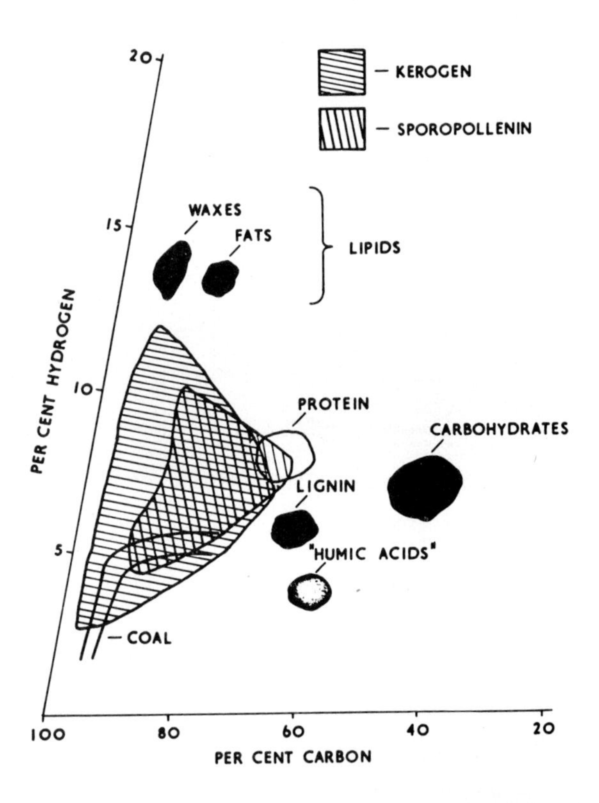
RELATIONSHIP OF COMPOSITION OF KEROGEN, SPOROPOLLENIN AND NATURAL PRODUCTS.
KEROGEN
SPOROPOLLENIN
WAXES
FATS
LIPIDS
PROTEIN
CARBOHYDRATES
LIGNIN
"HUMIC ACIDS"
COAL
PER CENT HYDROGEN
20
15
10
5
100
80
60
40
20
PER CENT CARBON

FIGURE 13

(Figure 13). The products from artificially degraded sporopollenin at various temperatures include, not only hydrocarbons but also aromatic compounds and fatty acids which are all known to be present in sediments (Calvin 1969). These results endorse the original suggestion of Steuart (1912) that sporopollenin is a probable contributor to kerogen and is a possible source of oil. Recent work (Combaz 1969) has also shown that similar spectra of products to those present in many crude oil accumulations can be generated by thermal degradation of *Tasmanites*, pollen grain and spore exines as well as from other sedimentary insoluble matter.

Preservation of Organic Matter in Sediments

The occurrence and nature of morphologically intact material varies considerably as one examines different rocks. With increasing age the fossil remains become less differentiated until in more ancient rocks virtually the only recognisable bodies are spore and spore-like organisms. These observations reflect the greater stability of the spore wall material and the lesser stability of other materials which make up the greater part of the plant.

Plants are mainly composed of cellulose, other polysaccharides and lignin which are subject to attack by a wide range of micro-organisms and are also susceptable to chemical attack, mainly hydrolysis. The rate of hydrolysis of such materials will vary with temperature but especially with pH of the aqueous phase. At low pH values (acid) the cellulose components will survive for only a brief period; in alkaline or neutral conditions the cellulose will be more stable, but would not be expected to survive the enormous time span into the Pre-cambrian in any but the most minute quantities. White (1933) has outlined how initially cellulose, then hemicelluloses are lignin are the first substances to undergo decomposition in aqueous media whereas spores show maximum stability under these conditions (Figure 12).

PROGRESSIVE DECOMPOSITION OF PLANT COMPONENTS UNDER NORMAL SEDIMENTARY CONDITIONS

(AFTER D.WHITE, ECON. GEOL., 28, 556, 1933)

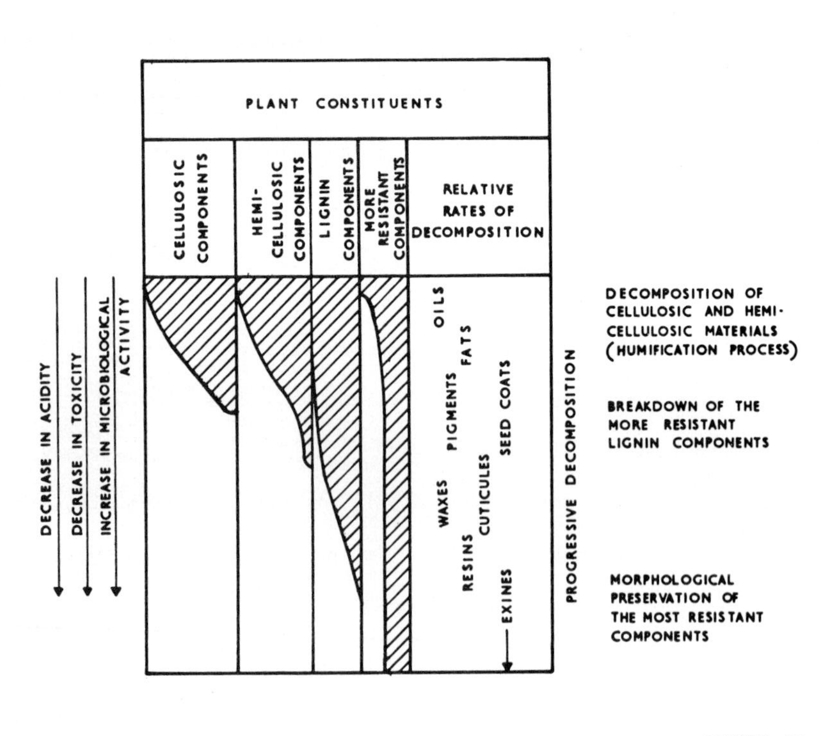

FIGURE 12

In addition to the preservation of fossil spore-like organisms in Pre-cambrian sediments, filamentous materials have also been described (Schopf and Barghoorn 1967; Pflug 1967; Schopf 1968; Schopf and Barghoorn 1969; Gutstadt and Schopf 1969) and raise the question are these filaments, which survive acetolysis, composed of sporopollenin in the same way as the associated spore-like material. It would be unusual if cellulosic filaments had survived into the Pre-cambrian, were resistant to acetolysis and survived the palynological extraction methods, since they are generally easily degraded. Carbohydrates are less stable than other organic matter and taking all the sources of information on the stability of sugars into consideration Degens showed that carbohydrates are more readily eliminated than other components of organic matter in the first stages of diagenesis (Degens 1967).

An explanation for the presence of these acetolysis-resistant filamentous materials in sediments may be provided by a process taking place in modern plants (Brooks and Shaw 1970). In modern plants during the process of sporopollenin deposition to form the exine, coating of the adjacent sporogeneous tissue occurs. This has been observed in various plants (Rosanoff 1865; Schanarf 1923; Kosmath 1927; Pettitt 1966; Banerjee 1967; Heslop-Harrison 1968) and provides the tissue with the same resistant properties as the exines, making them resistant to microbiological and non-oxidising chemical attack. The filamentous materials observed in ancient sediments may be preserved with the spores because they are coated with the same resistant material. Comparison of the resistant membrane investing the tapetum and sporogenous tissue and extra-tapetum membrane containing the spores in certain *Compositae* (Plates 1 and 2) (Heslop-Harrison 1968) with the structurally and organically preserved micro-organisms of the Pre-cambrian Bitter Springs Formation (Schopf 1968) and Beck Spring Dolomite of Southern California (Plate 3) (Gutstadt and Schopf 1969) shows the preserved spores encompassed in a resistant sheath, have many palynological similarities to modern spores and their associated resistant tissue and filaments.

Occurrence of Organic Matter in Pre-cambrian Sediments

Studies on sedimentary organic matter are providing new information of early terrestrial life and evidence that plant life existed in the late Pre-cambrian is now well established (Calvin 1969), but some doubt remains about the evidence for the presence of life in the early Pre-cambrian, especially the Onverwacht Series sediments (3,400 to 3,700 million years old) of Southern Africa.

To date the evidence for the presence of life having existed in sediments has rested on:-

a) histological examination of the sedimentary rocks and identification of morphological microfossils of a spore-like nature;

b) the presence of soluble and readily extractable organic compounds which have chemical structures (mainly hydrocarbon in nature) and occur in ratios characteristic of compounds one might expect to have arisen from biological systems;

c) the examination of the insoluble organic matter they contain which is quantitatively by far the major organic constituent of sediments.

The presence of fossil micro-organisms in sedimentary rocks

The last period of 600 million years has been well documented by palynologists and investigations during the past few years are leading to partial understanding of the nature of early biological organisation. Micro-organisms of the late Pre-cambrian have been identified in sediments as far apart as Northern Ontario (Tyler and Barghoorn 1954), Southern California (Gutstadt and Schopf 1969), Southern Africa (Schopf and Barghoorn 1967; Plug 1967) and Central and Southern Australia (Schopf 1968; Schopf and Barghoorn 1969). Although the presence of spore-like bodies in the Onverwacht Series Chert has recently been questioned (Nagy and Nagy 1969), recent work (Brooks and Muir 1970) shows that various samples of Onverwacht Chert are relatively rich in spore-like organisms.

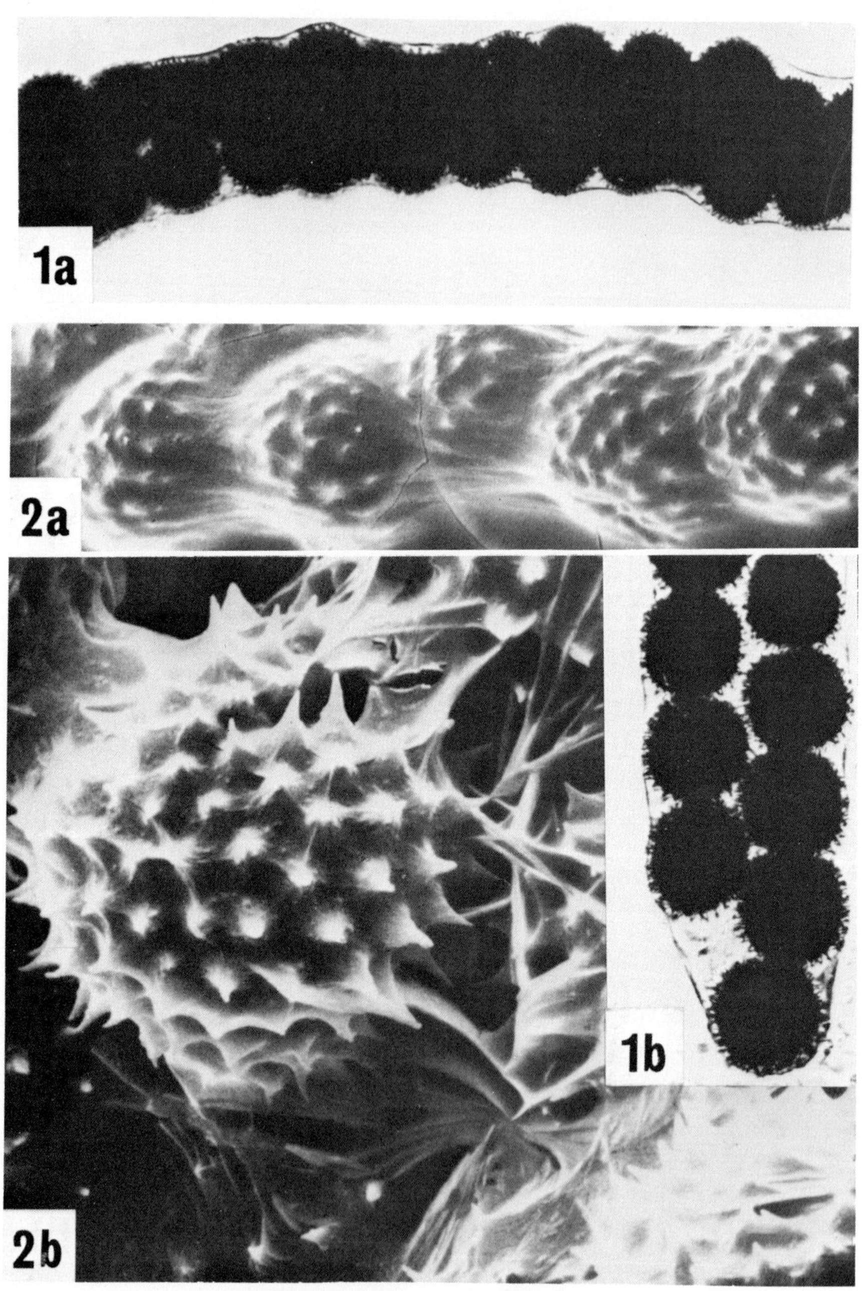

Plates 1 & 2

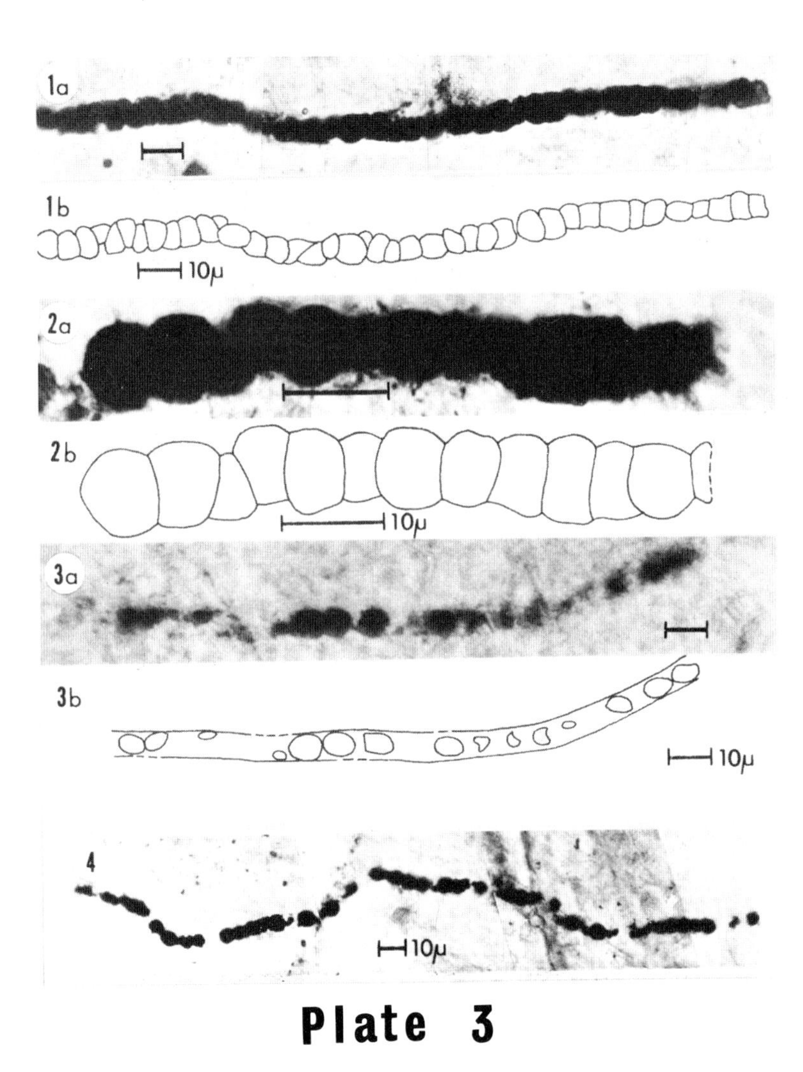

Plates 1 & 2. Spores of _Tagetes patula_ invested in an extra-tapetal membrane. Pl. 1 is an optical micrograph Pl. 2a & b are scanning electron micrographs. (By kind permission of J. Heslop-Harrison, 1968).

Plate 3. Possible organic sheaths on algal fossils from the late Precambrian of California. (By kind permission of J.W. Schopf [Gutstadt & Schopf, 1969]).

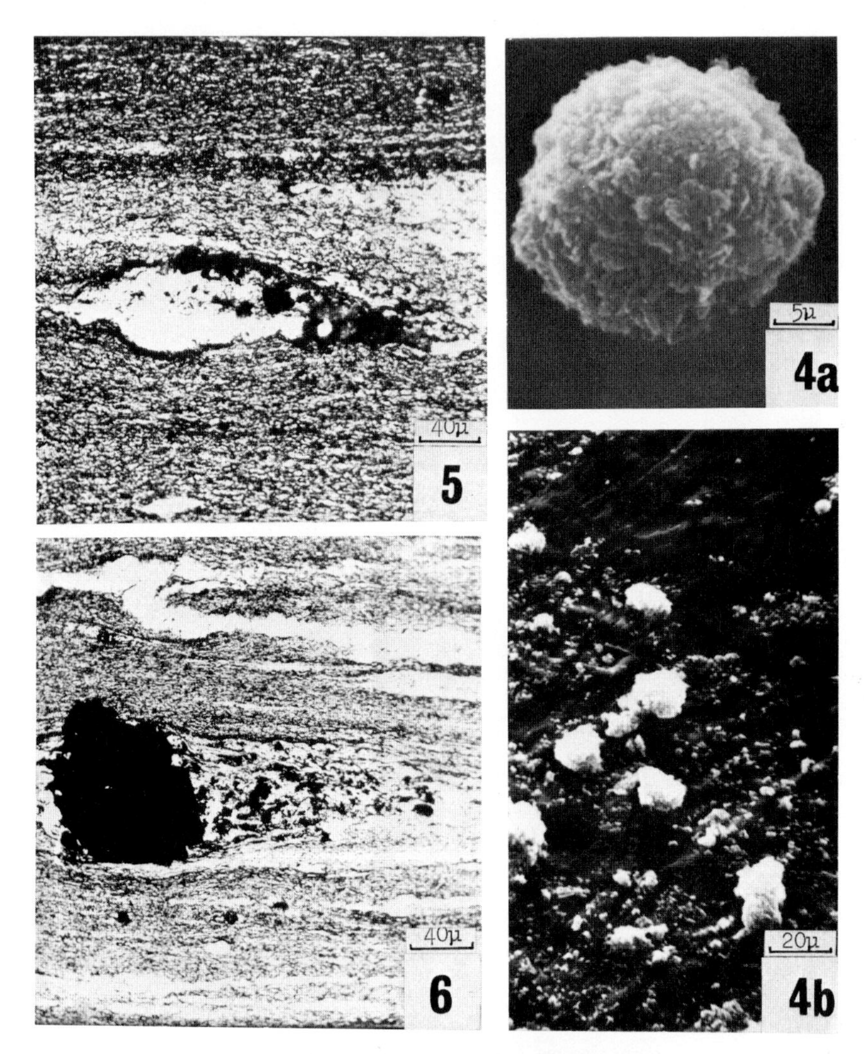

Plates 4,5 & 6

Plate 4. Photomicrographs of morphologically intact organic matter from the Onverwacht Chert, Swaziland, South Africa 3.4 - 3.7 $X10^9$ years old. The sphaeroidal bodies are strikingly similar to those described by Schopf, 1970, and also illustrated using a scanning electron microscope. Plates 5, and 6 are optical micrographs which show dispersed insoluble organic matter. Also from the Onverwacht Chert.(Brooks & Muir)

Some studies on the Onverwacht Chert

The cherts of the Onverwacht Series are, at present the oldest known sedimentary rocks on earth. They are part of the Swaziland System of Eastern Transvaal, Southern Africa. The Swaziland System is a folded synclinal belt forming the Barberton Mountain Land and is well exposed in the Barberton-Badplass area (Haughton 1969). The System consists of three rocks series; the youngest is the Moodies Series, the intermediate Fig-tree Series and the oldest Onverwacht Series. The Onverwacht rocks, which are the lowest of the strata, lie approximately 35,000 feet below the Fig-tree Series, are in contact with the surrounding granite terrain. There is some intrusion, but most of the beds are well preserved and little affected by heat (Han and Calvin 1969). The elemental analysis and mineral-type analysis of the rock show that the Onverwacht and Fig-tree sediments are composed mostly of α-quartz and various other minor mineral types (Brooks 1970). The Onverwacht Chert (0.05 per cent) contains less carbonate than the Fig-tree Chert (1.58 per cent), but the organic content of the Onverwacht sample (0.24 per cent) is similar to the organic content of the Fig-tree sample (0.22 per cent).

Attempts have been made to date the Onverwacht and Fig-tree Cherts using Rb-Sr isotope dating (Allsopp, Ulrych and Nicolaysen 1968), but interpretation of a precise geological data is difficult; using this method the Fig-tree Chert has been dated at $2.98 \pm 0.2 \times 10^9$ years old and the Onverwacht Chert to be older than 3.0×10^9

years. Recent stratigraphical dating (Han and Calvin 1969) has shown the Onverwacht Chert to be 3.4 to 3.7 x 10^9 years old.

Nagy (Nagy and Nagy 1968 and 1969) has reported that most of the Onverwacht Chert is metamorphosed (cf Han and Calvin 1969), but some of the sediment is not affected by too severe thermal metamorphism. The sample studied, which was collected from the same strata as Nagy's material (Nagy and Nagy 1969) has been examined for thermal history (Walls 1970) and results suggest the sample was never heated to much higher temperatures than 200^oC.

A section of rock with clean new surfaces from the central portion of the sample was allowed to dissolve in hydrofluoric acid for two weeks. The dark residue was examined using electron scan microscopy and spore-like microfossils were observed (Plate 4) (Brooks and Muir 1970). These micro-organisms have a constant size with an average diameter of 15μ and their morphology (spheroidal spore-like cells with reticulated surface texture) is similar to those described by Schopf (1970) as blue-green algae (*Myxococcoides minor* Schopf) from the Bitter Springs Formation of Central Australia.

It was postulated (Brooks and Shaw 1968; Brooks 1970) that there may be a relationship between the chemical nature of these spore-like micro-organisms and the more abundant amorphous insoluble organic matter present in the sediment (Plates 5 and 6) and examination of the chemical nature of this material may show it to be identical with biological material and thus confirm the presence of plant material in the sediment. The soluble organic matter (0.3

ppm) was extracted from the rock and analysis suggested it to be contamination (Brooks 1970). Recent studies (Nagy 1970) examined a representative sample of Onverwacht Chert and found it to have a porosity and permeability that could absorb at least 0.35 g of soluble alkanes dissolved in water in a period of 3.0 x 10^9 years. In contrast the insoluble organic matter (Table 11) is considered

ANALYSES OF SOME SEDIMENTARY ROCKS AND CARBON ISOTOPE VALUES

	AGE (10^6 Y.O.)	EMPIRICAL FORMULA	% SOLUBLE ORGANIC MATTER	δC^{13}	% INSOLUBLE ORGANIC MATTER	δC^{13}	% CARBONATE	δC^{13}
EL LAJJUN OIL SHALE	45	$C_{90}H_{126}O_{19}S_4$	2.82	—	17.80	—	12.4	—
GREEN RIVER FORMATION	50	$C_{90}H_{127}O_{20}S_6N$	3.40	—	12.30	—	17.4	—
NONESUCH SHALE	1000		0.06	−28.1	4.00	−28.1	13.1	−8.3
FIG-TREE CHERT	3100	$C_{90}H_{103}O_{13}SN_2$	0.4ppm	−27.5	0.22	−26.9	1.6	−2.2
ONVERWACHT CHERT	3400 TO 3700	$C_{90}H_{72}O_6N_{0.5}$	0.3ppm	−24.2	0.24	−15.8	0.05	(−3.2)

TABLE 11

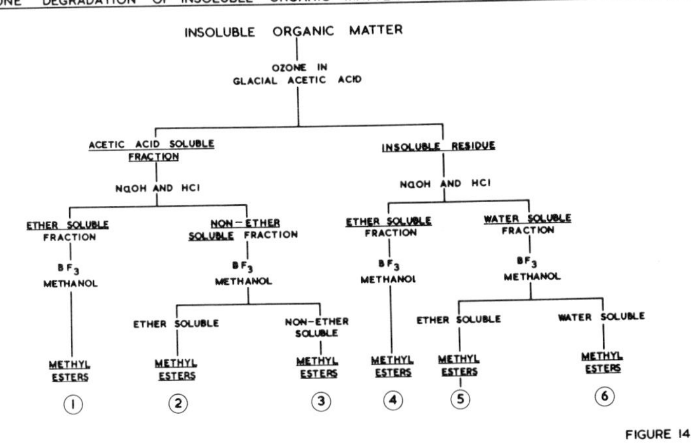

FIGURE 14

indigenous to the sediment. The extracted insoluble organic matter was degraded using ozone and the various fractions examined (Figure 14) using gas chromatography. These degradation products (Figure 15) were similar to those obtained from some other sedimentary insoluble organic matter and also to the ozone degradation products of sporopollenin (Brooks and Shaw 1968 and 1970).

Nagy (Scott, Modzeleski and Nagy 1970) discussed the presence of aromatic polymers in rocks, such as coals and kerogens and associated these with being derived from lignin, which is the case with black coals, but he rightly questions the presence of lignins in the early Pre-cambrian era and states that it is difficult to envisage other biochemicals that could be present in sufficient quantities to account for such large accumulation of aromatic matter.

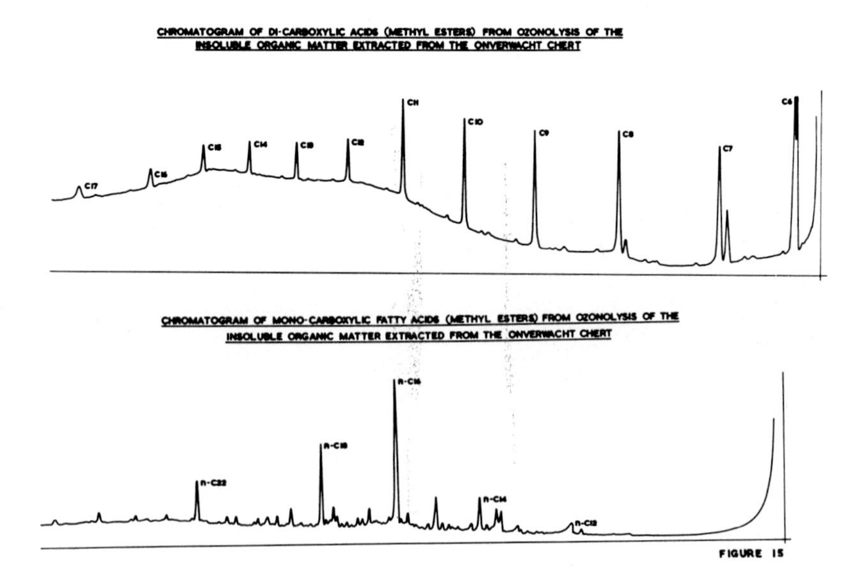

FIGURE 15

We have reported (Brooks and Shaw 1970) that aromatic compounds are characteristic of sporopollenin and synthetic polymers and are produced by heating (Potonié and Rehnelt 1969 and 1970). Therefore it is unnecessary to postulate a lignin precursor (which is unlikely in evolutionary terms) because the properties and chemical nature of the Onverwacht Chert insoluble organic matter and spore-like micro-organisms can be related to the biological material sporopollenin.

Stable carbon isotope studies on the Onverwacht insoluble organic matter

The stable carbon $\delta^{13}C$ value for the Onverwacht insoluble organic matter (Table 11) is significantly different from those of the Nonesuch and Fig-tree Pre-cambrian rocks. If one assumes that Nonesuch and Fig-tree deposits are examples of un-metamorphosed carbonaceous sediments, the $\delta^{13}C$ values for the insoluble organic matter suggest a marine deposit origin ($\delta^{13}C = -26$) and the $\delta^{13}C$ value for their carbonates are representative of a marine or oceanic deposit ($\delta^{13}C$ 0 to -5) (Figure 16). Although the $\delta^{13}C$ value (-3.2) for the Onverwacht Chert carbonate suggests a marine origin also, the insoluble organic matter $\delta^{13}C$ value (-15.8) is very unlike the value for the insoluble organic matter present in the associated Fig-tree sediment.

Silverman (1964) showed that heating organic matter results in the evolution of low molecular weight hydrocarbons (Figure 17) which show an enrichment of ^{12}C isotope (eg methane with $\delta^{13}C = -38$) which results in depletion of the $\delta^{13}C$ value (towards $\delta^{13} \rightarrow 0$)

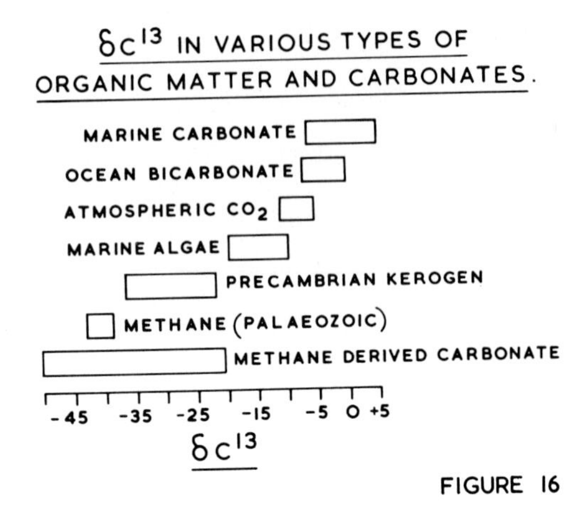

FIGURE 16

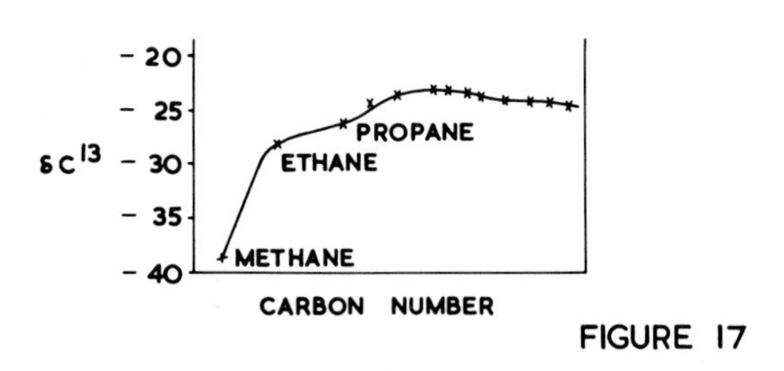

FIGURE 17

for the residual insoluble organic matter. The Onverwacht Chert has been heated to around 200° and low molecular weight hydrocarbons evolved. This evolution of hydrocarbons would result in the residual insoluble matter being slowly altered and the recorded

$\delta^{13}C$ value (-15.8) being the result of thermal alteration. The processes that have taken place in the Onverwacht Chert may be summarised as follows, insoluble organic matter, including some morphological spore-like organisms of biological origin similar to the material present in the younger Fig-tree sediment have been thermally altered resulting in evolution of hydrocarbons and aromatisation of the chemical structure of the residual organic matter.

Pre-cambrian Paleochemistry and Evolution

Although the earth is generally accepted to be about 4,500 million years old, the oldest known rocks (the Katarchean Series) have been dated about 3,600 to 3,800 million years old and this has led various workers (Sutton 1967) to suggest that the whole of the earth's surface was melted and re-worked at about 4,000 million years ago. This reworking of the surface is estimated to have taken place at temperatures in excess of 600°, at which temperature it would have been impossible for life, as we know it today, to have existed or even for the macromolecules that are considered precursors of life to have survived. Since recent studies have shown that the Pre-cambrian sediments contain evidence for plant life (Figure 18) including the oldest known sediments dated at 3,400 to 3,700 million years old, we reach the interesting question of not only how did life on earth begin, but we must ask ourselves when did it begin.

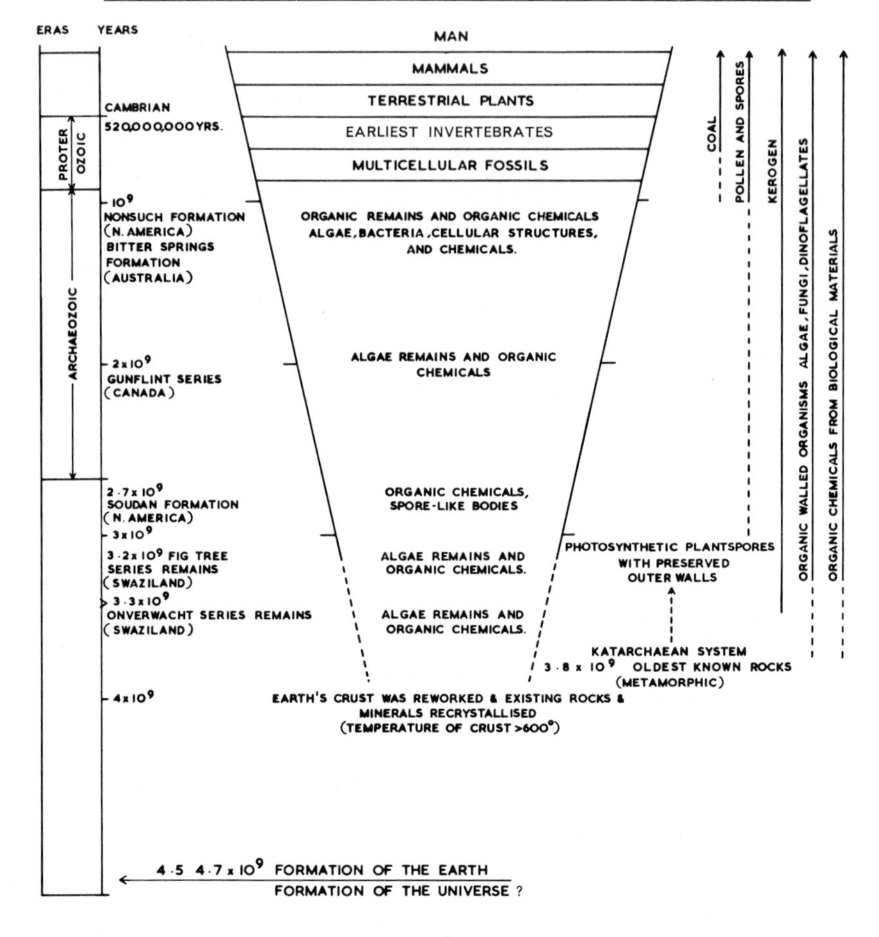

FIGURE 18

Acknowledgements

I wish to thank Dr. G. Shaw (University of Bradford) for his help and guidence during most of the work. Also appreciation is expressed to Mr. M.G. Collett (BP Research Centre, Sunbury-on-Thames) for reading the manuscript and for useful comments and suggestions and to the various people at the Research Centre who assisted in the preparation of the manuscript.

References

Allsopp, H.L., Ulruych, T.J. and Nicolaysen, L.O. (1968) *Can. J. Earth Sci.* 5, 605.

Banerjee, U.C. (1967) *Grana Palynologica*, 7, 365.

Bowen, H.J.M. (1960) *Intern. J. Appl. Radiation Isotopes*, 7, 261.

Braconnot, H. (1929) *Ann. chim. phys.* 2, 42.

Brooks, J. (1967) *M. Sc. Thesis* (University of Bradford).

Brooks, J. (1970) *Ph. D. Thesis* (University of Bradford).

Brooks, J. and Muir, M.D. (1970) unpublished results.

Brooks, J. and Shaw, G. (1968) *Nature*, 219, 532.
Grana Palynologica, 8, (2-3, 227.
Nature, 220, 678.

Brooks, J. and Shaw, G. (1969) *Nature*, 223, 754.
in *International Pullman Conference on Pollen 1969* (Ed. J. Heslop-Harrison) Buttersworth Publishers London, in Press.

Brooks, J. and Shaw, G. (1970) *Nature*, 227, 195.
Umschau, 18, 579.

Brooks, J. and Wood, D. unpublished results.

Burlingame, A.L., Wszolek, P.C. and Simoneit, B.R. (1968) in *Advances in Geochemistry 1968* Pergamon Press.

Calvin, M. (1969) *Chemical Evolution,* Oxford U.P. London.

Cane, R.F. (1968) *Proc. Roy. Soc. Tasmania,* 102, 85.

Collett, M.G. (1968) private communication.

Combaz, A., Marchand, A. and Libert, P. (1969) *Rev. Inst. Franc du Pêtrole,* 24, 3.

Correia, M. (1969) *Rev. Inst. Franc. du Pêtrole,* 24, 1417.

Dawson, J.W. (1871) *Amer. J. Ser., ser 3,* 1, 256.

Degans, E.T. (1967) in *Diagenesis in Sediments* (Ed. G. Larsen and G.V. Chilingar), Elsevier Publishing Co.

Eisenack, A. (1938) *Z. Geschiebeforsch,* 14, 1.

(1963) *Biol. Rev.,* 38, 107.

(1966) *Arch. Protistenk,* 109, 207.

Gutjahr, C.C.M. (1966) *Leidse Geologische Mededelingen,* 38, 1

Gutstadt, A.W. and Schopf, J.W. (1969) *Nature,* 223, 165.

Han, J. and Calvin, M. (1969) *Nature,* 224, 1082.

Haughton, S.H. (1969) *Geological History of South Africa* (Geol. Soc. S. Africa)

Heslop-Harrison, J. (1968) *Can. J. Botany,* 47, 541.

Hooker, J. (1852) *Quartz. J. Geol. Soc. (London)*, 9, 12.

John, -. (1814) *Journal für Chemie and Physik.* 12, 244.

Jeuniaux, C. (1965) *Bull. Soc. Chim. Biol.* 47, 2267.

Kosmath, L. (1927) *Ost. bot. Z.* 76, 235.

Kwaiatkowski, A. and Lubliner-Miahowska, K. (1957) *Acta Societatis Botanicorum Polonide*, 26, 5.

Larson, D.A. and Lewis, C.W. (1961) *Am. J. Bot.* 48, 934.

Macrae, J.C. (1943) *Fuel Sci. and Pract.* 22, 117.

Mazliak, P. (1968) *Progress in Phytochemistry, Vol. 1*, Interscience Press.

Muir, M.D. and Sarjeant, W.A.S. (1970) in Press.

Nagy, B. and Nagy, L.A. (1968) in *Advances in Geochemistry 1968* Pergamon Press.

Nagy, B. and Nagy, L.A. (1969) *Nature*, 223, 1226.

Nagy, B. (1970) *Geochim. Cosmochim. Acta*, 34, 525.

Newton, E.T. (1875) *Geol. Mag.* 12, 337.

Ostenfeld, C., Wandel, C.F. and Knudsen, M. (1899) in *Plankton* (Ed. C. Ostenfeld).

Park, R. and Epstein, S. (1960) *Geochim. Cosmochim. Acta*, 21, 110.

Pettitt, J.M. (1966) *J. Linn. Soc. (Bot)*, 59, 253.

Pflug, Hans D. (1967) *Rev. Palaeobotan. Palynol.* 5, 9.

Potonié, R. and Rehnelt, K. (1969) *Bull. Soc. Roy. Sc. Liége,* 38, 259.

Ralph, T.S. (1865) *Trans. Roy. Soc. Vic.,* 6, 7.

Rogers, H.D. (1965) *Proc. Phil. Soc., Glasgow,* 6, 48.

Rosanoff, S. (1865) *Jb. wiss. Bot.* 4, 441.

Schanarf, K. (1923) *Ost. bot. Z.* 72, 242.

Schopf, J.W. (1967) in *McGraw-Hill Yearbook Science and Technology*

Schopf, J.W. (1968) *J. Paleontol.* 42, 651.

Schopf, J.W. (1970) *J. Paleontol.* 44, 1.

Schopf, J.W. and Barghoorn, E.S. (1967) *Science,* 156, 508.

Schopf, J.W. and Barghoorn, E.S. (1969) *J. Paleontol.* 43, 111.

Scott, W.M., Modzeleski, V.E. and Nagy, B. (1970) *Nature,* 225, 129.

Shaw, G. and Yeadon, A. (1964) *Grana Palynologica,* 5, (2), 247.

Shaw, G. and Yeadon, A. (1966) *J. Chem. Soc.* (C), 16.

Silverman, S.R. (1962) in *Oil Sci. Session, Budapest 1962 - 25 years Hungarian Oil,* 308.

Silverman, S.R. (1964) in *Isotopic and Cosmic Chemistry* (Ed. H. Craig)

Smith, A.V.H. (1968) private communication.

Sprunk, G.C., Selvig, W.A. and Ode, W.H. (1938) *Fuel Sci. and Pract.* 17, 196.

Staplin, F. (1969) *Bull. Can. Petroleum Geology,* 17, 47.

Steuart, D.R. (1912) in *The Oil Shales of the Lothians*, H.M.S.O.

Sutton, J. (1967) *Proc. Geol. Ass.* 78, 498.

Tyler, S. and Barghoorn, E.S. (1954) *Science*, 119, 606.

Tschudy, R.H. (1969) in *Aspects of Palynology* Wiley-Inserscience Publishers New York.

van Gijzel, P. (1967) *Leidse Geologische Mededelingen*, 40, 263.

Wall, D. (1962) *Geol. Mag.* 99, 353.

Walls, R. (1970) private communication.

White, D. (1920) *Trans. Amer. Inst. Mining and Met. Engr.* 65, 176.

White, D. (1925) *Trans. Amer. Inst. Mining and Met. Engr.* 71, 282.

White, D. (1933) *Econ. Geol.* 28, 556.

Wickman, F.E. (1952) *Geochim. Cosmochim. Acta*, 2, 243.

Winterstein, A. and Kühn, R. (1932) *Ber. Deut. Chem. Ges*, 65, 1873.

Zetzsche, F. and Huggler, K. (1928) *Annalen*, 461, 89.

Zetzsche, F. and Kälin, O. (1931) *Helv. chim. Acta*, 14, 517.

Zetzsche, F., Kalt, P., Liechti, J. and Ziegler, E. (1937) *J. Prakt. Chem.* 148, 267.

DEMONSTRATION OF SURFACE FREE RADICALS ON SPORE COATS BY ESR TECHNIQUES

N.J.F.DODD and M.EBERT, Paterson Laboratories, Christie Hospital and Holt Radium Institute, Manchester 20.

Abstract

The free radical content of spore coats of *Osmunda regalis* and *Lycopodium clavatum* has been investigated by electron spin resonance (ESR) spectroscopy. The behaviour of spore coat radicals towards paramagnetic gases amd diphenylpicrylhydrazyl (DPPH) in benzene solution showed the presence of surface radicals. Van der Waals adsorption occurred with NO and NO_2 whereas O_2 was chemisorbed. As a result NO and NO_2 could be pumped off but O_2 had to be removed from the surface by chemical reaction with NO. Charcoal was used as a model for surface free radical behaviour. Exposure to ionizing radiation produced free radicals in the bulk of the spore coats, but did not add to the surface free radicals. Radiation produced radicals did not react with DPPH in solution and did not show reversible reactions with NO and O_2.

Introduction

The endogenous free radical population of spores of *Osmunda regalis* was examined during a comparative study of the biological radiation sensitivity of the spores and the free radicals produced in the spores by ionizing radiation (Dodd and Ebert 1969, in press). A large endogenous free radical population was located on the surface of the spores, which was influenced by certain paramagnetic gases, used to modify the biological radiation sensitivity. Endogenous free radicals showing a similar behaviour were also discovered using spores of *Lycopodium clavatum*. These radicals were characterized by comparison with the surface radicals of charcoal.

Materials and Methods

Osmunda regalis spores were harvested from plants growing in parks and gardens in and around Manchester. *Lycopodium clavatum* was bought commercially. The viability of the spores was unaffected by any of the treatments to be described, with the exception of NO_2 treatment. The gases used were commercially available and purified before use. Care was taken when NO and O_2 were used to flush with N_2 before admitting the other gas. A Hilger and Watts Microspin electron spin resonance (ESR) spectrometer with a dual cavity operating in the H_{014} mode was used to measure free radical concentrations. Absolute measurements involved double integration of the signal. Relative measurements were often possible by comparison of peak heights. The surface of the spores was examined using a Cambridge Instruments Stereoscan microscope.

Results

Morphology. Spores of *Osmunda regalis* have been shown (Erdtman 1957) to be highly sculptured. The present examination has shown the surface to be composed of ridges, deep hollows and sharp spines (fig.1). The surface area of the spores is consequently very large

A

B

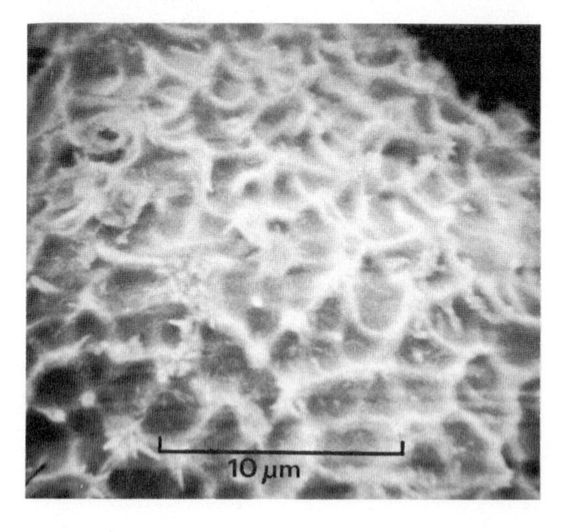

Fig.1. Scanning electron micrographs of an Osmunda regalis spore. (A). The whole spore. Its collapsed form is a result of drying. (B). The spore surface at 5 x greater magnification than in (A).

compared with their volume. A sectional micrograph (fig.2) shows

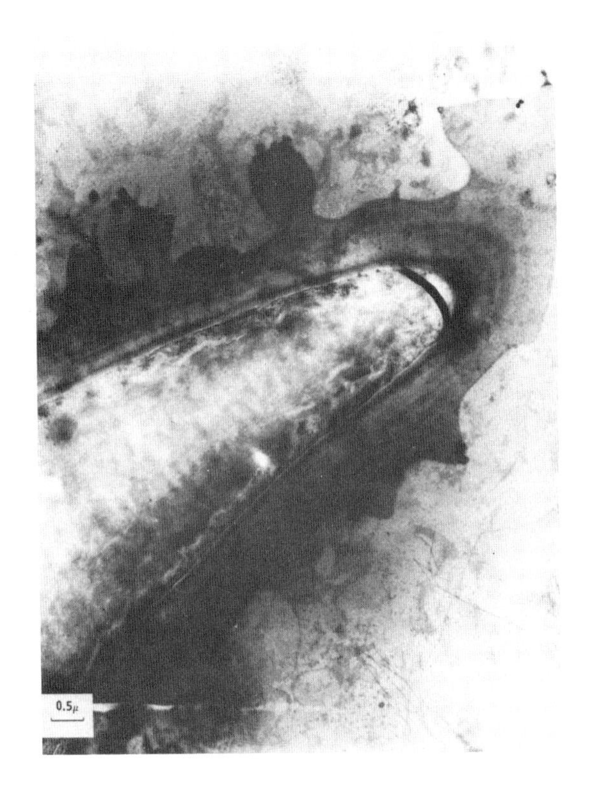

Fig.2. Sectional electron micrograph of a spore of Osmunda regalis.

striations in the coat. Electron dense spines are visible. In some sections narrow pores, of mean diameter about 50 nm, were visible passing through the coat. It is estimated that there are between 10 and 20 of these pores in each spore. Three germinal furrows were observed in the coats by optical microscopy. When the spores are soaked in water they split along these furrows and the spore coats are shed apparently unchanged as the spores grow and divide to form prothalli. We separated spore coats from the cell contents by mechanical disruption and washing with water. The separated coats were a pale yellow colour, in contrast to the bright green of the whole spores. Elemental analysis of these coats gave approximately 57% carbon, 7% hydrogen and 6% nitrogen by weight.

Demonstration of free radicals. Whole undried spores showed about 10^{10} paramagnetic metal ions and about 3×10^{9} free radicals per spore, when examined by ESR. These free radicals represented only about 10% of the total endogenous free radical population, the remaining 90% being masked, as will be shown below, by water and atmospheric oxygen. The ESR signal of these radicals (fig.3) is a

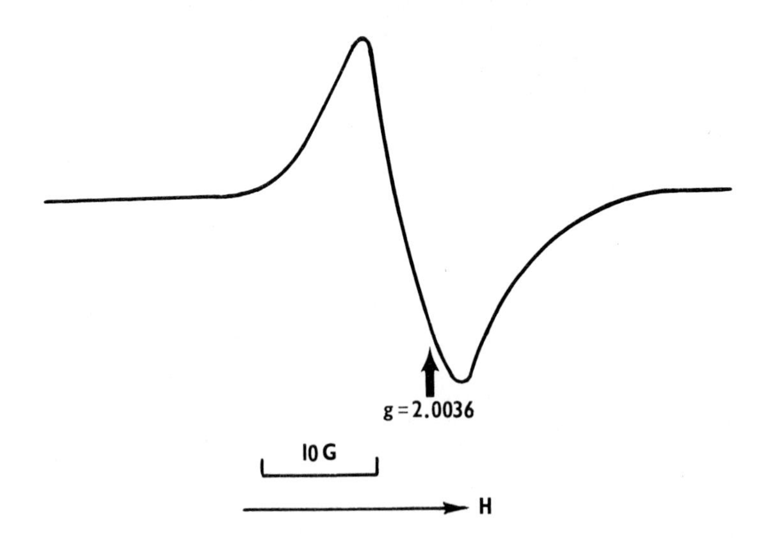

Fig.3. ESR signal of Osmunda regalis spores after the endogenous free radicals have been unmasked. The magnetic field increases to the right. The vertical arrow indicates the position of the DPPH resonance.

singlet of peak to peak line width 9 G and g-value 2.005. Free radicals giving an identical signal could be unmasked in the isolated spore coats.

Various treatments produced changes in the peak height of the ESR signal of whole spores (fig.4) without changing the shape of

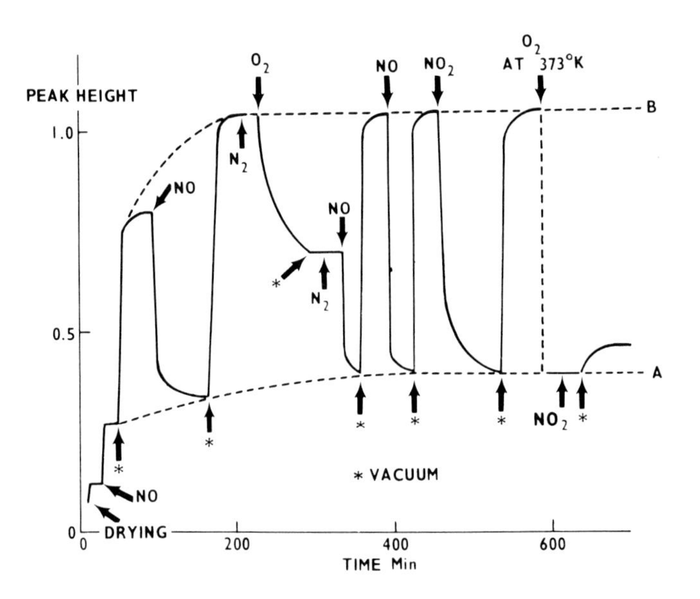

Fig. 4. The changes in peak height with time of the endogenous signal of spores of Osmunda regalis, produced by various treatments. Arrows indicate the times at which given treatments were started. See text for explanation of lines marked A and B. Abscissa: time, in min.; ordinate: peak height relative to an external Mn^{2+} standard.

the signal. Therefore in this case the peak height is proportional to the number of observable free radicals. When water molecules were removed by drying the spores by P_2O_5 in vacuo some of the endogenous radicals were unmasked. The number of observable radicals was increased by a factor of two. No change was produced by introducing N_2 or O_2 at this stage, whereas introducing NO increased the number of observable radicals by a further factor of 2. The largest proportion of endogenous radicals was unmasked on removing NO by evacuation, but could be masked again by re-introducing NO. Thus cyclic changes in the observable radical population could be produced by alternate exposure to NO and evacuation. Both the upper (B) and the lower (A) level of the cycle reached steady values after two or three cycles. The separation between the limits of the cycle represented a radical concentration of about 3×10^{10} radicals per spore or 5×10^{17} radicals per gram.

Oxygen and NO differ in the way in which they mask the radical population. Masking by NO was complete within minutes while O_2 took several hours, proceeding by approximately second-order kinetics, with a half time of 40 to 50 min. at room temperature. Oxygen masking was arrested but not reversed by evacuation. When spores were heated to 100° C in O_2, masking was complete within 10 min. In contrast, heating spores in NO produced an effect similar to evacuation, the masking effect was reduced and more free radicals were observed.

Free radicals fully or partly masked by O_2 could be unmasked using NO. Introduction of NO to spores partly masked by O_2 rapidly masked the remaining radicals (fig.4), but when NO was removed by evacuation the radicals masked by both O_2 and NO were unmasked. The cycle of masking the radicals by O_2 and unmasking them using NO could be repeated many times without changing the nature or concentration of the radicals. It can be inferred that in untreated spores endogenous radicals are masked by atmospheric oxygen.

Another paramagnetic gas which can mask the endogenous free radicals is NO_2. Its behaviour is similar to that of NO, masking is established rapidly and can be reversed by evacuation or heating. However, unlike NO, NO_2 is unable to remove the O_2 which masks the endogenous radicals.

Localization of free radicals. Spore coats separated from the spores showed, within experimental error, identical changes in free radical concentration when exposed to O_2, NO and NO_2. The radicals responsible for these changes are therefore located in the coats. The chemical nature of the coats is not precisely known, therefore the experiments were repeated using spores of *Lycopodium clavatum*, the coat material of which has been extensively studied (Afzelius 1956; Afzelius et al 1954; Brooks and Shaw 1968; Pettitt 1966; Shaw and Yeadon 1966; Zetzsche et al 1937; Zetzsche and Kalin 1931; Zetzsche and Vicari 1931) and is known to be composed mainly of sporopollenin. *Lycopodium clavatum* spores gave an endogenous free radical signal differing slightly from that of *Osmunda regalis* spores, but their behaviour towards NO and O_2 was identical. The signal observed after unmasking with NO (fig.5) has a peak to peak line width of 15 G and a g-value of 2.009. The similarity in behaviour of the endogenous spore coat radicals from both species suggests the presence of some common material. The difference in the shape of the ESR signals from the two species may reflect differences in composition of the sporopollenin.

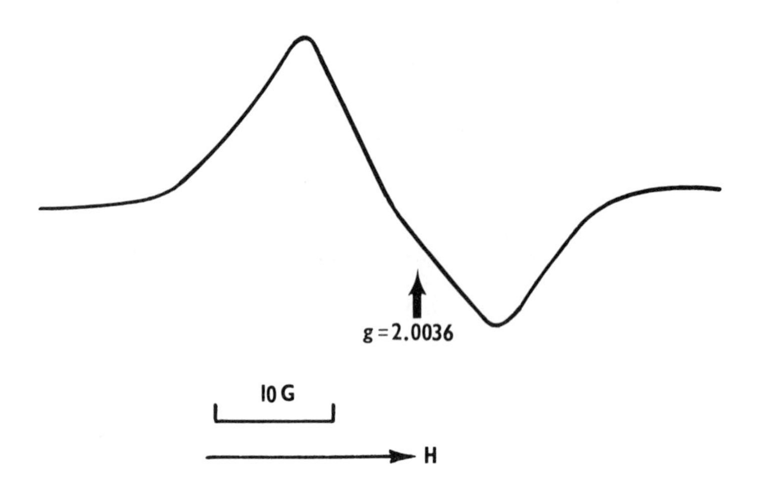

Fig. 5. ESR signal of Lycopodium clavatum spores after the endogenous free radicals have been unmasked. The magnetic field increases to the right. This signal was recorded under similar conditions of gain to Fig. 3. The vertical arrow indicates the position of the DPPH resonance.

The distribution of spore coat radicals was examined using a benzene solution of the radical scavenger diphenylpicrylhydrazyl (DPPH). This is a stable solid free radical giving an intense purple coloration in solution. Spores and spore coats of Osmunda regalis treated with NO to unmask the endogenous free radicals rapidly decolorized DPPH solutions. Untreated spores and spores or spore coats exposed to O_2 after NO treatment, i.e. those having endogenous radicals masked naturally or artificially by O_2 showed no such effect. These results show that the free radicals are available for reaction with the large molecules of DPPH and are therefore located on the surface of the spore coats. Electron micrographs showed that the size of the hollows in the spore coats was an order of magnitude greater than the molecular diameter of DPPH.

Free radicals are produced within the spore coats by ionizing radiation. These radicals, while giving identical ESR spectra to the endogenous radicals, did not react with DPPH or show the cyclic changes in NO and O_2. They decayed irreversibly in N_2, O_2 and NO (fig.6) and had no detectable interaction with the endogenous

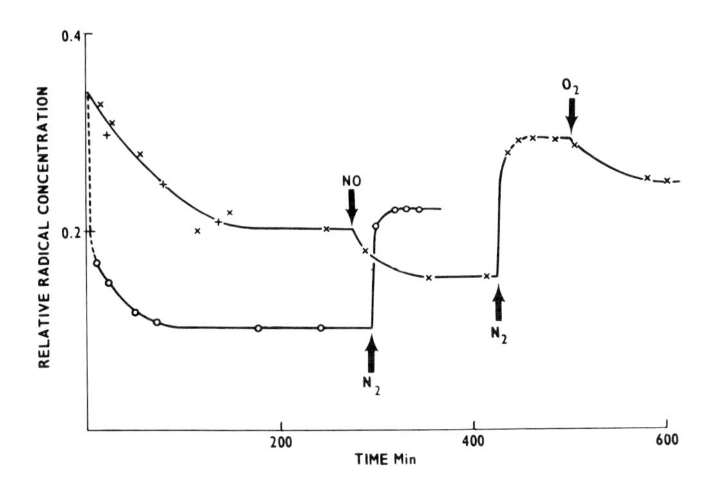

Fig.6. Decay of radiation produced radicals in Osmunda regalis spore coats. Arrows indicate the beginning of given treatments. Key: + spores irradiated and kept in N_2; x spores irradiated in O_2 and subsequently treated with (i) NO, (ii) N_2, (iii) O_2; 0 spores irradiated in NO and subsequently treated with N_2. Abscissa: time from mid-point of irradiation, in min.; ordinate: observable radical concentration relative to external Mn^{2+} standard.

radicals which could be unmasked by NO. The different reactivity of the radiation produced radicals is a consequence of their site of formation. Ionizing radiation produces radicals throughout the material. These are therefore largely inaccessible to DPPH radicals in solution in contrast to the endogenous radicals on the surface.

Discussion

In discussing the behaviour of the endogenous surface radicals of the spores, it is helpful to refer to ESR work on charcoal reported by other authors (Austin and Ingram 1956; Austin et al 1958) and repeated by ourselves. Mechanisms for the interaction of surface free radicals and paramagnetic gases have been proposed. Spore coat radicals and the surface radicals of charcoal behave similarly towards NO but differ in their reactions with O_2. Nitric oxide and NO_2 are readily adsorbed on the surface of spore coats and charcoal. These gases, being free radicals, mask the free radicals on the surface to which they are adsorbed. The gases are readily removed from the surface by evacuation as shown by the unmasking of the surface radicals. This suggests physical adsorption of the van der Waals type. Confirmation of this is provided by the effect of temperature. Van der Waals adsorption decreases with increasing temperature. Consequently when spores are heated in NO or NO_2 the extent of masking by the gases decreases, as during evacuation, and more radicals are seen by ESR.

Oxygen presents a different picture, behaving in one way with charcoal radicals and in another with spore coat radicals. The signal from charcoal is rapidly and reversibly masked by O_2 in a manner consistent with van der Waals adsorption. Oxygen will also be held on the surface of the spores by van der Waals adsorption. Interaction of O_2 with the surface radicals is demonstrated by the slow masking observed at room temperature. This interaction is accelerated by increased temperature, although the extent of van der Waals adsorption is reduced and is not reversed by evacuation. The O_2 held on the spore coats can, for want of a better term, be called chemisorbed, implying some chemical reaction requiring activation energy.

It has been shown above that chemisorbed oxygen is removed from the spore coats by NO. This can be explained if the oxygen molecule, which is a bi-radical, is bound by interaction of only one unpaired electron. The other interacts with the unpaired electron of NO to form an intermediate. This weakens the surface-oxygen bond with the result that the intermediate breaks down on evacuation, giving free spore coat radicals and probably NO_2, which is removed by evacuation. Nitrogen dioxide itself is unable to remove chemisorbed oxygen, probably because formation of an intermediate is energetically unfavourable.

Biological significance of the chemisorbed oxygen on spore coats was sought. The germination rates, in aerated and deaerated water, of _Osmunda regalis_ spores with and without chemisorbed oxygen on their coats were compared. Although the oxygen status of the water had a marked effect, the presence or absence of chemisorbed oxygen produced no detectable effect. A rough calculation showed that the total amount of oxygen available on each spore is small and would be utilized by a growing spore in about 10 min. Therefore while the chemisorbed oxygen does not play an important role in germination it may be utilized during this process or have some other biological significance.

It may be worth noting that empty sporangia of *Osmunda regalis* show endogenous free radicals giving signals and reactions that are identical with those of the surface free radicals of spore coats.

The presence of radicals on the spore coats may be a result of the process by which the coats are formed. It has been postulated (Brooks and Shaw 1968) that coats are formed by oxidative polymerization. Such a reaction may proceed by a free radical mechanism and some of the radicals involved may be trapped. The predominant occurrence of radicals on the surface of the spore coats may reflect the slow rate of polymerization in a biological system, which allows complete reaction of radicals within the polymerizing material.

Summary

The existence of free radicals on the surface of coats of *Osmunda regalis* and *Lycopodium clavatum* has been demonstrated using ESR. These radicals are, in the natural state, masked to a small extent by moisture and to a larger extent by O_2. The moisture can be removed by drying in vacuo whereas O_2, which is chemisorbed, requires chemical interaction with NO for its removal.

Acknowledgments

We thank Dr.R.D.Butler, Botany Dept., University of Manchester and Miss B. Lomas, Textile Technology Dept., U.M.I.S.T., who kindly provided the electron micrographs, and the Superintendent, Manchester

Corporation Parks Department for allowing us to collect spores.

References

Afzelius B.M. 1956 Grana Palynologica 1 22.

Afzelius B.M., Erdtman G. and Sjostrand F.S. 1954 Svensk.Botanisk Tidskrift 48 155.

Austin D.E.G. and Ingram D.J.E. 1956 Chem. Ind. 37 981.

Austin D.E.G., Ingram D.J.E. and Tapley J.G. 1958 Trans.Faraday Soc. 54 400.

Brooks J. and Shaw G. 1968 Nature 219 532.

Dodd N.J.F. and Ebert M. 1969 Nature 221 1245.

Dodd N.J.F. and Ebert M. Int.J.Radiat.Biol. In Press.

Erdtman G. 1957 Pollen snd Spore Morphology and Plant Taxonomy Vol. II p.84 Almquist and Wiksell, Stockholm.

Pettitt J.M. 1966 Bull.Brit.Mus. (N.H.) Geology 13 221.

Shaw G. and Yeadon A. 1966 J.Chem.Soc (C) 16.

Zetzsche F. and Kälin O. 1931 Helv.Chim.Acta 14 517.

Zetzsche F., Kalt P., Liechti J. and Ziegler E. 1937 J.Prakt. Chem. 148 267.

Zetzsche F. and Vicari H. 1931 Helv.Chim.Acta 14 58.

APPLICATION OF SCANNING ELECTRON MICROSCOPE TECHNIQUES AND OPTICAL MICROSCOPY TO THE STUDY OF SPOROPOLLENIN

M.D. Muir and P.R. Grant.

Geology Department, Royal School of Mines, Prince Consort Road, London, S.W.7., England.

Abstract

Although a large number of papers have recently been published using the scanning electron microscope to illustrate morphological details of spores and pollen grains, little work has been done to utilise the full potential of the instrument, which is capable of use as an analytical tool. Fern spores have been treated chemically with different standard maceration methods and acetolysis, and the results have been examined and assessed using the scanning electron microscope to investigate detailed morphological structures.

These results have been checked optically, and are illustrated. Polarisation microscopy has also been used.

The new technique of cathodoluminescence (the study of electron stimulated light) has also been applied to the same material, and the results of this study have been compared with the evidence supplied by U.V. fluorescence microscopy. Differences in properties between the endospore and perispore of several polypodiaceous ferns are described.

Several years ago, one of the authors of this paper was examining spores and pollen grains from the Pliocene of the Island of St. Helena in the Mid-Atlantic. The assemblage was of considerable interest, being the first pre-Pleistocene assemblage described from the islands of the Mid-Atlantic ridge, and, as well as that, it had a peculiar composition, in which fern spores were the dominant constituent of the assemblage. This was in marked contrast with the present-day flora of the island, and an attempt to explain this was made by Muir and Baker (1967).

One of the major difficulties encountered in describing this assemblage, was that 75% of the palynomorphs found were monolete polypodiaceous endospores, which appeared to possess no distinguishing characteristics at all.

Later, however, on examining the acetolysed assemblage using the scanning electron microscope, it became evident that there were at least two kinds of endospore, one smooth, and one with pronounced, albeit, small ornamentation (Muir, 1970). This led us to two conclusions: 1. that examination of polypodiacean endospores of living species using the scanning electron microscope might yield further details about the fine structure of the exterior, which might have taxonomic value and, consequently, stratigraphical value when fossil assemblages are examined; 2. that before such a study could have any real value, it would be necessary to investigate the morphological effects of different kinds of preparative techniques upon the surface structures of such spores. Martin (1969) investigated the effects of acetolysis on a variety of angiospermous and gymnospermous grains, and found that acetolysis often cleared debris from apertures of grains, and rendered details of surface ornament much more obvious than was visible in unacetolysed material. He found, however, that gymnosperm pollen such as the bisaccate pollen of *Pinus* was liable to show signs of collapse, and that although fine structure was revealed more clearly, the overall structure of the grains was obscured. Martin ascribed the changes caused by acetolysis to the removal of acid-soluble material, although he did not specify what this material was. Since by definition, sporopollenin is resistant to acetolysis, then the material that obscured the ornament and the apertures can not be regarded as sporopollenin. This material, may in fact, represent the final deposits from the tapetum (see Echlin, this volume) at least in the case of the angiosperm pollen grains. In view of the work of Dickenson (this volume) which suggested that the endexine of *Pinus* has a different composition from its ektexine, which is known to be sporopollenin, then, perhaps, the collapse of fresh *Pinus* pollen in Martin's experiment may be explained by the use of this information. Whatever the composition of the endexine, because it is not sporopollenin, it is not wholly resistant to acetolysis, and the treatment caused at least a partial solution of the endexine leading to the final collapse of the grain.

A further example of this removal of acid-soluble 'matrix' from spores and pollen grains has been found in some of the triradiate fern spores from St. Helena. On examining unacetolysed material, the abundant spores of Lygodium sp., Pteris sp. and the smooth triradiate spore tentatively assigned to Vittaria sp. (Muir & Baker,1967), we noticed that the contact areas of all these spores were smooth and featureless. However, on examination after acetolysis, all three species were characterised by the presence of small 'pores' on the contact areas. This phenomenon has also been described by Elsik (this volume), and he attributed these 'pores' to bacterial action. However, we would like to make the tentative suggestion that they may represent the removal of some acid-soluble material, and further that they may play some part in germination. Three examples of these are illustrated (figs. 1 - 3) and an optical micrograph of a Jurassic spore (Deltoidospora sp.) appears to show the same features (fig.4).

Thus we came to the conclusion that if fresh spores and pollen have walls that are not composed entirely of sporopollenin, but which have areas, or even partial mixtures where other substances are present, then it ought to be possible to determine the position of these substances in the spore wall by examining the variation in properties of the spore wall by as many different techniques as were available to us.

A further point which we wanted to examine, of particular interest to the palaeopalynologist, is that the perispores of the various monolete members of the Polypodiaceae (used in an old-fashioned collective sense) is never found fossil. Many of the structures described in the literature as perispores are, in fact, more reasonably interpreted as detached layers of the ektexine. This kind of separation has been found in spores of all groups of plants, although it appears to be particularly common in, for example, the Lycopsida. Furthermore, structures similar to the so-called perispores may be produced by compressing spores beneath the cover-slip of the preparation in which they are to be examined, and consequently, only where neither of these criteria apply can it be said with certainty that a true perispore is present. A further reason for doubting the presence of fossil perispores (at least in pre-Pleistocene material) is that in almost all cases where they have been described, they occur not on monolete spores, by far the most common occurrence today, but usually on triradiate spores. In any event, even such records as exist are few, and for the present purpose, we shall suppose that perispores are not fossilisable.

We decided to experiment with two different kinds of fern spores to begin with, and these are the only two species which have undergone all the tests applied. They are *Thelypteris palustris* , and *Dryopteris filix-mas*, both of which produce monolete spores, which are reported to possess a perispore.

We began our experiments with *Dryopteris*. Optically, this is a simple, smooth, monolete spore enclosed within a loosely enveloping perispore (fig. 5). Such spores have been examined on the scanning electron microscope (Jermy, 1970), and were shown to possess a noticeable fine structure on the perispore (see fig.6, our photograph) Now one of the optical properties of any anisotropic substance is its birefringence, and this can be examined using the polarising microscope. When Dryopteris spores are examined in this way, they show strong birefringence in the region of the perispore, and not in the general region of the endospore. However, the general poor resolution of images using cross polarised light is a limiting factor in such investigations (fig.7). Earlier work by Sitte (1959, 1960) and some others, although principally concerned with angiosperm pollen, using these techniques showed that the maximum birefringence was confined to the outer layers of the exine, and that the anisotropy of the nexine was demonstrably different from that of the sexine. However, the perispore of Dryopteris which has, at any rate, a quite different ontogenetic origin from the exine proper, has a much brighter birefringence than the exine of the angiosperm pollen grains described by Sitte, and this may reflect a difference in chemical composition or structure. This difference in composition may explain why such perispores are not normally found fossil.

Using the scanning electron microscope, we examined the morphology of grains that had been acetolysed, treated in an ultrasonic vibrator, and then washed in distilled water and mounted on a glass cover slip,and coated with gold in the usual way for examination. The ultrasonic treatment appeared to be the most drastic; the perispores of nearly every spore having been shaken free of the endospore, only a few tattered remnants being preserved. Acetolysis did not completely destroy the perispore, but caused it to collapse and apparently to thin.

Some examples of this are given in figs. 8, 9, and 10.

Thelypteris palustris, also a monolete polypodiacean spore which possesses, apparently, a tightly appressed perispore which separates from the endospore at regular intervals over the surface of the endospore forming conical projections above the general surface level. Optically, it is clearly distinct from *Dryopteris* (fig. 11), but when examined with cross-polarised light, it also shows that the strongest birefringence is confined to the outer layer of the spore (the perispore), and that the conical projections from the surface are markedly birefringent, showing up as bright spots (fig. 12). Thus, optically, *Thelypteris* shows morphological differences from *Dryopteris*, but the perispores of both appear to react the same way to cross-polarised light.

As a result, however, of the chemical and physical treatments, applied before examination with the scanning electron microscope, *Thelypteris* responded quite differently from *Dryopteris*. Even prolonged ultrasonic treatment (more than 10 minutes at high frequencies) was unable to cause the complete removal of the perispore; although corrosion of the points of the conical projections is marked (figs. 13, 14), the tightly appressed parts of the perispore are not readily removed - in fact we were unable to remove them at all. Acetolysis also had markedly less effect on *Thelypteris* than *Dryopteris* (fig. 15).

Thus the morphological differences clearly visible between the two species are reflected in their response to chemical and physical treatments, which responses must in themselves reflect compositional or molecular structural differences. Thus it seems possible that the suggestion made by Chaloner (this volume) that minor differences in composition between the sporopollenins of the higher plants ought to be looked for. Some such differences have already been determined(see Shaw, this volume table 6). These minor variations in composition may be employed as a kind of sure reflection of the species under investigation, and might, in future, be employable in determining, for example, evolutionary sequences in spores and pollen grains, and such important matters as whether the first distally germinating pollen grains were composed of a sporopollenin typical of polliniferous plants, or whether it has the kind of sporopollenin typical of sporogenous plants.

Plainly, there are methods for examining the physical properties of fern spores other than those described above. Van Gijzel has demonstrated (1967 a, b, and this volume) that the use of ultra-violet stimulated fluorescence can reveal a great many subtle differences between spores and pollen of different types, and from rocks of different ages. By the use of various sophisticated measuring instruments, he has been able to refine this technique greatly until it is capable of very subtle distinctions, such as determination of bonds present in a molecule, and the demonstration of the chemical differences between fossil and fresh sporopollenin.

A further method of, so to speak, *in situ* chemical study of sporopollenin is made available by the use of a new technique in the field of scanning electron microscopy. The normal image obtained by this method is the result of the collection of the secondary electrons (electrons emitted from the atoms comprising the surface affected by the incident beam), and in some cases, primaries (the higher energy electrons from the electron beam itself) which are elastically reflected back from the specimen surface. However, many other kinds of radiation are emitted and these can be collected and analysed and can be used to give some indication of the chemistry of the specimen under examination. X-rays characteristic of the elemental composition of the specimen are emitted and can be used most successfully for the analysis of most elements. In order to use the X-rays of the lighter elements, it is necessary to utilise a crystal spectrometer (wavelength dispersive) for the analysis. This system can analyse elements of atomic number down to boron (atomic number = 5), and in theory, at least, would seem to be an ideal system for palynological studies. However, although it is possible to use such a system to make direct analyses of the amount of carbon in sporopollenin, the method is inaccurate, and suffers from the disadvantage that in order to make a quantitative analysis, it is necessary for the specimen surface to be flat, which involves considerable preparational difficulties. Further, in order to stimulate sufficient X-rays, a high beam current must be used on the specimen, which has the undesirable side effects of heating, and sometimes destroying, the specimens. Thus, at the moment, this technique, although interesting, appears to be impractical for the study of sporopollenin.

The other type of X-ray analyser presently available (the energy dispersive type) is, at the moment, unable to analyse elements lighter than fluorine, which makes it entirely unsuitable for palynological work.

However, specimens excited by electron beams also emit ultra violet, visible, and infra-red radiation, and these signals may be collected, using a similar collection system to the normal one, but omitting the bias and the scintillator. An image of the light being emitted from the specimen may then be displayed on the normal display tube of the scanning electron microscope (see Muir, Grant, & Hodges, (in press); Grant, (in press). Further, it is possible to use a spectrometer, in the same way as did van Gijzel (1967a, this volume) to analyse the light emitted by the specimen (Muir, Grant & Hodges, (in press); Grant (in press). Although we know that sporopollenin emits electron stimulated light or cathodoluminescence (cathode luminescence), we do not yet know if it will prove to be as useful as the ultra-violet stimulated fluorescence described and illustrated by van Gijzel (1967a, b, this volume).

There are, however, several difficulties to be overcome in preparation methods. The most important of these is that if we attempt to analyse the light being emitted from the surface of a specimen, it is important that the material under examination is not coated with a metallic (or carbon)coating. Metallisation is important for the examination of spores and pollen, in order to produce a thin coating of a conducting substance which will carry away any electrical charge produced on the specimen by the electron beam. Since spores and pollen grains are more or less entirely non-conducting, they are particularly susceptible to charging effects, which frequently spoil the beautiful, clear, all-in-focus images that we have grown accustomed to expect. However, the intensity of cathodoluminescence is generally very low, and even a thin coating may effectively absorb any light emitted by the specimen. Coating with a metal such as gold, which when very thin is translucent, and may transmit light in various parts of the spectrum, will obviously alter the spectrum of light emitted from the spores and pollen grains, even if it does not entirely absorb it. So we have had to try to devise means of examining uncoated material: these include the mounting of the spores and pollen grains on a double sided conducting adhesive tape; on a metal-bearing adhesive; on silver dag (a suspension of silver particles in a volatile liquid); and by letting the grains dry directly on to the specimen holder. In fact this latter method has proved to be the simplest and best (fig. 16).

A further preparational difficulty is that we do not, as yet, know what effects different preparational methods will have on the spectra emitted by the spores and pollen under examination. Van Gijzel (1967a) has stressed the importance of preparation methods on the UV-fluorescence microscopical results, and our preliminary investigations have indicated that this is also true for cathodoluminescence. Untreated spores of Acrostichum give off quite a strong cathodoluminescence signal (fig. 17, 18), while specimens which have been acetolysed gave off only a feeble signal, from the region of the triradiate mark. Extensive experiments have not yet been attempted, but it seems likely that reagents which affect or inhibit UV-fluorescence will have the same effects on cathodoluminescence.

We have studied the cathodoluminescence images of various fern spores, including *Dryopteris filix-mas*, *Thelypteris palustris* and *Acrostichum aureum*. Since preparational effects have such drastic effects on the cathodoluminescence signals, we have confined our attentions to fresh, untreated material from herbarium specimens. At the present time, all that is evident is that light is emitted from the surface of the specimens, but until we are able to analyse the spectrum produced, we cannot make use of this as an analytical tool. If, however, it becomes possible to remove constituents of the spore wall, which are not sporopollenin, without damaging or altering the sporopollenin itself, then we will be able to make accurate estimations of the differences between the sporopollenin walls of various plants by the cathodoluminescence signals which they emit.

References

GRANT, P.R., 1971 (in press).

JERMY, A.C., 1970. *Watsonia*, 8, 3-15.

MARTIN, P.S., 1969. *Proc. 2nd.S.E.M. Symposium, I.I.T.R.I., Chicago, Illinois*, 89 - 103

MUIR, M.D., 1970. *Rev.Palaeobotan.Palynol.*, 10, 85 - 97.

MUIR, M.D., & BAKER, I., Palaeogeog. Palaeoclimatol. Palaeoecol., 5, 251 - 268.

MUIR, M.D., GRANT, P.R., & HODGES, G.M., 1971 (in press).

SITTE, P., 1959. Naturforschg. 14b, 575 - 582.

SITTE, P., 1960. Grana Palynologica, 2, 16 - 37.

VAN GIJZEL, P., 1967a. Leidse Geol. Meded., 40, 264 - 317.

VAN GIJZEL, P., 1967b. Rev. Palaeobotan. Palynol., 2, 49 - 79.

Acknowledgements

The authors are indebted to Dr. K.L. Alvin, of the Botany Dept., Imperial College for his help in obtaining and preparing material used in this study. The Natural Environment Research Council provided the funds for some of the equipment used. We are also grateful to Mr. K. Fowler, Portsmouth Polytechnic, for the material of *Acrostichum*.

FIGURE LEGENDS

Figure 1, 2 and 3. *Lygodium* sp., *Pteris* sp. and triradiate polypodiacean spore; Lower Pliocene; St. Helena. All three specimens show fine pitting on the proximal face in the contact areas. Such pitting does not occur on the contact areas of monolete spores. Scanning electron micrographs.

Figure 4. *Deltoidospora* sp.; middle Jurassic; Yorkshire. It is optical micrograph shows a similar pitting phenomenon. The specimen is probably a fern spore.

Figure 5. *Dryopteris filix-mas*. Optical micrograph showing the loosely enveloping perispore around the endospore.

Figure 6. *Dryopteris filix-mas*. Scanning electron micrograph showing the finely granular surface of the perispore (cf. Jermy, 1970).

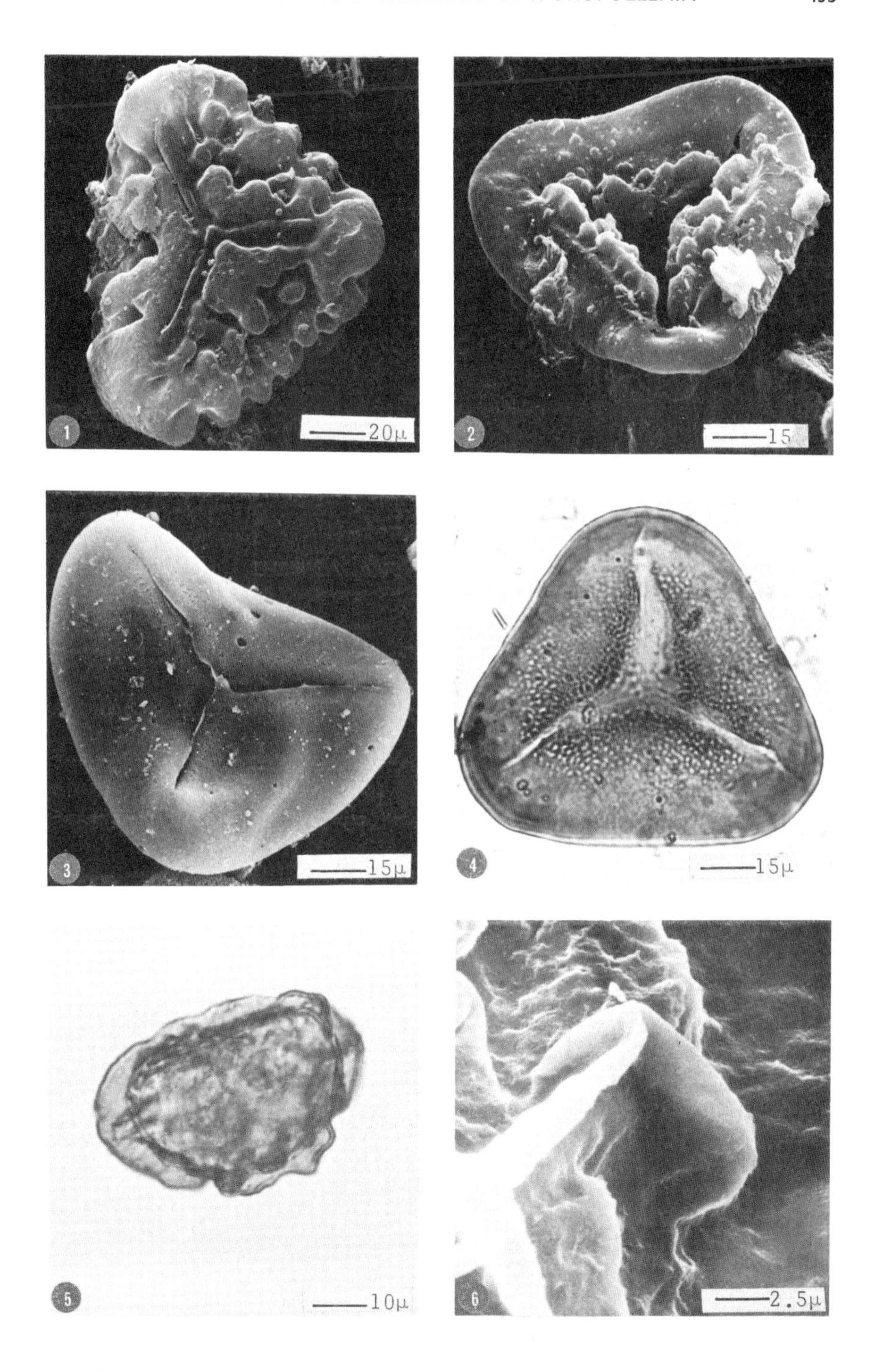
1
20µ
2
15
3
15µ
4
15µ
5
10µ
6
2.5µ

FIGURE LEGENDS

Figure 7. Dryopteris filix-mas. Optical micrograph under crossed-polarisers. The strongest birefringence is in the peripheral region, corresponding to the perispore (see fig. 5).

Figure 8. Dryopteris filix-mas. Scanning electron micrograph to show the perispore collapsed about the endospore.

Figure 9. Dryopteris filix-mas. Scanning electron micrograph of acetolysed specimen from the same plant as Figure 8. Note how the perispore appears to have shrunk and collapsed round the endospore.

Figure 10. Dryopteris filix-mas. Scanning electrom micrograph of fresh material treated in an ultra-sonic probe for 10 minutes. The perispore has been nearly completely removed (specimen on left), only a few remnants of it adhering to the endospore (specimen in centre). The endospore appears to have a certain amount of fine structure.

Figure 11. Thelypteris palustris. Optical micrograph showing the conical projections of the perispore in profile.

Figure 12. Thelypteris palustris. Optical micrograph of several specimens under cros-polarisers. The periphery of the spore is very bright giving a uniaxial figure; the conical projections of the perispore and visible as small, bright, dots.

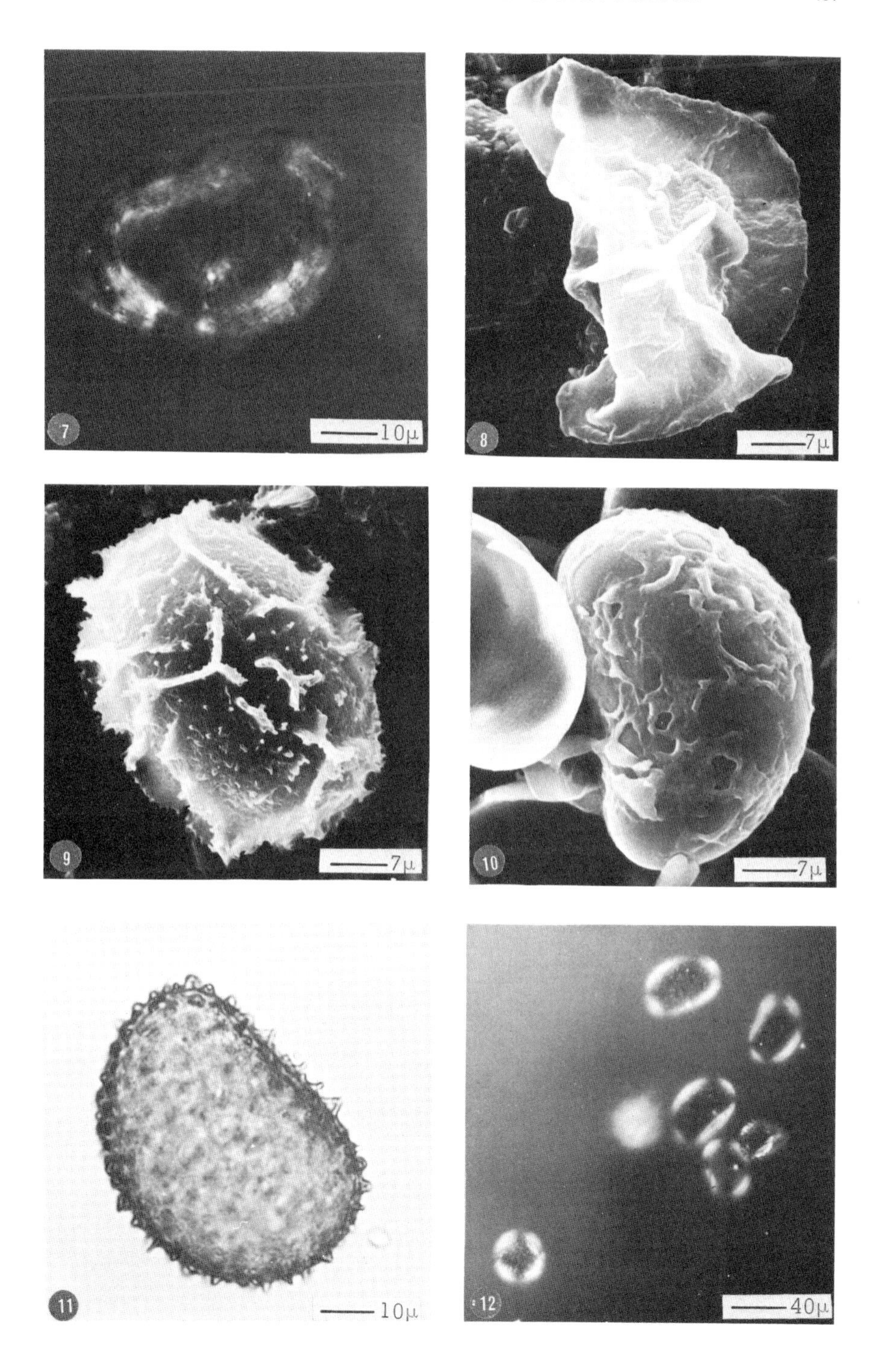
7
10μ
8
7μ
9
7μ
10
7μ
11
10μ
12
40μ

FIGURE LEGENDS

Figure 13. *Thelypteris palustris*. Scanning electron micrograph of untreated specimen. This specimen has *not* been coated with a conducting metal as is normal. It was photographed at 30 kV, after having been dried from an aqueous suspension on the metal specimen holder. The conical projections of the perispore have a series of holes round their bases.

Figure 14. *Thelypteris palustris*. Scanning electron micrograph of specimen treated in an ultra-sonic probe. The conical projections have all disappeared and the lower parts of the perispore adhere firmly to the endospore.

Figure 15. *Thelypteris palustris*. Scanning electron micrograph of acetolysed specimen. Some of the conical projections still remain.

Figure 16. *Thelypteris palustris*. Cathodoluminescence mode photograph of the specimen in Figure 13. Note that while the conical projections are still visible as bright, strongly luminescent signals, the holes are no longer visible.

Figure 17. *Acrostichum aureum*. Scanning electron micrograph of acetolysed specimen showing the curious surface structure produced by this treatment. Non-acetolysed specimens are smooth.

Figure 18. *Acrostichum aureum*. Cathodoluminescence mode photograph of an uncoated, unacetolysed specimen. This species emits much less light than *Thelypteris*.

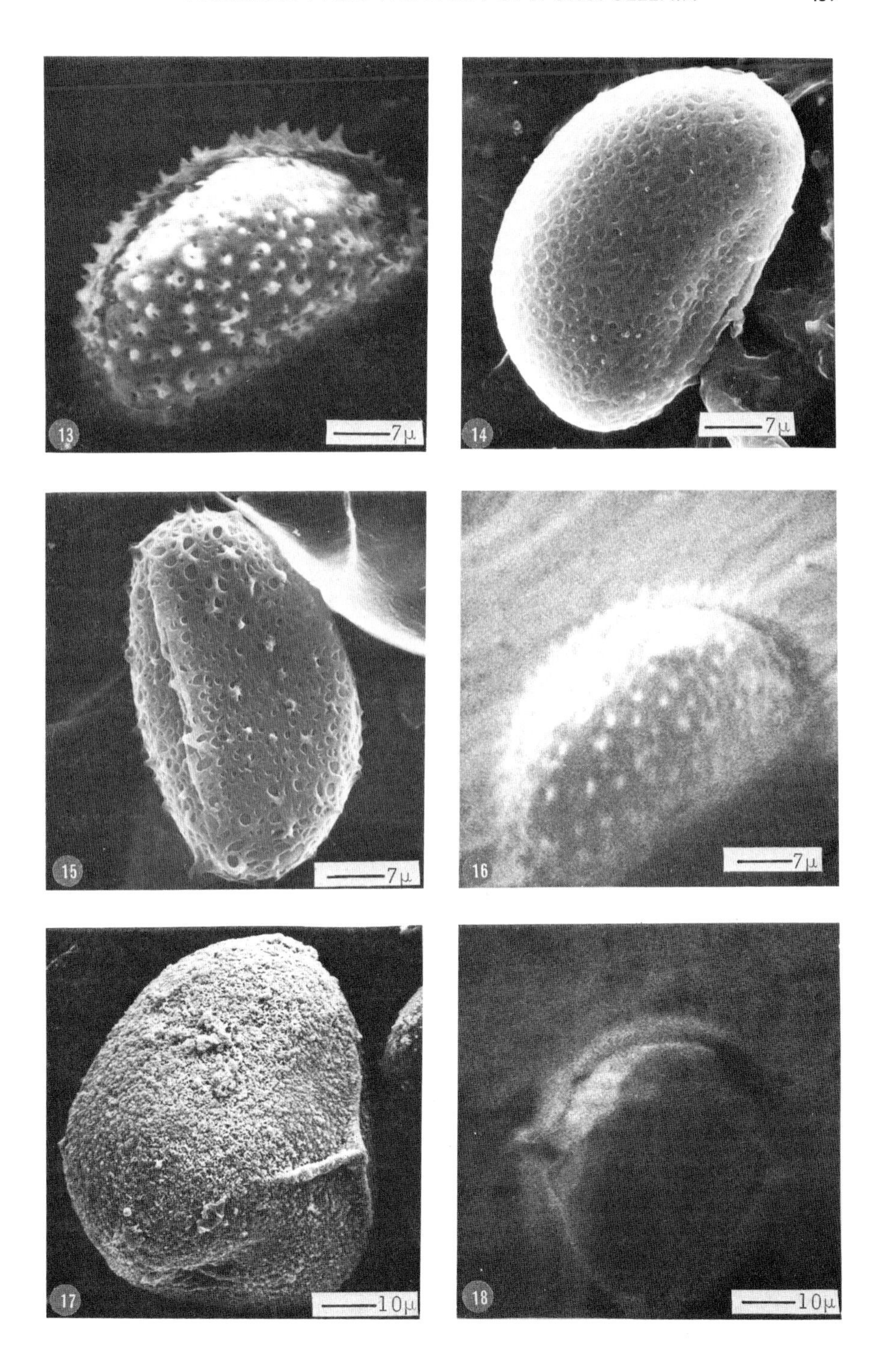
13
7μ
14
7μ
15
7μ
16
7μ
17
10μ
18
10μ

A Scanning electron microscope study of exine structure.

Michel HIDEUX (1)

Scanning electron microscopy has chiefly been used in palynology for the investigation of surfaces (exine and apertures). Ultrasound can break the pollen grains and cause the exine to rupture; then, structural study of the exine is possible. The great depth of field, the three dimensional observations, the localization of exine rupture with regard to the whole pollen grain, the easiness of preparation accentuate the advantages of S.E.M.

Pollen grains are previously treated by a chemical method, then they are treated by ultrasonic vibration, as it has been demonstrated in table 1, it is very important to empty the pollen grains of their cytoplasm and intine to observe them in very good conditions after ultrasound. Three chemical treatments have been used on the same sample : *Montinia caryophyllacea*, to compare them. The acetolysis of Erdtman has been used with good results except when exine is thin and flexible; collapsed pollen grains are obtained after the deposition of an evaporated layer of metal (gold-palladium) in a conventional vacuum unit ; the KOH treatment (10 % KOH) gives more collapsed pollen grains ; the best results with such exine have been obtained by using the acid-hydrolysis method (80 % H_2SO_4).

---

(1) Muséum National d'Histoire Naturelle : Laboratoire de Palynologie de l'E.P.H.E., Laboratoire de Biologie Végétale Appliquée, 61, rue de Buffon 75-Paris V°; Laboratoire d'Ecologie Générale, 91-Brunoy.

After chemical treatment, physical treatment by ultrasound takes place, using a generator from the "Société Piézo-Céram réf. 16 G" and 3 barium titanate transducers that vibrate at determined frequencies; only low frequencies close to the audible are effective to break the pollen grains ($\leqslant$ 40 kc/s); this equipment requires the use of a little airtight glass container with a flat bottom immersed in the transducer tank filled with water (a wetting agent is added to give a better result). Pollen grains can be treated in water or in acetic acid (dilute or concentrated) or in 70° ethyl alcohol. The power of emission is 5.7 watt/cm2.

The frequency of broken pollen grains depends on the physical aspect of the pollen grains. The physical treatment lasts from one minute (*Pinus*) to several hours (Umbelliferae or Saxifragaceae), but with shorter periods a good clearing of the exine surface is always obtained. The phenomenon of cavitation is responsible for the exine rupture, cavitation ceases beyond 70°C. After a short suspension in absolute ethyl alcohol to help the dehydration (equally valid for fresh material), pollen grains are transferred with a "Pasteur pipette" onto a 12 mm diameter cover slip fixed with "rubber cement" on a stub. If a short suspension is necessary to assist dehydration, it is sufficient, because longer periods can produce a softening of exine. The different physical and chemical treatments are summarized in table 1.

Acetolysis resistance was already a well-known microchemical test to detect sporopollenin (Shaw and Yeadon 1966 ; Brooks & Shaw 1968) ; Ultrasound action permits a physical approach to the knowledge of the sporopollenin molecular structure (exine resistance to rupture).

Appendix Plate 1 :

A few aspects of the various final stages in the ontogenesis of the exine : *Montinia caryophyllacea* THUN.

The material observed comes from both young and ripe anthers of flowers collected in South Africa (NBG)[1]. This material is not fixed but simply dried in the herbarium.

On an immature pollen grain (stage following the release of tetrad microspores) in which some microspores are still partially adhered (Fig. 5) an almost continuous superficial layer (Fig.4) lacking differentiation between the apertural zones (intercolpi) is differentiated as a partial tectum in the zones close to the apertures where the reticulum is already visible.

The exine sections of unripe pollen grains obtained by breaking after prolonged treatment with 80 % H_2SO_4 (100°C) show us the structure of the exine at an early stage (Figs.11, 12, 13) : the columellae are short and very dense, their heads are already fused and form a partial tectum. The pollen grains at the same stage of development, but not having undergone any chemical treatment, in situ in the anther (Figs.1, 2, 5,6) confirms the preceding findings : the differentiation of the lumina of the reticulum can clearly be seen (Fig.2). This material seems to play a part in the exine formation : it is very probably of a sporopollenic nature, since it is not affected by acid hydrolysis (Fig.4).

---

(1) NBG. Compton Herbarium abbreviation. Cape Town. South Africa

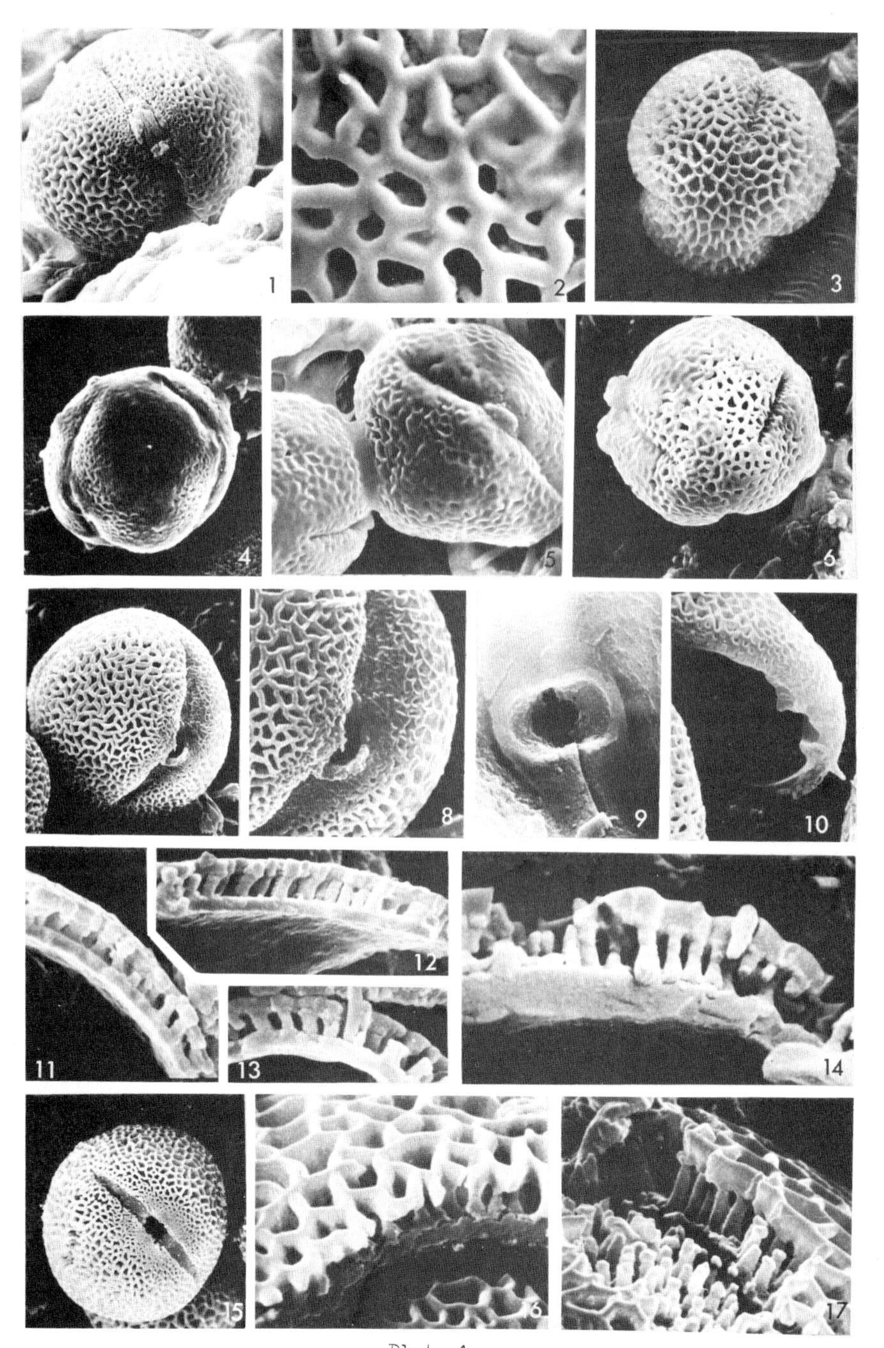

Plate 1

Table 1. Various pollen treatments

Chemical treatment	-	-	-	+ acetolysis	+ acetolysis	+ + prolonged acetolysis	+ 10 % KOH	+ 80 % H_2SO_4
Physical treatment (a)	-	-	+	-	+	optional	+	optional if prolonged action
Dehydration (b)	-	+	+	+	+	+	+	+
Results	(1)	(2)	(3)	(4)	(5)	(6)	(7)	(8)

(a) Ultrasound : 40 Kc/s (b) 70° or 90° ethyl alcohol

(1) Extrusion of intine at apertures. Variability of results: compressed apertures. Metallization not easy.
(2) Coating subtances of exine surface dissolved ; Metallization not easy.
(3) Structural study of exine impossible because expansion of inside material.Ultramicrotomed anthers are necessary (HESLOP-HARRISON I969).
(4) Pollen grains with softened exine are collapsed.
(5) Stereostructural study of exine possible except pollen grains with softened exine who have folded rupture.
(6) Method used for exine resistant to rupture by ultrasound.
(7) Results are mediocre. Exine softened ; many collapsed pollen grains.
(8) H_2SO_4 hardens exine and aspect of apertures remains turgescent like in fresh material ; very good conditions.

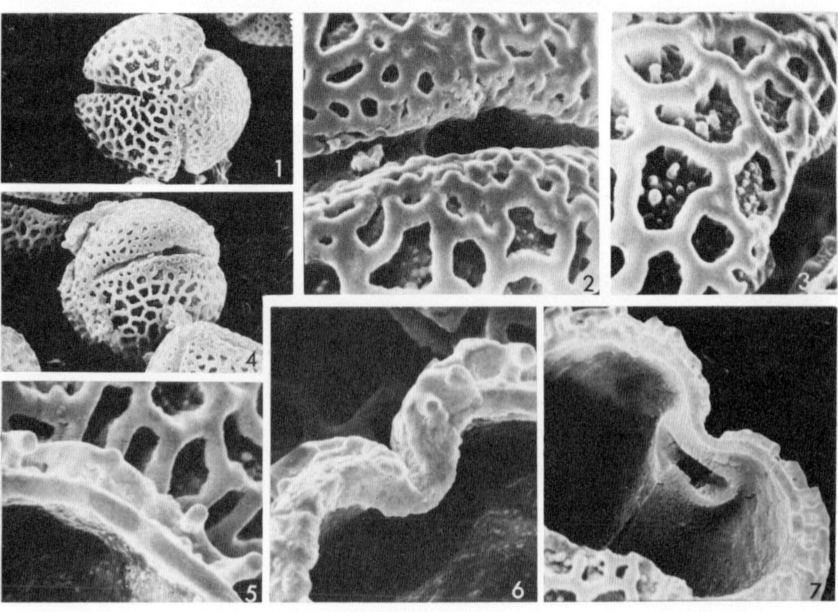

Plate 2

Acknowledgements.

I particularly thank Madame Cerceau who made it possible to carry out this work ; Monsieur Marceau whose long experience I was able to take advantage of and Mademoiselle Derouet who helped me with the photography. I also express my thanks to Professeur Laffitte and Mademoiselle Noël of the Geological Laboratory of the Museum where I was able to have acces to the Cambridge Stereoscan and avail myself of Monsieur Bossy's technical help.

Key to plates :

Plate 1. Montinia caryophyllacea THUN. (NBG)[1]

Scales : Figs. 1, 3, 5, 6, 7, 10, 15. —— = 10 μ
Figs. 8, 9. —— = 4 μ
Figs. 2, 11, 12, 13, 14, 16, 17. —— = 2 μ

Treatments :
fresh material : Figs. 1, 2, 3, 5, 6.
acetolysis and ultrasound : Figs. 14, 15, 16, 17.
prolonged acid hydrolysis : Figs. 11, 12, 13.
acid hydrolysis and ultrasound : Figs. 4, 8, 9.
basic hydrolysis and ultrasound : Fig. 10.

Plate 2. Grevea eggelingii MILNE-REDH. (EA)[2]

Scales : Figs. 1, 4. —— = 10 μ
Fig. 7. —— = 4 μ
Figs. 2, 3, 5, 6. —— = 2 μ

Treatments :
acetolysis and ultrasound : Figs. 1, 2, 3, 4, 5, 6, 7.

Photographs : Laboratoire de Géologie du Muséum National d'Histoire Naturelle.

(1) NBG. Compton Herbarium abbreviation. Cape Town. South Africa.
(2) EA. East African Herbarium abbreviation. Nairobi.East Africa.

Bibliography.

BROOKS, J. and SHAW, G. (1968), Nature Lond., 219, p. 532-533.
CERCEAU, M-T. ; HIDEUX, M. ; MARCEAU, L. and ROLAND, F. (1970), C.R. Acad. Sc. Paris (Ser. D), t. 270, p. 66-69.
HESLOP-HARRISON, J. (1969), Engis Stereoscan Colloquium, p. 89-95.
MARCEAU, L. (1969), Pollen et Spores, 11, 1, p. 147-164.
Mc INTYRE, D.J. and NORRIS, G., (1964), New-Zealand Journal of Science, 7, 2 p. 242-257.
SHAW, G. and YEADON, A. (1966), J. CHEM. (C), p. 16-22.

AN EXPERIMENTAL INVESTIGATION INTO THE DECAY OF POLLEN AND SPORES IN VARIOUS SOIL TYPES

A.J. Havinga, Laboratorium voor Regionale Bodemkunde, Duivendaal 10, Wageningen, Netherlands.

Summary

Mixtures of 19 pollen and spore species were deposited in various soils in 1964, in order to trace their preservation or decay in course of time in different environments.
Provisional results have already been reported at the pollen congress at Utrecht in 1966. Since then decay proceeded only slightly in Sphagnum peat, became more evident in Carex peat and in a podsolized sand soil, and caused a strong or extremely strong deterioration of many species in a river clay soil and leaf mould in a greenhouse. As a result the relative frequencies of the species in the two latter soils shifted considerably. Lycopodium clavatum appears very resistant to microbial attack, whereas Taraxacum and Polypodium show a fairly good resistance. Alnus, Corylus and Myrica disappeared nearly quantitatively. The frequencies of Acer, Carpinus, Fraxinus, Populus, Salix and Ulmus are strongly reduced. An intermediate position is taken by Fagus, Tilia, Juniperus, Quercus, Pinus, Betula and Taxus. Perforation type corrosion is the main cause of the decay of most species in these soil types. The sequence of increasing susceptibility to corrosion obtained from the experiments in the biologically very active soils, river clay soil and leaf mould, differs considerably from the sequence of increasing susceptibility to pure chemical oxidation and that

obtained from published evidence about corrosion in biologically poor soils. The two latter sequences show a good mutual resemblance.

In the podsolized sand soil and Carex peat a gradual thinning of the pollen or spore wall is the normal process, though in the latter soil corroded grains increased in number during the last years.

Besides corrosion and thinning other types of deterioration may be shown more or less frequently. In particular Polypodium shows a great variety in deterioration phenomena, though the spores are not readily destroyed.

Introduction

This paper is a continuation of the experimental investigation into pollen corrosion in different soil types; Sphagnum peat, Carex peat, podsolized sand soil, river clay soil and leaf mould in a greenhouse. A preliminary report was given at the International Conference on Palynology at Utrecht in 1966 (Havinga, 1967). Since then deterioration has strongly increased in the latter four soil types.

The experimental details are as follows: a mixture of pollen and spores of 19 plant species and an inert filling material (sand or clay) was placed in little pouches of nylon gauze. A series of these pouches was deposited in each of the soil types mentioned. These were subsequently sampled at half yearly intervals. For further details see Havinga 1967.

The results of the experiment up to and including November 1969 have been brought together in several tables which will be discussed separately below.

Results

Tables 1 and 2

Preceding the experiment, pollen counts were made from 24 different pouches. From these counts it appeared that mixing of the various pollen and spore species had not succeeded absolutely perfectly. In order to give an impression of the variations in the pollen and spore mixture, average values of 6 counts are given in the collective column 'before the experiment'. These average values are shown in columns 1, 2, 3 and 4. The mean values of all 24 counts are represented in column (1+2+3+4):4. (Table 1a).

The percentages, found after deposition of the pollen and spore mixture in the soils, are given per year, and are arranged according to soil type. The percentages found in only two of the five soil types, namely river clay soil (Table 1b) and leaf mould (Table 1c), are represented in the table because they are the only soil types in which the composition of the mixture has altered in the course of time. The figures shown are averages of two semi-annual counts. The total number of pollen grains and spores counted are given at the bottom of the successive columns (PT).

Representing the original percentages as directly calculated from the counts (PT=100%) would have given a table difficult to interpret.

In order to provide a clear presentation of the data, the original percentages were reduced as follows.

First the percentages were transformed into promilles to avoid decimals.

Then, in order to have a standard, the *Lycopodium* promilles (figures in brackets) were fixed at 100 in all columns of the three tables, and the promilles of the other species rationalized proportionally. *Lycopodium* is the only species fit for such use, because spores or pollen of all but this species became more or less detiorated in the course of the experiment. It may be assumed that no *Lycopodium* spores vanished or became indeterminable.

After this arithmetical treatment a homogeneous pollen and spore mixture before the experiment was "created" by another arithmetical manipulation, that shifting of the ratios during the experiment of any species might be easily compared with that of all other species. To this end all the mean values already rationalised in agreement with the mean value of *Lycopodium* = 100, were now also fixed at 100 (column (1+2+3+4):4), whereas the reduced values in the other columns representing the pollen and spore assemblages before as well as during the experiment, were rationalised again, namely in this way, that they remained proportional to the new mean value of 100.

With the aid of the original promilles for *Lycopodium* (figures in brackets) and the original promilles for the mean values (figures in brackets in column (1+2+3+4):4), the original promille or percentage of each species and data can be recalculated.

Example. The reduced promille of Polypodium in leaf mould in 1969 is 98 (see the Table 1c). This figure is arrived at as follows. The original promille (x) of Polypodium was first multiplied by (100:295) (see Lycopodium in Table 1c), and the original promille of the mean value of Polypodium before the experiment (7, see after Polypodium in column (1+2+3+4):4 of Table 1a) by (100:79) (see after Lycopodium in the same column). 7 x (100:79) was then fixed at 100, and correspondingly x x (100:295) multiplied by (79:7).
x x (100:295) x (79:7) = 98; hence x = 25 0/00 or 2,5%.

The number of Polypodium spores counted is (25:1000)x 2221 (see PT) = 55.

The author did not succeed in finding a suitable mathematical treatment for the data. The variations, as shown, per plant species and per year vary to such a degree, however, that a good impression of the course of the decay may be obtained by mere observation. It must be considered, however, that the fairly wide variations in the table are caused partly by the arithmetical reduction applied. For instance, the figures in column 4 (Table 1a) are considerably lower than those in the columns 1, 2 and 3. This is partly due to the fact that in column 4 the original promille (according to PT=1000 0/00) for Lycopodium is higher than in the other columns (see the figures in brackets). Lycopodium spores were distributed through the pollen and spore mixture less homogeneously than most other species.

In Table 2 the species are arranged according to their

susceptibility to decay in a more comprehensive way than could be done in Table 1. In some cases the position of a species may be argued. *Tilia*, for instance, is more resistant than *Juniperus* and *Quercus* up to and including 1968 (see the figures in Table 1). In 1969, however, its percentage became significantly lower. The percentage decline in the course of time may apparently vary from one species to another.

The sequences in river clay soil and leaf mould are much the same. This does not mean, however, that decay has been equally strong. The figures in Table 1 clearly show that more pollen is lost in the former soil type than in the latter. When discussing Table 5 this phenomenon will be dealt with in more detail.

Lycopodium and *Polypodium* attain the same position at the top of Table 2. Yet if we pay attention to the state of preservation (in 1969) of their spores as established by visual observation, it appears that the former species is by far the most resistant (see Table 4,) It should be mentioned here that we did not find it very difficult to classify *Lycopodium* or *Polypodium* spores (nor the very resistant *Taraxacum*) according to their state of preservation; this is in contrast to many of the other pollen species.

Table 3

As previously stated, shifting of the percentage share of the various species in the pollen and spore mixture took place only in the river clay soil and the leaf mould. It happened with part of the pollen gradually growing indeterminable, to be succeeded possibly, by total destruction. In the soils mentioned, perforation corrosion is prominent in contrast to the podsolised sand soil and peaty soils. In order to establish whether a relationship exists between the disappearance of certain pollen species and their susceptibility to perforation corrosion Table 3 was constructed to show the proportion of grains with a perforated exine in 1965 and 1969 for each species. The data for the other soil types are shown along with those for the river clay soil and leaf mould.

Comparing the figures for the river clay soil and leaf mould in Table 3 with those in Table 2 leads to the following conclusions. *Lycopodium* spores and *Taraxacum* pollen occurring in the upper part of the sequences in Table 2 do not show perforations at all; but *Polypodium* notwithstanding its high position in these sequences exhibits perforations in 1969. *Corylus*, and *Myrica* appear to be the most susceptible species according to both tables (c.f. below). The following differences between the tables are also worth noting. Table 3 shows that *Taxus*, *Salix*, *Fraxinus* and *Populus* are more resistant to perforation corrosion than *Pinus*, *Tilia* and *Fagus*. From Table 2 it is evident, however, that the former four species (i.e. *Taxus*, *Salix*, *Fraxinus* and *Populus*) have disappeared to a greater extent than the other three. This difference may be due to the fact that pollen grains of *Pinus* and *Fagus* are so massive, and to differential thinning effects which were so much more prominent in *Taxus*, *Salix*, *Fraxinus* and *Populus* than in the other two species.

Apparently the susceptibility to perforation corrosion determines the position in the sequences of Table 2 for only some species.

Compared with the situation in 1965, decay resulting from perforation corrosion has increased considerably in 1969. In both years we may establish a fairly good correspondence between the figures of many of the species, along with smaller or greater differences between those of the other species, in the two different soil types. In this connection it is worth mentioning Andersen's (1970) sequence of increasing susceptibility to perforation corrosion from pollen assemblages in forest moss humus, which is as follows: *Quercus*, *Fraxinus*, *Tilia*, *Betula*, *Ulmus*, *Alnus*, *Fagus*, *Carpinus*, *Corylus*. Apart from some differences it agrees fairly well with our experience. Between 1965 and 1969 the ratio of *Tilia* in the river clay soil to *Tilia* in the leaf mould, small in the former year, has increased to unity. Six species in river clay soil and 4 in leaf mould entirely unaffected in 1965 demonstrate a more or less extensive perforation pattern by 1969. Within this group *Polypodium* in leaf mould is most severely affected. *Lycopodium* and *Taraxacum* appear to be the only species entirely unaffected.

A striking feature in Table 3 is that since 1965 the percentages of grains showing perforation corrosion for *Myrica* and *Corylus* show a decline; yet decay proceeded from 1965, with the result that in 1969 only a few specimens of both species per 1000 pollen grains counted were found (see Table 1). Apparently the chance of becoming affected by corrosion lessened as the number of grains became greatly reduced.

In *Carex* peat, perforation corrosion is not an important degradation feature until 1969; as in the case of river clay soil and leaf mould, *Alnus*, *Corylus* and *Carpinus* pollen are very susceptible to corrosion. Some other species easily affected in river clay soil and leaf mould, however, are represented by only low percentages in Table 3 (e.g. *Myrica*). We may probably expect a strongly increasing deterioration of these species in the future.

The differences between river clay soil and leaf mould on the one hand, and *Carex* peat on the other, are so great that it may probably be assumed that different microorganisms have been at work.

Pollen grains with a perforated exine were only exceptionally found in the podsolised sand soil and *Sphagnum* peat. The grains affected belong to the same species as are most susceptible to perforation corrosion in river clay soil and leaf mould.

Table 5

It has been shown above that intensity and type of corrosion is strongly affected by soil type. The data given in Table 5 endorse the conclusion that intensity of corrosion is dependent upon soil type. The state of preservation is classified on visual observation as unaffected, intermediately affected and severely affected. Indeterminable pollen grains are included in the last class. The figures are given for the pollen and spore mixture as a whole at four different times.

Except for river clay soil and leaf mould in November 1969, the percentages in this table were calculated in such a way that possible effects of variations in the quantities of various species, having different susceptibilities to corrosion, were minimised (c.f. Havinga, 1967, p. 91). Such a calculation could not be made for the data from river clay soil and leaf mould in Nov. 1969 because of the very small number of pollen grains of susceptible species counted.

The data show that the proportions for the various preservation classes in the two peaty soils and the podsolised sand soil have altered little since 1965. The constancy in *Sphagnum* peat is in accordance with expectation. Yet it is unlikely that the corrosion process therein will stop in the future because care has been taken that the pouches containing the pollen and spore mixture are always under periodically dry conditions in the peat, this in contrast to pollen in *Sphagnum* peat under natural conditions.

Corrosion in *Carex* peat and podsolised sand soil in fact was more advanced in 1969 than might be suggested from the data in the table. Within the class "intermediately affected", a certain shifting into the direction of the "severely affected" class has occurred. This change is not made manifest in the table because of the criteria chosen in 1964 for the separation of the three preservational classes. The shift is more pronounced in the podsolised sand soil than in the *Carex* peat. The close agreement shown by the presentation of the data is therefore a little misleading.

The stronger deterioration in the podsolised sand soil is caused by thinning of the exine, a process which became significant in this soil after 1966. Perforation corrosion appeared only in *Carex* peat after 1966 and became more important in 1969.

The preservation of individual pollen or spore types often diverges in both kinds of soil. For instance, *Lycopodium* and *Tilia* are well-preserved in both the podsolised sand soil and in the *Carex* peat. *Alnus* and *Taraxacum* are well preserved in the podsolised sand soil; whereas *Acer* and *Myrica* are better preserved in *Carex* peat. In both soils, however, *Taxus* and *Polypodium* are strongly affected. *Quercus* is badly preserved in the podsolised sand soil.

The percentages in river clay soil and leaf mould in November 1969 were calculated directly from the counts, because at that time several species are so poorly represented in the pollen and spore mixture. These species, however, are not wholly unavailable for the calculations of the figures in Table 5, since they are probably represented in the "indeterminable pollen" group. The "indeterminable pollen" contributes to the "severely affected" pollen class in the table. Because pollen grains which have entirely disappeared cannot be included in the tables, the percentages for the "severely affected" class in fact give a more favourable picture of the preservation than is in fact the case.

With the aid of data in Table 1, an estimation of the quantity of the pollen totally destroyed in the course of the experiment can be made as follows.

Comparing the sum of the ratios of all species in the river clay soil in 1969 (100 + 98 + 60 + ... etc. = 427) (see Table 1b) with the sum of the ratios of all species before the experiment (19 x 100 = 1900) (see Table 1a) provides the proportion of the original pollen and spore stock preserved to the extent, that at least the grains could be identified. The proportion amounts to 427 : 1900 = 22%. In the same year a quantity equal to 43% of the total of all determinable pollen and spores (see column "Indet. in % PT" in Table 1b) is classified as indeterminable. This agrees with (43 : 100) x 22 = 9,5% of the original stock. Hence 100-(22+9,5) = 68,5% of the original stock in river clay was lost through complete decay. In the same way in leaf mould the loss was calculated to be 56%.

In agreement with the above conclusion, the pollen and spores, classified as "intermediately affected" appeared slightly more strongly affected in the river clay than in leaf mould. Thinning, in particular, had become more intense in the former soil than in the latter.

So that, as in the case of Carex peat and podsolised sand soil, the figures for river clay soil and leaf mould do not serve as a good basis for mutual comparison. However, they are significant, when compared with the figures for the other soil types.

Lycopodium spores are so very resistant against corrosion that even in the biologically very active soil types (river clay soil and leaf mould) they are not or only slightly affected; in

1969 more than 70% remained unaffected. For *Polypodium*, the only species besides *Lycopodium* showing no quantitative decline (see Table 1), the percentage for unaffected pollen amounts to no more than 10%.(See Table 4).

Table 6

Table 6 compares the time when in river clay soil the extremely corrosion susceptible species(*Alnus*, *Corylus*, and *Myrica*)quantitatively decline with the time where perforation corrosion becomes important. In this table, *Alnus*, *Corylus*, and *Myrica* are represented by their original unreduced percentages (PT = 100%).

As early as November 1965, 20 months after starting the experiment, the major part of the pollen grains of all three species already had perforated exines. At the same time, the amount of identifiable grains of *Corylus* and *Myrica* had declined strongly, in contrast to the group of *Corylus* and *Myrica* not mutually distinguishable. It is assumed that loss from this latter group as a result of decay was compensated for by deterioration of well preserved specimens of the still distinguishable species. The amount of grains belonging to the group *Corylus* and *Myrica* not mutually distinguishable declined in 1966. Supply apparently had slackened because the stock of unaffected pollen had greatly dwindled (cf. Havinga, 1967, p. 90). *Alnus* also became rarer in 1966. The results in leaf mould were similar to those in the river clay soil.

From the foregoing, it is evident that pollen not too severely affected by perforation corrosion may still be suitable for pollen analysis, although the distinction between *Corylus* and

Myrica may be difficult to make.

In *Carex* peat, the situation is quite different. In this soil, perforation corrosion did not appear before May 1968, and only in 1969 was fairly strong corrosion observed (see Table 3). Alteration of the proportional composition of the pollen and spore mixture had not occurred at this time, and the state of preservation of the mixture as a whole was still fairly good.

Corrosion phenomena

The corrosion phenomena observed during the experiment may be found in pollen and spores from a natural site. This is supported by published evidence.

Cushing (1967) studying corrosion phenomena and state of preservation in a lake deposit consisting of subsequent layers of silt, (algal) copropelium and *Depranocladus* peat, distinguished the following deterioration types: corroded (perforation corrosion), thinned, degraded, crumpled and broken. Cushing found that perforation corrosion was most significant in the peaty layer. Degradation dominated in the silt. The latter type, which has not been discussed in this article so far, implies, according to Cushing, "structural rearrangement, so that sculptural and structural details are resolved only with difficulty". In our experiment, obscuring of the structure of the exine (of *Corylus* and *Myrica* in particular) had frequently proceeded to such an extent that the endexine and ektexine had become undifferentiateable. This was observed in all soil types except *Sphagnum* peat.

Crumpled and broken grains were found by Cushing in a small

but fairly constant quantity throughout the profile. We seldom found crumpled grains, but ruptured or broken grains were not infrequent, especially in river clay soil and leaf mould. Lycopodium and Polypodium in particular appeared to be liable to this kind of deterioration. Rupturing was strongly promoted by applying some pressure to the cover slip.

Elsik (1966) describes and illustrates, amongst other things, simple circular to slightly irregular perforations about $\frac{1}{4}$ to 1μ in diameter, and 1 to 2μ in diameter in palynomorphs from older deposits. Similar perforations were fairly frequently found in Polypodium spores, and more sporadically in some pollen species.

Polypodium, in distinction from all other species, shows the widest range of corrosion phenomena. As well as the phenomena already mentioned, we also observed gradual desintegration, complete crushing, and the formation of ball-like protrusions.

In contrast to the Polypodium spores, spores of Lycopodium are extremely resistant to decay. Ruptured or broken specimens only appeared more frequently; besides only some slightly thinned grains were also found. (Only one grain was observed whose exine was locally intensely thinned).

Some of the corrosion phenomena described above are illustrated by the figures 1 - 12.

Some general remarks and conclusions.

Near the wall of a pouch, the spores and pollen grains are more strongly affected than in the centre. Apparently the microflora or fauna causing the deterioration intrudes but gradually. The following figures from two analyses in 1969 of the pollen and spore mixture in a pouch in the podsolised sand soil illustrate this clearly. 12% of the pollen and spores counted from samples near the wall of the pouch were classified as severely corroded, whereas from samples near the centre, the figure was never more than 7%.

Though the quantity of unaffected pollen and spores in the mixture declines to a large extent as time passes, some specimens of even the most susceptible species will always remain more or less intact, or even quite unaffected. For example, of 500 grains counted in May 1969, in a sample from a pouch in river clay, only three belonged to *Myrica*; two of these were classified as "intermediately affected", and the other as "unaffected". In November of the same year, only three *Carpinus* grains were found among 500 grains of other species, and two of these were classed as "intermediately affected", and the other as "severely affected".

All the *Carpinus* pollen in the experiment was gathered from a single tree, so that genetic differences between individual grains would be minimised. The phenomenon referred to above cannot be explained by the fact that the general state of preservation of the pollen mixture increases towards the centre of a pouch because the differences are too small. Variations depending upon the physical or chemical nature of the soil matrix, which might be considered a causative factor under natural conditions do not occur in the pouches.

In the author's opinion the differential susceptibility of individual grains within a species probably depends to a certain extent on the variations in the chemical structure of the sporopollenin, even in the case where the grains originate from the same anther. This is supported by the fact that differences in the state of preservation are sometimes shown by the grains in a reference slide with freshly gathered pollen. Differences in the chemical structure of sporopollenin might depend on the degree of maturity of the pollen grains, or possibly some other factor.

In samples from a natural site, containing affected pollen, the same phenomena may be found as in the pouches. For instance, the author analysed many samples from podsolised sand soils from different parts of the Netherlands. These, like the experimental pouches of this soil type, nearly all showed little or no perforation corrosion; the main process during the often severe deterioration of the pollen exine was thinning. As a result of this process, some of the *Quercus* pollen had even vanished, a fact which was borne out by the pollen analytical data. Whether the relatively intense thinning of the exine of this species in our experiment will have the same effect in future remains to be seen.

Sometimes the behaviour of the pollen in the pouches appeared to differ from that in natural soil. *Alnus* and *Corylus* show much less perforation corrosion in river clay soil than in the pouches. In natural river clay soil, thinning is a much more important process. In agreement with these phenomena is the fact that these species have disappeared much less in the natural clay than in the pouches, although the time of sedimentation in the natural soil was much earlier than the beginning of the experiment.

The fairly good preservation was inferred from a comparison of two different pollen spectra, one from the pollen in the clay near the pouches, which is well-drained and with strong biological activity, and the other representing the pollen flora from another part of the same sediment, occurring in a similar time-stratigraphic position, but where wet anaerobic conditions, guaranteeing good preservation, prevail.

It is worth repeating here, that during his investigation of the lake bottom sediment, Cushing (1967) found pollen strongly affected by perforation corrosion mainly in *Depranocladus* peat (see above). In the sequence *Populus*, *Quercus*, *Ulmus*, *Fraxinus*, *Alnus* and *Betula* the pollen was more strongly affected. This sequence is partly in agreement with, as well as, different from the sequences in Table 2.(c.f. Andersens sequence, earlier)

When analysing the pollen content of a *Carex* peat in the warm, and dry Middle European "Trockengebiet", the author found *Alnus* and *Corylus* far better preserved than *Pinus*, which result contrasts strongly with our experience during the experiment.

The author previously investigated the susceptibility of the same pollen and spore species as in this article to pure chemical oxidation in a slightly acid milieu (Havinga, 1967). The sequence of increasing susceptibility obtained appeared to agree well with a sequence obtained from published evidence on the behaviour of pollen in biologically poor natural soils. It differs considerably from the sequences in Table 2 (this article) which show the decay of pollen in river clay soil and leaf mould which have a rich micro-organic content.

Finally, the results of oxidising a mixture of _Alnus_, _Tilia_ and _Quercus_ pollen at different pH's should be mentioned (Havinga, 1963, p. 29). It has proved that _Quercus_ is most readily affected in an acid milieu, and _Alnus_ in an alkaline medium.

Decay of pollen and spores depends, among other things, on soil type, soil pH, and climate. Each of these factors varies according to place and time, thereby altering its effect on the processes of decay in a way which often varies according to the pollen or spore species. It will, therefore, probably be an impossible target to try to predict the effect of pollen corrosion on the pollen assemblages in all the different soil types used for palynological investigations at present.

Acknowledgements

The author is greatly indebted to Dr. J. Brooks, Dr. M. Muir, and Mr. P. R. Grant who kindly undertook the linguistic correction of the text.

References

ANDERSEN, S.T., 1970. _Danm. geol. Unders.,IIR_, 96,1-99.

ELSIK, W.C., 1966. _Micropaleontology_, 12,(4) 515-518.

CUSHING, E.J., 1967. _Rev. Palaeobot. Palynol._, 4, 87-101

HAVINGA, A.J., 1963. _Med. Landbouwhogesch. Wageningen_,63.1 - 93.

HAVINGA, A.J., 1964. _Pollen et Spores_, 2, 621-635.

HAVINGA, A.J., 1967. _Rev. Palaeobot. Palynol._, 2, 81-98.

Table Ia. Alteration of the pollen and spore mixtures in river clay soil and leaf mould in a greenhouse in the course of time. (1964= November 1964, 8 months after deposition of the mixtures in the soils.).

Pollen or spore species (c.f. Table II)	Ratios					
	before the experiment					
	1	2	3	4	(1+2+3+4):4	
Lycopodium	100	100	100	100	100	
	(67)	(79)	(75)	(97)		(79)
Polypodium	119	108	98	76	100	(7)
Taraxacum	120	104	98	77	100	(37)
Fagus	137	92	89	81	100	(61)
Tilia	132	89	100	78	100	(22)
Juniperus	118	105	106	71	100	(75)
Quercus	120	94	101	84	100	(45)
Pinus	112	100	104	83	100	(51)
Taxus	114	110	105	70	100	(44)
Betula	120	100	101	80	100	(52)
Populus	115	101	111	72	100	(49)
Salix	114	92	114	81	100	(53)
Carpinus	128	106	87	78	100	(46)
Ulmus	114	103	103	79	100	(62)
Fraxinus	111	105	109	74	100	(65)
Acer	128	100	100	71	100	(24)
Corylus	109	90	109	91	100	(85)
Alnus	109	101	103	87	100	(66)
Myrica	118	100	100	81	100	(76)
PT	5705	5919	5673	5881	23178	
Indet. in % PT	0	0	0	0	0	

The figures between brackets are promilles in accordance with PT = 1000 0/00.

Table Ib. Alteration of the pollen and spore mixtures in river clay soil and leaf mould in a greenhouse in course of time. (1964 = November 1964, 8 months after deposition of the pollen in the soils).

Pollen or spore species (c.f. Table II)	Ratios					
	river clay soil					
	1964	65	66	67	68	69
Lycopodium	100 (111)	100 (91)	100 (360)	100 (288)	100 (382)	100 (441)
Polypodium	76	108	140	130	76	98
Taraxacum	87	95	54	95	64	60
Fagus	84	81	29	49	36	42
Tilia	67	107	29	57	37	18
Juniperus	65	153	19	32	27	28
Quercus	46	125	12	36	23	23
Pinus	92	126	9	24	17	17
Taxus	110	81	13	17	13	9
Betula	66	92	16	25	7	6
Populus	83	106	18	10	8	5
Salix	49	83	15	10	5	5
Carpinus	58	50	10	5	5	5
Ulmus	63	90	10	13	3	2
Fraxinus	75	102	11	7	4	3
Acer	55	126	4	6	4	4
Corylus	77	20	9	30	2	1
Alnus	51	58	3	2	1	1
Myrica	44	21	1	1	1	0
PT	1329	955	528	1940	2902	2868
Indet. in % PT	1	16	35	45	39	43

The figures between brackets are promilles in accordance with PT = 1000 0/00.

Table Ic. Alteration of the pollen and spore mixtures in river clay soil and leaf mould in a greenhouse in course of time. (1964 = November 1964, 8 months after deposition of the pollen in the soils).

Pollen or spore species (c.f. Table II)	Ratios					
	leaf mould					
	1964	65	66	67	68	69
Lycopodium	100	100	100	100	100	100
	(83)	(96)	(230)	(268)	(278)	(295)
Polypodium	151	108	119	65	87	98
Taraxacum	104	124	77	54	71	72
Fagus	95	115	51	44	40	45
Tilia	96	150	96	42	46	25
Juniperus	114	149	41	33	38	43
Quercus	89	135	35	36	30	43
Pinus	140	135	36	55	37	29
Taxus	124	123	23	20	23	16
Betula	64	88	29	30	17	8
Populus	102	93	35	19	19	8
Salix	88	74	26	15	14	19
Carpinus	85	82	22	9	18	15
Ulmus	72	68	20	18	17	11
Fraxinus	100	97	24	16	12	9
Acer	100	135	19	26	23	15
Corylus	90	30	8	3	2	2
Alnus	82	47	9	5	3	2
Myrica	69	14	6	4	2	1
PT	894	874	1002	1921	2503	2221
Indet. in % PT	4	17	32	37	47	57

The figures between brackets are promilles in accordance with PT = 1000 0/00.

Table II. Sequences of increasing susceptibility to decay (from top to bottom). (c.f. Table I).

River clay soil	Leaf mould
Lycopodium, Polypodium	Lycopodium, Polypodium
Taraxacum	Taraxacum
Fagus	Fagus, Juniperus, Quercus
Juniperus, Quercus	Tilia, Pinus
Tilia, Pinus	Taxus, Salix, Carpinus, }
Taxus, Betula, Populus	Ulmus, Acer }
Salix, Carpinus, Ulmus, }	Betula, Populus, Fraxinus
Fraxinus, Acer }	Alnus
Corylus, Alnus	Corylus, Myrica
Myrica	

Table IV. Unaffected (u), intermediately (i) affected and severely (s) affected spores or pollen grains of Lycopodium, Polypodium and Taraxacum in river clay soil and leaf mould in a greenhouse, in 1969.

	Percentages					
	river clay soil			leaf mould		
	u	i	s	u	i	s
Lycopodium	77	23	0	72	28	0
Polypodium	10	35	55	13	49	38
Taraxacum	3	72	25	2	67	31

Table V. Unaffected (u), intermediately (i) affected and severely (s) affected pollen grains and spores in the various soil types at different times. (Nov. 1964 is 8 months after deposition of the pollen in the soils).

	Percentages														
	Sphagnum peat			Carex peat			Podsolized sand soil			River clay soil			Leaf mould in greenhouse		
	u	i	s	u	i	s	u	i	s	u	i	s	u	i	s
November 1964	92	8	0	84	16	0	63	37	0	26	61	13	23	60	17
March 1965	78	21	1	48	45	7	45	51	4	-	-	-	21	31	48
November 1965	-	-	-	-	-	-	26	61	13	17	36	47	16	25	69
November 1969	68	31	1	23	60	17	24	58	18	4	26	70	4	27	69

Table III. The proportions of the tested pollen and spore species in the various soil types at different times, the exine of which is affected by perforation corrosion. (1965 = 20 months after deposition of the pollen in the soils).

pollen or spore species.	Percentages						
	River clay soil		Leaf mould		Carex peat	Podsolized sand soil	Sphagnum peat
	1965	1969	1965	1969	1969		
Lycopodium clavatum	0	+	0	+	0	0	0
Taraxacum officinale	0	0	0	0	0	0	0
Polypodium vulgare	0	11	0	38	0	0	0
Juniperus communis	0	11	0	2	0	0	0
Taxus baccata	0	12	0	6	0	0	0
Salix sp.	0	12	2	9	0	0	0
Quercus robur	0	21	0	19	0	0	0
Fraxinus excelsior	0	25	4	14	0	0	0
Populus sp.	5	40	5	57	12	0	0

Acer pseudo-platanus	7	–	9	66	0	0	0
Pinus silvestris	12	57	8	64	2	0	0
Betula verrucosa	15	63	25	48	1	0	+ (1964,1967)
Tilia sp.	14	76	62	76	0	0	0
Fagus silvatica	22	63	24	75	6	0	+ (1967,1968)
Carpinus betulus	42	63	43	53	36	+ (1968)	0
Ulmus carpinifolia	58	79	82	89	1	+ (1964,1967)	+ (1965)
Alnus glutinosa	79	74	60	60	38	+ (1968,1969)	0
Myrica gale	88	–	97	73	8	0	0
Corylus avellana	92	54	96	61	34	+ (1967)	+ (1969)
Myrica or Corylus not mutually distinguishable	97	–	98	92	–	–	–

– = No reliable percentage available. + = 0,4 % or less.

Table VI. Decline of *Alnus*, *Corylus*, and *Myrica*, and the appearance of perforation corrosion, in river clay soil. (July 1964= 4 months after the deposition of the pollen in the soils.).

	Percentages						
	100% = sum of all pollen species				100%=total pollen of each species		
	Portion of the pollen mixture				Corroded pollen		
	J. 1964	N. 1964	N. 1965	N. 1966	J. 1964	N. 1964	N. 1965
Alnus	5.8	3.8	4.2	1.1	0	0	79
Corylus	8.5	8.7	1.8	3.4	0	0	92
Myrica	3.6	5.0	1.8	+	0	0	88
Cor. or Myr. not mutually disting.	3.4	3.5	4.3	0.9	0	0	97

J. = July; N. = November.

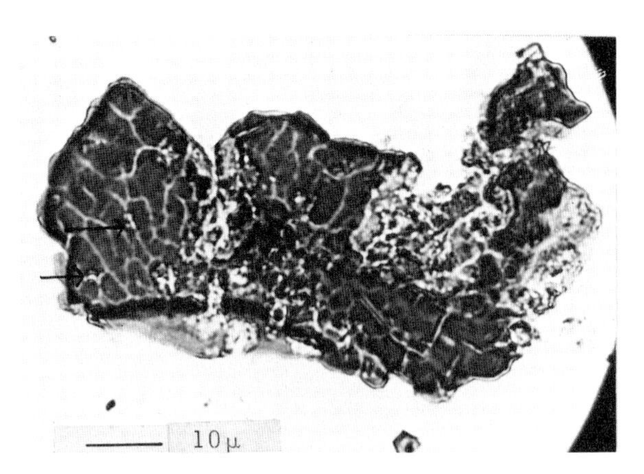

Fig. 1. *Polypodium* spore showing perforation corrosion.

The arrows point at spots where corrosion starts.

(In leaf mould)

Fig. 2. *Polypodium* spore showing moderate fragmentation.

(In leaf mould)

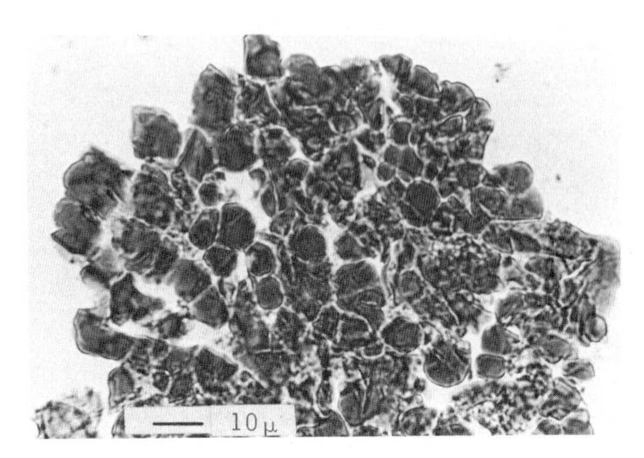

Fig. 3. *Polypodium* spore showing intense fragmentation and corrosion.
(In leaf mould)

Fig. 4. *Polypodium* spore with completely crushed appearance.
(In leaf mould)

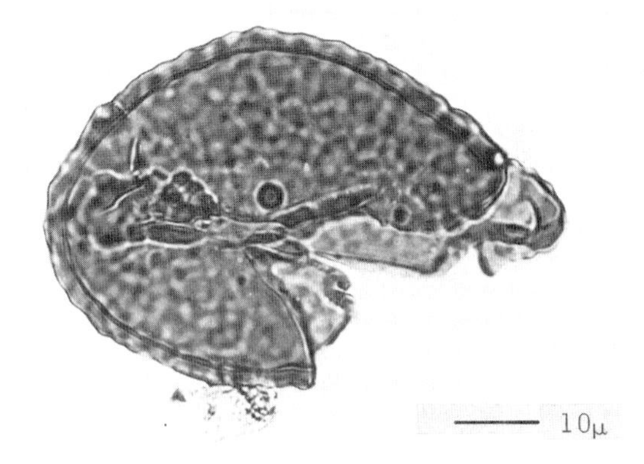

Fig. 5. *Polypodium* spore the wall structure of which is slightly degraded.

(In leaf mould)

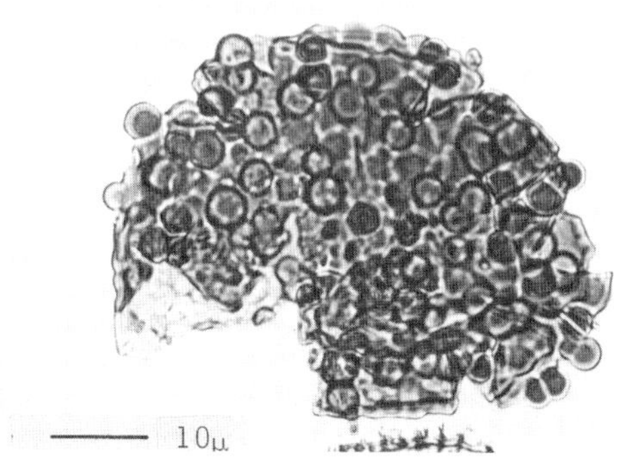

Fig. 6. *Polypodium* spore showing ball-like protrusions.

(In river clay)

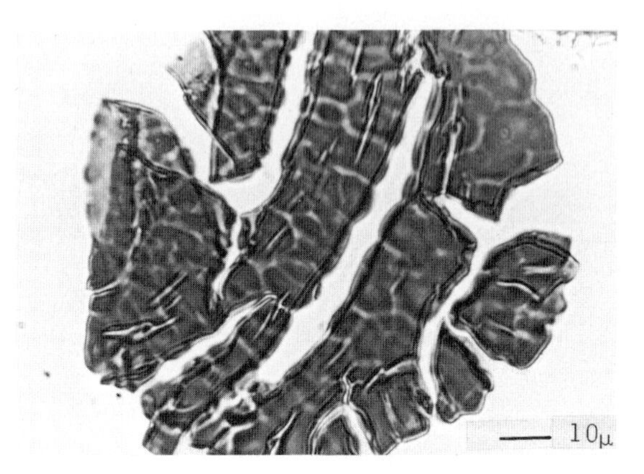

Fig. 7. *Polypodium* spore showing cracks.
(In leaf mould)

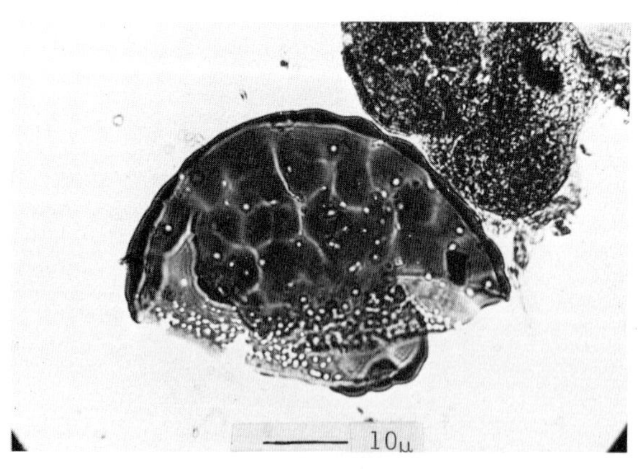

Fig. 8. *Polypodium* spore showing small circular perforations.
(In leaf mould)

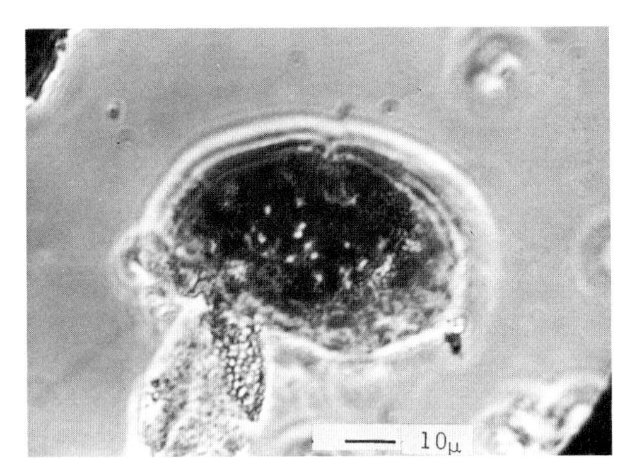

Fig. 9. *Tilia* pollen grain showing small circular perforations. (In leaf mould, phase contrast)

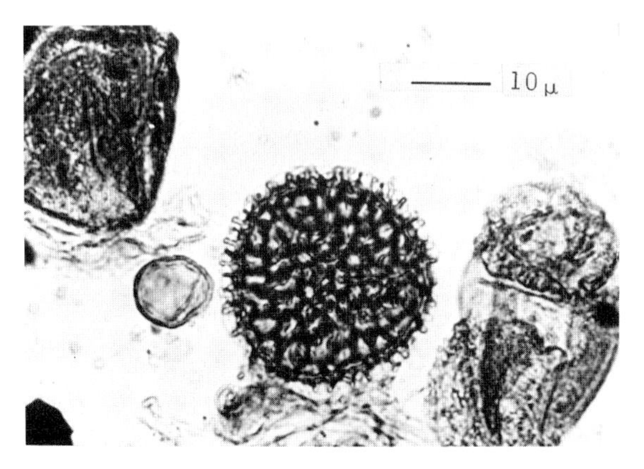

Fig. 10. From left to right. *Fagus* pollen grain showing extreme corrosion, unaffected *Lycopodium* spore, *Pinus* pollen grain showing extreme corrosion. (In leaf mould)

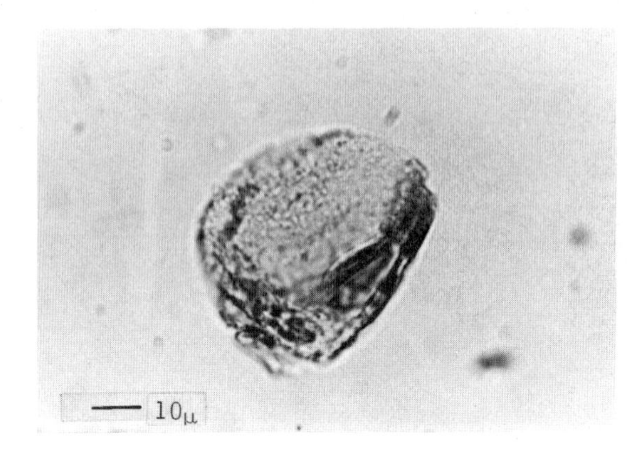

Fig. 11. *Corylus* pollen grain showing extreme corrosion.
(In river clay soil)

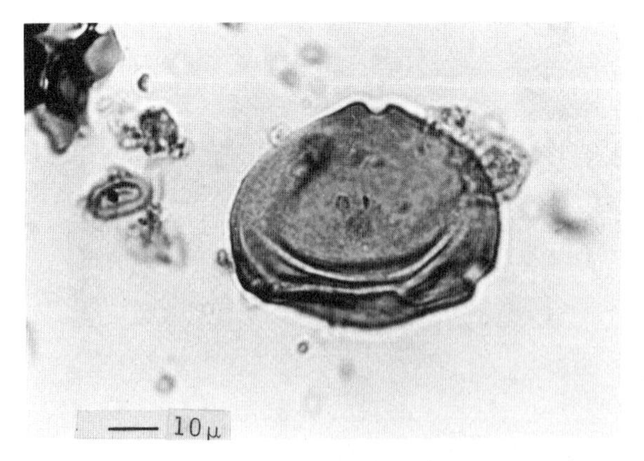

Fig. 12. *Corylus* pollen grain showing intense thinning.
(In podsolized sand soil)

DISCUSSION ON DR. A. J. HAVINGA'S PAPER

HESLOP-HARRISON: After seeing Dr. Elsik's slides, I would like to ask Dr. Havinga if he has also seen scars and rosette erosional structures in his material. If not, how does he explain this?

HAVINGA: Yes I have seen such structures, but I had not time in my lecture to show pictures of them.

HESLOP-HARRISON: Do you think you will be able to identify the causative organisms?

HAVINGA: I don't know, but I will work on it.

HESLOP-HARRISON: Good. It seems to me that we should try to obtain these organisms and culture them and see exactly what they do to pollen grains (what their sporopollinases are like).

CHALONER: Certain fungi commonly grow very easily on fossil spores. Dr. Elsik described how carbonisation appeared to halt microbial activity, but I have often experienced the growth of fungi (probably aquatic phycomycetes) on Carboniferous megaspores kept in distilled water after violent oxidative maceration.

ROWLEY: From Dr. Elsik's photographs, it appeared as if sometimes something had been added to the spore walls. Has Dr. Havinga noticed anything of this nature in his material.

HAVINGA: No. I don't think so.

HUGHES: What were the drainage arrangements for the soils in Dr. Havinga's experiments?

HAVINGA: It was conducted in wet soil in a normal river bank, but not always flooded.

HUGHES: Fossil material was probably flooded all the time.

HOLLEYHEAD: Are the soils acid or alkaline? In leaf mould which is mildly alkaline, the most degradation seems to occur. The soils are in the atmosphere, and along with the alkaline conditions chemical degradation due to oxidation is likely to occur. What are your comments on these points?

HAVINGA: The following pH's occur: Sphagnum peat, 4.1; Carex peat, 6.2; podsolised sand soil, 4.6; leaf mould, 6.5; river clay soil, 7.2. In river clay soil, also slightly alkaline, deterioration is most intense. Deterioration due to pure chemical/biochemical oxidation probably occurs along with perforation corrosion in the biologically active soils. Oxidation is more important in river clay soil than in leaf mould, where thinning is less intense.

MICROBIOLOGICAL DEGRADATION OF SPOROPOLLENIN

W. C. Elsik, Humble Oil & Refining Company, Houston, Texas

Abstract

Microbiological activity may result in definite patterns of removal of the exine from spores, pollen and other organic microfossils. Chemical or physical degradation does not produce these patterns, except in the case of crystal impressions onto or through the exine. Microbiological degradation of sporopollenin which results in definite patterns or scars is attributable to the higher bacteria (Actinomycetes) and true fungi. Other bacteria apparently decompose the exine in an orderly fashion but with no set or recognizable microscopic pattern.

Several types of degradation patterns are demonstrable. More than one type of pattern may be found on an individual exine specimen. Basic research into the exact nature of the attacking organisms and characteristic degradation scars produced by each is lacking. The stratigraphic and environmental significance of their occurrence is not yet evident; but it is obvious that the presence of degraded fossil microspores should be taken into account in any evaluation of a fossil assemblage, especially quantitative studies.

Introduction

Degradation of microspore exines and the sporopollenin which constitutes a large portion of the exine involves the decomposition by biological, chemical and physical means; all of which are often inseparable to some extent in the deterioration of the exine. Biological degradation is that degradation produced or affected by living organisms. The objective of this paper is to discuss and illustrate the degradation of the exine of microspores (pollen and spores) due to microbiological forces which results in definite patterned scars.

The degradation of organic material by micro-organisms has been well documented (Wehmer, 1915, 1925; Rege, 1927; Waksman, 1938; Weston, 1941; Barghoorn, 1942; Sparrow, 1943; Barghoorn and Linder, 1944; ZoBell, 1946; and Gaumann, 1952). Braun's study (1855) of *Rhizophidium pollinis* emphasized the occurrence of that aquatic phycomycete as a saprophyte on modern pollen. Goldstein (1960) reviewed the subsequent discovery of aquatic and soil Phycomycetes which are saprophytes on plant spores in nature. Fossil Phycomycetes from the Triassic of Arizona were described by Daugherty (1941). No clear cut patterns of degradation were produced in the fossil material.

Kirchheimer (1933) described three types of preservation of fossil *Myrica* sp. supposedly found in the same fossil inflorescence: 1) "korrodiert", 2) "desorganisiert", and 3) "hyalin". Although not illustrated in detail, the corroded type had ± patches of exine removed (op. cit., figs. 7 and 11). Disorganized microspores seemed to have layers of the exine separated (op. cit., fig. 12), and hyaline types supposedly retained only a thin, ± smooth vestige of the exine (op. cit., figs. 9 and 13). We are not concerned here with the latter two types; the hyaline type appears not to have been *Myrica* at all, at least not with greatly protruding pores like the other figured specimens. It is obvious that bacteria may have been responsible for the degradation of the triporate pollen in Kirchheimer's paper identified as *Myrica* (op.

cit., fig. 11). Likewise the *Lycopodium* drawn in his figure 30 has patches of exine supposedly completely removed (but impossible to determine from the illustration). Kirchheimer attributed this corrosion to factors occurring during the carbonization process of brown coal, especially in the presence of sulfur.

Kirchheimer (1935) reviewed the literature on degradation or preservation of fossil and modern microspores and added his own observations and interpretations of corrosion found on microspores in brown coals, including the observation that chytrids were also responsible for some of the degradation. Corrosion of *Corylus* pollen was also attributed to microbes. No clear illustrations of the type of corrosion being discussed was given. Even for *Pinus* pollen exhibiting relatively large holes in the exine (op. cit., pl. 8, figs. 1a-c) it is impossible to discern whether the degradation illustrated might have been of the scar type.

Goldstein (1960), and Sangster and Dale (1961, 1964) commented on bacterial degradation of fresh pollen. No definite patterns of degradation were described nor were the exact loci of bacterial attack noted, other than that the bacteria covered the exine and were also found inside the body cavity of the pollen. Goldstein (op. cit.), in his study of degradation due to Phycomycetes, failed to note the shape of the scars produced by the fungi. Goldstein did report that "smoothness of the pollen wall in the region immediately surrounding the penetrating rhizoids and discharge tubes of these organisms suggests that they digest rather than puncture the wall." Goldstein also noted the presence of

Actinoplanes (Actinomycetes) on his pollen material subjected to soil extracts.

Moore (1963b) began a series of reports on the fossil evidence of organic degradation. He refers to an early note of fungi on microspores by Renault in 1900. Moore described fossil coccoid, diplococcoid and bacillary bacteria, singly and in chains, as well as filamentous fungi occurring in degraded microspores. He noted that the distribution of the attacking organisms was controlled to some extent by the structure and ornament of the Carboniferous microspores examined. A degradation of the spore coat along the lines of attack implied definite degradation patterns, but the fact that the degradative organism was fossilized in place obscures the exact morphology of the scars.

The susceptibility of microspores to degradation, either biologically or chemically, was summarized by Moore (op. cit. and 1966) and Havinga (1964). Havinga noted that the susceptibility of microspores to corrosion, either bacterial or through chemical oxidation, is inversely proportional to the amount of sporopollenin contained in the exine. No definite degradation patterns are noted for the bacterial attack; two illustrations of the type of degradation described in this paper were referred to as perforation-type corrosion. Havinga credits the mention of this "perforation-type corrosion" to Andersen; the specimen illustrated (op. cit., fig. 2) however is actually a branching or rosette type of degradation scar. The perforations further appear to have some thin portion of the exine left in their floor and also very faint ridges of ektexine remain.

Coscarelli (1964) noted that cellulose fibers submerged in a marine environment were degraded by bacteria and fungi. The species were not identified. An illustration (op. cit., fig. 4) was published but does not show more than a suggestion of a definite pattern of attack. Shape of the scars was not described.

Elsik (1966a) illustrated the structural control of the arci of *Alnus* in the degradation of that angiosperm pollen. Elsik (1966b) discussed degradation scars found on fossil microspores and speculated on the biologic affinity of the causative organisms; both simple perforations and branching patterns were described.

Havinga (1967) reviewed the literature on the preservation of pollen and discussed the significance of corrosion and susceptibility of pollen to corrosion. Branching degradation patterns were again referred to as "perforation-type corrosion".

Degradation of sporopollenin is accomplished by micro-organisms, in particular the higher bacteria and true fungi. The bacteria and fungi may utilize the components of exine of a microspore and in so doing leave degradation scars. The scars produced are very regular and are of two types according to function. These are 1) perforate scars, produced by organisms primarily degrading the microspore cell contents, and 2) pit, rosette or branching scars produced by micro-organisms primarily involved with degradation of the exine. Some of the latter types, particularly branched scars, may be further restricted to the ektexine, or outermost layers of the exine. The pits, rosettes or more loosely branching scars are regular negative patterns resulting from the removal of sporopollenin.

The degradation scars produced microbiologically should be contrasted to degradation due to crystal growth, the only other means by which definite patterns are imparted or impressed upon the exine. Fig. 1 illustrates patterns produced by mineral crystallization, in this case pyrite. The pyrite has been removed with acid, leaving the degraded exine. Note that the growth of the crystals does not remove the exine, but in fact pushes the exine into folds around the outline of the crystals. This was pointed out earlier by Neves and Sullivan (1964) who also noted that it is generally accepted that pyrite is precipitated by bacteria; the pyrite crystals are formed in the exine at loci of bacterial decay. "The crystals will grow within the spaces formed by the bacterial decay and, in so doing, will penetrate the exine" (op. cit., p. 448). This explanation may be true for crystals growing outward from within the microspore body, but is not acceptable for those originating in the exine; otherwise there need be no exine preserved to be pushed into folds around the crystals. Other minerals as well may be responsible for this type of crystal deformation. Although not as common, rhomboidal forms similar to calcite have been observed, as well as acicular forms of other minerals; these are not impressions as in some cases they are restricted to the microspore exine.

All microbiological degradation scars discussed in this paper are preserved on fossil microspores recovered through normal palynological processing techniques, which is basically treatment of the sample with concentrated hydrochloric acid to remove carbonates, hydrofluoric acid to remove excess silicates (including

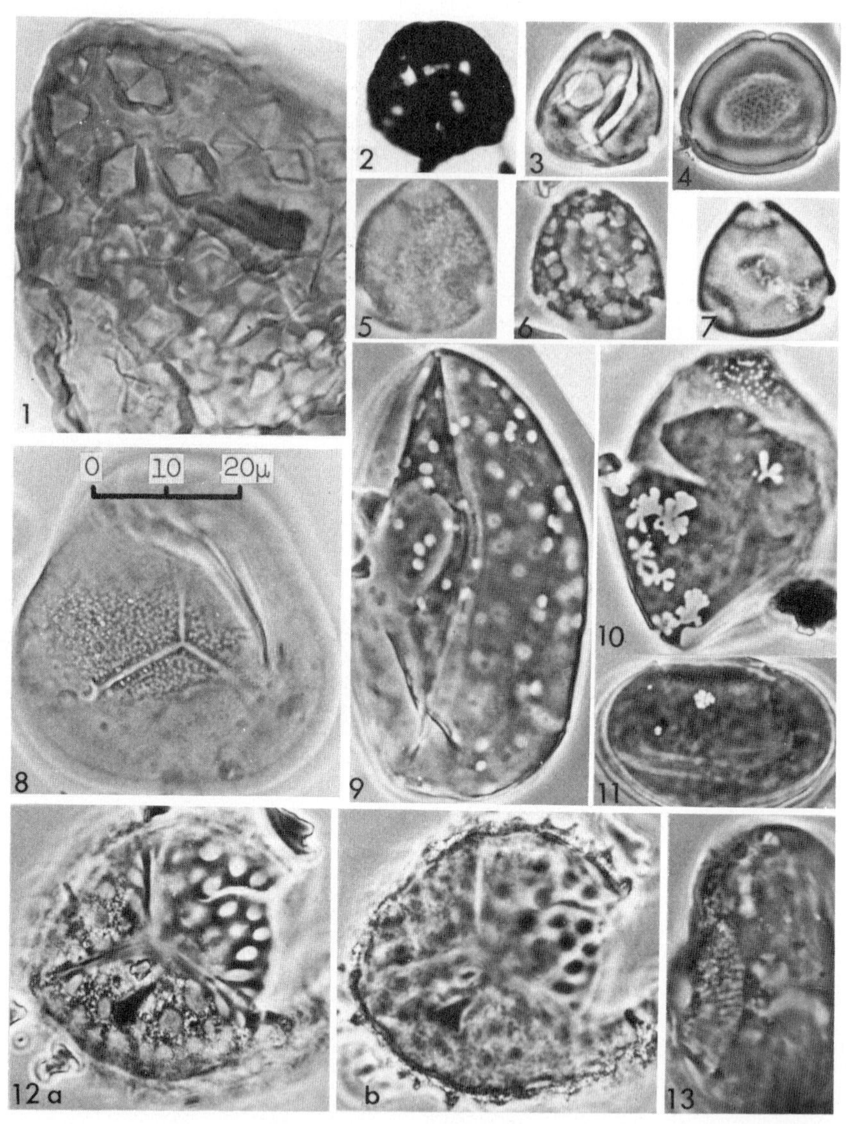

Fig. 1. Crystal degradation of microspore exine, identification impossible due to dense crystal scars. Fig. 2. Carbonized, slightly degraded specimen of *Platycarya*. Figs. 3, 5-13. Microbiological degradation of fossil microspores. Fig. 4. Non-degraded *Engelhardtia* pollen. Note perfect preservation. Ektexine is tectate smooth with scattered granules or papillae.

cements), concentrated nitric acid to remove pyrite and render excess organics more soluble, and concentrated ammonia to decant the soluble humics. Oxidation with Schulze's solution is used only when the organic material is carbonized to such an extent that more drastic treatment is required. Bleaching is hazardous to any sample and is used only to make opaque grains partially translucent to light for identification purposes.

Degradation Scars

The degradation scars described in this paper are found on fossil microspores of many form genera. Examples used are mainly the psilate genera *Laevigatosporites* and *Deltoidospora* and the ± smooth tectate *Engelhardtia* (Fig. 4). This is simply because the scars are much more obvious on such forms and does not imply (as in Havinga, 1964) that they are more common on psilate microspores.

Three types of scars were described by Elsik (1966b): 1) simple circular to slightly irregular perforations, ± ¼ to 1 micron in diameter (op. cit., pl. 1, figs. 1-3) and 1 to 2 microns in diameter (op. cit., pl. 1, fig. 4); 2) wedge-shaped perforations arranged in basic rosettes of four or more wedges with their narrow ends joined, often modified by further branching from the broad end of each wedge (op. cit., pl. 1, figs. 5-16); and 3) a meandering, branching groove ± ½ micron wide, with no particular symmetry (op. cit., pl. 1, fig. 17).

Four types were described by Elsik (1968a): Type 1, "Circular perforations through monolete and trilete microspore walls. Diameter from 0.2 to 1.2 microns. Perforations taper in longitudinal section and are widest at their outer end"; Type 2, "Circular to irregular perforations 0.6 to 2.0 microns in diameter. Both types 1 and 2 may occur on the same spores"; Type 3, "Irregularly branching systems composed of wedge shaped elements or grooves cut into sporomorph walls. Wedges ca. 0.5 micron at narrow end, 0.6 to 1.6 microns wide at opposite end and 1.6 to 2.4 microns long. Branching systems developed from basic rosettes of wedges joined at their narrow (ends). Systems randomly dichotomizing. Individual systems to 12.5 microns long. Entire spore walls may be involved in the distribution of these branching systems"; and Type 4, "Similar to type 3 but wedges never wider than 0.5 micron. Length of individual wedges to 1.5 microns. Entire branching system may be denser than type 3. Generally only one layer of the exine is involved. Limited to angiosperm pollen."

Basically, however, there are two types of scars produced by degradative organisms: 1) ± circular perforations of very small size through one or more layers of the exine, and 2) scars produced by branching or rosette individuals or colonies of bacteria and fungi. The circular perforations are ± 2.5 microns or less in diameter and generally penetrate the entire exine. These are due to penetrating rhizoids and exit tubes of Phycomycetes. Perforations 1 micron or smaller may also be ascribed to the bacteria, in which case only one layer of the exine may be affected. These smaller scars may be very densely developed over the exine or layer affected; endexine, ektexine, or separate layers thereof. This type of degradation grades in size downward to include gradual fading of the ektexine and endexine, which may be due not only to bacterial action but oxidation or solution due to other causes (Fig. 5). The specimen in Fig. 5 is a Paleocene *Triatriopollenites* exhibiting obscure to definite pitting of the ektexine and endexine. The pits do suggest bacterial degradation due to their circular shape.

Layers of the exine may exhibit differential susceptibility to degradation as in Fig. 8, which illustrates a perforation of a middle layer of the exine of an originally psilate trilete spore. The regular size and spacing of the cavities suggest actual ornamentation, but the surface layer remains psilate. Another example of this type of degradation is the *Anthoceros* microspore illustrated in Figs. 12a-b. This specimen is degraded only over two-thirds of the proximal portion of the microspore (Fig. 12a). Further, the innermost layer which corresponds to the endexine is not affected (Fig. 12b). Similar types of degradation occur in fossil Schizaeaceous grains (Elsik, 1968a, pl. 12, figs. 7-9; pl. 13, figs. 1-2).

Another mode of occurrence of the perforation scar is in the area of the laesurae of monolete and trilete spores. The upper portion of the specimen illustrated in Fig. 10 represents a broken trilete spore. Perforations are developed in an irregular area between the laesurae. Since these areas often have different staining characteristics as well, it is probable that exine composition plays a controlling role in the location of the perforations. This is especially evident in some species of fossil microspores referable to *Polypodium* (op. cit., pl. 7, fig. 6), where the perforations are developed in the grooves between the verrucae.

The second type of scar which is more easily related to microbiological degradation is generally limited to one layer; in the specimens illustrated here, the ektexine. If the endexine is very thin or absent, the scars appear to perforate the exine. Study subsequent to Elsik (1966a) has led to a re-evaluation of the original concept of ± circular perforations. The larger circular to irregular perforations illustrated in figure 4 (op. cit.), and equivalent to the type in Fig. 9, are upon close examination at high magnification revealed to be the scars of very small patch or rosette type. The edges of the scars are irregular corresponding to the wide ends of closely spaced ± wedge shaped elements much more obvious in larger scars of this type (Fig. 19). Further, very faint ridges radiating from a common center of raised material are sometimes ascertained (Figs. 6 and 14); this material is all that remains of the ektexine in the area of the scar. In fact, many of these degraded exines are not perforate. The exine is not completely perforate unless the wall is homogenous or the inner, more resistant layers weakened by enzymes destroying the ektexine, which results in further oxidation or bacterial decay of the endexine. In other words, Elsik (1966b) referred to at least two types as "perforations", when in fact the scars here do not fully penetrate the exine, but only one or two layers. This was corrected in part (Elsik, 1968a). The lack of endexine in a microspore thus attacked would result in an irregularly circular perforation produced by an organism that would have otherwise produced a rosette type scar.

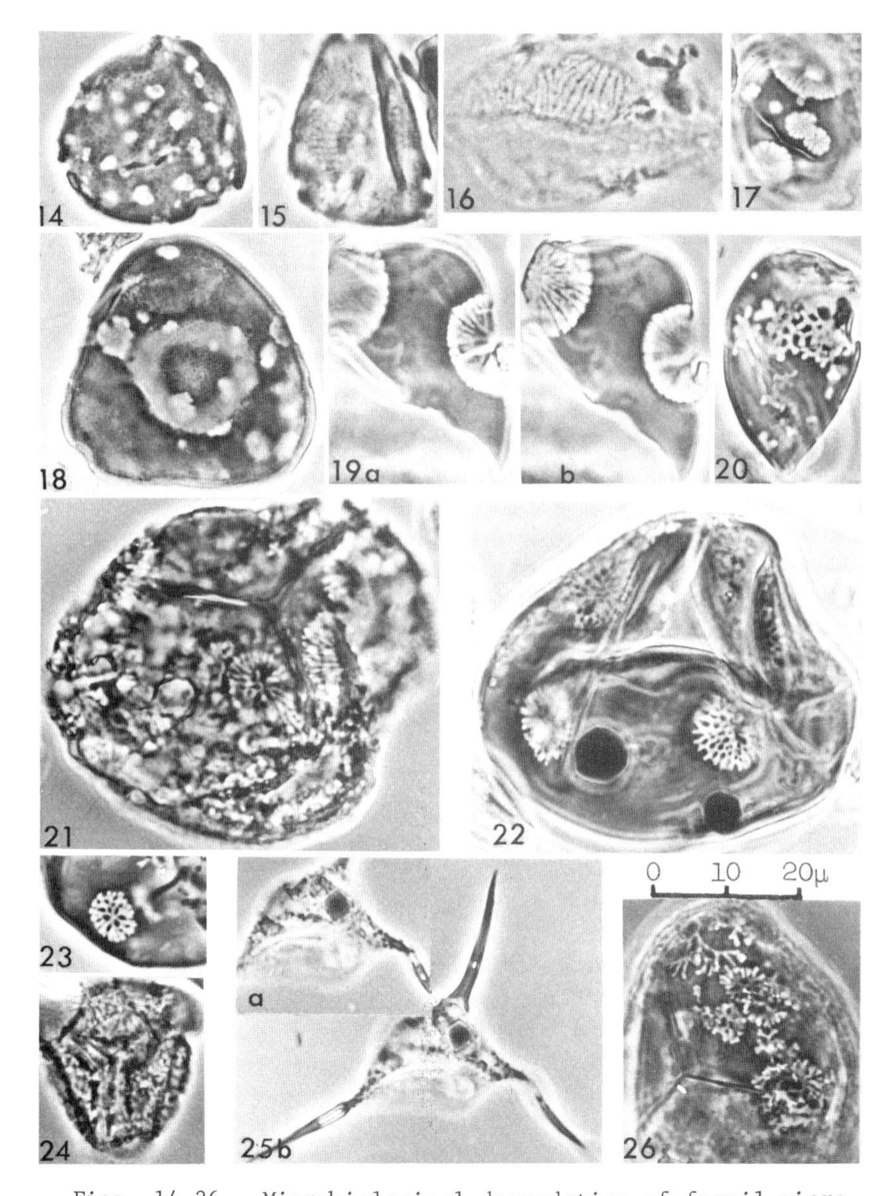

Figs. 14-26. Microbiological degradation of fossil microspores. Patch or rosette type scars due to Actinomycetes or fungi Fig. 14. Triatriate *Engelhardtia* pollen degraded to appear polyforate. Fig. 15. *Engelhardtia* pollen completely degraded except for ridges of ektexine and relatively undisturbed endexine.

The overall size of the rosette or branching scar may be relatively constant, or variations may be exhibited as in Figs. 11, 17 and 18. The shape of the individual wedges also varies to some extent within each rosette; greater variation is exhibited by rosette type scars which may be used to separate them morphologically into specific types (Figs. 15-16 vs. Fig. 17 vs. Fig. 19 vs. the wide wedge type in Figs. 10 and 28). These may or may not have a positive center of raised material. More openly branched types sometimes have a common center of origin (Figs. 20, 22-23) or are linear in development (Fig. 7; top of Fig. 26; Fig. 29; and Elsik, 1968a, pl. 1, figs. 10-11). Individual elements may or may not be straight or curved and wider at the terminal ends of each segment.

One type of scar actually seems to have been produced by a colony of individual microbes (Fig. 30). This illustration is of a specimen of leiosphaerid type with a relatively thick wall. Examination of the scars at different levels of focus reveals that the base of the scar is composed of material removed in a ± radiating manner from a common center; the grooves end at different positions at various levels of focus. This type has been seen only on this one specimen.

That apparently different type scars on the same specimen may actually represent arrested growth stages is suggested by Fig. 11. In Fig. 11 the upper left scar is a small ± circular area. The lower left scar is about twice as large and is irregular with a hint of a faint ridge of ektexine across the bottom of the scar. The larger scar on the right is definitely lobed with faint ridges of exine between some of the lobes. The ridges join and circumscribe a centrally located vacant area, perhaps corresponding to the original initiating scar.

Fig. 11 may on the other hand represent a situation similar to that of Fig. 13; at least two types of perforations and two types of branching rosette scars are present on this specimen of *Laevigatosporites*. Two types of rosette scars are exhibited by another *Laevigatosporites* (Figs. 32a-b). On one portion of the microspore only linear ridges of the ektexine remain (Fig. 32b). Although only faintly discernible, an endexinous layer is present in undamaged condition, for otherwise a microspore with this state of degradation would not retain its shape, nor would the thin ridges be held in place.

Identity of Causative Organisms

Positive identity of causative organisms for the scars is lacking as there are no "fossils" occupying the scars. This may be due to 1) the maceration process or 2) the nonresistant chemical nature of the organisms. A clue has been furnished by Moore (1963b), however, in that both 1) single cells and chains of cells of bacteria and 2) fungal-like mycelia have been found occupying similar type scars in Carboniferous material. We have only 1) the shape of the scars and 2) the observed position of growth of extant bacteria and fungi as further clues.

The true bacteria are not considered to be causative agents due to 1) small size and 2) nature of deterioration produced. The exine in several studies has been noted gradually to "fade", no perceptive pattern being produced. If no set pattern is obvious, or if

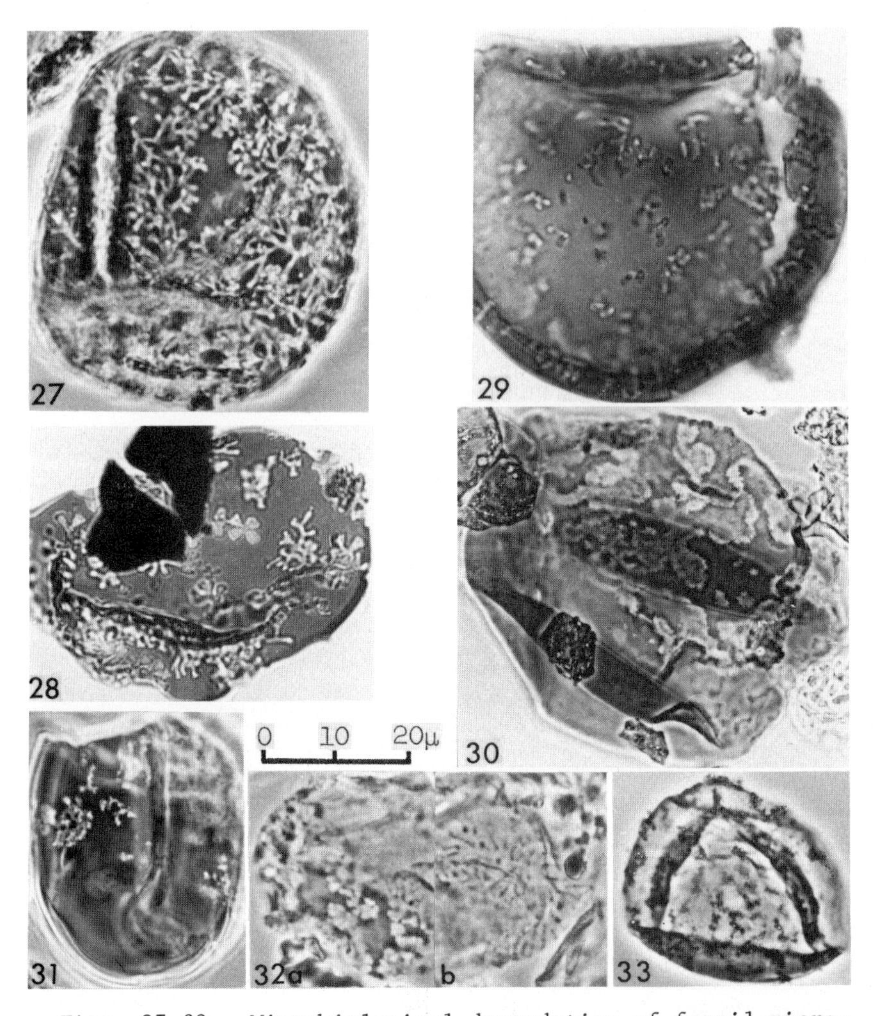

Figs. 27-32. Microbiological degradation of fossil microspores. Figs. 27-28. Predominantly linear branching scars. Figs. 29 and 31. Linear branching scars developed only in the ektexine of psilate spores. Fig. 30. Scars due to bacterial colony? Magnification 1/2 that of other figures. Fig. 32. Two types of scars on the same microspore. Note patches of untouched psilate exine. Fig. 33. Coccoid bodies attached to outer surface of Taxodiaceous microspore.

punctae or structurally controlled perforations ensue, the nature of the deterioration could be ascribed to either bacterial or chemical agents (as in Fig. 5). The origin of all perforate scars is therefore not obvious from microscopic examination. It is also possible that we are seeing the result of differential oxidation of rods of less resistant material in the exine in some cases, particularly where the degradation is limited to one layer of the exine (Fig. 8).

The penetrating rhizoids and exit tubes of Phycomycetes and other penetrating fungi and algae are undoubtedly responsible for some of the perforations. Miller (1955, 1961, 1965) in a continuation of his studies of fresh water Phycomycetes has shown spores encysting on the wall of host cells (often pollen grains), but in all cases only simple punctae at the point of entry are indicated. Further, an extremely small diameter for the entry point (punctae) is suggested. The zoospore of *Micromyces grandis* averages 2.3 microns and the penetrating rhizome is much thinner (Miller, 1955). The chytrids generally develop a thallus within the host cell, and when discharge sporangia are formed the host wall is again penetrated. These latter perforae may be ± 0.5 to 2.5 microns in diameter, judging from the drawings of Miller (1961). *Liquidambar styraciflua* L. was used as bait and was seen to be infected by several different fungi; the fungi in turn were infested with parasitic fungi (Miller, 1965)! Miller suggests that the penetration of a host wall by the discharge tube of some sporangia are in fact "a combination of puncture and digestion" (op. cit.). The entry rhizome is also usually more swollen over the surface of the pollen grain, suggesting the possibility of enzymatic attack and attachment not normally visible on specimens in living media under light microscopy.

The rosette and branching scars probably represent the differential degradation of microspore exines by rhizoids of Actinomycetes and true fungi. The rhizoids may prefer a sporopollenin substrate (species may be specific for sporopollenin or related exine components; perhaps more commonly ektexinous specific). The branching, radial scars may also be the result of direct growth of a bacterial or fungal colony.

Gwynne-Vaughan and Barnes (1937) made several observations about fungi, which are pertinent to this discussion: 1) Endophytic fungi may send short branches of hyphae into the cells being attacked. The branches may "become specialized as haustoria of limited growth and definite form" (op. cit., p. 20). 2) Ectophytic fungi may also send haustoria into the cells of the host, and in the case of *Stigmatomyces* a short, pointed haustorium penetrates only a short way into the chitinous covering of the parasitized insect, the suggestion being made that chitin is decomposed and used as food (op. cit., p. 276). 3) Hyphae of Ascomycetes and Basidiomycetes enter also the non-living elements of wood, penetrating the walls of the cells and delignifying them. The middle lamella of the wall "is dissolved...The enzymes responsible for these changes spread in a plane parallel to the walls" (op. cit., p. 14).

Modern studies of Actinomycetes have furnished evidence for the range in size and morphology of this group of higher bacteria.

Unfortunately, most of the forms studied are important from the standpoint of animal pathogenicity, and as such, effect upon host tissue is not applicable in this discussion. Further, the organisms are studied from cultures grown on artificial growth media.

Simply branched *Actinomyces* were illustrated by Howell et al. (1959) similar to the scars in Fig. 29. Maximum size for individual branched specimens was 8 microns for *A. israelii* and 10 microns for *A. naeslundii*. Each segment exhibits ± equal width for its entire length. Relative length of individual segments is variable. Pine and Hardin (1959) illustrated *Actinomyces israelii*, of perhaps a different strain, which possessed somewhat wider elements in a linear, simply branched system up to 7 microns long. Most of the specimens were bacillary in shape, but this may have been due to the growth stage studied. In addition, entire colonies with smooth margined, circular shapes up to 200 microns in diameter were illustrated.

Nocardia salivae, another Actinomycetes, forms similar branching filaments 0.5 to 1 micron wide and several microns long (Davis and Freer, 1960). Further, the colonies of this species, which may reach 725 microns, exhibits different colony form depending upon the type of constituent cells. The smooth and granular colonies are composed mainly of coccoid and short types; the extremely rough colonies which may be sunk into the substrate are composed of more filamentous cells. This latter type as illustrated by Davis and Freer is similar to Fig. 17.

It is apparent that the cells of different strains of the same species will exhibit different growth forms. Umbreit (1962) noted that the same organism grown in different growth media would adapt different growth forms, all of which would revert back to the original form when replaced in the original habitat. Examples given were *Butyribacterium rettgeri*, single bacilli of which in a changed medium developed budded and even bifid branches. This branched form was similar to another form identified only as "bifid bacteria" (op. cit., p. 197) which formed branched mycelia with swollen tips very similar to the wide wedge-shaped scars in Fig. 28.

Pine, Howell and Watson (1960) illustrated colonies of *Actinomyces bovis* 17 to 77 microns and larger in diameter. The growth form of the larger colonies in particular is very similar to that of the scars in Figs. 15, 16 and 19. The ektexinous ridges left in the scars may therefore be unoccupied areas analogous to the linear depressions on the upper colony surface. That this type scar is produced by a colony and not a radially branched individual organism is suggested by the fact that some of the ektexinous ridges are seen to be branched; if an individual all of the grooves should be connected.

Poindexter (1964), in a review of the *Caulobacter* group of bacteria, illustrates radial colonies of bacteria connected at the center by the base of their stalks. The colonies may be very dense and reminiscent of Fig. 30 (op. cit., fig. 5). Poindexter notes that when "stalks are present, they are characteristically all of equal length....component cells of a mature rosette adhere to it uniquely through attachment at the tip of the stalk to the common holdfast material" (op. cit., p. 258). Whether or not the stalks are of equal length is debatable, as one illustration (op. cit., fig. 1) has individual stalks attached to other stalks at a distance from the holdfast material greater than one stalk length; a branched colony is exhibited. This type of growth with a central holdfast might explain radial scars on sporopollenin in which there is a centrally located knob.

The Actinomycetes have been reviewed by Lechevalier (1964), who is of the opinion that "Actinomycetes are bacteria with fungal morphology" (op. cit., p. 233). The lack of a nuclear membrane is cited as evidence of strong affinity to the bacteria. However, the blue-green algae (Cyanophyta) also lack nuclear membranes (Smith, 1951). Basic generic classification of the Actinomycetes is based upon 1) presence or absence of sporangia, 2) whether the sporangia are borne on substrate or aerial mycelia, and 3) types of spores. Shape of substrate mycelial elements apparently are either not important in the classification of Actinomycetes, or perhaps are too variable within a genus under this classification; they are therefore not discussed by Lechevalier.

Other Bacteria

The fossil coccoid bacteria illustrated in Fig. 33 require some discussion. Small coccoid and bacillary objects attached to Taxodiaceous microspores, including *Exesipollenites* and *Spheripollenites*, are common in assemblages of Jurassic and Early Cretaceous age from the Gulf Coast of Texas. The coccoid form is most common in Gulf Coast material and is irregularly distributed over the microspore surface, sometimes forming very densely crowded areas. The bacillary form is less common and may be found singly or in limited patches on the same fossil microspore as the coccoid form.

There is no shape or size gradient between the two forms. Further, both types have been found attached to a specimen referable to the incertae sedis *Schizosporis*, of no genetic relationship to the Taxodiaceae. Any degradation scars that might have been produced by these bacteria would be very small and either 1) hidden underneath the bacterial cell or 2) not detectable with the light microscope. Further, the possibility exists that these fossil bacteria used the exine only as a place of attachment, deriving nutrients from other than the sporopollenin. If true bacteria, these objects represent types with very resistant walls, having survived the maceration process.

The coccoid type have been referred to in the literature as Ubisch bodies or orbicules when found on Taxodiaceous and related gymnosperm pollen. Gamerro (1968) described Ubisch bodies (orbicules) from male cones of Lower Cretaceous gymnosperms which include pollen of the form genera *Classopollis*, *Applanopsis* and *Trisaccites*. The bodies attached to the individual pollen grains appear somewhat irregularly spherical in shape and are perhaps analogous to the Gulf Coast material. The bodies preserved on the tapetal membrane of the cones, however, exhibit too diverse a range in size and shape and are not the same as the bodies illustrated in Fig. 33. Evidence that the bodies illustrated here are actually bacteria are 1) both coccoid and bacillary types may occur on the same affected sporomorph and 2) both types, and especially the bacillary bacteria, have been found on other than Taxodiaceous microspores, including the cysts of marine dinoflagellates.

Discussion

All microspores are subject to both bacterial and fungal degradation. Definite degradation scars are produced only by the higher bacteria (Actinomycetes) and fungi. These scars are not restricted to microspores and have been observed on microfossils by other investigators (Elsik, 1968a). Other fungal spores, algal spores and cysts (including dinoflagellates), bryophyte and pteridophyte spores, gymnosperm and angiosperm pollen, plant cuticle, incertae sedis (including acritarchs and animal egg cases) and chitinous linings of microforams may all be microbiologically degraded and scarred to some extent.

It is possible that soil organisms produce these scars on fossil material exposed at the surface but equal numbers of scarred specimens have been recovered from subsurface well cuttings and cores. Further, the scars have been observed on highly carbonized pollen (Fig. 2). Carbonization of microspores requires some degree of heat and pressure, generally due to depth of burial. It is believed that microbiological degradation would occur only before carbonization of the microspores. That the patterns may be caused by thermal alteration alone may be discounted as 1) the scars may be found on relatively fresh grains exhibiting no color change due to carbonization (Fig. 3) and 2) brownish orange to brown and darker colored microspores show no increase in presence of scars over that found in unaltered material and in fact some pollen carbonized to an opaque black condition exhibit no such scars.

The problem presented by the presence of degraded microspores in a fossil assemblage is two-fold: 1) misidentifications may be made or morphology misinterpreted by the unwary and 2) an additional bias must be allowed for in quantitative evaluations; the actual occurrence of an important constituent of the fossil flora may be masked. Examples of the former are *Punctatisporites granulatus* Roche (1969) which is a psilate trilete spore exhibiting patches of scarred ektexine similar to that in Figs. 21 and 26, and *Laevigatosporites micropunctatus* Roche (1969) which is a specimen of a psilate monolete microspore with similar scars (see also Figs. 27 and 31); both retain patches of unscarred ektexine. The mimicry of ornamentation is compounded by the confinement in most cases (at least of rosette and branched scars) of the degradation to the ektexine of the microspore. The *Gleicheniidites* in Fig. 24 is completely scarred. Some ± radiating ridges of ektexine remain. Fig. 25a-b is of a degraded *Veryhachium*. At least two layers of the otherwise nondifferentiated wall is suggested as the scars are not perforate. The three spines, although not in focus, exhibit degradation to different lengths up their bases. If examined only under bright field microscopy, these specimens might be described by the unwary as having a granulose or punctate ornamentation.

Selected layers may be partially (Elsik, 1968b, pl. 37, fig. 2) or completely (op. cit., pl. 36, fig. 13) destroyed. The layer attacked and scar produced depends on the type of attacking organism and its food preferences. This again points out a significant difference in endexine and ektexine composition, suggested earlier (Faegri and Iversen, 1964) after the differential staining results and electron transmission characters discussed by Larson and collaborators (Larson and Lewis, 1961; Larson and Skvarla, 1961).

The ready acceptance of stain by the ektexine of chemically treated pollen and the relative non-acceptance of stain by the endexine does suggest a basic chemical difference. But which is the most resistant? Opinions are contradictory. Sangster and Dale (1961), after an oxidizing chlorine experiment on various pollen, noted that the ektexine of *Typha* was stripped off leaving the endexine intact. They later (1964) implied that this effect was "similar to that found frequently in sporodermal breakdown" in the natural environment. This is supported by observation of degraded *Thomsonipollis* spp. from the Paleocene of the Texas Gulf Coast which have lost the ektexine. Contrasted to this is the observation (Rowley, this Symposium) that heavy metal stained nexine (endexine) is less resistant to certain chemical treatments than the sexine (ektexine). The same effect is noted (Erdtman, this Symposium) for certain fossil pollen grains recovered from a mineral soil. Resolution of this apparent contradiction hinges on further evaluation of sporopollenin along the lines suggested by Chaloner (this Symposium): "the designation of the exine substance as a single entity may be underestimating minor differences in composition which are of biological and systematic importance", which, if carried to an ultimate goal, would be to ascertain not only sporopollenin differences between groups of organisms but also between major layers of the exine and among species of related plants.

The intensity of attack of the degradative organisms varies from microspore to microspore, even within the same species in the same sample of sediment. Havinga's (1967) study of the relative susceptibility of microspores of 19 plant species placed in river clay and leaf mould for a period of several months indicated a series with *Lycopodium clavatum* as the most resistant and *Alnus glutinosa*, *Corylus avellana* and *Myrica gale* as the least resistant to degradative organisms producing scars. Three other habitats evidenced no degradation of this type in the same period of time. Study of the test microspores in the river clay and leaf mould after another time lapse (Havinga, this Symposium) indicates that the pollen of *Alnus*, *Corylus* and *Myrica* almost disappeared quantitatively and there was a shift towards higher infection of all types except *Lycopodium clavatum*, which still appeared to be very resistant to degradation. In addition, the microspores in those environments which before evidenced no microbiological scarring after a longer period of time began to exhibit degradation. The results of Havinga's study suggests several important points: 1) time of exposure must be recognized as a relative unknown factor, 2) the degradational organisms causing scars may be different in one environment as contrasted to another environment, and the differing degradative organisms may have different time lapses required before microspores are attacked (upon loss of inhibitory chemicals?), or 3) different organisms may be specific for types of microspores (sporopollenin), and still react in a general way parallel to Havinga's susceptibility series, but after a period of time (still undetermined) the effects are cumulative, and if carried to the extreme, all fossil types could reflect equal infection or almost complete removal from the sediment.

In some fossil assemblages, environments of deposition not determined, approximately 50 to 90 percent of all microspores may exhibit degradation scars. In other fossil assemblages, i.e. some lignites, most of the degradation may be limited to one microspore type, i.e. punctate *Laevigatosporites*, with varying degrees of infection of individuals of that type. A susceptibility scale (cf. Havinga, 1967) is therefore not as evident in fossil material. Further, the supposed more resistant types may be degraded to a greater extent, as in some south Texas Upper Eocene assemblages composed predominantly of *Cicatricosisporites* and *Engelhardtia*; the thick walled psilate trilete microspores are more often degraded.

The scars may be classified and described morphologically in a manner similar to other fossil micro-organisms. Morphological studies of minute details may reveal specific differences eventually to be related to extant causative agents. The identification of the extant causative agents in microbiological degradation will give us another means of tracing those taxa in the fossil record, similar to "trace" megafossils. Geographic, stratigraphic and environmental significance of occurrences can then be evaluated.

In the identification process we must restrict ourselves, obviously, to the use of microspores and material similarly affected as bait or substrate for the studies of growth and development of the degradative organisms. We can deduce how the sporopollenin is attacked at this point but we do not know what enzymes are involved. Pfeiffer (1964) notes that there "are more than 700 known enzymes - biological catalysts...." How many types are required by exine degradative organisms? Are oils, waxes and other chemicals present in the fresh microspores inhibitive to certain types of microbiological degradation? Is the sporopollenin a source of nutrients? What are the fermentation products and how do they chemically counter into the environmental scheme? At what point does diagenesis (carbonization) bar further microbiological degradation?

In connection with the entire spectra of events occurring in the fossilization of sporopollenin, we can also ask ourselves in a wider sense "What is the relation of this microbiologic activity to the maturation of all plant material in sediments and the production of hydrocarbons?" Is the origin of petroleum tied simply to a thermo-pressure gradient applied to organic source material or does microbiologic activity play a role? Can hydrocarbons be fermented by bacteria and lower fungi?

Acknowledgments

This paper is published with the permission of Humble Oil & Refining Company, and represents a continuing interest in the physical state and meaning of degraded modern and fossil microspore exines. My thanks to Dr. Jim Brooks, Exploration and Production Research Division of British Petroleum Company Limited, for the opportunity to express my views through the medium of this Symposium volume. My special appreciation goes to Mr. Robin Hollyhead, of the Organic Research Labs, School of Chemistry, University of Bradford, for his delivery of this paper to the participants of the Symposium.

References

Barghoorn, E. S. (1942). *Science* 96, 358-359.

Barghoorn, E. S., and Linder, D. H. (1944). *Farlowia* 1, 395-467.

Braun, A. (1855). *K. Preuss. Akad. Wiss. Berlin, Monatsber.*, 378-384.

Coscarelli, W. (1964). *In* "Principles and Applications in Aquatic Microbiology" (H. Heukelekian and N. C. Dondero, eds.), p. 113-147. John Wiley & Sons, Inc., New York.

Daugherty, L. H. (1941). *Carnegie Inst. Washington Pub.* 526, 1-108.

Davis, G. H. G., and Freer, J. H. (1960). *J. Gen. Microbiol.* 23, 163-178.

Elsik, W. C. (1966a). *Nature* 209, 825.

Elsik, W. C. (1966b). *Micropaleontology* 12, 515-518.

Elsik, W. C. (1968a). *Pollen et Spores* 10, 263-314.

Elsik, W. C. (1968b). *Pollen et Spores* 10, 599-664.

Faegri, K., and Iversen, J. (1964). "Textbook of Pollen Analysis", Hafner Publishing Co., Inc., New York.

Gamerro, J. C. (1968). *Ameghiniana* 5, 271-278.

Gaumann, E. A. (1952). "The fungi. A description of their morphological features and evolutionary development" (F. L. Wynd, trans.), Hafner Publishing Co., New York.

Goldstein, S. (1960). *Ecology* 41, 543-545.

Graham, A. (1962). *Jour. Paleontology* 36, 60-68.

Gwynne-Vaughan, H. C. I., and Barnes, B. (1937). "The Structure and Development of the Fungi", 2nd ed. (reprinted 1965), Cambridge at the University Press.

Havinga, A. J. (1964). *Pollen et Spores* 6, 621-635.

Havinga, A. J. (1967). *Rev. Palaeobotan. Palynol.* 2, 81-98.

Havinga, A. J. (1968). *Acta Bot. Neerl.* 17, 1-4.

Howell, A., Jr., Murphy, W. C., III, Paul, F., and Stephan, R. M. (1959). *J. Bacteriol.* 78, 82-95.

Kirchheimer, F. (1933). *Bot. Archiv* 35, 134-187.

Kirchheimer, F. (1935). *Beih. Bot. Centralbl.* 53, 398-416.

Larson, D. A., and Lewis, C. W., Jr. (1961). *American Jour. Bot.* 48, 934-943.

Larson, D. A., and Skvarla, J. J. (1961). *Pollen et Spores* 3, 21-32.

Lechevalier, H. (1964). *In* "Principles and Applications in Aquatic Microbiology" (H. Heukelekian and N. C. Dondero, eds.), p. 230-253. John Wiley & Sons, Inc., New York.

Miller, C. E. (1955). *Elisha Mitchell Sci. Soc. Jour.* 71, 247-255.

Miller, C. E. (1961). *Elisha Mitchell Sci. Soc. Jour.* 77, 293-298.

Miller, C. E. (1965). *Elisha Mitchell Sci. Soc. Jour.* 81, 4-9.

Moore, L. R. (1963a). *Vid. Akad. Oslo, M. N. Kl. Skr.*, n. s., no. 9, 1-14.

Moore, L. R. (1963b). *Palaeontology* 6, 349-372.

Moore, L. R. (1964a). *Yorkshire Geol. Soc. Proc.* 34, 235-292.

Moore, L. R. (1964b). *Cinquieme Congr. Intern. Strat. Geol. Carbonifere* (Paris, 1963), 587-592.

Moore, L. R. (1966). *Adv. Sci.* 22, 313-330.

Neves, R., and Sullivan, H. J. (1964). *Micropaleontology* 10, 443-452.

Pfeiffer, J. (1964). "The Cell", Time Inc., New York.

Pine, L., and Hardin, H. (1959). *J. Bacteriol.* 78, 164-170.

Pine, L., Howell, A., Jr., and Watson, S. J. (1960). *J. Gen. Microbiol.* 23, 403-424.

Poindexter, J. S. (1964). *Bacteriological Reviews* 28, 231-295.

Rege, R. D. (1927). *Ann. Appl. Biol.* 14, 1-44.

Roche, E. (1969). *Bull. Belg. Ver. Geol., Paleont., Hydrol.* 78, 131-146.

Rowley, J. R. (1963). *Grana Palynologica* 4, 25-36.

Sangster, A. G., and Dale, H. M. (1961). *Canadian Jour. Botany* 39, 35-43.

Sangster, A. G., and Dale, H. M. (1964). *Canadian Jour. Botany* 42, 437-449.

Smith, G. W. (1951). *In* "Manual of Phycology" (G. W. Smith, ed.), p. 13-19. Chronica Botanica Co., Waltham, Massachusetts.

Sparrow, F. K. (1937). *Biol. Bull.* 73, 242-248.

Sparrow, F. K. (1943). "Aquatic Phycomycetes (exclusive of the Saprolegniaceae and *Pythium*)", Univ. Michigan Press, Ann Arbor.

Umbreit, W. W. (1962). "Modern Microbiology", W. H. Freeman and Co., San Francisco.

Waksman, S. A. (1938). "Humus", 2nd ed., Williams and Wilkins Co., Baltimore.

Wehmer, C. (1915). *Deutsche Chem. Ges., Berlin* 48, 130-134.

Wehmer, C. (1925). *Brennstoff-Chem.* 6, 101-106.

Weston, W. H. (1941). *In* "A Symposium on Hydrobiology", p. 129-151. Univ. Wisconsin Press, Madison.

Yamazaki, T., and Takeoka, M. (1962). *Grana Palynologica* 3, 3-12.

ZoBell, C. E. (1946). "Marine Microbiology", Chronica Botanica Co., Waltham.

Discussion

Linskens: The isolation of degrading organisms could contribute to the elucidation of the structure of sporopollenin. Do you think it possible that the biosynthesis of sporopollenin could be studied *in vitro*?

Elsik: Yes. Existing monitoring systems could be adapted to record the types and amounts of dissolved organic substances that might be produced in an isolated culture media. In the October 1970 issue of Natural History, Bruce C. Parker mentions the use of such a device for measuring the amounts of the vitamins B_{12}, biotin and niacin dissolved in water.

Rowley: Do the organisms causing scars leave anything in or on the wall or do they only excavate the exine?

Elsik: Specimens of degraded exine examined to date have been processed through a harsh palynological treatment, including oxidation and sonic. In this material there are no remaining substances left by the degradative organisms. Recent advances in preparation technique allow examination of microspore exines after HCl and HF but before other chemical or mechanical treatment. Perhaps study of thin sections of untreated fossil exines *in situ* will be required to elucidate the presence or absence of material added by the degradative organism. From the present evidence, however, it appears that the exine is excavated with no replacement of material.

Holleyhead: You note that Ubisch bodies have been reported in the literature to be coccoid or bacillary bacteria and you agree with this. From the evidence of this symposium alone, Ubisch bodies certainly appear not to be related in any way to bacteria.

Elsik: Ubisch bodies examined *in situ* are indeed not related in any way to bacteria. Fossil bodies *similar* in morphology to Ubisch bodies do occur on dispersed microspores. The fossil material has been described as bacillary and coccoid bacteria. Their uniform morphology and occurrence on diverse types of microspores suggests a bacterial origin. Reyre illustrates similar bodies on fossil microspores referable to the Taxales and Cupressales in the August 1970 issue of Palaeontology. Both illustrated specimens (op. cit., pl. 59) possess clumps of coccoid bodies referred to in the case of the Cupressales specimen as glomerules. The point is raised here only to emphasize that there are the two suggested origins for attached, ± spherical bodies on certain fossil exines.

LIPID COMPONENTS IN FRESH AND FOSSIL POLLEN AND SPORES

G. DUNGWORTH

Department of Exobiology, Catholic University, Nijmegen

A. McCORMICK

Atomic Weapons Research Establishment, Aldermaston

T.G. POWELL and A.G. DOUGLAS

Organic Geochemistry Unit, Department of Geology,
University of Newcastle upon Tyne

ABSTRACT

Components of the lipid fraction of fresh spores and pollen and of fossil spores, isolated from the exinite fractions of three coals of Carboniferous age (circa 300 x 10^6 yr), have been examined and, in part, identified. Extracts of both fresh and fossil materials were analysed for, *inter alia*, hydrocarbons, alcohols and carboxylic acids by thin layer chromatography, column chromatography, gas liquid chromatography, mass spectrometry and combined gas chromatography-mass spectrometry.

The fresh pollen and spores contained only traces of aliphatic hydrocarbons whereas the fossil spores contained a series of n-alkanes. A series of isoprenoid alkanes were tentatively identified, by gas chromatographic retention data, in the fossil spores. A series of high molecular weight cycloalkanes, present in two of the fossil hydrocarbon fractions were thought to be triterpane hydrocarbons.

Saponification of the solvent extracted fresh pollen yielded a neutral fraction of predominantly n-fatty alcohols and sterols. The acid fraction was almost all n-C_{16} and n-C_{18} saturated and unsaturated fatty acids. Saponification of the exinite concentrates yielded acid fractions, two of which had distributions that paralleled the hydrocarbon distribution; they contained normal, isoprenoid and triterpenoid acids.

The solvent insoluble and unsaponifiable residues remaining after analysis were partially degraded by hydrogenolysis at elevated temperature and pressure and some preliminary results for these degradations are presented.

INTRODUCTION

Microscopic examination of polished coal surfaces in reflected light indicates that certain basic constituents can be distinguished. These constituents may be considered as organic equivalents of rock minerals; they can be related to the plant structures from which they were derived, and grouped accordingly. Three basic organic constituent or maceral groups are recognised, and named vitrinite, inertinite and exinite; this last group is believed to represent the fossil remains of cuticles, resins, algae and spores. For a discussion of this topic, see, for example Stach (1968) and Cooper and Murchison, this volume.

Non-destructive density separation methods as used in palynology (Gray, 1965) have been developed and applied to the further separation of the maceral groups. Thus the exinite group can be concentrated by flotation in inorganic salt solutions, or in heavy organic liquids, of appropriate densities in which vitrinites and inertinites sink (Cooper and Murchison, this volume). Complete separation is not usually possible due, among other things, to the associations of the macerals. A good example showing the association of fine-grained micrinite with microspores has been published by Stach (1968).

Work on both extracts, and tars obtained by low temperature pyrolysis of bulk coals, has shown that a wide variety of chemical types are present. To our knowledge, however, little has been published on the chemical constitution of the lipid fraction of the exinite group of macerals. A recent review of the Kauri resins by Thomas (1969), contains information on some of the diterpene constituents of fossil resins. Alginites, occurring as Tasmanites, have been investigated by Burlingame _et al._ (1969) who found a wide range of aliphatic and aromatic carboxylic acids in two samples of different age. A Carboniferous alginite, torbanite, has been shown to contain a series of normal and branched alkanes and also a wide range of aliphatic carboxylic acids (Douglas _et al._ 1969a, 1970). A recent alginite, coorongite, has also been shown by Douglas _et al._ (1969b) to contain normal and isoprenoid hydrocarbons and a series of terminal olefins.

Preliminary results by Hunneman and Eglinton (1969) indicate that dicarboxylic acids are present in the oxidation products of the extracted cutinite of Indiana Paper Coal. Surprisingly, there appear to be no reports on the chemical types present in the material extractable from sporinite, the dominant member of the exinite group.

Kroger _et al._ (1964) have analysed tars, produced by the

low temperature carbonisation of exinite concentrates, and shown that they contain a homologous series of normal paraffins and a wide range of olefins and branched paraffins. In addition, polycyclic aromatic hydrocarbons were found containing up to five condensed rings which were, no doubt, mainly produced by dehydrocyclisation reactions of aliphatic molecules.

The lipid constituents of some modern pollens and spores have been determined, and a review is in preparation. (Stanley and Linskens, 1970). Saturated fatty acids ranging from C_6 to C_{22}, and C_{18} mono- and polyunsaturated acids have been isolated from dandelion pollen by Standifer (1966), and from some conifer species by Ching and Ching (1962). In *Pinus ponderosa* the major acids of the phospholipid fraction were shown to be palmitic, oleic and linoleic by McIlwain and Ballou (1966), while Hoeberichts and Linskens (1968) indicated that palmitic acid was the major component of the C_{12} to C_{18} acids present in *Petunia* pollen. Brooks and Shaw (1968) have shown that the major acid of *Lilium henryii* pollen esters (as carotenoid esters) was palmitic acid, which occurred together with a wide range of other normal and branched-chain acids.

Sterols are found in the pollens of a wide variety of plants and a mass spectrometric survey by Standifer *et al.* (1968) has shown that cholesterol, 24-methylenecholesterol, β-sitosterol and stigmasterol are present. Little is known of the triterpenoid content of pollens and we are only aware of the recent identification by Devys *et al.* (1969) of three triterpene alcohols, related to cycloartenol, in the pollen of dandelion.

Chemical studies aimed at identifying the nature of the tough exine of pollen and spores has been reviewed by Shaw (1970). Earlier work by Zetzsche was not completely rewarding; nevertheless he assigned to the pollen exine, sporopollenin, a polyterpenoid structure, (Zetzsche *et al.* 1937). Shaw, in his most recent work, is in agreement with this, describing sporopollenin as an oxidative polymer of carotenoids or carotenoid esters.

In the following, we describe some preliminary results regarding the lipid content of some modern and fossil pollens and spores, and also of the hydrogenolysis products of both. Hydrogenolysis is one of the techniques that has been successfully used by, for example, Hubbard and Fester (1959) to degrade kerogen, a term used to describe the organic macrostructure which occurs in most sediments and which is insoluble in common organic solvent.

The three exinite rich Carboniferous coals discussed here, contained negligible amounts of cuticle, resin and algal bodies as determined microscopically; two were bituminous coals whose exinite fraction contained microspores and megaspores (High Hazles, from the East Midlands and Harvey-Beaumont, from Northumberland); the other was a bituminous coal containing mainly megaspores from Donibristle, Fife. The fresh pollen and spores were from *Pinus pinaster* and *Lycopodium clavatum* respectively.

RESULTS AND DISCUSSION

I. EXINITE LIPIDS

Exinite concentrates were prepared of the crassidurain band

of the Harvey Beaumont seam, and the High Hazels seam, using Thoulet's solution (equal parts of potassium iodide and mercuric iodide in distilled water). Normal procedures for clean working were used to ensure that contamination was kept to a minimum; these procedures for cleaning and handling of samples, apparatus etc. had been detailed previously (Eglinton et al. 1966). The cleaned powdered (through 240 mesh) coal samples were stirred to a slurry in Thoulet's solution of specific gravity 1.25 and centrifuged; the floating exinite concentrate was removed, washed and dried. The concentration of exinite measured microscopically was about 70%, the bulk of the remaining material being inertinite. Palynological examination of the Harvey Beaumont exinite concentrate showed the presence of Densosporites (75%); Laevigatosporites (7.5%); Lycospora (5%); Columnospora (2%); Punctatisporites (2%); Dictyotriletes (1%) and Granulatisporites (1%). The exinite from Donibristle colliery was concentrated to 93% simply by grinding and sieving; this simpler procedure was possibly due to the large content of megaspores.

Extracts of the exinite concentrates were chromatographed on alumina and the hydrocarbon eluates, after analysis by thin layer chromatography (TLC) and infra-red spectroscopy (IR) were separated into normal and non-normal fractions; they were then analysed by gas chromatography (GLC) and combined gas chromatography-mass spectrometry (GC-MS). The analytical procedure is outlined in Fig. 1.

The amounts of exinite extracted, the yields of extract,

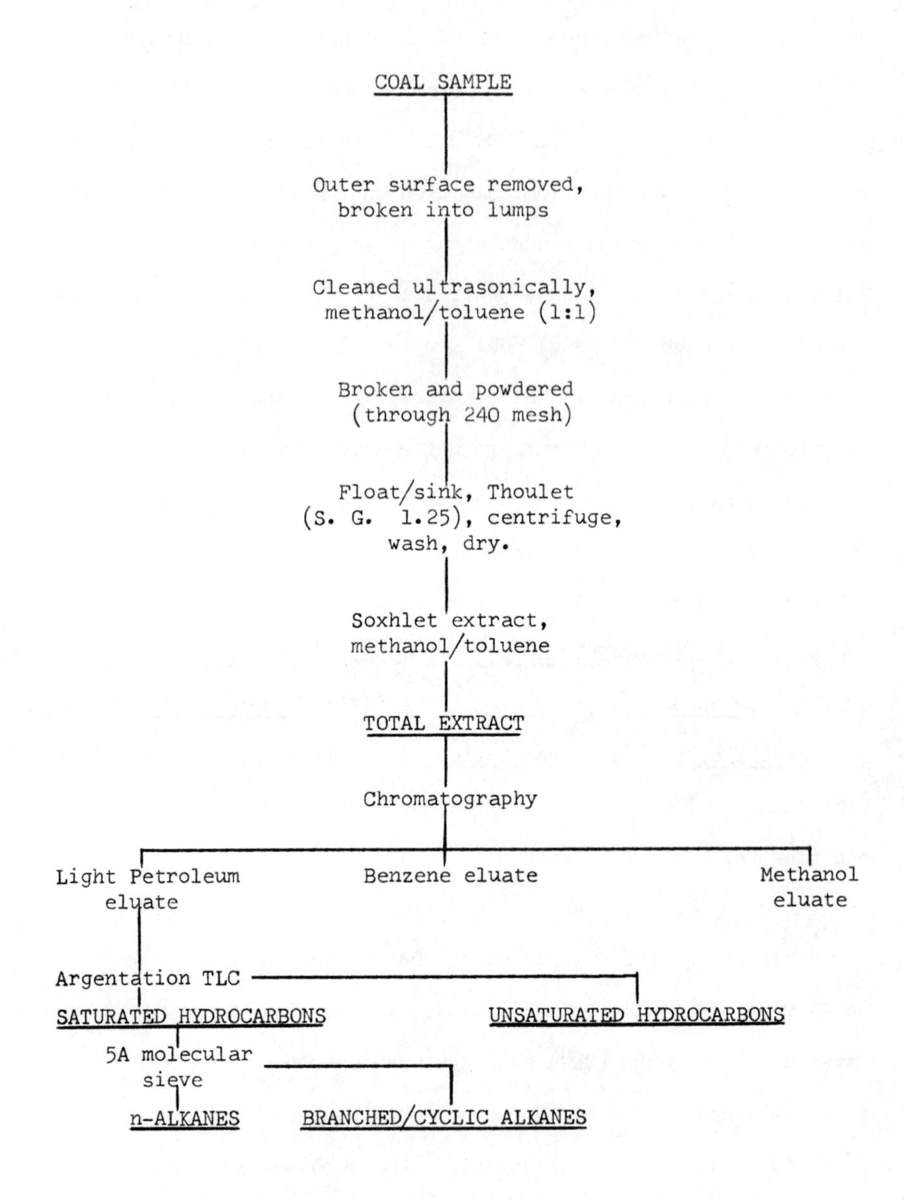

FIG. 1. Analytical procedure for preparing, and extracting, exinite concentrates.

and the saturated hydrocarbon fractions (light petroleum eluates) after column chromatography, are shown in Table 1 below.

TABLE 1. Yields of extract and hydrocarbons from exinite concentrates.

	Quantity extracted (g)	Total extract (mg/g)	Light petroleum eluate (mg/g)
Harvey Beaumont exinite	8	33.8	5.6
High Hazles exinite	10	39.3	3.3
Donibristle exinite	46	30.9	0.8

Hydrocarbons

The light petroleum eluates were shown, by argentation TLC and IR, to be contaminated with unsaturated material. Removal of this by preparative argentation TLC, followed by separation of the normal from branched/cyclic alkanes with 5A molecular sieve, gave the results shown in Table 2.

TABLE 2. Saturated and unsaturated hydrocarbons in exinite concentrates.

	n-Alkanes (mg/g)	Branched/cyclic alkanes (mg/g)	Unsaturated fraction (mg/g)
Harvey Beaumont exinite	1.3	2.6	1.1
High Hazles exinite	0.1	0.4	2.7
Donibristle exinite	0.01	0.4	0.3

Gas chromatograms of both the normal and branched/cyclic alkane fractions were recorded using high resolution packed columns; the

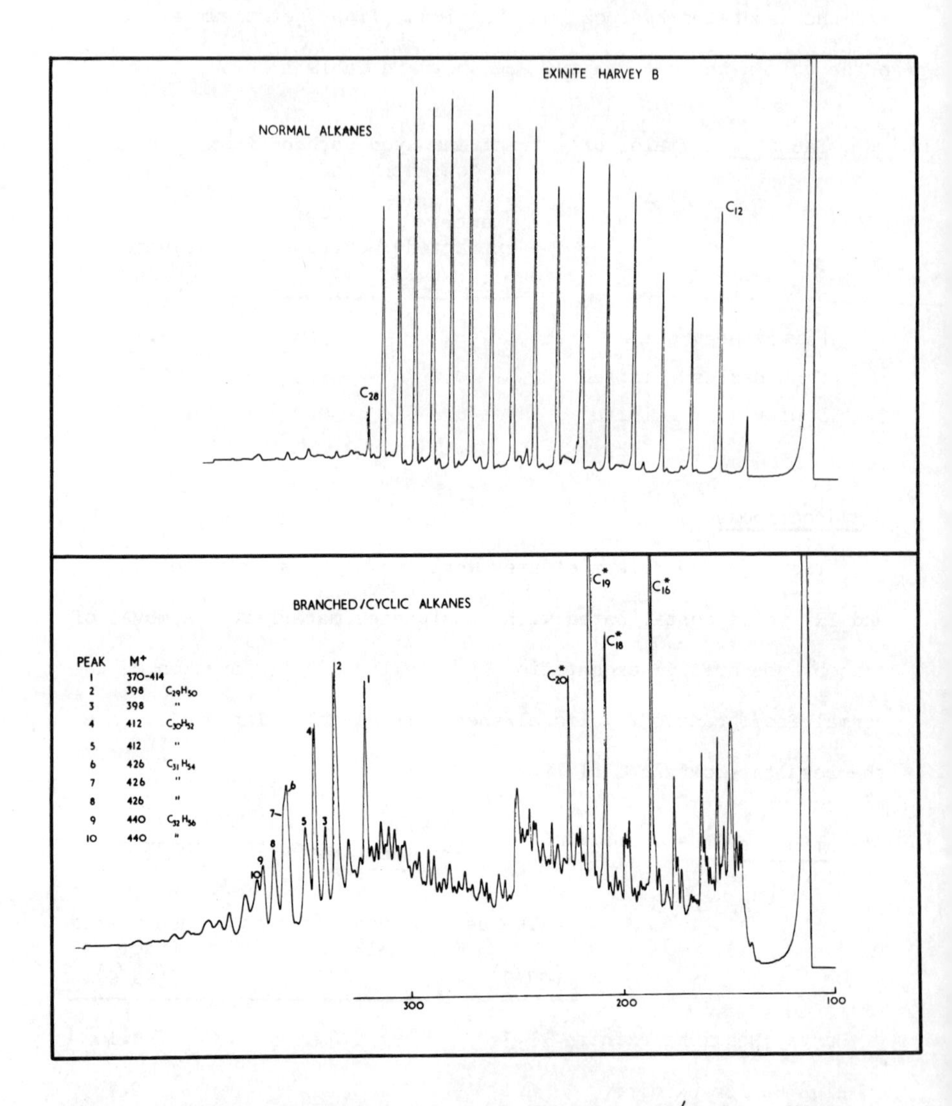

FIG. 2. Gas chromatograms of normal and branched/cyclic alkanes obtained from Harvey Beaumont exinite. Conditions, 20 ft. x 0.040 (i.d.) column containing 3% OV-1 on 100-120 mesh Gas Chrom Q; temp. programmed from 100-300° at 4°/min.; injector 300°, detector 300°. The nitrogen flow rate in this and Figs. 3, 4, 6, varied between 5 and 12 mls/min.

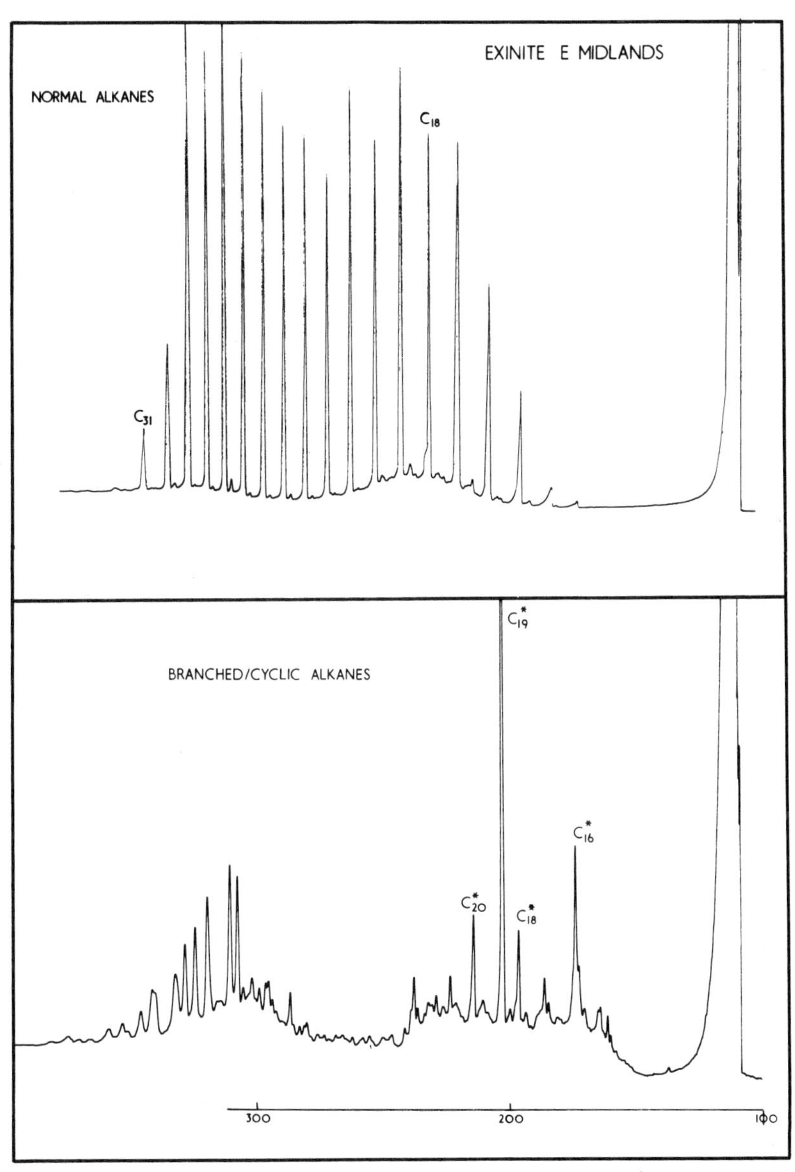

FIG. 3. Gas chromatograms of normal and branched/cyclic alkanes obtained from High Hazles (East Midlands) exinite. Conditions as for Fig. 2.

chromatograms are shown in Figs. 2, 3 and 4 and the chromatographic conditions are noted in the figure legends.

Harvey Beaumont exinite shows a distribution of normal hydrocarbons (determined by coinjection of standards), ranging from C_{11}-C_{28} with the odd carbon numbered members predominating, the ratio of the amount of odd to even alkanes (Carbon Preference Index, CPI) being 1.09. The High Hazles sample shows a distribution in the range C_{13}-C_{31} with a still greater odd predominance (CPI = 1.30). Rank determinations of vitrinite partings, associated with these exinites, showed that the Harvey Beaumont sample had a higher rank (83.6% carbon) than that from High Hazles (81.8% carbon). These rank/CPI relationships are in agreement with the findings of Brooks and Smith (1967) who noted similar relationships in a variety of coals. It is possible that the increase in the amount of n-alkanes obtained from the Harvey Beaumont exinite is due, in part, to its higher rank and consequently greater maturation.

Donibristle exinite provided a much smaller hydrocarbon fraction than the other two exinites (Tables 1 and 2). The distributions of normal and branched/cyclic alkanes in the extract, Fig. 4, are unlike those of High Hazles and Harvey Beaumont exinites, and also of others, currently being examined in this laboratory. The nature of the constituents of the branched/cyclic fraction of lower average molecular weight is being investigated; attempts to confirm the n-alkane distribution, using fresh samples of coal, will be undertaken.

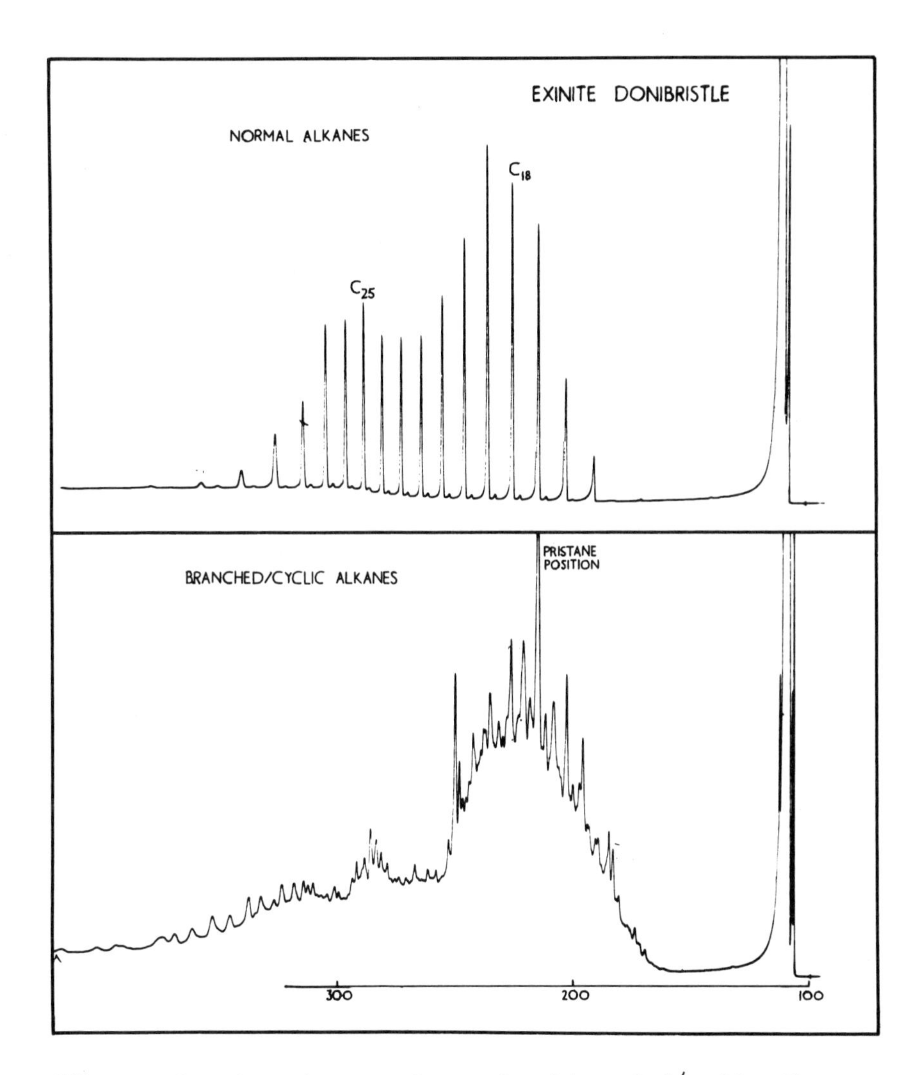

FIG. 4. Gas chromatograms of normal and branched/cyclic alkanes obtained from Donibristle exinite. Conditions as for Fig. 2.

Gas chromatograms of the branched/cyclic fractions from the Harvey Beaumont and High Hazles samples are also shown in Figs. 2 and 3. The components of both mixtures are concentrated predominantly in two ranges equivalent of about n-C_{20} and below, and to n-C_{28} and above respectively. In the lower molecular weight part of both fractions, the acyclic isoprenoid hydrocarbons 2,6,10-trimethyldodecane (farnesane); 2,6,10-trimethyltri -, tetra -, and pentadecane; 2,6,10,14-tetramethylpentadecane (pristane); 2,6,10,14-tetramethylhexadecane (phytane) and 2,6,10,14-tetramethylheptadecane, are present; some are numbered, and starred in the figures. These compounds were identified by comparing their retention indices with those in the literature (Douglas _et al_. 1970), and also with those of isoprenoid alkanes isolated by us from a Carboniferous shale (Plessey Mussel Band, Northumberland), for which mass spectra had been obtained that accorded well with the mass spectra of authentic compounds. Furthermore, coinjection with a mixture of authentic C_{15}-C_{21} isoprenoid hydrocarbons reinforced the appropriate peaks.

An increase in the pristane/phytane ratio with increasing rank in coals was noted by Brooks _et al_. (1969). In agreement, the ratio increases from 4:1 in High Hazles exinite (rank equivalent to 81.8% carbon) to 5:1 in the higher rank (83.6% carbon) Harvey Beaumont exinite. This increasing ratio, with concomitant increases in the total amounts of pristane and phytane, they attribute to diagenetic changes in the soluble wax esters, and in the chloroform insoluble part of the coal, during coalification.

In that part of the gas chromatograms equivalent to about n-C_{28} to n-C_{34} a number of large peaks appear. One of the samples (Harvey Beaumont) was examined by combined gas chromatography-mass spectrometry, spectra being recorded at the points numbered 1 to 10 in Fig. 2. These all had m/e 191 as base peaks and, apart from spectrum number one which had a number of apparent molecular ions, each had predominantly one molecular ion in the C_nH_{2n-8} series, with a small contribution from the next highest homologue. The most characteristic fragment in the mass spectra of pentacyclic triterpenes is at m/e 191 (Budzikiewicz et al. 1963, 1964) whereas m/e 217 is characteristic of steranes (Biemann, 1962, Schnoes, 1970). Spectrum number one was dominated by a species of molecular weight 370, with the next three members of this C_nH_{2n-8} series, and three of the series C_nH_{2n-6}, beginning at m/e 386, as major contaminants. The base peak of this spectrum was at m/e 191 but there was a fairly intense peak at m/e 217 suggesting that the C_nH_{2n-6} compounds are steranes. Three of the hydrocarbon spectra are reproduced in Fig. 5. In addition to the base peak at m/e 191, these spectra all have prominent peaks at m/e 95, 109, 123, 137, 177 and 205, an intense molecular ion and a prominent M-15 fragment. All of these features are characteristic of pentacyclic triterpanes (Hills and Whitehead, 1966, 1970). The fragment at m/e 191 arises by fission across ring C of the triterpane molecule and the positive charge may be retained by either fragment.

In C_{30} compounds, e.g., lupane, cleavages 1 and 2 both result in an ion of m/e 191, whereas other members of the homologous series

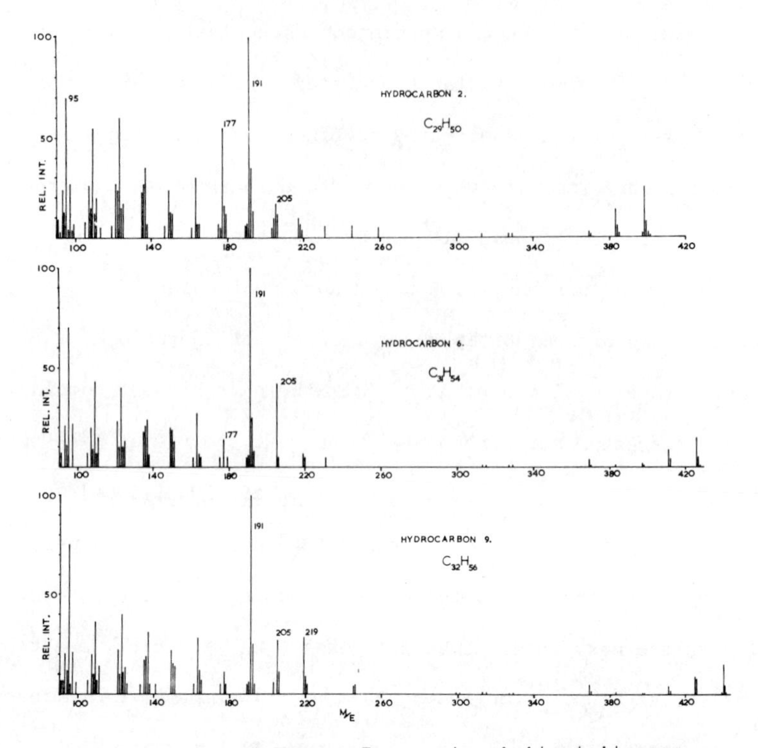

FIG. 5. Mass spectra of Harvey Beaumont exinite triterpanes, obtained from GC-MS scans of peaks 2, 6 and 9 in Fig. 2.

will give 191 and 177, 191 and 205, 205 and 205, 191 and 219 etc., by these processes. Other minor fragmentation pathways also give rise to these ions, e.g., lupane has an ion of low abundance at m/e 205 in its mass spectrum (Hills and Whitehead, 1966). Hydro-

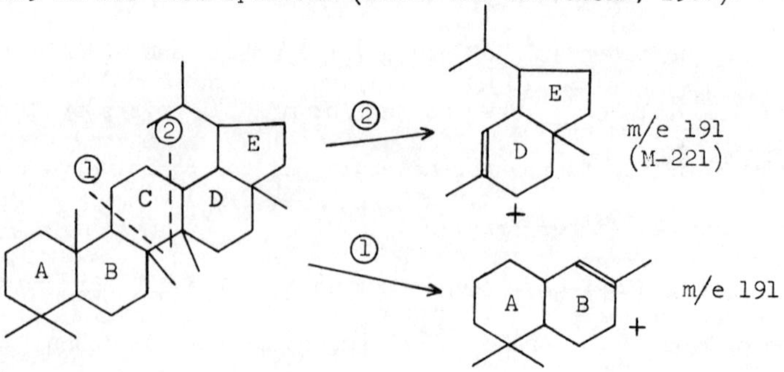

A B C D E ① ② m/e 191 (M-221) m/e 191

carbon 9, Fig. 5, has base peak at m/e 191 and an intense peak at m/e 219; from this it would appear that the two "extra" carbon atoms are on the same side as the molecule. The nor-triterpane, hydrocarbon 2, has a mass spectrum very similar to that of adiantane (Henderson *et al.* 1969; Hills and Whitehead, 1970). Apart from a type classification, and general conclusions such as those outlined, the similarity of the spectra of the triterpanes precludes their use for unequivocal identification even where a pure specimen, and an authentic reference standard, are available. Much further work is therefore required before the identity of the hydrocarbons in this exinite can be established; this might include measurement of gas chromatographic retention data, on high resolution columns of different polarity, or where sufficient material is available by X-ray analysis, as achieved by Hills *et al.* (1968). A summary of the formulae of the major species found in the GC-MS analysis is given in Table 3.

TABLE 3. Triterpane components of Harvey Beaumont exinite.

Scan No.	M^+	Empirical formula of major triterpane
1	370-414	$C_{27}H_{46}$
2	398	$C_{29}H_{50}$
3	398	"
4	412	$C_{30}H_{52}$
5	412	"
6	426	$C_{31}H_{54}$
7	426	"
8	426	"
9	440	$C_{32}H_{56}$
10	440	"

Carboxylic Acids

Acids were isolated from the Harvey Beaumont and High Hazles exinites in the following manner. The crushed concentrates were saponified with methanolic potassium hydroxide, acidified and extracted. The bulk of the non-acidic constituents was removed on potassium hydroxide impregnated TLC plates and the recovered acid salts were methylated. Pure methyl ester fractions were prepared by preparative TLC and analysed by GLC and GC-MS without attempting further separation into normal and branched/cyclic fractions. The acids in Donibristle exinite were obtained by saponifying the total extract and then proceeding in the manner described above. Interestingly, no acids were obtained when this thoroughly extracted exinite concentrate was saponified. Table 4 below indicates the content of acids in the exinite concentrates.

TABLE 4. Carboxylic acids in exinite concentrates

	Wt of sample (g)	Acids (as methyl esters) (mg/g)
Harvey Beaumont exinite	70	0.11
High Hazles exinite	86	0.33
Donibristle exinite	46	0.01

Gas chromatograms of Harvey Beaumont, High Hazles and Donibristle exinite carboxylic acid esters are shown in Fig. 6 in which the range of normal acids is n-C_{12} to n-C_{24}; n-C_{14} to n-C_{29} and n-C_{12} to n-C_{26} respectively. Some of the members of this homologous series are clearly evident in the middle part of the chromatograms which also show the even/odd character of the distribution. These acids were characterised by coinjection with authentic

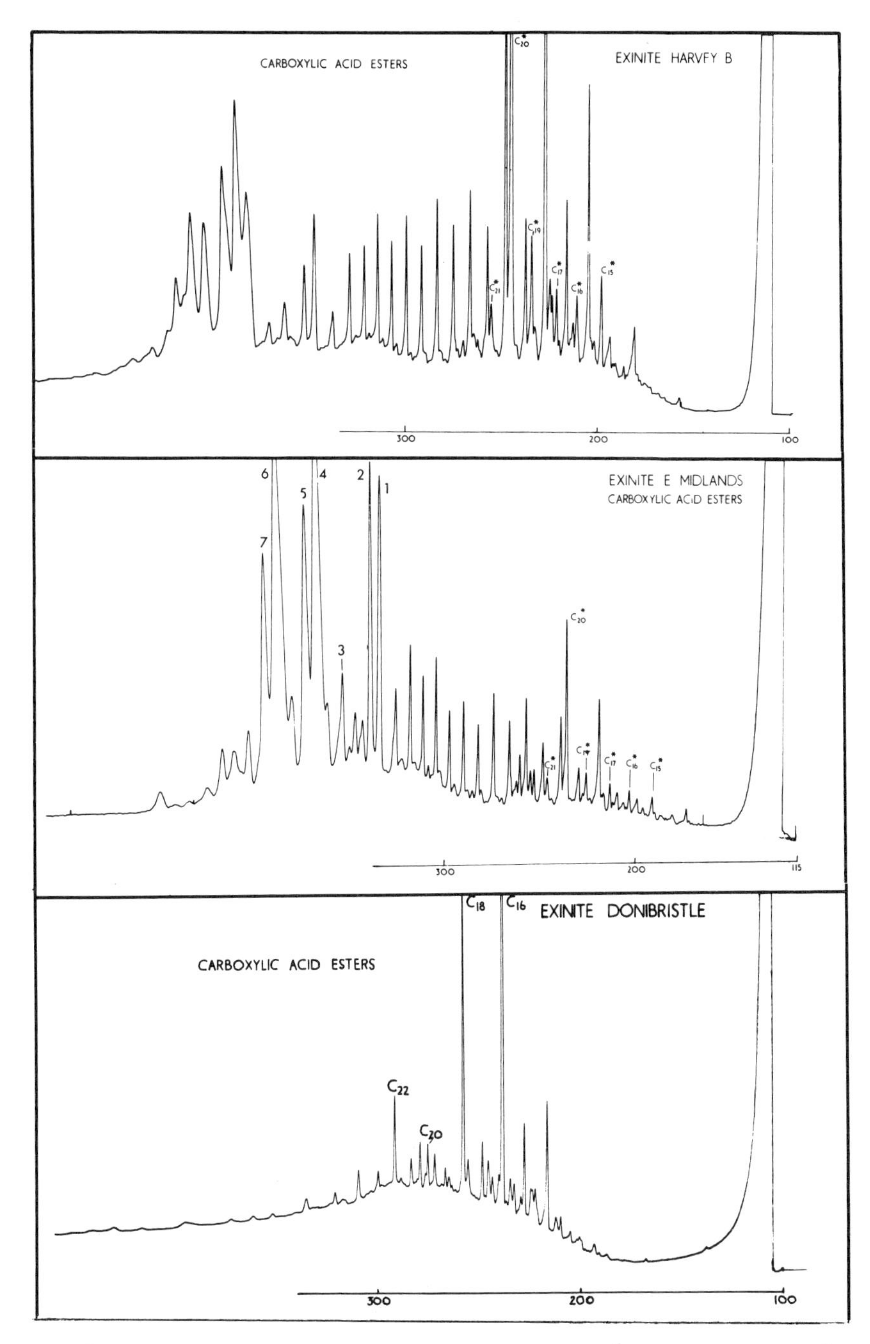

FIG. 6. Gas chromatograms of carboxylic acids (as methyl esters) obtained from Harvey Beaumont, High Hazles (East Midlands) and Donibristle exinites. Conditions as for Fig. 2 except initial oven temperature, as shown.

$n\text{-}C_{12,14,16,18,20,23,24,25,26,28}$ and $_{30}$ methyl esters. The very strong predominance of the normal C_{16} and C_{18} acids in the Harvey Beaumont and Donibristle samples is disquieting, since it is not reflected in the acids above $n\text{-}C_{18}$. Neither is the marked predominance of these two acids evident in most ancient sediments for which data have been published, although van Hoeven _et al_. (1969) do find a similar distribution in one Precambrian sample. Handling in the laboratory is unlikely to have contaminated the samples, and the nature of the solid coal blocks, from which the samples were prepared, does not encourage the view of contamination by, for example, percolating ground waters. There may be cause for concern with the Donibristle sample, which has lain as a solid block on a museum shelf for many years. If these acids are indeed present in the exinite concentrates in such large amount, it is possible that coalification has generally reduced the even/odd predominance. At the same time, if large amounts of saturated or unsaturated even acids, free or as glycerides, were present in the original material they might well persist sufficiently as saturated acids after diagenetic reduction, to give chromatograms as shown. The possibility that these predominating acids in Harvey Beaumont exinite are attached to the insoluble organic matrix as esters, where they may be partly 'protected' from geochemical change, is currently being investigated; we have shown above that they are not thus attached into the Donibristle exinite.

Isoprenoid acids ranging from C_{15} to C_{21} are shown numbered, and starred, in the chromatograms. They were identified by comparing their equivalent chain lengths with those of the isoprenoid acids in an extract of Green River Shale; these data were obtained on columns containing OV-1 and Apiezon L. The ECL values obtained were in very good agreement with values published by Douglas _et al_. (1970) for some authentic isoprenoid acids, and those in Green River and Serpiano Oil shales. Interestingly, isoprenoid acids are absent or at most are minor constituents of the Donibristle exinite.

Large amounts of acids above n-C_{28}, and not homologous with the normal acids, are evident in both Harvey Beaumont and High Hazles exinite, but not in Donibristle exinite. Since the corresponding hydrocarbon fraction of the first two contain triterpane hydrocarbons, it appeared likely that those substances might be triterpenoid acids. The acid esters from High Hazles exinite were examined by GC-MS and mass spectra were recorded for the seven peaks labelled in this part of the chromatogram. These spectra all had m/e 191 as base peak, intense M and M-15 ions and a prominent fragment at M-221. This latter fragment is analogous to that produced by cleavage 2 above is lupane. By introducing a sample of the mixture into the mass spectrometer via the direct insertion probe, the exact masses, and therefore formulae, of some parent molecular ions and of the m/e 191 fragment were established as in Table 5.

TABLE 5. Mass measurements of peaks in the carboxylic acid methyl ester fraction of High Hazles exinite.

Mass measured	Requirements
428.3647	$C_{29}H_{48}O_2$ requires 428.3654
470.4115	$C_{32}H_{54}O_2$ requires 470.4124
191.1799	$C_{14}H_{23}$ requires 191.1800

From these and the GC-MS spectra, three of which are reproduced in Fig. 7, it can be confidently stated that the acids are monobasic

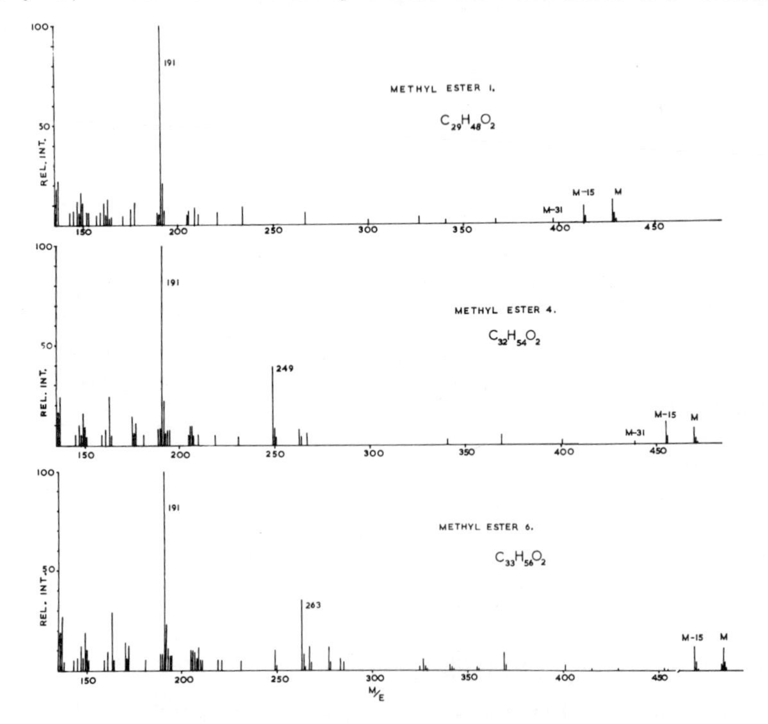

FIG. 7. Mass spectra of High Hazles (East Midlands) exinite triterpenoid acids, obtained from GC-MS scans of peaks 1, 4 and 6 in Fig. 6.

triterpenoids and, like the hydrocarbons, probably have the same structure for rings A and B, the differences being in the substitution of rings D and E. Data for the acids are collected in Table 6 below.

TABLE 6. Triterpenoid acids in High Hazles exinite.

Scan No.	M^+	Empirical formula of major component
1	428	$C_{27}H_{45}CO_2CH_3$
2	428	"
3	456	$C_{29}H_{49}CO_2CH_3$
4	470	$C_{30}H_{51}CO_2CH_3$
5	470	"
6	484	$C_{31}H_{53}CO_2CH_3$
7	484	"

II. FRESH POLLEN AND SPORE LIPIDS.

Pinus pollen and *Lycopodium* spores were extracted ultrasonically to give a lipid fraction. The lipid extract and the residual pollen and spores were saponified, and separated into neutral and acid fractions according to the scheme shown in Fig. 8. The extract afforded free lipids, the residual pollen and spores bound lipids.

Neutral Lipids:

The neutral lipid fraction was chromatographed (alumina, grade 2) using light petroleum, benzene, benzene/methanol (9:1) and methanol as eluants. Fractions were monitored by TLC and combined appropriately, to give a hydrocarbon fraction (light petroleum eluate) and an alcohol fraction (benzene/methanol and methanol eluates), consisting of long chain alcohols and sterols. Gas chromatographic analysis of the hydrocarbon fraction, obtained only from the free lipids, and shown by argentation TLC to be free of olefinic material, indicated that n-alkanes ranging from C_{13} to C_{25} were present. The surprisingly low CPI value of 1.50 would have been lower still but for the predominance of the C_{21}

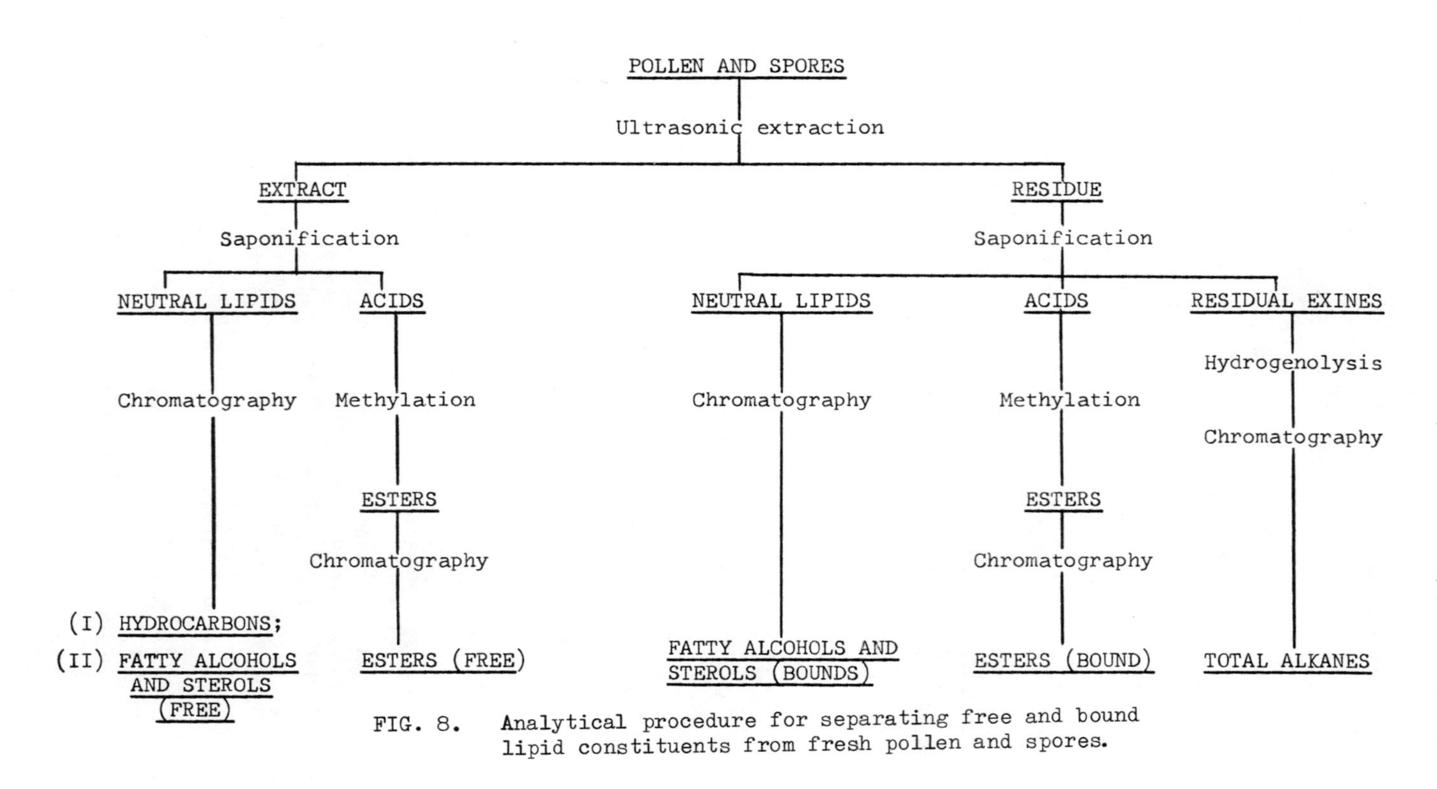

FIG. 8. Analytical procedure for separating free and bound lipid constituents from fresh pollen and spores.

compound, heneicosane.

The free and bound fatty alcohols were analysed as silyl derivatives, by gas chromatography, and shown to be predominantly the n-C_{18} and n-C_{20} compounds, as determined by coinjection with authentic compounds. The sterols were not identified, but analysis of other Pinus species by Standifer et al (1968) has shown that there the principal sterol is β-sitosterol. Quantitative data for the analyses of the neutral lipid fractions are included in Table 7.

Carboxylic Acids:

The acids were methylated, chromatographed and analysed by argentation TLC at -20° (Morris et al. 1967) using light petroleum/ether (9:1) as developer; this showed that both species contained saturated, mono- and diunsaturated esters. Mass spectra of the bound ester mixtures were obtained Fig. 9 at low (20eV) electron energies, to enable the easier recognition of molecular ions. Prominent molecular ions at m/e 298 ($C_{18:0}$); 296 ($C_{18:1}$); 294 ($C_{18:2}$); 270 ($C_{16:0}$) and 268 ($C_{16:1}$) together with intense ions at m/e 74 and 87 indicated the presence of the above acid esters. The prominent ion at m/e 264, Fig. 9, could indicate the presence of a $C_{16:3}$ ester, but this is not substantiated by TLC or GLC data; it is possible that it represents the intense M-32 ion noted by Ryhage and Stenhagen (1960) in the mass spectrum of methyl oleate. Further evidence was obtained for the presence of these particular esters by measuring their gas chromatographic retention data on high resolution packed columns containing variously OV-1,

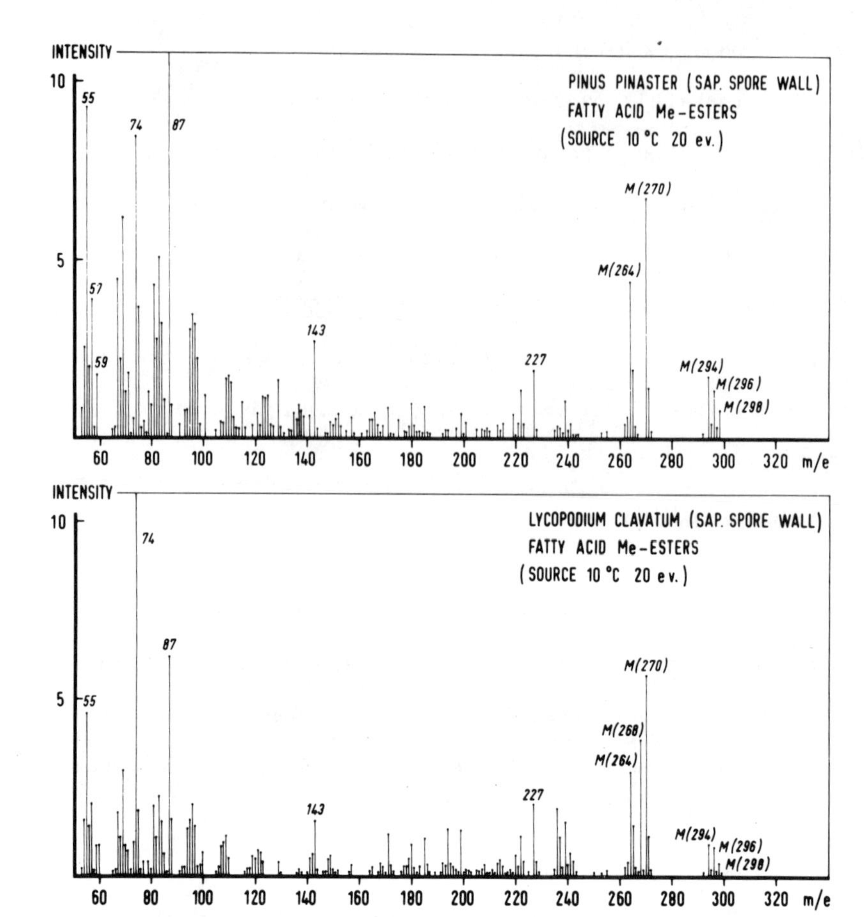

FIG. 9. Mass spectra of mixed fatty acid methyl esters obtained by saponification of the exines of P. pinaster pollen and L. clavatum spores.

Apiezon-L and Polysev. Individual acids were identified by comparing the equivalent chain lengths of their methyl esters with those of authentic esters; excellent agreement was obtained. Gas chromatograms of the free acid esters, from both species, are shown in Fig. 10. The free and bound acids in Lycopodium gave chromatograms that were almost superposable, the bound Pinus acids do not contain the n-C_{12}, n-C_{14}, n-C_{19}, n-C_{20} or n-C_{22} compounds

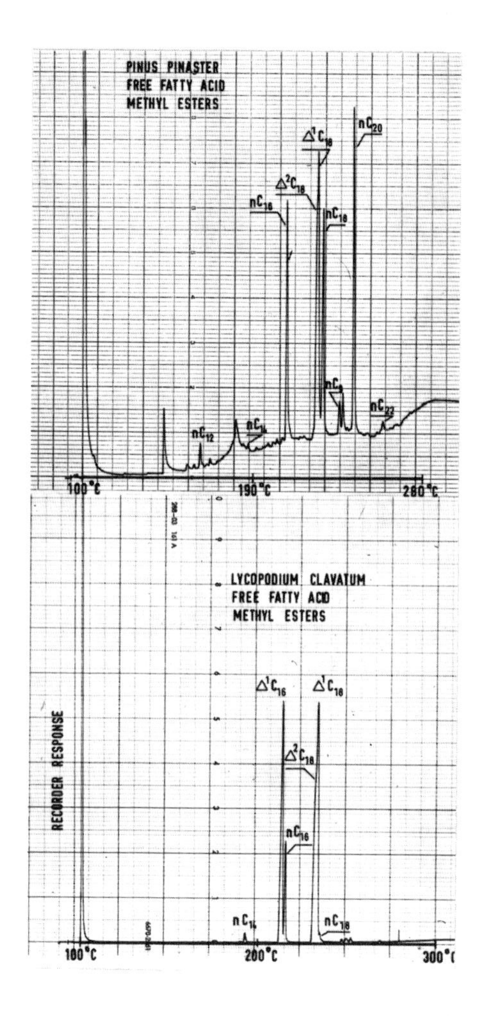

FIG. 10. Gas chromatograms of free fatty acids (as methyl esters) of P. pinaster pollen and L. clavatum spores. Conditions, 12 ft. x $\frac{1}{8}$ in (o.d.) column containing 3% OV-1 on 100-120 mesh Gas Chrom Q, temp. programmed as shown at 4°/min.; injector 300°, detector 300°, nitrogen 17 ml/min.

present in the free acids.

TABLE 7. Hydrocarbons, acids and alcohols in the free and bound lipids of fresh pollen and spores.

Quantitative Results	Lipid concentration *Pinus pinaster* pollen (mg/g)	Lipid concentration *Lycopodium clavatum* spores (mg/g)
Soluble lipids	108	216
Total hydrocarbons	1.2	0.1
Fatty alcohols and sterols (free)	43.8	75.0
Fatty alcohols and sterols (bound)	52.0	84.9
Total fatty alcohols and sterols	95.8	159.9
Fatty acids (free)	9.8	94.2
Fatty acids (bound)	20.2	96.0
Total fatty acids	30.0	190.0

III. HYDROGENOLYSIS OF FRESH AND FOSSIL POLLEN AND SPORE EXINES

Hydrogenolytic degradation of the thoroughly extracted fossil exinite, and the extracted and saponified fresh exines, was accomplished by heating a mixture of the exinite with stannous chloride (ratio 2:5) at 380°, for 90 minutes at 2,000 p.s.i. Benzene extraction of the reaction mixture, followed by chromatography (alumina, grade 1), afforded a saturated hydrocarbon fraction (light petroleum eluate); yields of extract and hydrocarbon are given in Table 8. Preliminary experiments had shown that very little hydrocarbon material was produced at 320° (*L. clavatum*) and 230° (Harvey Beaumont exinite). The hydro-

TABLE 8. Hydrocarbons from the hydrogenolysis of fresh and fossil pollen and spore exines.

	Quantity extracted (g)	Benzene soluble material (B) (mg/g)	Total alkanes (A) (mg/g)	A/B %
Donibristle exinite	1.03	838	160	19.0
Harvey Beaumont exinite	0.75	255	4.7	1.8
Pinus pinaster	0.55	909	8.4	0.9
Lycopodium clavatum	0.52	175	19.0	10.9

carbon fractions from *P. pinaster* and *L. clavatum* had infra-red bands which indicated the presence of terminal olefins (910 and 990 cm^{-1}), and *trans* disubstituted olefins (970 cm^{-1}), but not aromatic compounds. Gas chromatograms of both hydrocarbon fractions, Fig. 11, indicate that a homologous series of normal alkanes, extending from about C_{11} to C_{28} are present. Coinjection of a mixture of standard n-C_{14} to n-C_{20} alkanes confirmed the presence of these compounds. Tentative identification of terminal olefins, which can be seen partially resolved from the n-alkanes in the chromatograms, was made by comparing their retention indices with those of a standard mixture of the authentic n-C_{14} to n-C_{20} compounds.

Gas chromatograms of the fossil exinite alkanes are very complex. In Fig. 12, the large 'hump' in the Harvey Beaumont alkane chromatogram is due to unresolved components in the mixture and not to instrumental conditions. Nevertheless, superposed on this complex mixture there are normal alkanes ranging from about C_{12} to C_{28} in which n-C_{14}, n-C_{15} and n-C_{16} are particularly

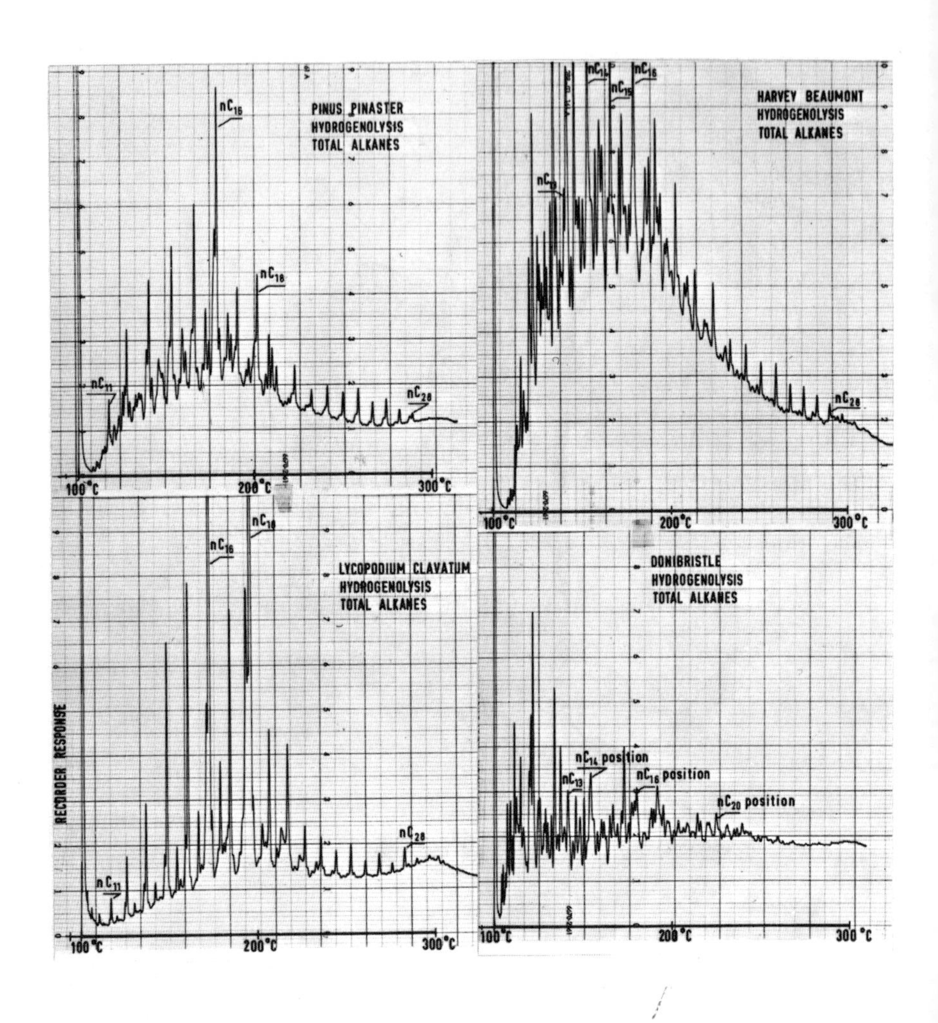

FIG. 11. Gas chromatograms of alkanes (and alkenes) obtained from the hydrogenolysis products of P. pinaster pollen and L. clavatum spores. Conditions, as for Fig. 10.

FIG. 12. Gas chromatograms of alkanes obtained from the hydrogenolysis products of Harvey Beaumont and Donibristle exinite concentrates. Conditions as for Fig. 10.

abundant. The hydrocarbon fraction from the Donibristle exinite is somewhat less complex than the above, and shows no prominent normal alkanes. Thus, preliminary results show that the alkane distributions, obtained by hydrogenolysis of two fossil exinites, are markedly different, although their distribution in one of the exinites (Harvey Beaumont, Fig. 12) is similar, in some respects, to that in the degraded _P. pinaster_ and _L. clavatum_ exines.

Results of the hydrogenolysis of fresh and fossil pollen and spores may be summarised as follows. The exine of _P. pinaster_ pollen has been almost completely converted to benzene soluble substances (91%) whereas that of _L. clavatum_ has not (18%). Of this, _L. clavatum_ has more than double the amount of alkanes, which constitute about 11% of the benzene soluble material compared with about 1% for the _P. pinaster_ exine. Since the extracted pollen and exine had been vigorously saponified, the hydrocarbons are unlikely to represent alkyl moieties derived from alcohols, or acids, esterified to the organic macrostructure; this conclusion is at variance with that inferred by Brooks and Shaw (1968). The possibility that the _L. clavatum_ exine is partly composed of polymers of unsaturated fatty acids, cross-linked in a manner that renders them susceptible to hydrogenolysis, is currently being studied. The amount, type and distribution of some hydrogenolysis products of the fossil exinites are also consistent with their having different structures.

REFERENCES

Biemann, K., Mass Spectrometry, (McGraw-Hill, New York, 1962).

Brooks, J.D., and Smith, J.W., Geochim. et. Cosmochim. Acta, 31, 2389, (1967).

Brooks, J., and Shaw, G., Nature, 219, 532, (1968).

Brooks, J.D., Gould, K., and Smith, J.W., Nature, 222, 257,(1969).

Budzikiewicz, H., Wilson, J.M., and Djerassi, C., J. Amer. Chem. Soc., 85, 3688, (1963).

Budzikiewicz, H., Djerassi, C., and Williams, D.H., Structural Elucidation of Natural Products by Mass Spectrometry, Vol.2, (Holden-Day, San Francisco, 1964).

Burlingame, A.L., Wszolek, P.C., and Simoneit, B.R., Advances in Organic Geochemistry 1968, edit. by Schenck, P.A., and Havenaar, I., 131 (Pergamon Press, Oxford, 1969).

Ching, T.M., and Ching, K.K., Science, 138, 890,(1962).

Devys, M., Andre, D., and Barbier, M., C.R. Acad. Sci. Paris, 269, 798, (1969).

Douglas, A.G., Eglinton, G., and Maxwell, J.R., Geochim. et. Cosmochim. Acta, 33, 569, (1969a); 33, 579 (1969b).

Douglas, A.G., Blumer, M., Eglinton, G., and Douraghi-Zadeh, K., Tetrahedron, in press (1970).

Douglas, A.G., Douraghi-Zadeh, K., Eglinton, G., Maxwell, J.R., and Ramsay, J.N., Advances in Organic Geochemistry 1966, edit. by Hobson, G.D., and Speers, G.C., 315, (Pergamon Press, Oxford, 1970).

Eglinton, G., Scott, P.M., Belsky, T., Burlingame, A.L., Richter, W., and Calvin, M., Advances in Organic Geochemistry 1964, edit. by Hobson, G.D., and Louis, M.C., 41, (Pergamon Press, Oxford, 1966).

Gray, J., Handbook of Palaeontological Techniques edit. by Kummel, B., and Raup, D., 530, (W.H. Freeman, London, 1965).

Henderson, W., Wollrab, V., and Eglinton, G., Advances in Organic Geochemistry 1968, edit. by Schenck, P.A., and Havenaar, I., 181, (Pergamon Press, Oxford, 1969).

Hills, I.R., and Whitehead, E.V., Nature, 209, 977, (1966).

Hills, I.R., Smith, G.W., and Whitehead, E.V., Nature, 219, 243, (1968).

Hills, I.R., and Whitehead, E.V., Advances in Organic Geochemistry 1966, edit. by Hobson, G.D., and Speers, G.C., 89 (Pergamon Press, Oxford, 1970).

Hoeberichts, J.A., and Linskens, H.F., Acta Bot. Neer. 17, 433, (1968).

Hoeven, W. van ,Maxwell, J.R., and Calvin, M., Geochim. et Cosmochim Acta, 33, 877, (1969).

Hubbard, A.B., and Fester, J.I., Bureau of Mines Report of Investigations No. 5458 (1959).

Hunneman, D.H., and Eglinton, G., Advances in Organic Geochemistry 1968, edit, by Schenck, P.A., and Havenaar, I., 131, (Pergamon Press, Oxford, 1969).

Kröger, C., Goebgen, H.G., and Klusmann, A., Brennstoff-Chemie, 45, 170, (1964).

McIlwain, D.L., and Ballou, D.L., Biochemistry, 5, 4054, (1966).

Morris, L.J., Wharry, D.M., and Hammond, E.W., J. Chromatog., 31, 69, (1967).

Ryhage, R., and Stenhagen, E., J. Lipid Res., 1, 361, (1960)

Schnoes, H.K., and Burlingame, A.L., Advances in Analytical Chemistry and Instrumentation, edit. by Burlingame, A.L., 369, (John Wiley and Sons, London, 1970).

Shaw, G., Phytochemical Phylogeny, edit. by Harbourne, J.B., 31, (Academic Press, London, 1970).

Stach, E., Coal and Coal-Bearing Strata, edit. by Murchison, D.G., and Westoll, T.S., 127, (Oliver and Boyd, Edinburgh, 1968).

Standifer, L.N., Ann. Entomological Soc. Amer., 59, 1005, (1966)

Standifer, L.N., Devys, M., and Barbier, M., Phytochemistry, 7, 1361, (1968).

Stanley, R.G., and Linskens, H.F., Biochemistry of Pollen, in preparation, (1970).

Thomas, B.R., Organic Geochemistry, edit. by Eglinton, G., and Murphy, M.T.J., 599, (Springer-Verlag, Berlin, 1969).

Zetzsche, F., Kalt, P., Liechti, J., and Ziegler, E., J. Prakt. Chem., 148, 267, (1937).

ACKNOWLEDGEMENTS

We are much indebted to Dr. B. Owens, Institute of Geological Sciences, Leeds, for the palynological analysis, Mr. J. Allan for preparing the Donibristle exinite concentrate, the Curator of the Hancock Museum, Newcastle upon Tyne, for a gift of Donibristle spore coal, Mr. M. Willemse for the fresh pollen sample and Dr. P. Albrecht for the standard isoprenoid hydrocarbons. We thank the Natural Environment Research Council for grants to A.G.D. and T.G.P.

THE PETROLOGY AND GEOCHEMISTRY OF SPORINITE

B. S. COOPER AND D. G. MURCHISON

Organic Geochemistry Unit, Department of Geology,
University of Newcastle upon Tyne.

ABSTRACT

Sporinite is a term normally used to describe an organic constituent (maceral) that is widespread in coals, particularly those of Palaeozoic age. The maceral is formed from spore exines, which, over a range of coal rank, but primarily in the bituminous coals, display a substantial and continuous variation of chemical and physical properties with increasing coalification. The properties of sporinite eventually become similar in the low-volatile bituminous coals to those of the dominant coal maceral, vitrinite, which is derived primarily from lignin-cellulose-humic complexes in the original coal-forming peat. These property changes with rank can be mainly attributed to the influence of relatively small increases of temperature ($<200^{\circ}C$) acting over widely varying periods of geological time. Sporinite can also show considerable range in its character, even in single coals, because of biochemical effects early in its depositional history. Variations in the composition of sporinite, whose relationship to sporopollenin is also discussed, may help in elucidating the geochemical history of coals.

INTRODUCTION

The internationally accepted term 'Sporinite' (International Committee for Coal Petrology, 1963), is used to describe an organic constituent (maceral) that is particularly widespread and, depending on the original depositional environment, sometimes abundant in coals of Palaeozoic age. The maceral is derived from the outer membranes of spores and from pollen grains, but the term sporinite does not include any contribution from fungal spores which, over the entire coalification (rank) range from peats to low-volatile bituminous coals, display properties that are in marked contrast to those of sporinite.

RELATIONSHIPS AND DISTRIBUTION OF SPORINITE

Sporinite belongs to the 'Exinite' maceral group, a term which encompasses three other macerals of similar properties; resinite, cutinite and alginite, which form respectively from resins, cuticles and algae. The relationship of the exinite macerals to other coal macerals is shown in Table 1.

The morphographic nature of the term sporinite must be clearly understood. It is a collective term that can be used by petrologists without any implication to either its physical or chemical composition. However, it is worth emphasising that it is possible to measure certain physical properties (particularly optical) of spore exines in coals and thus define sporinite for any specific level of rank. Because correlations between the physical and chemical properties of the maceral are available, conclusions

Table 1.

Group	Maceral	Origin
Exinite	Sporinite	Spore and pollen exines
	Cutinite	Cuticles
	Resinite	Resins
	Alginite	Algae
Vitrinite	Collinite Telinite	Wood and bark tissues
Inertinite	Fusinite Semifusinite Micrinite Sclerotinite	Primarily from wood and bark tissues which have suffered a different biochemical history to the tissues forming the vitrinite macerals; a small contribution from fungal tissues in certain coals

can also be drawn from the physical measurements on the general chemical constitution of sporinite at different levels of coalification.

The underlying bases of the definitions of sporinite and the term 'Sporopollenin' (Zetsche, 1930) are quite different. Sporopollenin is not a morphographic term and current usage indicates a solely chemical basis for its definition. For example, Brooks and Shaw (1968) have shown sporopollenin of recent spores and pollen to be an oxidative polymer derived from carotenoids and carotenoid esters. But the term is apparently being even more widely and collectively applied than sporinite, without the benefit of any morphographic control, to describe not only pollen and exines of vascular plants, but also fossil-spore material and the polymerised cell contents of aquatic algae and bacteria. In fact, such a usage, even when limited by compliance with properties derived from studies on chemical composition, solubility, pyrolysis and oxidation products means that sporopollenin as defined will also include fossil cuticle and fossil resin, as well as many kerogens which can be shown to be mixtures of plant remains of various origin.

In autochthonous Palaeozoic coals resinite and cutinite form only a small fraction of the exinite group, sporinite being dominant; alginite is confined to 'drift' coals, particularly to the rich algal coals termed torbanites, which are relatively rare in occurrence. It should also be noted that a considerable deposition of spores, along with other degraded plant material, occurs in many sediments associated with coal seams. There would appear to be no objection to extending the use of the term sporinite to cover spore and pollen occurrences in these primarily inorganic sediments.

The distribution of spores in the modified Carboniferous peats that now form a large bulk of the more mature coal resources of the world has been considered by Smith (1962), who was able to recognise several spore phases within coal seams. Each phase was characterised by a particular spore assemblage and each could be related to slightly different levels of water-table in the coal-forming swamp. Spore-rich coals, crassidurains, are associated with the Densospore phase. The investigations of the physical and chemical properties of sporinite have almost exclusively begun with the isolation of exinite from crassidurains in which the initial spore concentration was often greater than 50 per cent by volume of the whole coal.

MORPHOLOGY OF SPORINITE

Coals are normally examined with the microscope using incident-light, oil-immersion objectives on relief-polished surfaces. While spores and pollens from peats and soft brown coals are little distorted from their original form in life, in coals of higher rank the bodies are much compressed, because of weight of sedimentary overburden, to form flat, circular or ellipsoidal discs parallel to the coal banding. In sections perpendicular to the banding, the individual exines possess an elongated form (Fig. 1), often

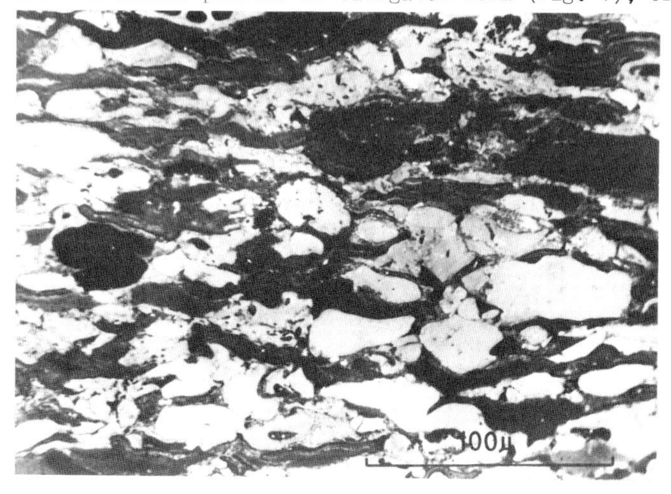

Figure 1. Low reflecting, elongate, thin-walled microspores in coal; the higher reflecting material mainly represents polymerised cell infillings: reflected light, oil immersion, Jurassic, Maghara, Sinai.

with the interior of the spore appearing as a thin line (Fig. 2), which sometimes contains traces of other modified plant materials or inorganic matter, for example, clay minerals. Less frequently the cavity of the spores is more open (Fig. 3).

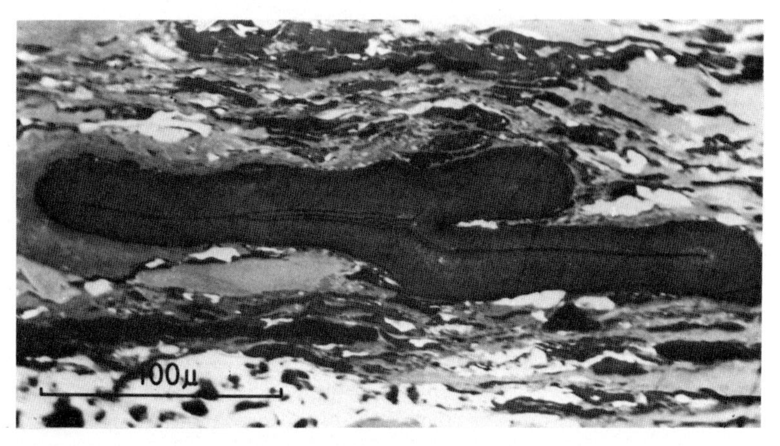

Figure 2. Megaspore in low-rank bituminous coal showing low reflectivity and high relief: reflected light, oil immersion, Stone Coal, Northumberland.

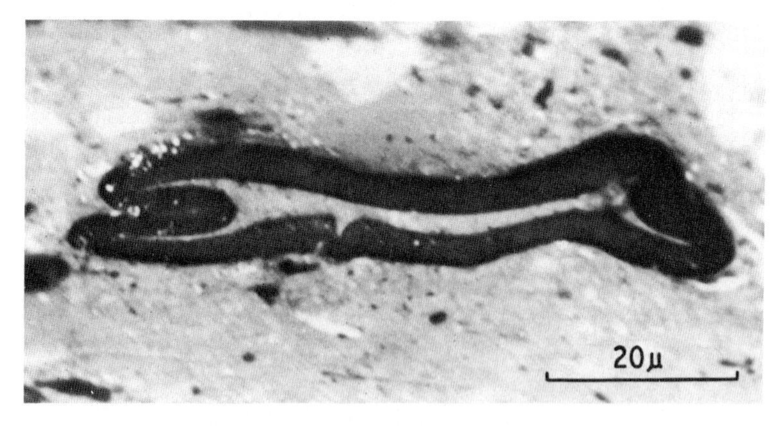

Figure 3. Microspore with partly compressed cavity now containing vitrinite and flecks of micrinite: reflected light, oil immersion, Jurassic, Scotland.

Spore exines show a wide variety of sculpturing and ornamentation. Individual exines may display a two-fold layering, which is sometimes apparent by conventional reflected-light microscopy, but which also may be enhanced or even revealed by fluorescence microscopy. More detailed consideration of variations in spore morphology in coals can be found in articles by Stach (1964) and Potonié, Rehnelt, Stach and Wolf (1970).

Spores range in length from less than 10 microns to greater than 3000 microns. In coal microscopy a distinction based on size is made between megaspores and microspores, the former usually being much larger. It is acknowledged that 'microspore' is a collective term, since small megaspores, isospores and pollen grains, as well as true microspores, are likely to be grouped under this heading.

PROPERTIES OF SPORINITE

Physical

The physical properties of sporinite, as they are modified by geochemical coalification processes, are well known and have been summarised by the International Committee for Coal Petrology (1963). Potonié, Rehnelt, Stach and Wolf (1970) have recently given a more detailed consideration of the physical properties and the same authors have described the chemical properties of the maceral and suggested chemical structures for the constitution of Carboniferous spore exines.

Most of the variation in the physical properties of sporinite that can be observed with the microscope takes place within the bituminous-coal rank range and involves a reduction in contrast between the sporinite and other organic constituents, for example, in polishing relief, which is reduced, and in density and micro-indentation hardness, which both rise with rank increase. In peats and brown coals spore exines are a blue-grey colour in reflected light and they change to black or dark-grey in the hard lignites and high-volatile bituminous coals. Frequently the exines in low-rank bituminous coals display reddish internal reflections (Fig. 4). With further rise in rank the exines become lighter grey in colour and increasingly similar in optical properties to the main organic constituent of coals, vitrinite. The change in colour of the sporinite in the bituminous-rank range is associated with a rise in the maximum oil reflectivity of the

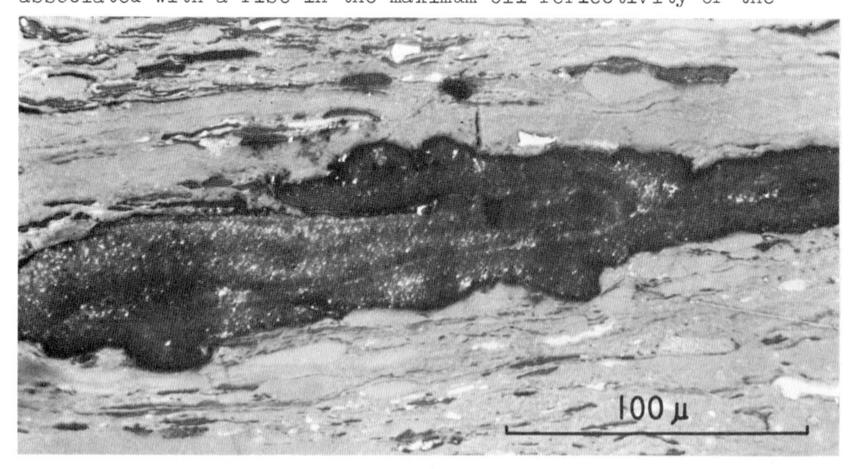

Figure 4. Megaspore in low-rank bituminous coal showing low reflectivity; the bright flecks are due to internal reflections within the exine: reflected light, oil immersion, Top Yard Seam, Northumberland.

maceral from 0.10 per cent to 1.50 per cent, corresponding to a reduction in the volatile matter of the whole coal from approximately 40 to 20 per cent. Van Gijzel (1968) has shown that the spectral ratio, Q, which gives an estimate of the fluorescence value of spore exines, falls with increase of reflectivity, i.e. with rising coalification.

Chemical

Any study of the chemical properties of sporinite ideally requires that the maceral should be separated from all other organic and inorganic constituents of the coal. Preparation of pure sporinite concentrates is, however, extremely difficult and the most successful separations achieved have been of the exinite group macerals rather than of sporinite (e.g. Kröger, Pohl and Kuthe, 1957, Dormans, Huntjens and van Krevelen, 1957; Ladam, Iselin and Alpern, 1958). These separations were mainly achieved by initial fine grinding of the coals followed by immersion of the pulverised coal in a liquid of suitable specific gravity, for example, zinc chloride or a mixture of carbon tetrachloride and toluene, to separate the exinite macerals from other constituents of the coals. Such procedures yielded high exinite concentrates, sometimes as great as 97 per cent by volume. While these concentrates would have to be regarded as exinite (sensu stricto), their composition would, however, be essentially sporinite with small amounts of contamination by the macerals resinite and cutinite. As indicated earlier, it is unusual to find substantial concentrations of either resinite or cutinite in Palaeozoic coals.

Further separation of the macerals within the exinite concentrates by 'sink and float' methods is impossible, because of insufficient density differences between sporinite, resinite and cutinite of individual coal samples. A different approach is required. While Murchison and Jones (1963) were able to prepare pure concentrates of resinite on a scale sufficient for micro-analysis by isolating the maceral under a stereoscopic microscope with probes, the same method is not satisfactory for spore exines. Individual megaspores can certainly be removed from the bedding planes of coals and carbonaceous shales, but because the compressed central cavities of the spores have been filled with a variety of inorganic and organic matter at the time of deposition, there is still the problem of removing these contaminants. 'Sink and float' methods for secondary purification cause losses that are too great to be countenanced with the small amounts of maceral that can be isolated with probes.

Another alternative, recently employed by Potonié, Rehnelt, Stach and Wolf (1970), is to isolate the spores in part mechanically and partly by chemical treatment using nitric acid or hydrogen peroxide to remove organic matter associated with the spores. The possibility of chemical attack on the exines by such reagents is obvious, even when these are carefully used in dilute form. The conclusion must be that no method yields an entirely satisfactory separation. 'Sink and float' techniques and chemical separations result in high yields of spore exines, but in selecting a method, the dangers of modifying the exines during the severe ball-milling prior to specific-gravity separation has to be balanced against possible oxidation of the exines by chemical reagents.

Fig. 5 illustrates the generalized coalification tracks for the dominant coal macerals based on a plot of atomic ratios H/C:O/C using data from van Krevelen (1950, 1961, 1963). The implication of the diagram is that the different macerals, which form from progenitors of widely varying compositions, follow convergent development lines that reflect their response to diagenetic and post-diagenetic processes.

For a British high-volatile coal, the average composition of its sporinite would lie close to $C_{100}H_{110}O_{10}$, a composition which

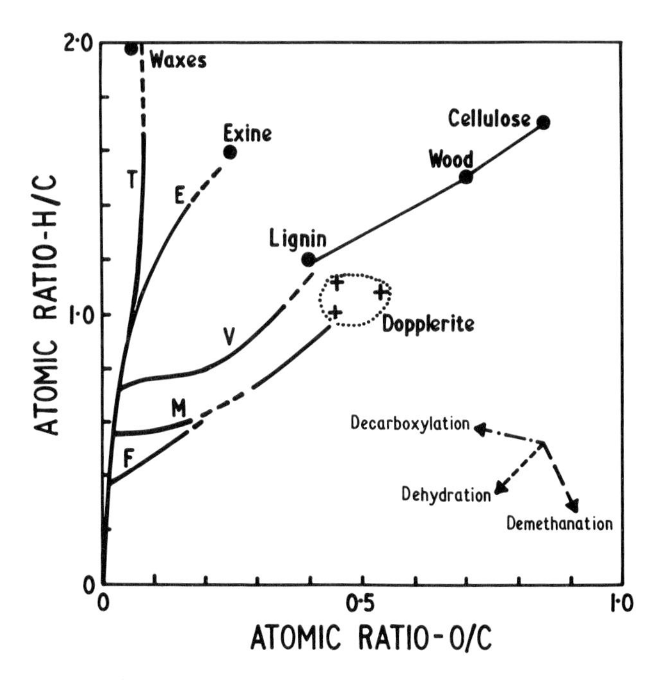

Figure 5. Coalification tracks for the dominant macerals and group macerals (after D.W. van Krevelen, 1950, 1961, 1963).

E - Exinite
F - Fusinite
M - Micrinite
V - Vitrinite
T - Alginite (Torbanite)

agrees substantially with data given by Potonié, Rehnelt, Stach and Wolf (1970) for individual sporinite concentrates produced by a variety of separatory techniques. At higher levels of rank, the development lines for sporinite and vitrinite coincide in the semi-anthracite range, where the two macerals are regarded as being very similar and possessing a composition that approximates to $C_{100}H_{75}O_5$.

Extensive investigations of macerals and maceral groups, including spore-rich exinites, were made during the period 1955-1965, employing conventional solvent extraction techniques, X-ray diffraction, infra-red and NMR spectroscopy. The work of the earlier part of this period has been well reviewed by Brown (1959), Tschamler and de Ruiter (1963) and Dryden (1963). The summarised data show that the oxygen content of sporinite must be mostly contained in hydroxyl and ether groupings, since carbonyl and carboxyl groups form less than 1 per cent of the oxygen present. The basic structure of sporinite is cyclic, partially aromatic with peripheral aliphatic substituents and linkages. Rise of rank causes an increase in aromaticity, a loss of hydroxyl content and an increase in solubility. Dryden (1963) in his analysis showed that in comparison with vitrinite of the same coal, spore-rich exinite possessed a) a higher atomic H/C ratio, b) a lower ratio of aromatic to aliphatic hydrogen, c) the same number or fewer aromatic rings per 'cluster' (aromatic nucleus) and, d) approximately the same number of aromatic carbon atoms per cluster.

More recently, Tschamler and de Ruiter (1966) have shown that for each 100 carbon atoms of spore-rich exinite in a high-volatile bituminous coal, 62 of these are aromatic, 26 occur in CH_2 groups, 6 in CH groups and 6 in CH_3 groups. There were a substantial number of aliphatic cyclic rings present in the separation. X-ray measurements showed that the ring size of the aromatic nuclei, including peripheral aliphatic carbon atoms, lay between 2.2 and 3.5, suggesting that the nuclei are probably built up from benzene or napthalene structures.

Products of the pyrolysis of spore-rich exinites might be expected to indicate the most frequently occurring aromatic group present in the maceral. While Kröger, Goebgen and Klusmann (1964) demonstrated that 1-, 2-, and 3-ring aromatic compounds were present in exinite tar, benzpyrene and 4-ring compounds occurred in much larger quantities, suggesting that condensation and dehydration had taken place during pyrolysis.

In relation to vitrinite, sporinite has a low solubility in such solvents as pyridine and dimethyl formamide. The maceral can, however, display a much greater solubility than vitrinite in simple organic solvents such as benzene, methanol or acetone; extracts of whole coal samples in these solvents have been attributed to exinite. No total analyses of such extracts are available, but they contain alkanes, which have mainly been studied, aromatics and oxy-compounds.

Brooks, Gould and Smith (1969) have shown that alkanes are generated during coalification after the brown-coal rank stage.

With rise of rank there is a progressive change in alkane composition in which the odd-carbon preference of the n-alkanes is lost and both pristane and phytane increase in abundance. Similarly, Leythaeuser and Welte (1969), in a survey of Westphalian coals, describe changes in extractibility with rank in which there is a maximum in extractibility close to a volatile-matter content of the whole coal of 30 per cent and a carbon content of 85 per cent. There were also changes in the distribution of n-alkane numbers which coincided with maximum extractibility. Similar effects have been noted in coals intruded by a dolerite sill and related to variations in spore properties (Cooper and Murchison, 1970). All these phenomena can be related to a radical molecular rearrangement in the sporinite, which leads at first to maximum adsorptive capacity (and hence to extractibility) and then, as rank increases, causes the molecular structure to approach that of vitrinite with its low extractibility.

RESISTANCE OF SPORINITE IN THE GEOLOGICAL ENVIRONMENT

Biochemical Stage

The methods of isolation of sporinite forced upon the chemist by the heterogeneous nature of coal means that, with a few rare exceptions, chemical data on sporinite result from bulk analyses which can take no account of possible variations in the maceral isolated from a single coal. On the other hand, microscopical examination can indicate, through differences in optical properties of the maceral, that chemical variations are likely within sporinite concentrates, although the amount of altered organic material involved may be relatively small. It is true that qualitative and quantitative microscopical studies on a wide variety of spore-bearing coals suggest that, in any individual coal sample, large numbers of exines have approximately the same properties. However, systematic investigation of possible chemical 'type' variations within the spores of a restricted sample have not been attempted. Type variations in vitrinites are recognised as one source of chemical variation within single coal samples, particularly those of low rank.

The difficulty of demonstrating type variations in spores lies in the problem of isolating sufficient spores of precisely similar optical properties which would act as a control for chemical examination. More severe alterations to spore exines of single samples are, however, recognised. These alterations are the result of processes occurring during the biochemical stage of coal formation or prior to it before deposition of the exines. Stach (1962) has illustrated the effects of biochemical coalification on the exinite macerals, demonstrating the occurrence of both vitrinized and fusinized miospores, cuticles, resins and algae, in coals of varying rank. Figs 6 and 7 are composite photographs of typical fusinized megaspores, the brittleness and fragmentation of the spores being quite characteristic of such modified exines, which are normally much tougher and elastic. Severe corrosion of the exine surfaces is usual (Fig. 8). In contrast, the spores may be vitrinized (Fig. 9) and it is not uncommon to find

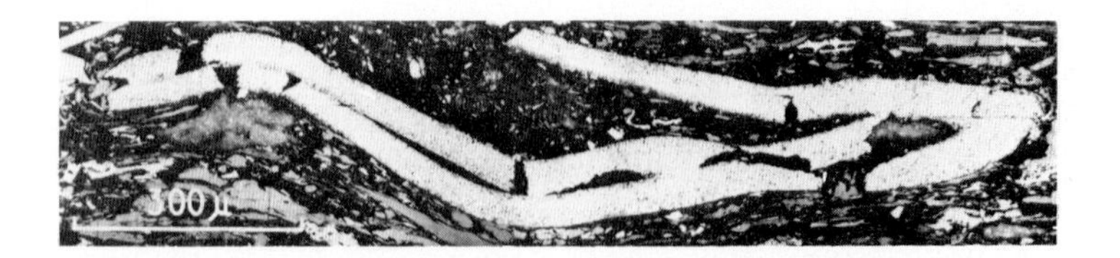

Figure 6. Fusinized megaspore in carbonaceous shale illustrating the brittleness of the spore: composite photograph from 7 separate negatives, reflected light, oil immersion, Beaumont Seam, Northumberland (reproduced with permission from the Editor of Fuel).

Figure 7. Unbroken and relatively uncompressed fusinized megaspore in carbonaceous shale: composite photograph from 4 separate negatives, reflected light, oil immersion, Beaumont Seam, Northumberland.

gradations within single miospores from exinite to vitrinite or exinite to semifusinite or even fusinite (Figs 10 and 11) (Bell and Murchison, 1966). The correlations that have been established between the optical and chemical properties of the coal macerals (see, for example, International Committee for Coal Petrology, 1963), indicate that the optical variations shown in Figs 6 - 11

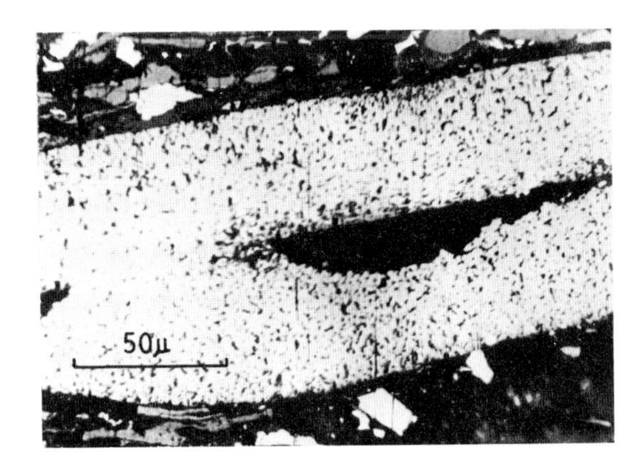

Figure 8. Corrosion of a fusinized spore: reflected light, oil immersion, Beaumont Seam, Northumberland.

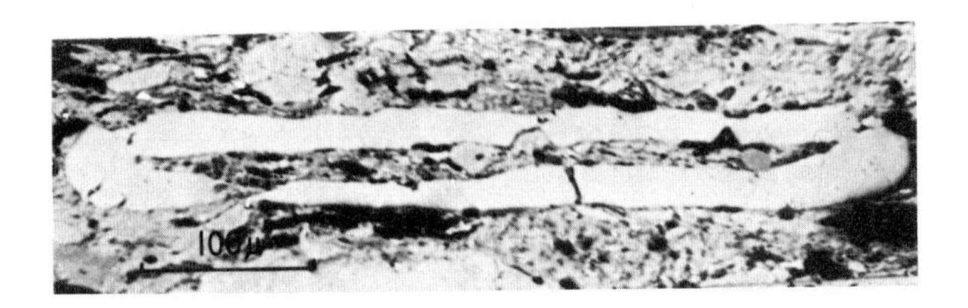

Figure 9. Vitrinized megaspore: reflected light, oil immersion, Jurassic, Maghara, Sinai.

represent substantial chemical alteration of the entities away from the norm for the main mass of exinite in each of the coals.

Such occurrences in coals are widespread, although the number of entities involved is small. The vast majority of spores reposing in coal seams have probably been freed from the parent organism and deposited in the coal-forming swamp area relatively rapidly, then being protected from further degradative processes by stagnant conditions in which no violent fluctuations of pH develop. On the

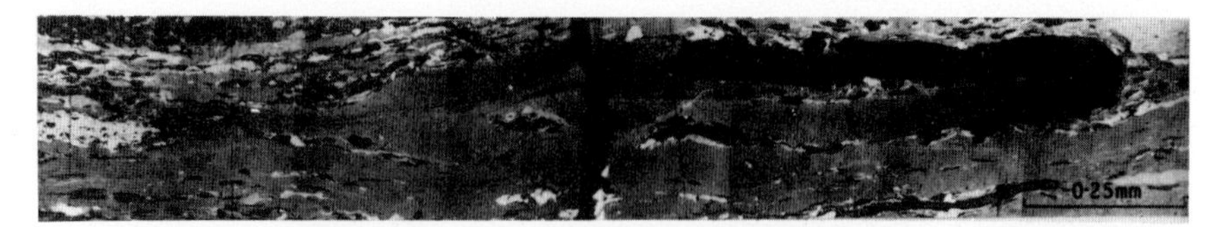

Figure 10. Partially vitrinised megaspore in low-rank coal; composite photograph from 16 separate negatives, reflected light, oil immersion, Beaumont Seam, Northumberland (reproduced with permission from the Editor of Fuel).

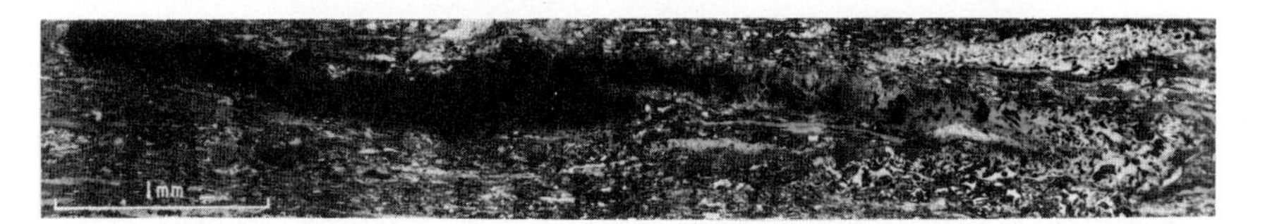

Figure 11. Exinite material, possibly a miospore, showing a transition in properties from exinite to semi-fusinite; composite photograph from 150 separate negatives, reflected light, oil immersion, Beaumont Seam, Northumberland (reproduced with permission from the Editor of Fuel).

other hand, occurrences such as those referred to above, must be explained. Despite claims that spores are highly resistant to degradation, in the laboratory or elsewhere, there is ample evidence from the microscopy of coals and associated sediments that spores can be markedly affected in the biochemical stage of coal formation or even earlier in their history. Temperature and pressure effects during the later geochemical coalification of the macerals must be excluded as a cause for the property variations referred to in Figs 6-11.

Some of the fusinized and vitrinized miospores will have been blown or washed into the coal-forming peat, having already suffered the marked alterations that are now visible. Likewise, the inclusion from other sources of what were whole fragments of 'foreign' peat in which all the contained macerals now differ from those of the surrounding coal could similarly occur (Cooper and Murchison, 1970). However, it appears likely that many of the more radical changes to the spores have occurred near to or at their site of deposition. It also seems possible to relate the amount of alteration of the exine to the type of organic sediment in which it is enclosed. Thus, while the proportion of spores showing biochemical alteration in the microlithotypes of normal coals is relatively low, in carbonaceous shales the proportion of strongly corroded, semi-fusinized and fusinized spores is much higher. Burghardt (1964) has related the corrosion of sporinite in carbonaceous shales directly to the alkalinity of the deposit. Although exposure to atmospheric oxygen might also have played some part in the modification of the exines, it was not felt to be a dominant influence in the modification of the spores.

Geochemical Stage

The influence of the depositional environment and the biochemical stage of coalification on sporinite can be very striking. During the later, and generally much more extended, geochemical or 'metamorphic' stage, modifications to the properties of sporinite are considerable, if more gradual and less obvious than the changes that may affect individual exines during their early history. Since regional coalification is a relatively low-temperature phenomenon, which takes place over varying periods of geological time, sporinite, notwithstanding that it has probably the lowest thermal decomposition point of all the macerals at any rank level, never shows the evidence of plasticity or vacuolation that would be expected had the maceral approached its softening point. However, as a major volatile producer, particularly in the low-to medium-rank bituminous coals, even regional coalification processes may produce some evidence of the removal of large amounts of primarily aliphatic material during rank increase. Stach (1964) has illustrated mega-spore exines which display transverse cracks which are attributed to release of methane during coalification. This condition is, however, quite different to the appearance produced by the much shorter term and more rigorous environment in which an igneous intrusion has thermally metamorphosed a coal and gas bubbles from the devolatilising exines have accumulated around their margins.

The conditions which produce a certain rank of coal are variable. In the geochemical stage prolonged low-temperature heating or shorter-lived and higher temperatures may result in the same degree of coalification being attained. The changes in the composition of vitrinite during coalification may be envisaged as simple reactions in which carbon dioxide, water and methane are eliminated; sporinite probably yields the same breakdown products, but in different proportions. However, since sporinite will alter at a different rate to vitrinite, there may be unique combinations of vitrinite and sporinite for each temperature-time combination.

The atomic ratios H/C:O/C from the analyses of three sets of separated spore-rich exinites (Given, Peover and Wyss 1960; Kröger and Bürger, 1959; Alpern 1956) are plotted in Fig. 12. The

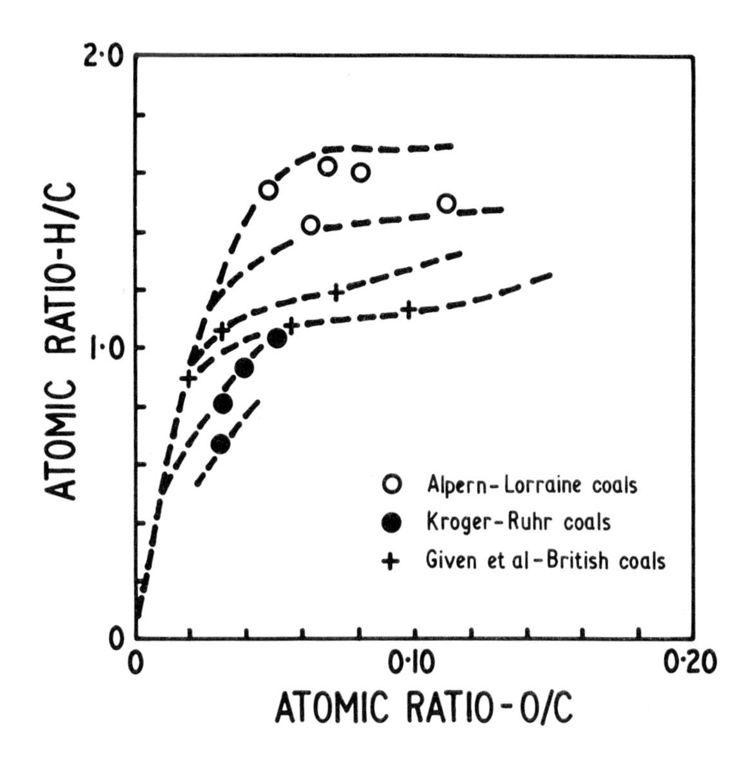

Figure 12. Plots of the atomic ratios H/C:O/C for spore-rich exinites from three different coalfield areas.

sporinites from the Lorraine coals lie well away from those of the Ruhr and the British coals; the Lorraine samples are perhydrous. The differences in the plots might be ascribed to the geochemical settings of the three sets of coals. The Lorraine coals were deposited in a limnic basin, whereas the Ruhr and British coals (mainly from the Yorkshire-Nottinghamshire coalfield) formed in paralic basins. There is then the possibility of slight differences in ground-water chemistry affecting the coals.

A more interesting speculation is that differences in the tectonic and coalification histories are the cause of the divergences between the sets of data. Comparison of the tectonic structures and the rank profiles of the Ruhr and Saar-Lorraine coalfields show that the lines of equal rank cut across the fold structures and are sub-parallel to unconformable beds above, indicating completion of coalification after folding which also took place at the end of Carboniferous times. The Lorraine sporinite compositions may then be the result of long exposure to moderate temperatures, but the Ruhr sporinites, and possibly those of the British coals also, have possibly been taken to higher temperatures for shorter periods of time. The British sporinites do, however, raise some difficulties. Suggate (1970) has shown that the coalification in the Yorkshire-Nottingham coalfield was post-folding. It might be expected therefore that coalification would have proceeded slowly at moderate temperatures as the Mesozoic cover gradually accumulated. On the other hand, there is evidence that the mineralisation, which gave rise to the Derbyshire deposits, was continuing throughout Permo-Triassic times and also that there was post-Permian intrusion of igneous sills in the region. Both these factors might account for a short-lived but substantial rise in the geothermal gradient.

REFERENCES

Alpern, B., Rev. ind. minérale, 38, 170-81, (1956).

Bell, J.A., and Murchison, D.G., Fuel London, 45, 407-15, (1966).

Brooks, J.D., Gould, K., and Smith, J.W., Nature, 222, 257-9, (1969).

Brooks, J., and Shaw, G., Nature, 220, 678-9, (1968).

Brown, J.K., B.C.U.R.A. Monthly Bull., 23, 1-17, (1959).

Burghardt, O., Fortschr. Geol. Rheinld. u. Westf., 12, 421-8, (1964).

Cooper, B.S., and Murchison, D.G., in Organic Geochemistry - Methods and Results (edit. by Eglinton, G. and Murphy, M.), 720 (Springer-Verlag, Berlin, 1969).

Cooper, B.S., and Murchison, D.G., Unpublished data (1970).

Dormans, H.N.M., Huntjens, F.J., and van Krevelen, D.W., Fuel London, 36, 321-39, (1957).

Dryden, I.G.C., in Chemistry of Coal Utilisation (edit. by Lowry, H.H.), 232-95 (John Wiley, New York, 1963).

Gijzel, P. van, Leidse geol. Meded., 40, 263-317, (1968).

Given, P.H., Peover, M.E., and Wyss, W.F., Fuel London, 39, 323-340, (1960).

International Committee for Coal Petrology, International Handbook of Coal Petrography, 2nd Edn., (C.N.R.S., Paris, 1963).

Krevelen, D.W. van, Fuel London, 29, 269-84, (1950).

Krevelen, D.W. van, Coal, 116 (Elsevier, Amsterdam, 1961).

Krevelen, D.W. van, in Organic Geochemistry, (edit. by Breger, I.A.), 200 (Pergamon Press, London, 1963).

Kröger, C., and Bürger, H., Brennstoff-Chem., 40, 76-85, (1959).

Kröger, C., Goebgen, H.G., and Klusmann, A., Brennstoff-Chem., 45, 170-8, (1964).

Kröger, C., Pohl, A., and Kuthe, F., Glückauf, 97, 122-37, (1957).

Ladam, A., Iselin, P., and Alpern, B., Brennstoff-Chem., 39, 58-61, (1958).

Leythaeuser, D., and Welte, D.H., in Advances in Organic Geochemistry 1968, (edit. by Schenck, P.A., and Havenaar, I.), 432, (Pergamon Press, London, 1969).

Murchison, D.G., and Jones, J.M., Fuel London, 42, 141-58, (1963).

Potonié, R., Rehnelt, K., Stach, E., and Wolf, M., Fortschr. Geol. Rheinld. u. Westf., 17, 461-98, (1970).

Smith, A.H.V., Proc. Yorks. geol. Soc., 33, 423-74, (1962).

Stach, E., Brennstoff-Chem., 43, 71-8, (1962).

Stach, E., Fortschr. Geol. Rheinld. u. Westf., 12, 403-20, (1964).

Suggate, R.P., Personal Communication, (1970).

Teichmüller, M., and Teichmüller, R., in Coal Science (edit. by Gould, R.F.), 133-155 (American Chemical Society, Washington, 1966).

Tschamler, H., and de Ruiter, E., in Chemistry of Coal Utilisation (edit. by Lowry, H.H.), 35-118 (John Wiley, New York, 1963).

Tschamler, H., and de Ruiter, E., in Coal Science (edit. by Gould, R.F.), 332-341 (American Chemical Society, Washington, 1966).

Zetsche, F., Mitt. naturforsch. Ges. Bern, (1930).

DIAGENESIS OF SPOROPOLLENIN AND OTHER COMPARABLE ORGANIC SUBSTANCES: APPLICATION TO HYDROCARBON RESEARCH

Michel Correia *

Abstract

An attempt is made to evaluate the various parameters of the diagenetic evolution of sporopollenin and other associated organic elements. This study has been statistically treated with a C.D.C. 6400 computer.

The diagenetic evolution of sporopollenin appears to depend largely on its maximum depth of burial and on the temperature attained at this depth. However, some parameters like the lithology, and age of the studied geological formation also have some influence on the evolution.

The organic elements, (vegetable remains, amorphous organic matter) encountered in association with the spores and pollen seem to be less affected by thermal evolution. This is probably due to the physico-chemical conditions prevailing at the time of sedimentation.

The genesis of hydrocarbons pre-supposes certain essential constituents such as adequate quantity and suitable type of organic matter, as well as a moderate thermal maturity. Such characteristic factors may be determined by physico-chemical analysis and geochemical interpretation. Moreover, the microscopical examination of sporopollenin and associated organic elements offer valuable supplementary information to these analyses.

In some oil-producing countries, it was possible to define

* Much of the work done here was completed when the author was working at the Institut Français du Pétrole.(Present address: Société Nationale des Pétroles d'Aquitaine, Research Centre, PAU)

the relationship between the quality and degree of evolution of the microscopically observed organic matter of the mother-rock on the one hand, and on the other, the various types of hydrocarbons encountered.

It is therefore possible to evaluate oil potential possibilities from the microscopical study of sporopollenin.

Introduction

This study concerns the diagenesis of sporopollenin and associated organic substances which generally accompany spores and pollen grains. Later, we shall show that there may be possibilities of applying microscopic observations of different morphological types to determine the degree of diagenesis of the organic matter which may be used in petroleum predictions.

Organic matter observed under the microscope

The organic matter in sediments represents a very small proportion of the total compared with the mineral matter; whereas with coals, the proportions are reversed.

There are three types of organic matter present in sediments:-

1. Those which are insoluble and have definite morphological shapes, of more or less known biological affinity.

2. Insoluble organic matter of less well defined morphology.

3. Soluble organic matter.

In general, in palynological preparations, it is possible to recover the first two types of organic matter.

The insoluble organic matter with definite morphologic shape is a large group, in which can be placed both properly defined microfossils, microorganisms, and organic debris.

Fossil microorganisms are very diverse: viz. spores and pollen grains, dinoflagellates, acritarchs, chitinozoans, scolecodonts, microforaminifera. These microorganisms have been described by various workers (MULLER, 1959; STAPLIN, 1961; de JEKHOWSKY, 1963; TAUGOURDEAU, 1966 and 1968; WILLIAMS and SARJEANT, 1967).

These diverse forms have different origins, and different distributions. Thus, the numerical distribution of spores and pollen is a function of proximity to the coast, c.f. recent sediments where spores are more abundant on the coast-line, and rarer in the open sea. Dinoflagellates require marine conditions and their abundance is usually associated with the occurrence of very small quantity of spores and pollen. The acritarchs perhaps were related to the Dinoflagellates, although they seemed to occupy a less saline environment. The chitinozoa, like the scolecodonts, are of animal origin, the first are found in marine deposits, although the scolecodonts appears to be associated with a reef facies.

Debris of organs or organisms. Generally plant debris predominates over that produced by animals. It appears that the abundance of cuticles, tracheids, sporangial debris and other such remains are inversely proportional to the quantity of spores and pollen (CORREIA, 1969).

Insoluble organic matter with poorly defined outlines includes fine organic matter which may be colloidal. This kind of debris is usually encountered when microorganisms are rare and plant debris less abundant.

Eventual chemical composition of organic elements seen under

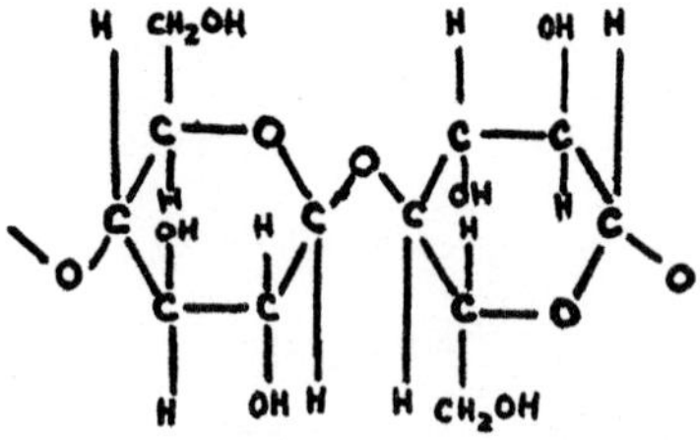

Fig.1. Molecular formula of cellulose

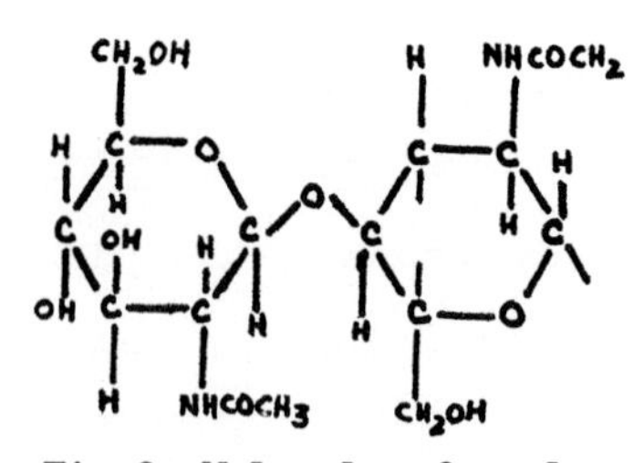

Fig.2. Molecular formula of chitin

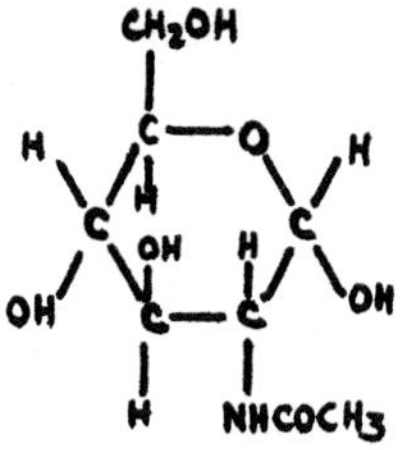

Fig.3. Molecular formula of N-acetylglucosamine

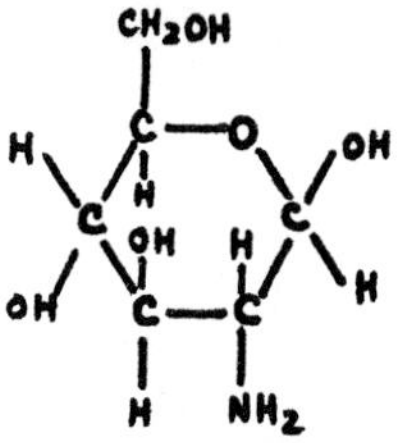

Fig.4. Molecular formula of glucosamine

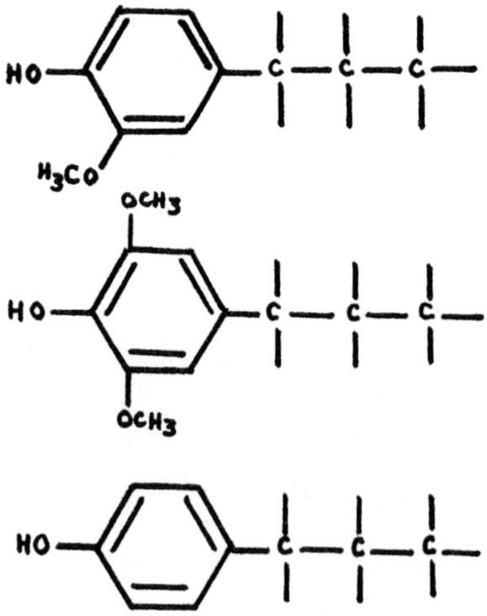

Fig.5. Possible molecular formulas for lignin. (after MANSKAYA & DROZDOVA, 1963)

the microscope. The diagenesis of microscopic organic matter may be partly related to their chemical composition. We will attempt to define the various organic and mineral constituents which compose them, in the living state; then follow the diagenetic modification of this matter.

These various microorganisms are composed essentially in the living state of carbohydrates, lignin, proteins, lipids, pigments and often mineral constituents.

The cellular plasma is generally composed of simple sugars whereas the membrane is often impregnated with polysaccharide material of which cellulose is the most abundantly occurring constituent. The membrane of plant cell-walls, spores and pollen, and even recent Dinoflagellates all contain cellulose in variable proportion. In contrast, fossil forms appear to be bereft of cellulose (DEFLANDRE, 1938). The linking β-glucoside C-O-C which binds molecules of glucose (Fig.1) may be disrupted by a specific enzyme emitted by bacteria or by oxidation, and so the absence of cellulose may be explained in fossil forms by bacterial action, or oxidation during the depositional process (CHIAVERINA, 1962; HUGEL, 1964; VAN KREVELEN, 1964).

Chitin is the polysaccharide of the animal kingdom. Its molecular formula is similar to that of glucose, but it is differentiated from cellulose by the presence of the $NHCOCH_3$ radical (Fig.2). The β-glucoside link may be broken either by bacterial action or by oxidation, and leads to the formation of acetylglucosamine (Fig.3) or glucosamine (Fig.4). The product may be further degraded to produce a stable compound comparable with 'melanoidine' (MAILLARD, 1912), made evident on the graptolites of the Dictyonema shale (MANSKAYA and DROZDOVA, 1963). Chitinozoa and scolecodonts do not react to the tests for chitin (EISENACK, 1931).

Lignin is an important constituent in plant cell walls, cuticles, and tracheids. The total proportion of lignin in wood is variable, and the gymnosperms have a higher proportion than the angiosperms. Its molecular formula is derived from a phenolic ring with a short, 3-carbon chain (Fig.5) (CHIAVERINA, 1962). It is thought that lignin may be partly, or fully combined with cellulose or hemi-cellulose. The chemical stability of lignin has proved difficult to study; since with the normal methods used, it is possible to obtain many different products. Thus, after solubilisation of cellulose and the hemicelluloses with a strong acid (CHIAVERINA, 1962) dark coloured humic substances are formed. In contrast, if lignin is treated with H_2SO_4 or KOH, and the product can be precipitated, a number of lignins, differing in composition and properties, are produced (CHIAVERINA, 1962). The term lignin certainly covers a complex group of compounds which may not be more precisely defined. It may, however, be concluded, that during the course of sedimentation, the celluloses and hemicelluloses are degraded, and that the lignin fraction of the plant assemblage is fossilisable.

Proteins may comprise 10-35 per cent of recent spores and pollen grains (HUGEL, 1964), and play an important biological role. Spores and pollen grains contain amino acids, amongst which we can find the most stable: lysine, glycine, isoleucine and leucine

(Table 1) (CALIFET and LOUIS, 1965; HUGEL, 1964) and fossil graptolites also show the presence of amino acids (MANSKAYA and DROZDOVA, 1962). It may therefore, be possible in the future to find such amino acids in fossil spores and pollen grains.

Lipids may constitute from 1-20% of recent pollen grains (HUGEL, 1964). It is known that after saponification and elimination of the acid fraction, hydrocarbons, alcohols, and sterol products remain. Now the polymerisation products of alcohols and fatty acids are found in plant cuticles and the external membrane (exine) of spores. It is believed that some types of organic matter with similar structure to that of certain living algae (e.g. Botryococcus-type) contain lipids in great quantities. There is not a great deal of analytical or bibliographical information on the relationships between lipids and the various fossil organisms that are the subject of this investigation.

Pigments are numerous: flavonoids, chlorophyll, carotenoids. Chlorophyll is changed during the course of deposition; it loses its magnesium, and produces stable complexes, by combining with either vanadium or nickel. Most living plants, and certain microorganisms contain chlorophyll, which in the fossil state is no longer detectable, and only the stable degradation products are found. It now appears that the carotenoids are responsible for the formation of sporopollenin (BROOKS and SHAW, 1968). Because this is the subject of this Symposium, we will not repeat here all the ideas about this substance. It should be remembered, however, that the proportion of sporopollenin is variable between genera and species of spores and pollen. On the other hand, it

TABLE I

Composition and amino-acid content of three pollen grains (after Nielsen,1955)

Values are given in g/100 g. of proteins.

Amino acids	Pollen grains		
	Zea mays	*Alnus glutinosa*	*Pinus montana*
Alanine	+	+	+
Arginine	6.3	9.8	6.4
Glutamic acid	+	+	+
Glycine[*]	+	+	+
Hydroxy-proline	+	+	+
Isoleucine[*]	+	+	+
Leucine[*]	7.6	6.0	6.5
Lysine[*]	5.0	4.7	5.1
Phenyl-alanine	2.9	2.3	2.1
Proline	+	+	+
Serine	+	+	+
Valine	+	+	+

+ = traces [*] = mentioned in text.

TABLE II

Mineral content determined by ash analysis (after Hugel,1964)(%)

Potassium	20 to 40
Magnesium	1 to 12
Calcium	1 to 19
Phosphorus	1 to 20
Silicon	2 to 10
Sulphur	1
Chlorine	0.8
Copper	0.05 to 0.08
Iron	0.01 to 0.3

seems that certain organisms like Dinoflagellates also possess sporopollenin. In fact, sporopollenin, however it be defined, could be the coloured substance in the external membrane of certain microorganisms.

The most abundant mineral constituents of living spores and pollen grains appear to be potassium, magnesium, calcium, silicon, and phosphorus (HUGEL, 1964), but minor quantities of sulphur, chlorine, copper, and iron (Table II) are also encountered. To our knowledge, the other microorganisms and remains of organs, have not been subjected to analysis for their mineral content.

Evolution during sedimentation

During the process of sedimentation, physico-chemical and mechanical conditions may occur that can modify the chemical composition and physical aspect of various components in the organic matter. The physical degradation undergone by the various microorganisms seems to be the most important process of fossilisation. Various authors (HAVINGA, 1964) have elsewhere reported on various aspects of degradation during the time of sedimentation. In this article, which concerns the corresponding chemical evolution undergone by these organisms, it is reasonable to assume that the degradation takes place by a progressive elimination of the less stable products and a resulting concentration of the most stable types (melanoidine, humic acids, amino acids, degradation products of lignin etc.). Sporopollenin seems to be very resistant and remains extremely stable throughout the period of deposition. It seems equally likely that certain morphological types can be transformed during aerobic phases when the plant cell walls are corroded and finally produce an organic substance which looks smooth or finely granulated under the microscope (BARGHOORN, 1952).

Diagenetic evolution

We have thus tried to place in order, on the basis of their physico-chemical properties, sporopollenin and the other chemical constituents of insoluble organic matter. Our aim is to show how these components evolve during the course of diagenesis, and which have diagenetic properties that are most noticeably modified. In fact, we will compare the diagenetic evolution of sporopollenin with that of other organic constituents.

Method of study

Organic particles show diagenetic evolution by becoming progressively more opaque and the colours darken, passing from transparent yellow, to orange, red, reddish brown and finally, to black. This colorimetric evolution is progressive, but manifests itself differently, on various parts of individual microfossils or organic debris. For example, appendages, or membranous folds become brown more quickly than smooth areas. This increasing opacity can be measured by different methods: the light absorption method (GUTJAHR, 1966) and the light reflectance method to give a calculated reflectance power (ALPERN, 1964 and 1967).

CORREIA (1969) in a recent publication described the methods used to transfer the various geological and palynological data to punched cards. We will here described the principal characteristics of this transfer. It is essential that all the samples should be treated by the same standard extraction technique:

TABLE III

Characteristics of different states of preservation as seen in different groups of microfossils (after Correia, 1967 & 1969).

Spores & Pollen Grains	Dinoflagellates Hystrichospoeres Acritarchs	Chitinozoa	Plant Debris	Amorphous Organic Matter	State of Preservation
Wall of central body, transparent clear yellow Trilete mark clear, distinct	Transparent, clear yellow Insertion of appendages clearly visible	Clear yellow to brown Neck & body translucent Internal structure clearly visible. Appendages transparent	Clear yellow to brown Cuticles & tracheids well-preserved	Clear yellow to red	1
Wall of central body, orange to reddish Trilete mark slightly opaque	Transparent, but tending to brown Base of appendages becoming opaque	Identical to state 1	Identical to state 1	Reddish brown	2
Central body brown. Trilete mark opaque, becoming obscure	Yellowish brown Appendages becoming more opaque	Identical with state 2 or 4	Becoming opaque Fragmentation	Brown	3
Central body opaque. Trilete mark hard to see	Opaque, appendages dark - black	Dark, neck opaque Splitting between body & neck. Brittle	Between states 3 and 5	Black	4
Brown spherules	Polygonal forms but indeterminate	As above	Brown & black small fragments	Black	5

Central body here means main part of the spore or pollen grain.

$HNO_3 + H_2O + NaCl$

$HF + H_2O$

$HCl + H_2O$

mounting in glycerine jelly between slide and cover slip.

Preliminary observations have made it possible to draw up a chart of preservation taking into account at the same time morphology and colorimetry of each group of organisms and debris (CORREIA, 1967). Each of the groups represented (spores, pollen grains, acritarchs, dinoflagellates, hystrichospheres, chitinozoans, scolecodonts, plant debris) the commonest species have been chosen for observation; fifty of each were examined. It was then possible to give each group a certain colour value and in the final analysis to define, in relation to a chart, the state of preservation for the whole assemblage with all its different constituents (Table III).

These observations were made on samples coming from formations of different ages, and from different regions. The regions were the Paris Basin, the Aquitaine Basin, the Saxe Basin, the northern border of the Hoggar massif (Polignac, Ahnet Mouydir, Erg Chech), Madagascar (Morondava Basin and Majunga Basin), and Chile (Magellan Basin). We have, therefore, examined many hundreds of samples and have retained for this paper 1328 core-samples. Taking into account the total of 37 parameters (Fig. 6 and 7) for each sample of organic matter, we have available 49,000 facts, which were coded (i.e. they were attributed to different variables in value). Each region and each of the core samples, has been numbered and the age of the sample transcribed into millions of years using the absolute chronologic scale of SMITH and WILCOCK (1964). Further, we have indicated

Fig.6. List of the 37 codes of variables for the computer programme

PALYNO	Palynological number
REGION	Region of the basin
FORAGE	Number of the bore-hole
AGE	Age in millions of years
PROFAC	Actual depth
PROFM1	Maximum depth of burial (1)
PROFM2	Maximum depth of burial (2)
PROFM3	Maximum depth of burial (3)
GRADIE	Temperature gradient (°C)
EPENM1	Epoch of maximum burial (1)
EPENM2	Epoch of maximum burial (2)
EPENM3	Epoch of maximum burial (3)
TEMPAC	Present-day temperature
TEMPM1	Temperature of maximum burial (1)
TEMPM2	Temperature of maximum burial (2)
TEMPM3	Temperature of maximum burial (3)
SPORPO	State of preservation of spores & pollen
DEBVEG	State of preservation of plant debris
MATORG	State of preservation of amorphous organic matter
CLAGEN	General state of preservation
QSPORP	Relative quantity of spores & pollen
QDEBVE	Relative quantity of plant debris
QMATOR	Relative quantity of amorphous organic matter
DINOFL	State of preservation of dinoflagellates
ACRITA	State of preservation of acritarchs
CHITIN	State of preservation of chitinozoa
QDINOF	Relative quantity of dinoflagellates
QACRIT	Relative quantity of acritarchs
QCHITI	Relative quantity of chitinozoa
ARGILE	Percentage of argillaceous matter
CALCAR	Percentage of calcareous matter
SILICE	Percentage of siliceous matter
COULEU	Colour of the rock
CARORG	Organic carbon
CARMIN	Mineral carbon
HUILE	Proximity to trace of oil
GAZ	Proximity to trace of gas

Fig.7. Example of the data on cards concerning one sample
(1 line = 1 card)

```
PALYNO+0.8428E+04   REGION+0.2000E+01   FORAGE+0.2260E+01   AGE   +0.3730E+03   GE046401
PROFAC+0.1519E+04   PROFM1+0.2130E+04   PROFM2+0.2130E+04   PROFM3-0.0000E+04   PR046402
GRADIE+0.2600E+01   EPENM1+0.3000E+03   EPENM2+0.0700E+03   EPENM3-0.0000E+03   GR046403
TEMPAC+0.5400E+02   TEMPM1+0.7000E+02   TEMPM2+0.7000E+02   TEMPM3-0.0000E+02   TE046404
SPORPO+0.2000E+01   DEBVEG+0.2500E+01   MATORG-0.0000E+01   CLAGEN+0.2250E+01   CO046405
QSPORP+0.2000E+01   QDEBVE+0.2000E+01   QMATOR+0.    E+01                       QO046406
DINOFL+0.0000E+01   ACRITA+0.1500E+01   CHITIN+0.0000E+01                       OR046407
QDINOF+0.0000E+01   QACRIT+0.1000E+01   QCHITI+0.0000E+01                       QU046408
ARGILE+0.4000E-00   CALCAR+0.0000E+00   SILICE+0.6000E+00   COULEU-0.0000E+01   PO046409
CARORG-0.0000E+00   CARMIN-0.0000E+00   HUILE +0.2000E+01   GAZ   +0.1000E+01   GO046410
```

one, two and three assumptions of maximum burial. Our knowledge of the present geothermal gradient has enabled us to calculate the present temperature of the various sediments, and what they would have been during their period of maximum burial. Finally, it was necessary to take account of the petrological nature of the sample: the percentages of clay, lime, and silica have been indicated as a function of their lithology.

A numerical classification has been used to transfer the colours observed in the organisms (yellow = 1, orange = 2, reddish brown = 3, black = 4). The quantity of organisms present in each assemblage has been classified thus:

0 = absent; 1 = rare; 2 = present; 3 = intermediate;

4 = abundant; 5 = very abundant.

The proximity of traces or fields of hydrocarbons (gas and oil) has been indicated by the value 0 when there is no trace, 1 for a showing, and 2 for an oil or gas field.

It was necessary to handle all these data with a computer. All the information was recorded on data sheets, then on punched cards, and finally stored on magnetic tapes (de JEKHOWSKY, 1968).

The statistical calculations carried out on the CDC6400 computer by the program of treatment were classical: distribution, correlation and regression (BIENNER, de JEKHOWSKY, PELET and TISSOT, 1968).

<u>Diagenetic alteration factors affecting preservation</u>

The essential parameters for the transformation of organic matter contained in the rocks are temperature, pressure and the presence of mineral catalysis. The increasing effects of burial subsequent to subsidence results in rises of temperature and increase of pressure. It is important to note that the temperature reached will be possibly much higher than the local geothermal gradient. By a catalytic process, the lithology may affect the results and favour a thermal change.

Burial

Burial affects the evolution of organic material by the increase in temperature and results in the selected constituents of organic matter themselves being characterised.

If we examine the samples studied in the Paris Basin (Trias and Lias) we notice, while comparing the general state of preservation with the present depth of burial and with the depth of maximum burial, that the coefficients of correlation are identical, i.e. 0.73 - 0.72 (Table IV). It is the same in Chile (Magellan Basin) where the coefficient of correlation with the present depth of burial and with the maximum depth of burial are very close : 0.83 - 0.85 (Table V). The close similarity of these coefficients of correlation may be explained by the phenomena of a regular, homogeneous subsidence in each of these basis.

In contrast to this, in other regions (Polignac Basin, Madagascar, Aquitaine), if we only examine the actual depth of burial, the correlations are not significant (Aquitaine) or very poor (Table IV). The geological history of these basins has been characterised by important tectonic phases which led to uplift of certain areas, and initiated important erosional phases. The present day burial does not, therefore correspond to the maximum subsidence. This is particularly the case in the Polignac Basin

TABLE IV

Comparison of correlation coefficients between the general state of preservation and the depth of burial.

BASIN	ACTUAL DEPTH OF BURIAL	MAXIMUM DEPTH OF BURIAL
Aquitaine	0.09 (not significant)	0.46
Magellan	0.83	0.85
Morondava (Madagascar)	0.48	0.65
Paris	0.73	0.72
Polignac	0.37	0.59

TABLE V

Variation of the correlation coefficients between the general state of preservation and the depth of burial as a function of the age of the formation.

BASIN	Aquitaine	Morondava	Paris	Polignac
Total	0.46	0.65	0.72	0.59
Cretaceous	0.17*			
Jurassic	0.86	0.67 0.54 (Lias) (Trias)	0.77(Lias)	
Triassic	0.25	0.69 (Permo-Trias)	0.42	
Carboniferous				0.00*
Devonian				0.32
Silurian				0.67

* Correlation not significant

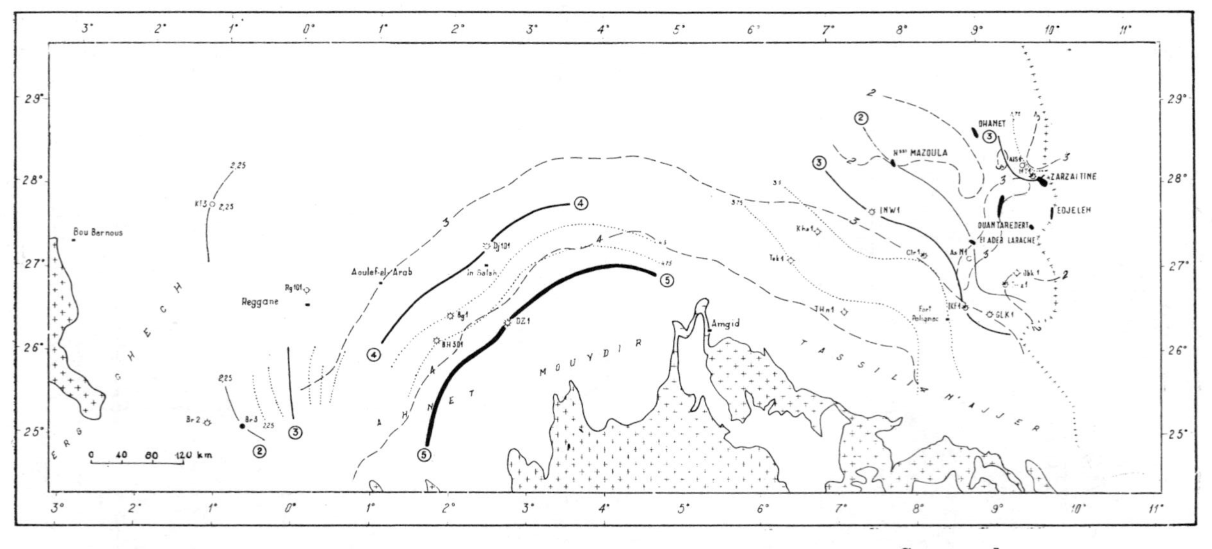

Dry hole — Trace of oil — Trace of gas — Oil producer — Gas producer

② ③ ④ ⑤ Limit of basement — Isogradient geothermal curve

Changing state of preservation

Figure 8. Distribution of the geothermal gradient and states of preservation north of the Hoggar Massif Polignac. Ahnet-Mouydir. and Erg Chech Basins. (after CORREIA.1969)

where the Hercynian tectonic phase uplifted the western part, and was followed by intensive erosion. It is possible to estimate the maximum depth of burial reached by the formations sampled: the correlations are significant in every case, and the coefficients of correlation between the general state of preservation and maximum burial are improved.

We conclude therefore, that burial, and particularly the depth of burial reached at the point of maximum subsidence has an influence on the state of preservation of the elements of the organic matter studied.

Temperature

Increasing depth of burial causes increase of temperature. Thus, for a given depth, the temperature is a function of the local geothermal gradient. Only the present day geothermal gradient is known, and it is assumed that it must always have been the same throughout geological time.

In the case of the Basins of Polignac, Ahnet Mouydir and the Erg Chech, states of preservation agree with the distribution of the geothermal gradient (Fig.8). In effect, the regions with a weak gradient ($2^{o}C$ per 100 m.) give well preserved elements, while those with stronger gradients (3.5 to $4^{o}C$ per 100 m.) give extremely altered microfossils.

The correlation between the values of the temperature at maximum burial and those of the preservation of the Silurian microfossils from the Polignac Basin is better than that obtained from the zone of maximum burial (i.e. 0.67 for maximum burial, and 0.77 temperature at maximum burial (Table IV)).

Age

Since burial and age act in the same direction, in order to use the evidence of burial properly, it is necessary to compare samples of the same age. Better coefficients of correlation are obtained from older and more deeply buried formations. This is true of the Polignac Basin where we concluded that for the Silurian, the coefficient of correlation (0.67) is better than that in the Devonian (0.32) and that of the Carboniferous where the correlation is not significant (Table V).

On the contrary, the other basins are different; generally, the coefficients of correlation are better for some formations than for others of different ages. Thus, in the Jurassic (Aquitaine, Madagascar, Paris Basin) the coefficients of correlation are good and much better than those of the Trias (Table V). Thus it seems that the influence of burial may be modified by the age of the deposit and/or by the characteristics of the deposits.

Lithology

We have tried to determine the influence of this factor on the 1328 samples examined.

Clay minerals are considered to be good catalysts, but in fact, while some are (e.g. montmorillonite), others are not so efficient. The clay environment is extremely heterogeneous in composition, and its behaviour as a catalyst may vary. In the carbonates, the environment is mineralogically constant, and their catalytic behaviour, although more unfavourable than the clays, is more homogeneous. Let us consider among the 1328 samples, those where the percentage of clay is greater than 50% (Group A), and those where the percentage of calcareous matter is more than 50% (Group B). Then we will examine the evolution of the general state of preservation as a function of the depth of burial in each of these groups. In Group A (+ 50% clay) the coefficient of correlation is of the order of 0.68, while for Group B (+ 50% of calcareous matter) it is fairly constant at 0.83. It is therefore demonstrated again in these coefficient values that the first idea is correct: the clay medium is heterogeneous and the correlation is not as good as in the calcareous environment which is more homogeneous from the point of view of their actions as catalysts.

It is important to know if, in a clay environment, alteration is stronger than in a calcareous environment. Considering the samples from the Paris Basin (Lias), whose state of preservation is better than 2, i.e. to the stage which should have been produced by alteration (Table III). The assemblage may be subdivided into two groups (Table IV); a group of samples whose depth of burial is less than 1,500 m, and a second group buried to greater than 1,500 m. If we examine the distribution of lithologies from each of these two groups, we notice that:

a) for burial of less than 1,500 m., 80% of the samples are clays and the organic components have passed stage 2 on the preservation scale; there are only 15% of calcareous samples, and

b) for burial of more than 1,500 m., there are no more than 16% clay samples where the state of preservation is altered, and most of the samples (72%) in this category are calcareous.

It seems therefore, that microfossils are more readily altered in the clay environment than in a calcareous one.

Diagenetic behaviour of the various organic elements studied.

The organic elements used for this study are very diverse. The value for the state of preservation is defined for the whole assemblage of microfossils studied, which makes it possible, if a group is absent, or poorly represented, nonetheless to define a state of preservation; the state of preservation takes account both of the morphology and the colour (Table VII). The behaviour of each of the groups has been studied on the basis of colorimetric observation. This is in fact the equivalent of the methods of measurement by absorption (GUTJAHR, 1964) or, in part, the reflectance measurements (ALPERN, 1964).

As a first attempt, we studied the colorimetric behaviour as a function of burial on the total assemblage of spores, pollen, plant debris, and organic matter. The study was made on samples from the Paris Basin (Trias and Lias). We have totalled the colours of the microfossils measured, and calculated a median value:

$$\text{C.V. median} = \frac{\text{C.V. spores and pollen grains} + \text{C.V. plant debris} + \text{C.V. organic matter}}{3}$$

(where C.V. = colorimetric value).

We have compared the evolution of this value with the increasing burial. The coefficient of correlation is equal at 0.45; but, the coefficient of correlation between the state of preservation and depth of maximum burial for the Lias/Trias assemblage from the Paris Basin is 0.72. Thus, taking into

TABLE VI

Comparison of correlation coefficients between colorimetric evolution and increasing depth of burial and temperature for certain groups of organic elements.

REGION	State of preservation	Spores and Pollen grains	Plant debris	Amorphous organic matter
Polignac Basin (Silurian)				
- PROFM*	0.67	0.40	0.30	0.52
- TEMPM°	0.77	0.61	0.49	0.55
Paris Basin (Lias)				
- PROFM*	0.77	0.47	0.24	0.42
- TEMPM°	0.75	0.48	0.21	0.38
Aquitaine Basin (Jurassic)				
- PROFM*	0.85	0.87	0.27+	0.65
- TEMPM°	0.85	0.87	0.32+	0.63
Magellan Basin (Chile)				
- PROFM*	0.84	0.77	0.59	0.57
- TEMPM°	0.83	0.77	0.59	0.55

PROFM* = depth at time of maximum burial. + = correlation not significant.
TEMPM° = Temperature at maximum burial.

TABLE VII

Influence of lithology on the state of preservation of microfossils(Paris Basin, Lias; samples where the state of preservation index in greater than 2)
(see Table III)

MAXIMUM DEPTH OF BURIAL (m)	NUMBER OF SAMPLES	PETROGRAPHY		
		50% silica	50% lime	50% clay
<1,500	(41)	(2) 5%	(6) 15%	(33) 80%
>1,500	(51)	(6) 12%	(37) 72%	(8) 16%

account only the colorimetric state, the correlation is not so good.

If we now take the colorimetric evolution of each of the groups as a function of burial, we conclude that spores and pollen grains have the best coefficient of correlation (from 0.40 to 0.87) (Table VI). For the other organic constituents this coefficient is much worse; plant debris gives 0.21 to 0.59; organic matter gives 0.38 to 0.65. It is noteworthy that these coefficients of correlation are always worse than those obtained using states of preservation (0.67 to 0.84). These differences in value are explained by the fact that the state of preservation takes into account several factors (morphological factors as well as colorimetric ones) while the colorimetric criterion is exclusive.

Electron microprobe study of the observed diagenetic evolution.

The structure of insoluble organic matter (kerogen) is little known. It appears to be a macromolecule with naphthene-aromatic rings bound by linkages of a hetero-aromatic type (TISSOT, 1969a). With the increasing temperature resulting from increasing depth of burial, these bonds of the type C-S, C-O, and occasionally C-C are broken and polycondensation is accentuated. The breaking of these bonds are accompanied by the expulsion of oxygen and light constituents (CO_2, H_2O, CH_4, etc.....).

The evolution of the states of preservation towards alteration has to be explained by reference to analogous systems. At the present time, it is difficult to confirm them. We can only give an account of our attempts to use the electron probe micro-

Table VIII

Characteristics of samples where the microfossils have been subject to electron probe microanalysis.

Sample Number	Actual Depth(m.)	State of Preservation
1	331	Good - 1.25 - Yellow
2	311	Medium -2.75- reddish-brown
3	2,716 3,500	Bad - 4 - Black

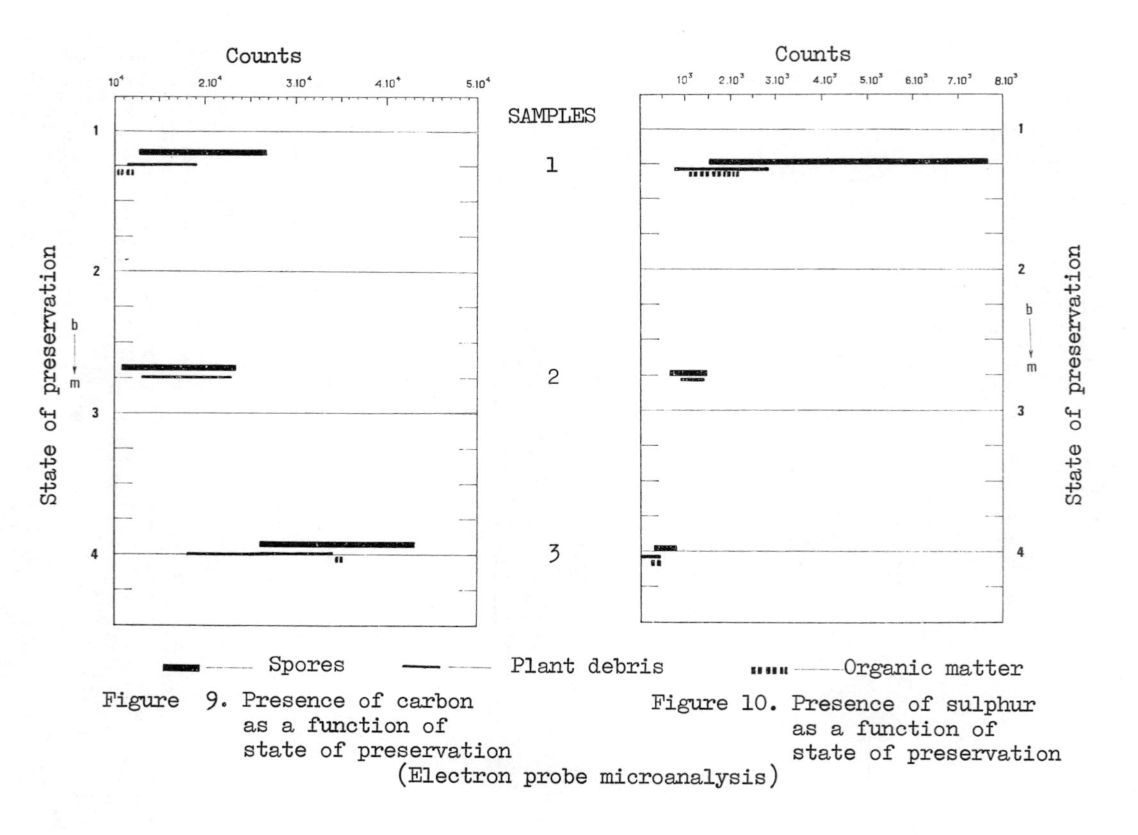

Figure 9. Presence of carbon as a function of state of preservation

Figure 10. Presence of sulphur as a function of state of preservation

(Electron probe microanalysis)

analyser to study this.

From the residues obtained after chemical treatment of the Jurassic sediments of Madagascar, we extracted various groups of microfossils: trilete spores; plant debris and amorphous organic matter in various states of preservation (Table VIII). The following elements were detected in different individual specimens of the three groups listed above:

C, N, O, F, Na, Mg, P, S, K, Ca, Fe.

(Hydrogen cannot be detected by an electron probe microanalyser).

The relative quantities of these different atoms are very variable, in the groups considered, but generally are consistent with the state of preservation. In effect, we could see that the different fossil microorganisms were systematically richer in carbon than in sulphur, oxygen and nitrogen. Other than that, well-preserved organisms (sample 1, Table VIII) are poorer in carbon than those which have degraded (sample 3, Fig.9). In contrast, the unaltered microfossils are rich in sulphur, while in those which are altered the sulphur content is very low (Fig.10).

It seems that there has been polycondensation of the molecules and a loss of light degradation products like CO_2 and H_2S. It is generally at the end of stage 4 (in state of preservation) that the important physico-chemical modifications become important. Stage 4, is about equivalent to anthracite rank (CORREIA, 1969).

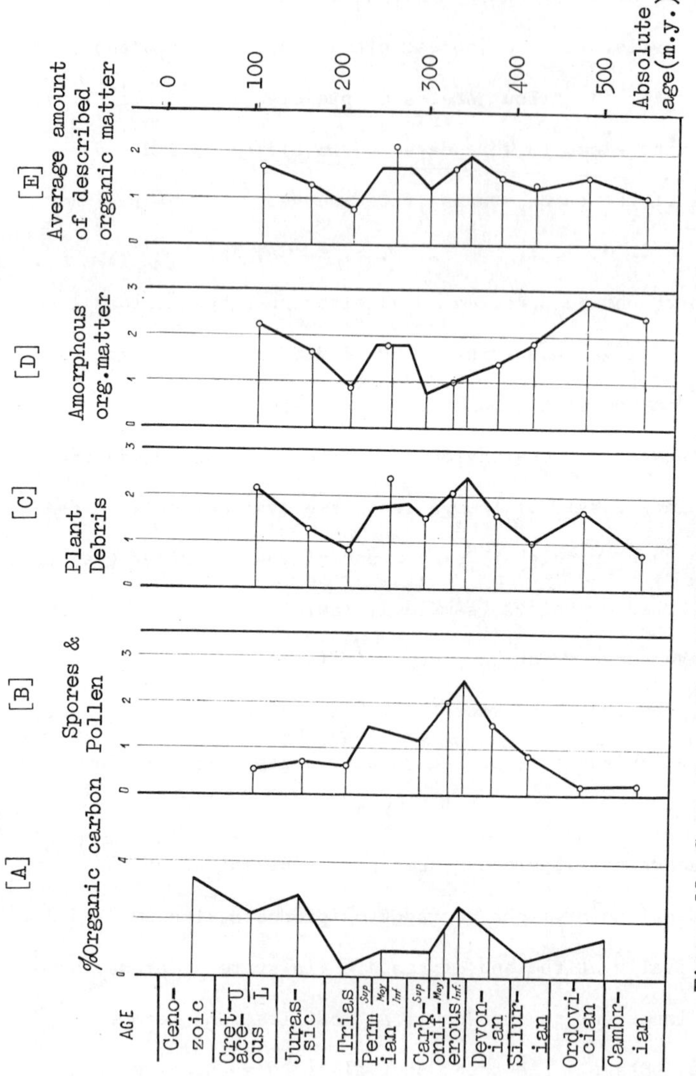

Figure 11. Comparison of quantitative evolution of organic matter and the described groups through geologic time. (after CORREIA, 1969).

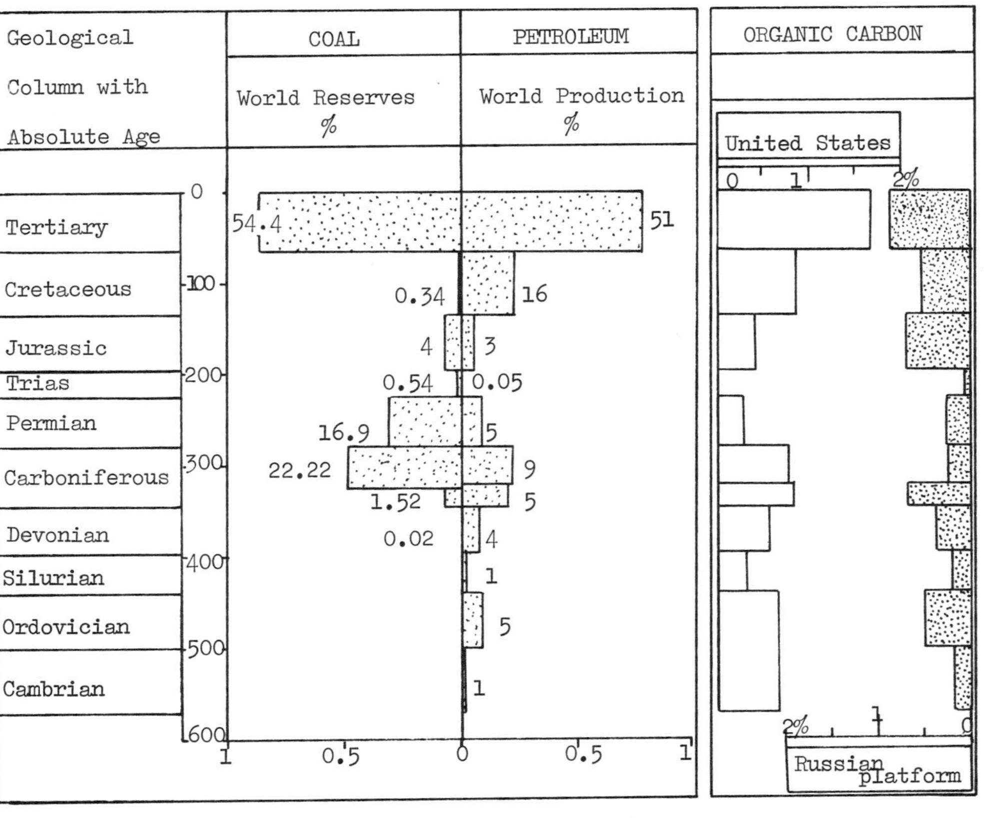

Fig. 12 Geologic distribution of coal and petroleum in the world; also a stratigraphic comparison of organic carbon in the petroleum provinces of the United States, and the Russian platform. (After J.DEBYSER and G.DEROO, 1969.)

Quantitative evolution of sporopollenin and other organic matter in the course of geological time - comparison with coals, petroleum and organic carbon.

In dealing with a number of samples from all the geological epochs from the Palaeozoic and Mesozoic, we have calculated (see above) the average quantity of various organic constituents in each stage. We have thus considered successively the quantity of spores and pollen, plant debris and amorphous material in the total organic matter under investigation. In one epoch, the Carboniferous, spores, pollen and plant debris are very abundant, whilst in the Ordovician, the debris and amorphous organic matter are the more prominent. The three categories are quantitatively important in the Permian, the Silurian and Trias represent two epochs with the smallest quantities.

Thus the world reserves of Triassic coal are very poor (Fig.12). The world production of petroleum is poor both in the Trias and in the Silurian (Fig.12). Finally the amount of organic carbon in the sedimentary rocks of the U.S.A. and on the Russian platform are at a minimum in the Trias and the Silurian (Figs. 11 and 12) (DEBYSER and DEROO, 1969).

Thus, in the 1328 samples studied with reference to their geological history, we found a quantitative evolution of organic components which generally matched the geological distribution of coals and oil deposits of the world. In particular the quantities of organic carbon present matched the petroliferous provinces, especially those of the Russian platform.

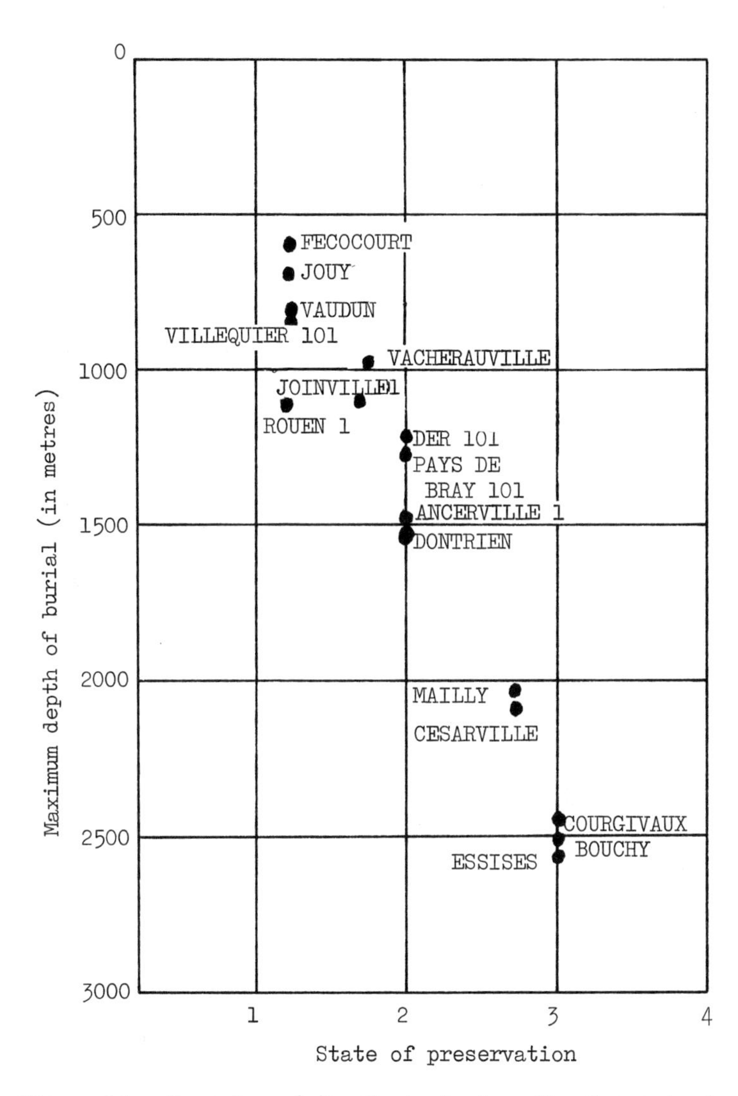

Figure 13. Toarcian of the Paris Basin. Development of the state of preservation of the organic matter as a function of the maximum depth of burial.

Relationships between the diagenetic evolution of the organic matter and the transformation of organic matter into hydrocarbons.

We will first examine two cases which are relatively simple because of their homogeneity, i.e. the chemical and lithological composition of the series studied are monotonous, there has been no metamorphism to add to the effects of diagenesis, and the series were deposited in a relatively short time. The two series are the Toarcian of the Paris Basin, and the sedimentary series of the Upper Cretaceous and Paleocene of the Douala Basin (Camerouns, Equatorial Africa). These two models have been the subject of relatively complete geochemical studies (LOUIS and TISSOT, 1967; ALBRECHT, 1969). Our observations lead to the conclusions that diagenesis is least in the Paris Basin; state of preservation = 3; and a maximum burial of the order of 2,500 m. (Fig.13) (TISSOT, 1969a). At Logbaba (Douala Basin) the depth of burial was 4,000 m. and the state of preservation reached stage 4 (Fig.14). Now, in the case of the Paris Basin, it was noticed that the ratio of totally saturated hydrocarbons to total organic carbon is augmented from 1,300 m. (equivalent to 60°C) downwards (Fig.15) and its maximum depth of burial is too small to see the ratio diminish in the latter depths as was the case in Logbaba (Fig.16). Indeed, we see that below 2,800 m. the ratio is not significant. Now these different levels at 1,300 m. (60°C) in the Paris Basin, and with only a very slightly different temperature at Logbaba (65°C), correspond to slightly different states of preservation: about 2 for the Toarcian, and 2.75 for Logbaba. The second level, i.e. the

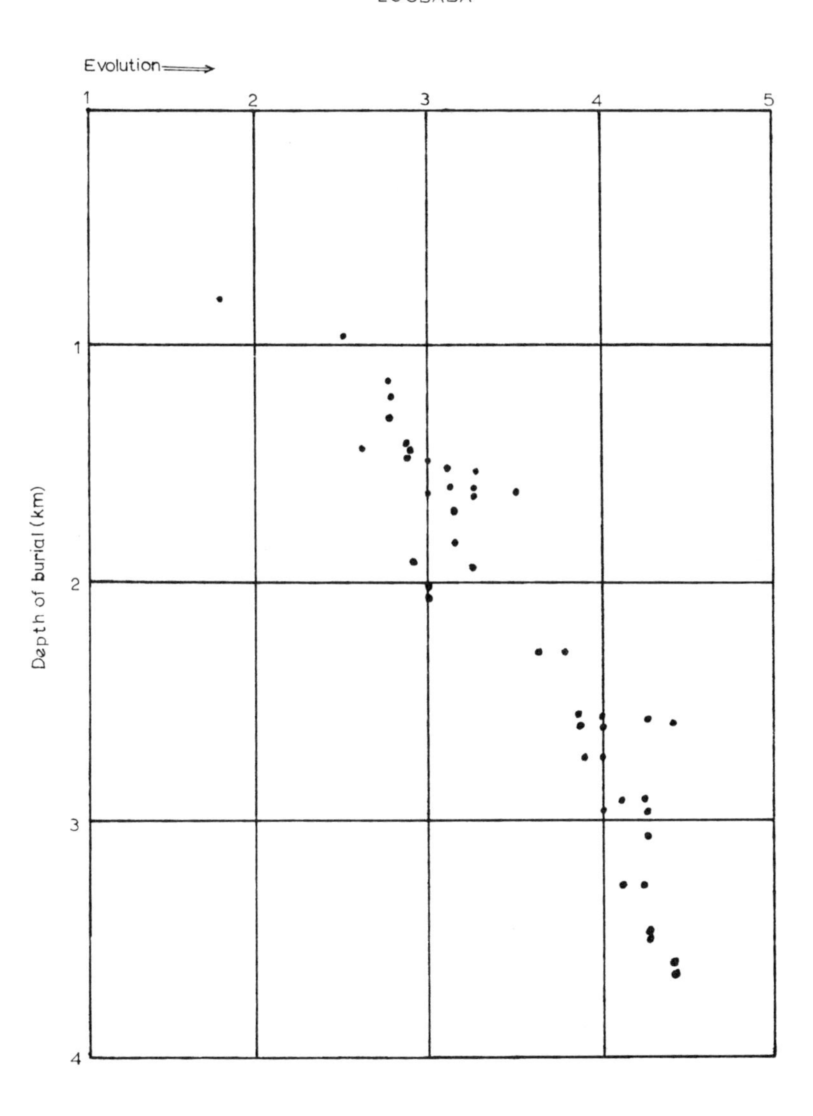

Fig 14

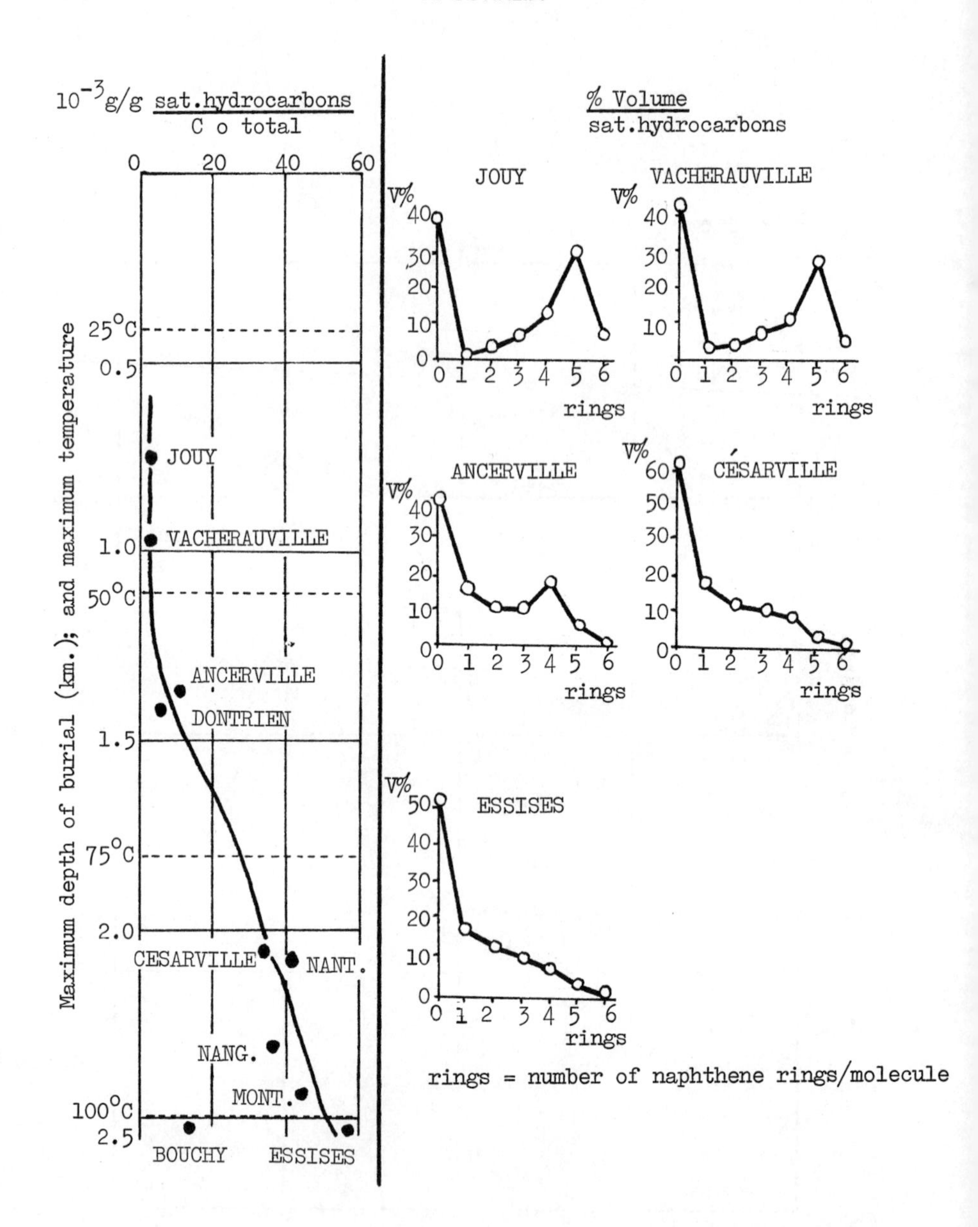

Figure 15. Variations of content and composition of saturated hydrocarbons as a function of the maximum depth of burial (after LOUIS & TISSOT, 1963)

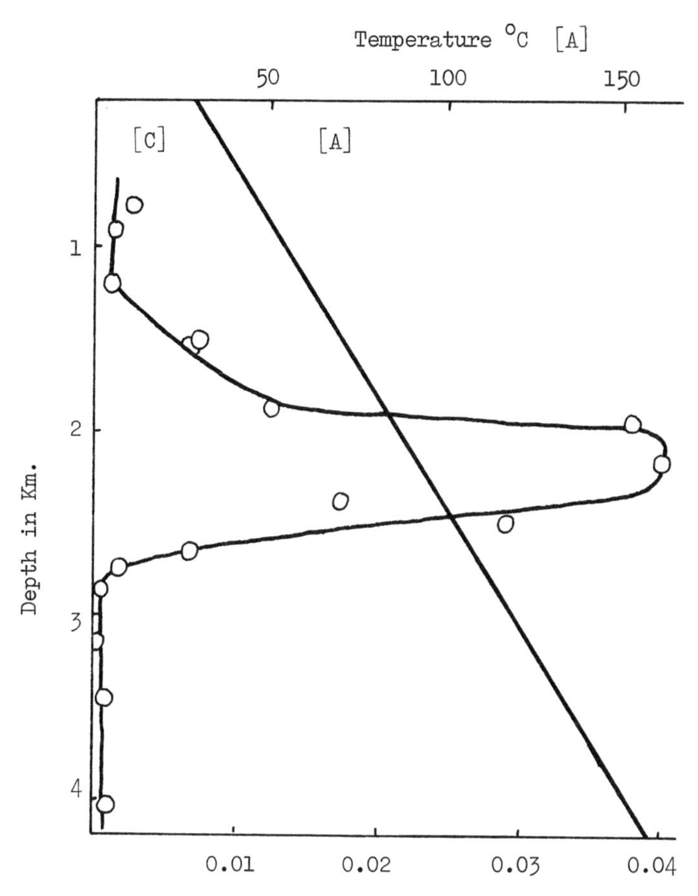

Figure 16. Variation of the relationship of saturated hydrocarbons/total organic carbon with depth in the borings at Logbaba(after ALBRECHT,1969)

one where the ratio has become very low, (2,800 m.) corresponds to states of preservation of 3.75 - 4.00. Are such states of preservation in accord with those encountered in the oil-fields?

Relationship between the diagenetic state of sporopollenin and other organic elements and the presence of hydrocarbons.

The genesis of hydrocarbons pre-supposes favourable conditions for the accumulation and preservation of organic material. Later, increases of temperature and pressure, for which burial is responsible, alter this organic matter (SOKOLOV, 1966 in TISSOT, 1969). We conclude that the quantities of oil produced are important at a particular level (about 1,300 m.), but deeper than 4,000 m., there will be no more oil produced, whereas gas will continue to be produced beyond this level (Fig.17) (TISSOT, 1969b).

The state of preservation of the studied organic matter allows us to determine the thermal maturity and consequently the maturity of the petroleum in a sediment. Further observations in a number of additional basins will either prove or disprove this hypothesis. If we take the Paris Basin as our first example, we know that the Dogger (M.Jurassic, Ed.) acts as cap-rock for the field in the centre of the basin. If we assume that the mother-rocks are the marine shales of the Bajocian, then migration only occurred over short distances, and the state of maturity of the organic matter may be related to the presence of hydrocarbons. The more stable the organic matter the better is the chance of finding hydrocarbons. The study of the organic constituents from the Lias and Trias allow

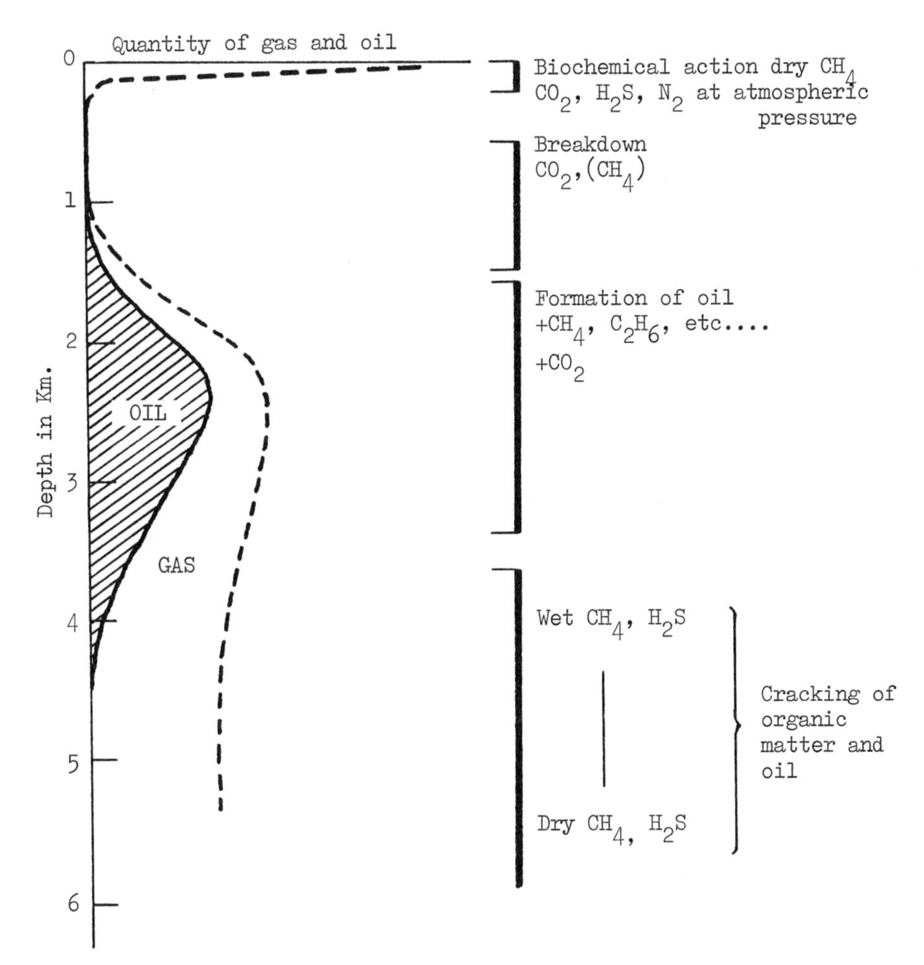

Figure 17. Schematic diagram showing variations in quantities of oil and gas as a function of depth. (After TISSOT, 1969b).

us to establish iso-preservation curves (Fig.18). We have found that for states of preservation with indices equal to or larger than 2, we find traces of oil,and oil fields occur when the state of preservation corresponds to an index greater than 3.

The basins situated on the northern edge of the Hoggar massif (Polignac, Ahnet Mouydir, Erg Chech) give us another opportunity to observe the relations between the states of preservation and the presence of hydrocarbons. In the Polignac Basin, we can see a relationship between good preservation and the presence of hydrocarbons (Fig.8). The hydrocarbon fields are localised by the upper zones where the microfossils are hardly altered, i.e. with a state of preservation corresponding to an index less than 3; when they are altered (index greater than 3) there are no traces of hydrocarbons (CORREIA, 1967). It appears as though these results are a contradiction of those obtained in the Paris Basin, but in fact, this is not so. Actually, the Polignac Basin has been the recipient of crude oils formed in the neighbouring subsident zones by migration (CORREIA, 1969). The north-west region, as a result of Hercynian tectonic uplift, has undergone an active hydrodynamism with, as a result, the washing out of hydrocarbon products. Thus it can be understood why the state of maturity of the organic matter, and more particularly of the studied organic components cannot easily be related to the presence of hydrocarbons.

In Ahnet Mouydir we found that the palynoplanktonic material was heavily altered, and only gas was found (Fig.8).

In the Aquitaine Basin, the oil fields are localised in regions

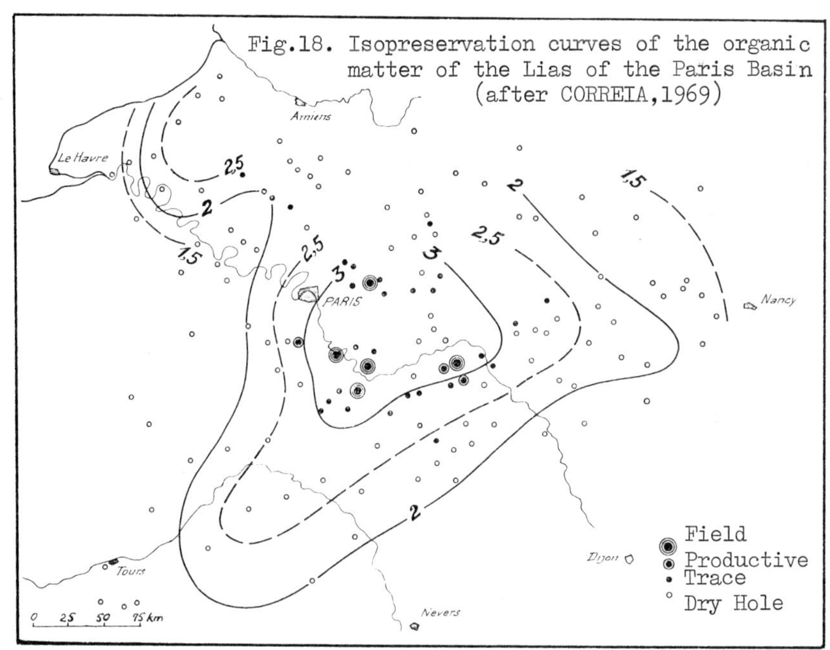

Fig.18. Isopreservation curves of the organic matter of the Lias of the Paris Basin (after CORREIA,1969)

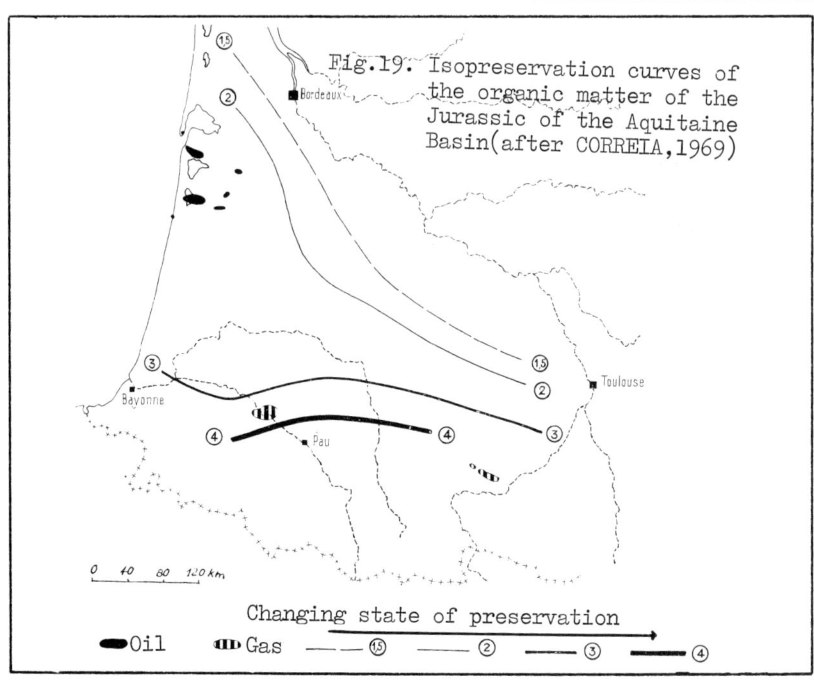

Fig.19. Isopreservation curves of the organic matter of the Jurassic of the Aquitaine Basin(after CORREIA,1969)

where the state of preservation index is greater than 2, while above a state of preservation index of 4, only gas is found (Fig.19). From these various observations we have deduced that the states of preservation of palynoplanktonic microfossils are related to the alteration of organic material into hydrocarbons.

Relationships between the different types of organic matter and the nature of the hydrocarbons formed.

The diagenesis of the three components, sporopollenin, amorphous organic matter, and other plant debris shows differences in behaviour.

Some of these components evolve hydrocarbons corresponding with the increase in temperature; these are mainly the spores and pollen; others appear to be affected by other factors of degradation, probably related to their original chemical composition and their thermal stability during diagenesis (e.g. debris of lignin, such as tracheids). Finally, a third group, amorphous organic matter appears to be affected both by diagenesis and by some other parameters. The difference in the diagenetic level seen in the Toarcian of the Paris Basin and at Logbaba is related to the type of organic matter present and to the temperature necessary to produce degradation of the various components. In the Toarcian, the organic material is more of a colloidal type, but at Logbaba, the material is of ligno-cellulosic nature (plant debris of lignins and celluloses). It seems that the first of these may be more easily transformed than the second, because the one transforms at about 60°C (Fig.15) (LOUIS and TISSOT, 1967), while the second requires a temperature in excess of 65°C (Fig.16) (ALBRECHT, 1969).

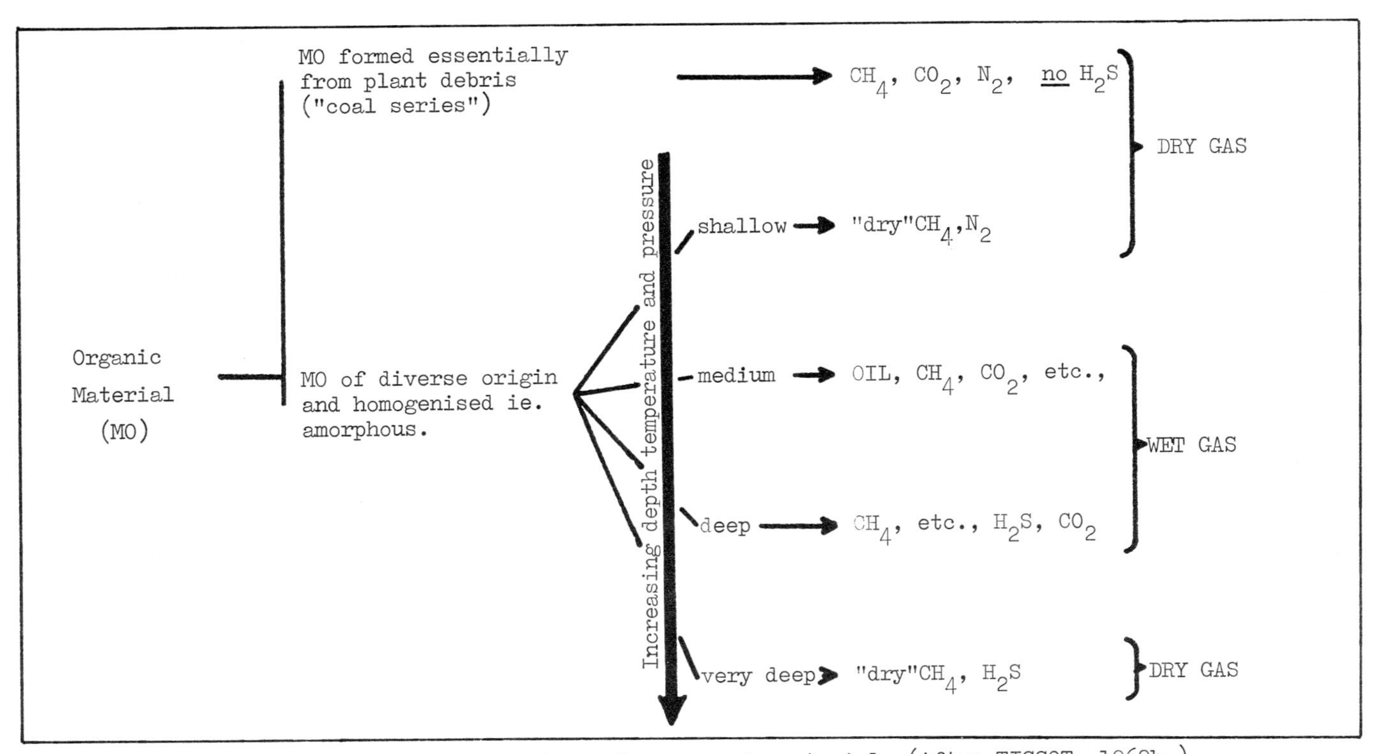

Fig. 20 Schematic of the formation of gas from organic material. (After TISSOT, 1969b.)

While it is admitted that sporopollenin is rare in these two cases, it seems more abundant in the second, where the proportion of plant debris is greater.

Thus we can observe, that diagenetic interpretations can be influenced by the type of the organic matter present. The inherent nature of the hydrocarbons formed seems to be related to the nature of the organic material present in the source rocks. VAN OYEN and his team, (also COMBAZ, STAPLIN, TISSOT, RADCHENKO, etc.) have shown that this relationship is possible. TISSOT (1969b) explains the formation of gas by considerable burial (Fig.20), or else by the presence of large quantities of organic matter similar to those found in coals. In such rocks, the organic matter is comparable in part, to those which we classed, as the group 'plant debris' (i.e. tracheids, lignin, cell walls, cuticles, etc.) whereas organic material that is of different origin and homogeneous, only produces on average maturation, oil.

The different categories of organic material used for the definition of the characteristic qualities of the source rock, and correspondingly the type of oil generated, can be classed together in two large groups:

1. lignitic material, and
2. sapropelic organic material.

The lignitic material comprises part of the plant debris described above, particularly the lignitic components (MOL - organic lignitic material) and tracheid debris (MOT - organic tracheid material).

Organic lignitic material: is characterised by its irregular polygonal outline with clear sharp edges, and is often brown but mostly black. Infra-red studies suggests these particles to be fragments of tracheids whose bordered pits appear to have been lost by secondary infillings.

Organic tracheid material: these are vessels from wood with pits, with scalariform elements being easily recognised. The elements are generally broken and are usually dark-brown to black in colour.

The *sapropelic material*: comprises organic material of very different origin (microscopic algae, soft tissues of plants and animals, and various microorganisms degraded by bacterial action) which has been homogenised. This kind of organic matter is in general amorphous. The sapropelic material is composed of colloidal organic matter (MOC), and found in marine sediments.

The colloidal organic matter has the appearance of being granular in the mass, and without true structure. The colour is variable with the degree of maturation, and associated with it, can be found globules of oil, in addition to pyritic inclusions.

Colloidal organic matter may be derived from microscopic algae of the type *Gloeocapsamorpha*, *Botryococcus*, and also from degradation products of certain plant tissues especially those rich in cellulose (MOV - vegetable organic matter).

As well as these four principal types of organic matter there are associated different organic material:

Smooth membranes (MOM), more or less transparent, fine, amorphous and with indefinite outlines;

Fine organic matter (MOF) which may be the result of the degradation of pre-existing organic components.

The morphologic variety is accompanied, as was indicated

earlier, by chemical differentiation. These chemical differences are shown in the study of the thermal diagenesis of the different organic types. The natural 'heating' of the different categories of organic matter cited above will lead therefore, to different hydrocarbon degradation products.

If the material observed is only of the lignitic type (MOL and/or MOT) the hydrocarbons produced will be generally of the dry gas type. Undoubtedly a certain amount of this type of organic matter appears to be required for this formation. In contrast, if the material observed is of the sapropelic type (MOC in particular), then under thermal influences the hydrocarbons produced may be of the following types:

i) Little developed maturation - nothing or traces of oil or wet gas.

ii) Average maturation - field of oil, or wet gas.

iii) Well matured - dry gas only.

iv) Very strong maturation - little gas, or nothing.

Finally, if the material is mixed, we can have relative proportions of each types and following the thermal alteration have oil, wet gas, or dry gas.

This scheme of hydrocarbon formation as a function of the nature of the organic matter, is verified by considering the two previously cited examples, the Toarcian of the Paris Basin, and the boreholes at Logbaba. In fact, the Toarcian of the Paris Basin is rich in sapropelic organic matter (MOC), and one finds liquid hydrocarbons in the centre of the basin, where the thermal effects have not been too strong. At the Logbaba boreholes, the organic matter is of the lignocellulose type, that is to say we find at the same time lignitic material (tracheids, and lignitic organic matter) and possible precursors of sapropelic matter (cellular and plant organic matter), and associated with this we find oil and gas, with a predominance of the gaseous products when the diagenesis is strong.

Sporopollenin is considered rather little in our scheme of the formation of hydrocarbons. We may suppose however, that the cellular products of degraded spores and pollen during the course of sedimentation and at the beginning of diagenesis are transformed and may contribute in part to the organic matter, described by TISSOT as the diverse and homogenised organic matter.

Conclusions.

Sporopollenin, as the substance which constitutes the external membranes of spores and pollen grains, appears to be a good thermal marker for diagenesis, compared with other associated organic matter. Sporopollenin appears to be relatively chemically very stable during the early and middle stages of diagenesis. The colorimetric variations undergone by it during this thermal evolution and which may be measured and evaluated by different methods (fluorescence, absorption, reflectance, evaluation of state of preservation) are at the moment little understood as physical and chemical processes. The various papers presented at this Symposium have shown it is now possible to begin an explanation of these observed phenomena.

Sporopollenin does not appear to be solely responsible for the formation of hydrocarbons. In fact, its importance is quantitatively reduced in comparison with the other organic matter which becomes predominant, in the generation of petroleum. This is mainly due to the stability of sporopollenin which appears to be difficult to transform. It must be admitted, however, that by degradation, sporopollenin may contribute in a small way to the formation of the organic matter of diverse and homogeneous origin, which appears to be the origin of the formation of hydrocarbons.

The morphologic study of the external membrane of spores and pollen has largely been used for chronostratigraphy, but it offers equal opportunities for the determination of the geological environment. Our studies show that sporopollenin may be an excellent guide for petroleum prospecting in delimiting zones favourable for the genesis of hydrocarbons.

Acknowledgements

I wish to thank the Institut Francais du Pétrole (I.F.P.), l'Entreprise de Recherches et d'Activités Pétrolières (ERAP-ELF-R.E.) and the Société Nationale des Pétroles d'Aquitaine (S.N.P.A.) who permitted the study of these samples coming from their sphere of operations, and allowed me to publish the results obtained. The relationships of the different types of organic matter and their eventual relationships with hydrocarbons has been included here thanks to G. Peniguel and his colleagues (Service Palynoplanktologie) here in S.N.P.A. I also wish to thank C.A.M.E.C.A. for the use of their electron probe, and M. Abstruc for his technical assistance.

References

ALBRECHT, P. (1969). These de doctorat, Universite de Strasbourg.

ALPERN, B. (1964). Advances in Organic Geochemistry. Proceedings of the International Meeting, Rueil-Malmaison.

ALPERN, B. (1967). Communication au Colloque International 'Le charbon en tant que roche et matiere premiere', Freiburg.

BARGHOORN, S. (1952). J. Sed. Pet., 22, 1, p.34-41.

BIENNER, F., de JEKHOWSKY, B., PELET, R., and TISSOT, B. (1968). Origin and Distribution of the elements. Pergamon Press, Oxford and New York.

BROOKS, J., et SHAW, G. (1968). Nature, 220.

CALIFET, Y., et LOUIS, M. (1965). C.R. Acad. Sci. Paris.

CHIAVERINA, J. (1962). These de doctorat, Grenoble.

COMBAZ, A. (1964). Rev. de Micropaleontologie, 7, n4.

CORREIA, M. (1967). Rev. Inst. Franc. du Petrole, XXII-9, p.1285-1306.

CORREIA, M. (1969). Rev. Inst. Franc. du Petrole, XXIV - 12, p.1417-1454.

DEBYSER, J., et DEROO, G. (1969). Rev. Inst. Francais du Petrole, XXIV-I, p.21-48; XXIV-2, p.151-174.

DEFLANDRE, G. (1938). C.R. Acad. Sci. Paris, 206, p.854-856.

EISENACK, A. (1931). Paleont. Z., 13, p.75-118.

HAVINGA, A.J. (1964). Pollen et spores, 6, 2, p.621-637.

HUGEL, M.F. (1964). These de doctorat, Paris.

GUTJAHR, C.C.M. (1966). Leidse geol. Mededel, 38, p.1-29.

JEHOWSKY, B. de. (1963). C.R. Somm. Sci. Soc. Biogeographie, Paris, 349.

JEHOWSKY, B. de. (1968). Rev. Inst. Franc. du Petrole, XXIII-5, p.595-607.

LOUIS, M., et TISSOT, B. (1967). VII° Congres Mondial du Petrole, Mexico.

MAILLARD, L.C. (1912). C.R. Acad. Sci., 154, p.66 and 155, p.1554.

MANSKAYA, M., et DROZDOVA, T.V. (1962). Geochimika, II.

MANSKAYA, S.M., et DROZDOVA, T.V. (1963). International series of Monographs in Earth sciences, Pergamon, 28.

MULLER, J. (1959). Micropaleontology, 5, p.1-32.

NIELSEN, N., GROMMER, J., et LUNDEN, P. (1955). Acta. Chem. Scand., 2, p.1100-1106.

RADCHENKO, O.A. (1968). Doklady Akad. Nauk. SSSR., 183, n 1, p.193-196.

SMITH, G., et WILCOCK, S. (1964). The Phanerozoic Time Scale Suppl. Quarterly Journal Geol. Soc., London, 120.

STAPLIN, F.L. (1961). Palaeontology, 4, part 3, p.392-424.

STAPLIN, F.L. (1969). Bull. of Canadian Petroleum Geology, 17, n 1, p.47-66.

TAUGOURDEAU, Ph. (1966). Memoire Soc. Geol. de France, XLV, fasc.I, memoire n 104, p.1-64.

TAUGOURDEAU, Ph. (1968). Rev. Inst. Franc. du Petrole, XXIII-10, p.1219-1271.

TISSOT, B. (1969a). Rev. Inst. Franc. du Petrole, XXIV-4, p.470-501.

TISSOT, B. (1969b). Rev. A.F.T.P. n 198, p.167-171.

VAN OYEN et al. (1965). Le Domaine de la Palynoplanctologie (Note interne SNPA-1965).

WILLIAMS, D.B., and SARJEANT, W.A.S. (1967). Marine geol. 5, n 5-6, p.389-412.

WILSON, L.R. (1964). Grana Palynologica, 5, n 3, p.425-436.

PLATE I

Figs. 1, 2, & 3: Lignitic organic matter (MOL)

Figs. 4, 7, & 8: Tracheids with bordered pits (MOT)

Figs. 5, 6, & 9: Scalariform vessels (MOT)

PLATE II

Fig.1 : Sapropelic organic matter (colloidal (MOC)

Fig.2, 3 : As above, but with several drops of oil

Fig.4, 5 : Microscopic algae (sapropel precursors ?)

Fig.6, 7 : Well preserved plant cellular organic matter (MOV)

Fig.8, 9 : Early stage of degradation in cellular matter

Fig.10 : Advanced stage of degradation of cellular organic matter (sapropel precursors ?)

PLATE III

Figs. 1, 2, & 3 :Fine organic matter (MOF)

Figs. 4, 5, 6, &7:Membranes (MOM)

Figs. 8, & 9 :Mixed assemblage with lignitic matter and sapropelic matter.

THE SCALE LINE ON EACH FIGURE REPRESENTS 20 microns

PLATE I

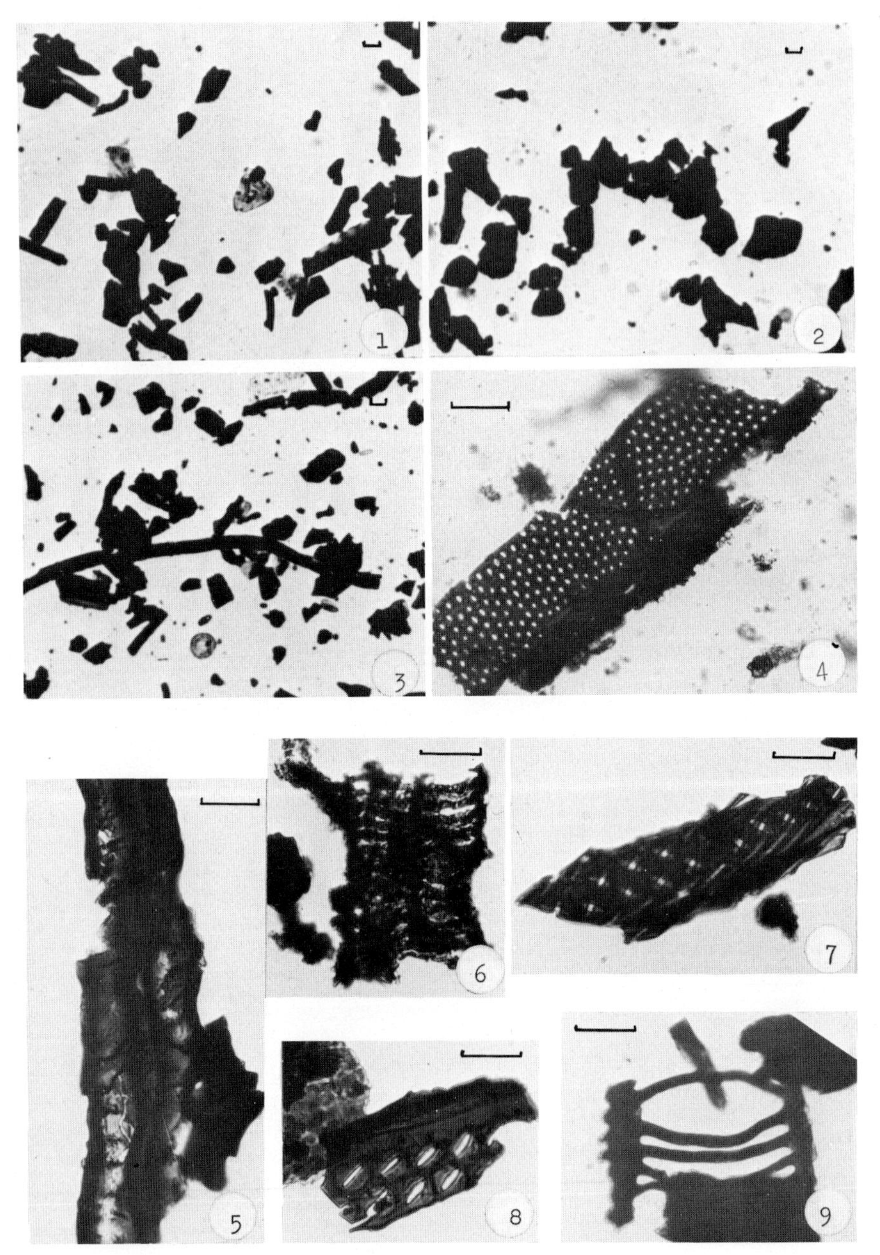

PLATE II

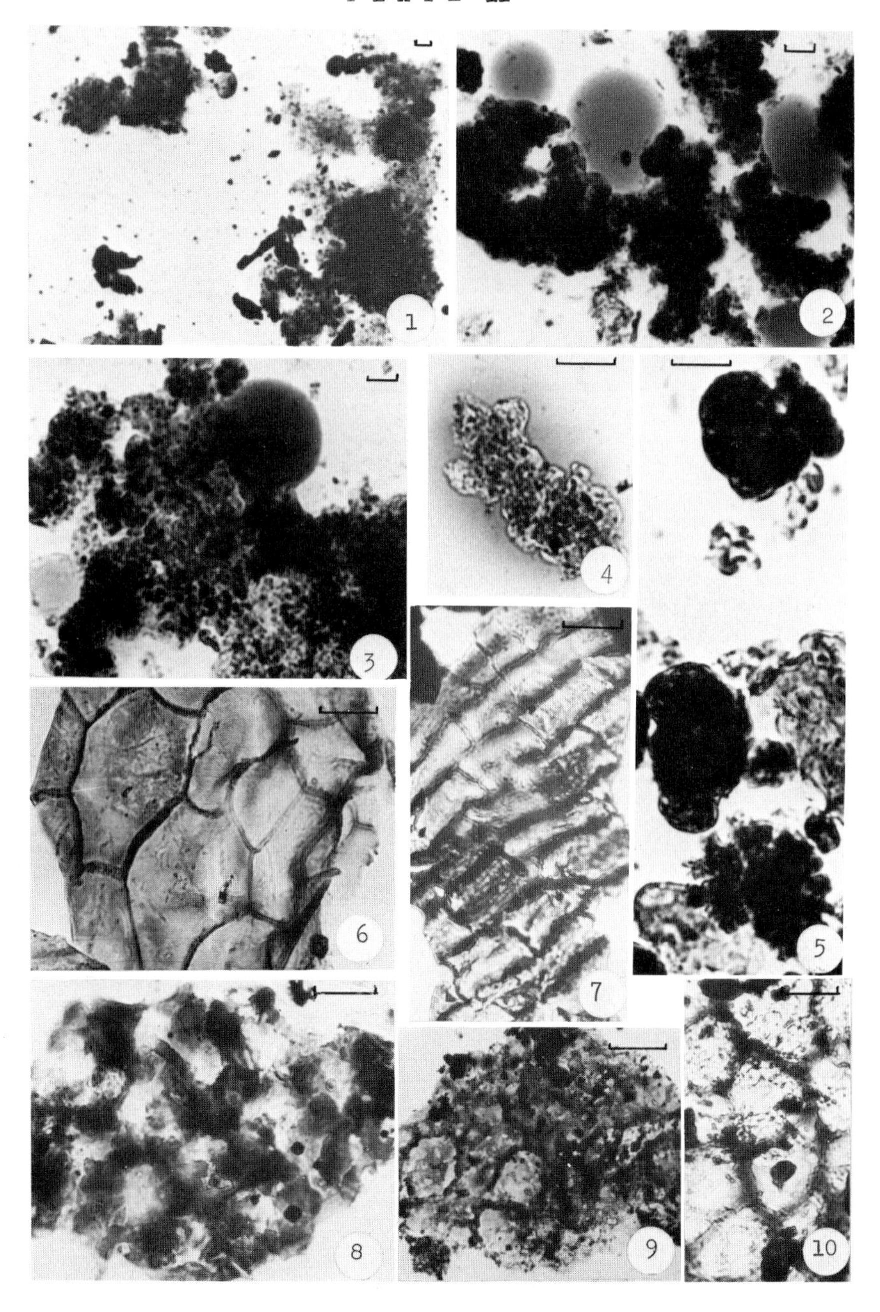

PLATE III

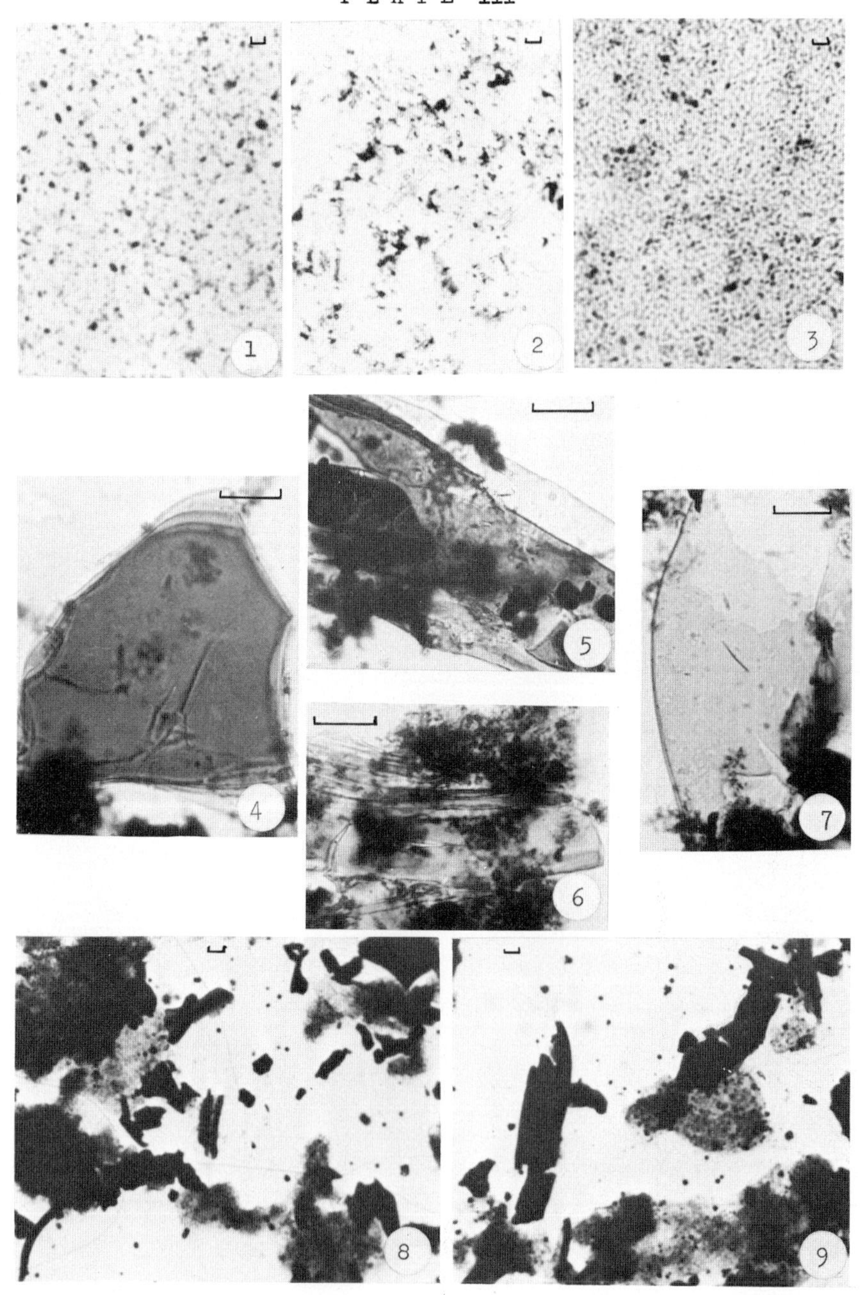

THERMAL DEGRADATION OF SPOROPOLLENIN AND CENESIS OF HYDROCARBONS

A. Combaz

Compagnie Francaise de Petroles, 20 Rue Jean Jaurs,
92 Puteaux, Hauts de Seine, France.

Comparison with some similar substances

ABSTRACT

It is possible to see a similarity of composition between sporoderm and various related substances thanks to our knowledge of their chemistry. Studies of the thermal degradation of fresh conifer pollens, selected or concentrated microfossils, spores and pollens, *Tasmanites*, and rock samples containing organic matter, have shown diverse reactions. Most of them have shown varying yields during successive thermolysis treatments, the products of which were analysed. The productivity curves show a bimodal profile, and the hydrocarbons obtained are often rich in the heavier C_{10} to C_{19} fraction.

Microscopic examination, after several treatments at 180°C for 10 minutes, showed that the morphological details are little changed although there is a darkening of the colour.

A small quantity of the material is converted into hydrocarbons.

The recorded Paramagnetic Electronic Resonance (R.P.E) signal increases sharply after several thermolysis treatments.

INTRODUCTION

The substance forming the exine of spores and pollen grains has, by its remarkable resistance to physical and chemical action, the characteristics of a natural 'plastic'. It is not by chance that exines have evolved such a material for the protection of the genetic heritage of plants. Although this substance has often been reported, it has remained little understood until recent years and few researchers have given much time to its investigation.

One can sum up the present knowledge by saying that the exine appears to consist of:

- the main part, originating from Ubisch granules, formed by condensed lipids, which on degradation produce fatty acids (eg palmitic acid) and phenolic acids.

- and the ligneous fibres, which are rare, and produce phenolic acids by chemical oxidation.

The whole thing makes a structure which is very resistant to attack by non-oxidizing chemical reagents. However, these very unsaturated substances are capable of fixing O, S or N during the fossilisation process. This fixation is accompanied by important changes in chemical structure.

The simplified formula of sporopollenin varies from $C_{90}H_{134}O_{20}$ to $C_{90}H_{142}O_{35}$ (BROOKS and SHAW, 1968) and is allied to the composition of many varied substances found in the vegetable kingdom. The anatomical and functional homogeneity of exine disguises a certain chemical heterogeneity. Some natural fossilized substances have similar properties and overall related molecular composition to exines.

Those fossil substances that are cited in particular by BROOKS and SHAW, (1968), are: the Tasmanites ($C_{90}H_{136}O_{17}$) of Permian age, kukersite, ($C_{90}H_{136}O_{18}$) of Ordovician age, and even the Green River Formation kerogen (Eocene) with its $C_{90}H_{134}O_{25}$ formula and containing 0.66% of nitrogen and 0.90% of sulphur.

EFFECTS OF THERMAL DECAY

A series of thermolytic[1] treatments were carried out on pollen material of *Pinus sylvestris* and other related materials. This was accompanied by chromatographic analysis of the decay products obtained in normal operating conditions. (BORDENAVE, COMBAZ et GIRAUD, 1966; GIRAUD, 1970).

A. Operating methods

Several milligrams of the material were heated, to 280°C for 10 minutes, in a low inertia thermal oven. The resulting products were first collected using a liquid nitrogen cooled trap, and then injected into a gas chromatograph by a current of helium (GIRAUD, 1970). The main advantage of this system, compared with the classic pyrolysis rests in the low heating temperature, and the simultaneous collection of the products before injecting them all together into the chromatograph. The present method differs slightly from that initially used by GIRAUD. The baffle, that caused a perturbation of the composition of the flux entering the chromatograph, has been suppressed, resulting in all the products passing through the capillary tube. The sensitivity of the apparatus allows less than 1 milligram of material to be studied.

B. Studies of Pinus sylvestris (collected at BORDEAUX (France) 1968)

Preparation of membranes: The method used was that of ZETSCHE and HUGGLER (1928) as modified by SHAW and YEADON (1966).

The content of insoluble carbon of these membranes is 51,1% ± 3,06 or 6% relative error (average of 2 readings).

Results: It was found useful to run a series of readings on the same material in order to compare the results after 10, 20, 30 minutes heating, etc ... These operations were continued until finally there were no longer useful readings. The figures used represent the number of micrograms of pure constituent per gram of total organic insoluble carbon (C_T).

---

[1] - In accordance with ANDREEV et al, 1955 we prefer to use this term for thermal decay at relatively low temperatures and reserve pyrolysis for that at high temperatures above (≯ 500°C)

Exptl. sample	1st thermolysis	2nd thermolysis	3rd thermolysis	4th thermolysis	5th thermolysis
Sample size	4,83 mg	4,83 mg	4,83 mg	5,09 mg	5,09 mg
C_T	51,1	51,1	51,1	51,1	51,1
TOTAL GAS	81,02	67,0	28,1	28,2	9,3
TOTAL OIL	45,9	18,6	23,8	13,02	15,57
G/O	1,75	3,6	1,2	2,14	0,59
A/O	4,4	5,4	2,0	2,35	0,98
TOTAL MONO-ARO	201,7	101,4	48,4	30,89	15,40
Exptl. sample	6th thermolysis	7th thermolysis	8th, thermolysis	9th thermolysis	10th thermolysis
Sample size	5,09 mg	5,09 mg	4,2 mg	4,2 mg	4,2 mg
C_T	51,1	51,1	51,1	51,1	51,1
TOTAL GAS	12,7	14,15	21	17,77	17,26
TOTAL OIL	18,21	51,81	35,01	30,58	35,96
G/O	0,7	0,28	0,6	0,58	0,48
A/O	1,22	0,56	0,72	0,63	0,62
TOTAL MONO-ARO	22,23	28,82	25,36	20,40	29,05

Table 1 Pine pollen GF976

1. Gas: The quantity of gas produced is small and decreases rapidly after the first 50 minutes treatment, then increased slightly to a maximum at the 8th period of thermolysis.

2. Oil: Overall there is little oil produced. There are two maxima, the first during the first 10 minutes, and the second after 1 hour's heating at 280°C. Normal light paraffins are the most abundant, with C_8 dominating in the early stages, then after 40 to 100 minutes the C_7 component dominated. The C_{14} to C_{17} normal paraffins

Fig 1 - Pollen of *Pinus sylvestris*

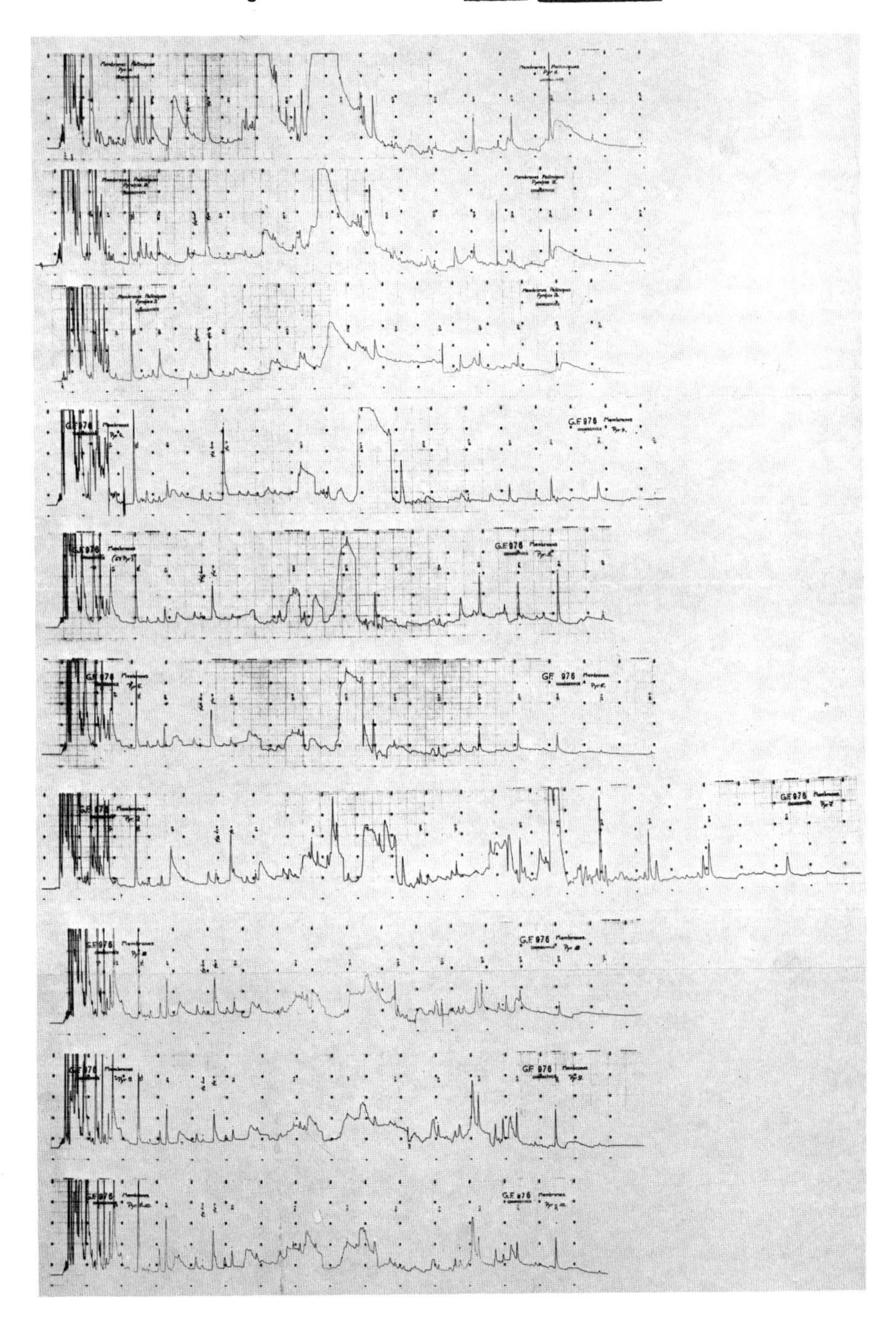

are well represented at the first thermolysis, but then little or nothing until at the 7th thermolysis, when there is a marked appearance of C_{15}. These normal paraffins represent 80% of the total oils produced.

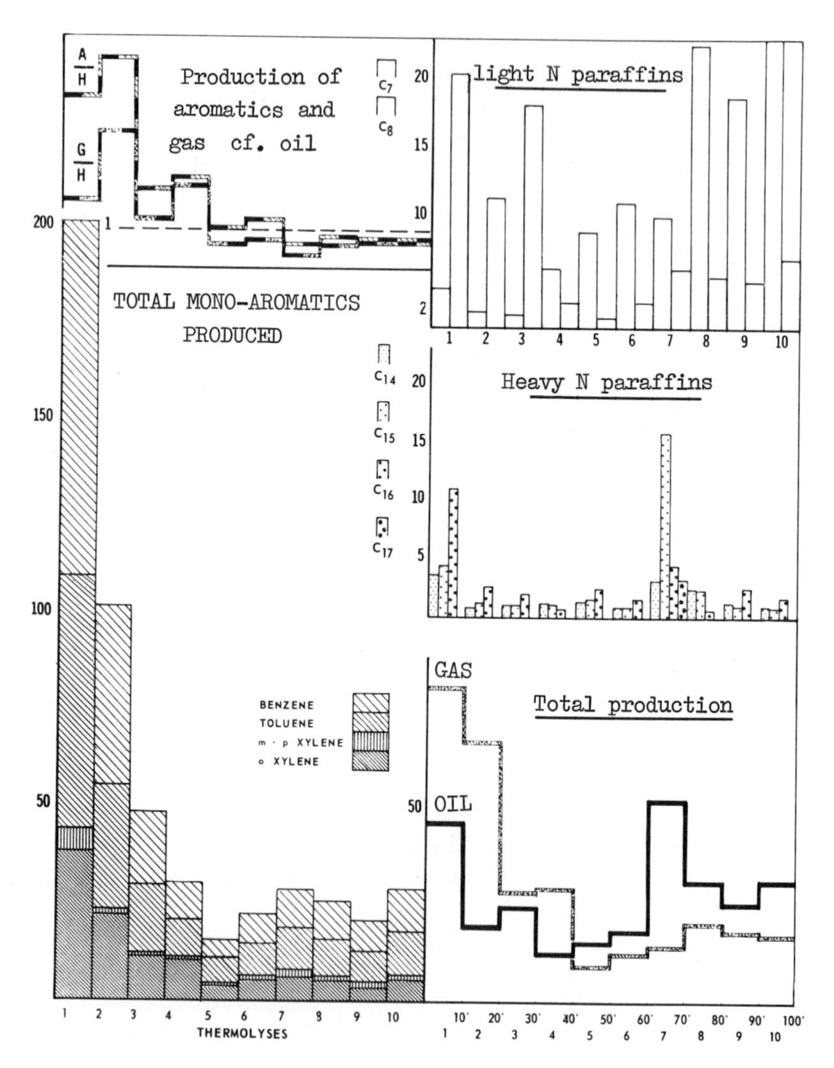

Fig 2 - Pollen of Pinus sylvestris

3. Aromatics: These are produced in far greater quantities than the oil constituents during the first 20 minutes (up to 4 times as much). In the same way that gases are produced, the production of aromatics decreases sharply up until the 5th thermolysis when there are two minor maxima at the 7th and 10th thermolysis. At each benzene is the predominant product. As with the production of gas this A/O relationship tends to ward unity and then gradually decreases after the 5th thermolysis.

4. Miscellaneous: Other products formed in large quantities were present on the chromatogram between C_9-C_{10}, and C_{10}-C_{11}, and to a lesser extent between C_8-C_9 and C_{16}-C_{17}. These components were not identified, but possibly could be fatty acids. The absorption of the solvents, used in the preparatory washing of the samples, masks the chromatogram readings below benzene with the result that C_4, C_5 and C_6 are excluded from the interpretation.

5. Loss of weight: The upper part of figure 3 shows the cumulative curve of percentage weight loss, based on two runs, with successive thermolysis. The decrease is regular except for the 1st and 7th thermolyses where the loss is greater. This phenomenon is even clearer on the following histograms:

- total loss of weight in mg to 100 g of product,
- hydrocarbons produced in γ/g of C_T.

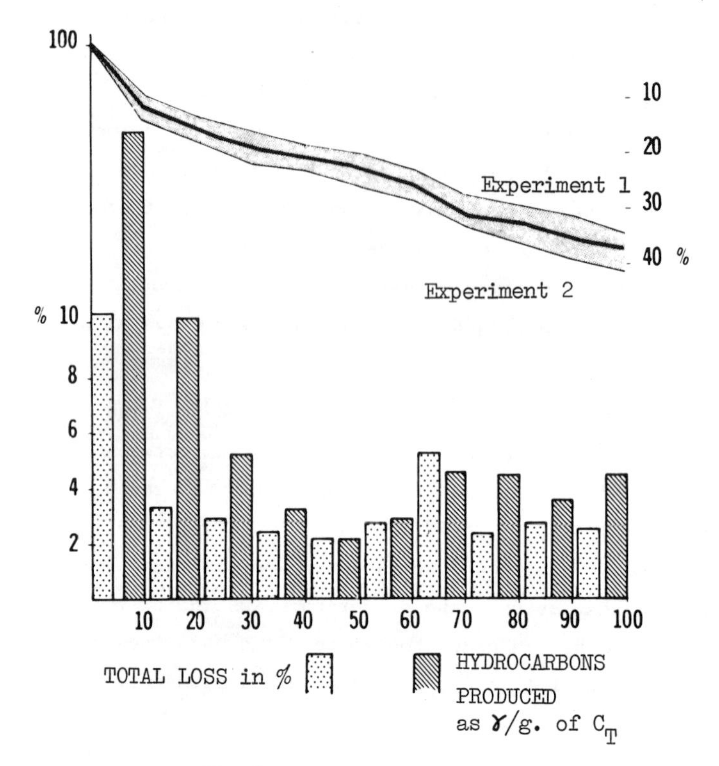

Fig. 3 - WEIGHT LOSS

C. Fossil pollens and spores

The material studied consisted of spores and pollens concentrated from the organic residue of macerated palynological material.

They were prepared using standard procedures which included, after crushing of rocks of different ages from different places, leaching with hot hydrochloric acid, then hydrofluoric acid, and finally separation of the neutralised residue by centrifuging in a dense fluid (d = 2.2). These residues were roughly sorted, using a stereomicroscope, in order to concentrate their microfossil content. The residue was then examined using the same thermolysis and chromatographic technique described above, with upto 4 successive treatments of the same material.

Spores and pollens from the 'Upper Westphalian' of England (GR 420 - R 7081): The spores which are very abundant and well preserved have a predominance of the Lycospora and Densosporites genera, and many Monosaccates. The concentrate also contained a certain quantity of cuticle fragments and lignitic debris.

The quantity of material used was 0.93 mg (C_T 75%). The yields were very low.

No. Experimental sample	1st thermolysis	2nd thermolysis	3rd thermolysis	4th thermolysis
Sample size	0,93mg	0,93mg	0,93mg	0,93mg
C_T	75 %	75 %	75 %	75 %
TOTAL GAS		10,72	3,0	3,6
TOTAL OIL	22,58	19,53	43,20	10,5
G/O		0,54	0,07	0,34
A/O	3,6	0,47	0,06	0,21
TOTAL MONO-ARO	81,97	9,15	2,7	2,2

TABLE 2 - Fossil pollens and spores from Westphalian of England GR 1420, R 7081.

1. Gas: Very small decreasing quantities.

2. Oily constituents: Poor yields with a sharp increase of lighter products at the 3rd thermolysis, but still in very small quantities, with C_7 as the predominate product. The medium products have their maximum at the 3rd thermolysis when C_{14} is accompanied by the heavy products in the form of C_{15}, C_{16} and C_{18}.

3. Aromatics: Their initial yield is the most important but decreased rapidly after the 2nd thermolysis. The A/O relationship is much less than 1.

4. Miscellaneous: Several unidentified products appear

Fig. 4 – RESIDUE OF POLLEN AND SPORES FROM THE WESTPHALIAN OF ENGLAND

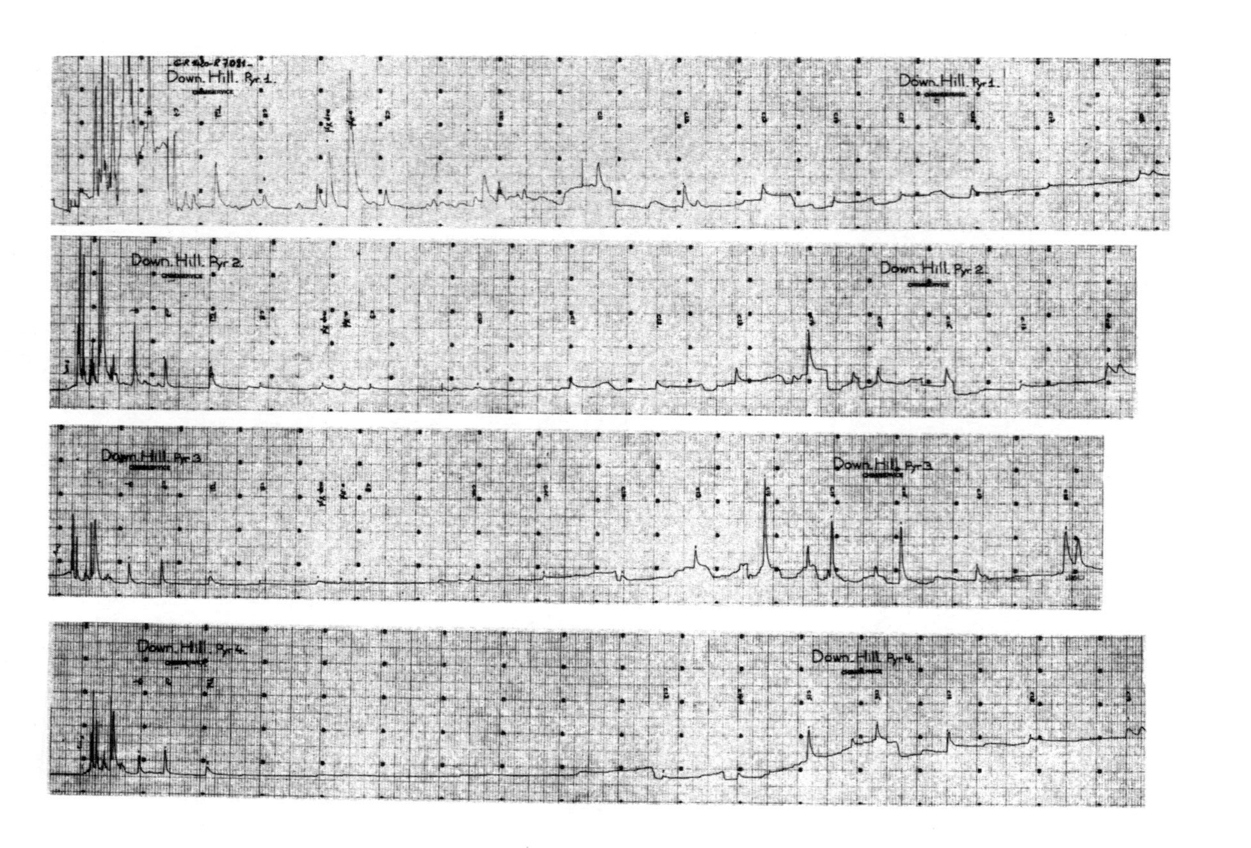

in very small quantities at the 2nd and 3rd thermolyses. They correspond to the C_{13} to C_{16} range. There is a persistence of a probable isoparaffin up to and just beyond the C_{18} range.

Saharan Upper Devonian (Famennian) Spores G1414, R 3440:

The residue used consisted of abundant, little altered, deep yellow to dirty-amber coloured spores. The main genera were *Hymenozonotriletes* (including *H. lepidophytus*), *Verrucosisporites*, *Leiotriletes*, etc.. There were also several Acritarchs and many fragments of altered cuticles.

Quantity of material used : 0.92mg - C_T : 75%.

No. Experimental sample	1st thermolysis	2nd thermolysis	3rd thermloysis	4th thermolysis
Sample size	0,92mg	0,92mg	0,92mg	0,92mg
C_T	75 %	75 %	75 %	75 %
TOTAL GAS	48	13,3	6,7	12,7
TOTAL OIL	1104	128,3	68,9	60,6
G/O	0,04	0,10	0,09	0,21
A/O	0,240	0,16	0,14	0,20
TOTAL MONO-ARO	265	20	10	12

TABLE 3 - Saharan Devonian Spores - GR 1414, R 3440

1. Gas: Low yield.

2. Oily constituents: The first thermolysis gave a large yield, after which it decreased regularly. The initial large yield consisted mainly of the lighter components, C_7 to C_{11}, but also contained normal paraffins up to C_{17}. Whereas the C_{11} products decreased regularly after the 2nd thermolysis with an increasing yield for C_{14} at the 2nd and 4th thermolysis, and for C_{16} and C_{18} at the 3rd thermolysis.

3. Aromatics: Although these are quantitively of much less importance than the normal paraffins, these are produced in

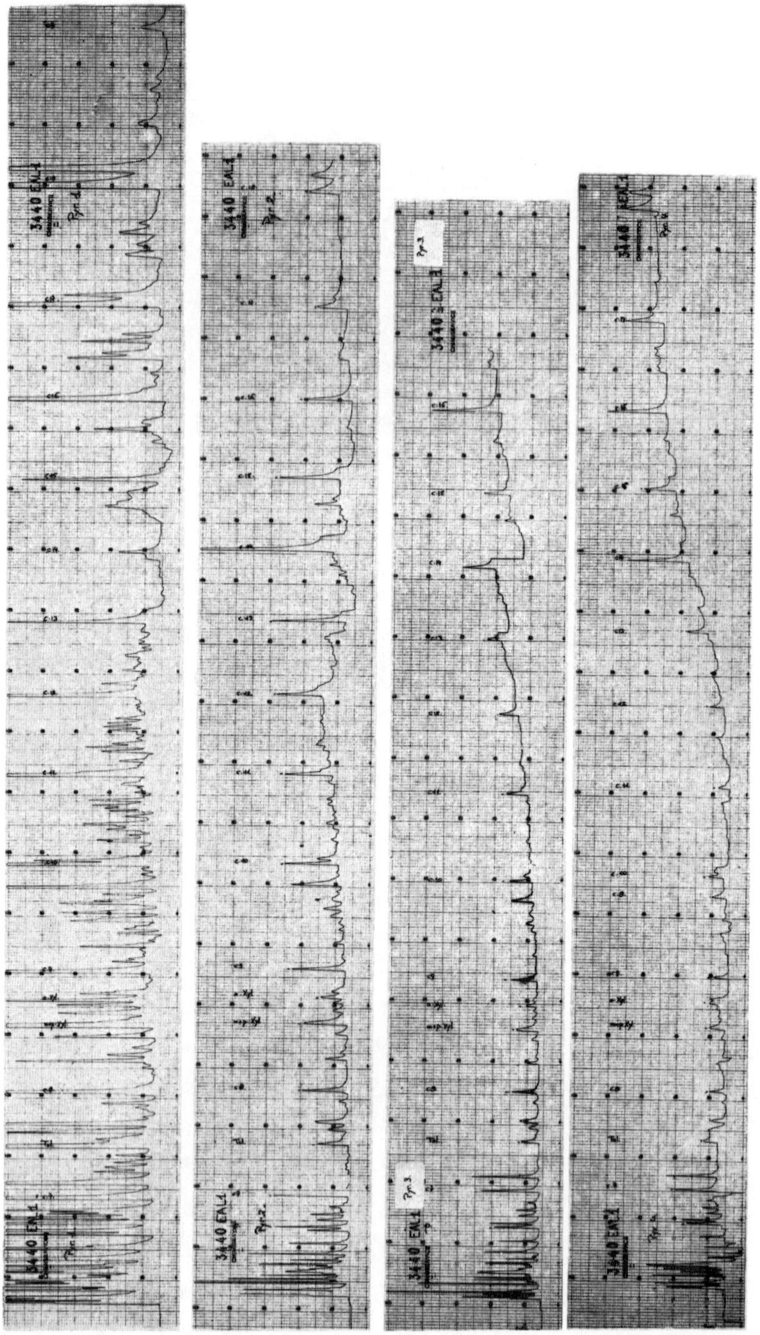

Fig. 5 – SPORE RESIDUES – UPPER DEVONIAN OF THE SAHARA

fair yield with the 1st thermolysis followed by a brusque diminuation as with the gases, with possible resumption at the 4th thermolysis.

4. Miscellaneous: There were many isomers in the light paraffins to C_{13} range and were mainly produced during the first 10 minutes. Following these became less and showed a strong maximum in the middle zones.

Upper Devonian (Fammenian) Spores from Libya GR 1413 - R 4082:

This residue is very similar to the previous examples, except that it is better preserved showing a clear amber colour. The dominant genus are *Densosporites*, *Verrucosisporites*, *Leiotriletes*, with common cuticle fragments, trichoma, some resinous globules, and rare Acritarchs.

Quantity used : 0.72mg - C_T: 75%.

No. Experimental sample	1st thermolysis	2nd thermolysis	3rd thermolysis	4th thermolysis
Sample size	0,72mg	0,72mg	0,72mg	0,72mg
C_T	75 %	75 %	75 %	75 %
TOTAL GAS	1	0,6	0,3	0,3
TOTAL OIL	54,1	18,75	59,19	9,86
G/O	0,02	0,03	0,006	0,03
A/O	0,05	0,05	0,013	0,06
TOTAL MONO-ARO	2,5	1	0,7	0,6

TABLE 4 - Devonian Spores from Libya - GR 1413, R 4082

1. Gas: Insignificant yields.

2. Oily constituents: Low yield with only traces of the normal paraffins C_{13}. The component being C_{14} only found during the first 10 minutes of thermolysis. The C_{15} to C_{18} range, (especially C_{16}) are the most representative with maximum concentration at the 1st and 3rd thermolysis.

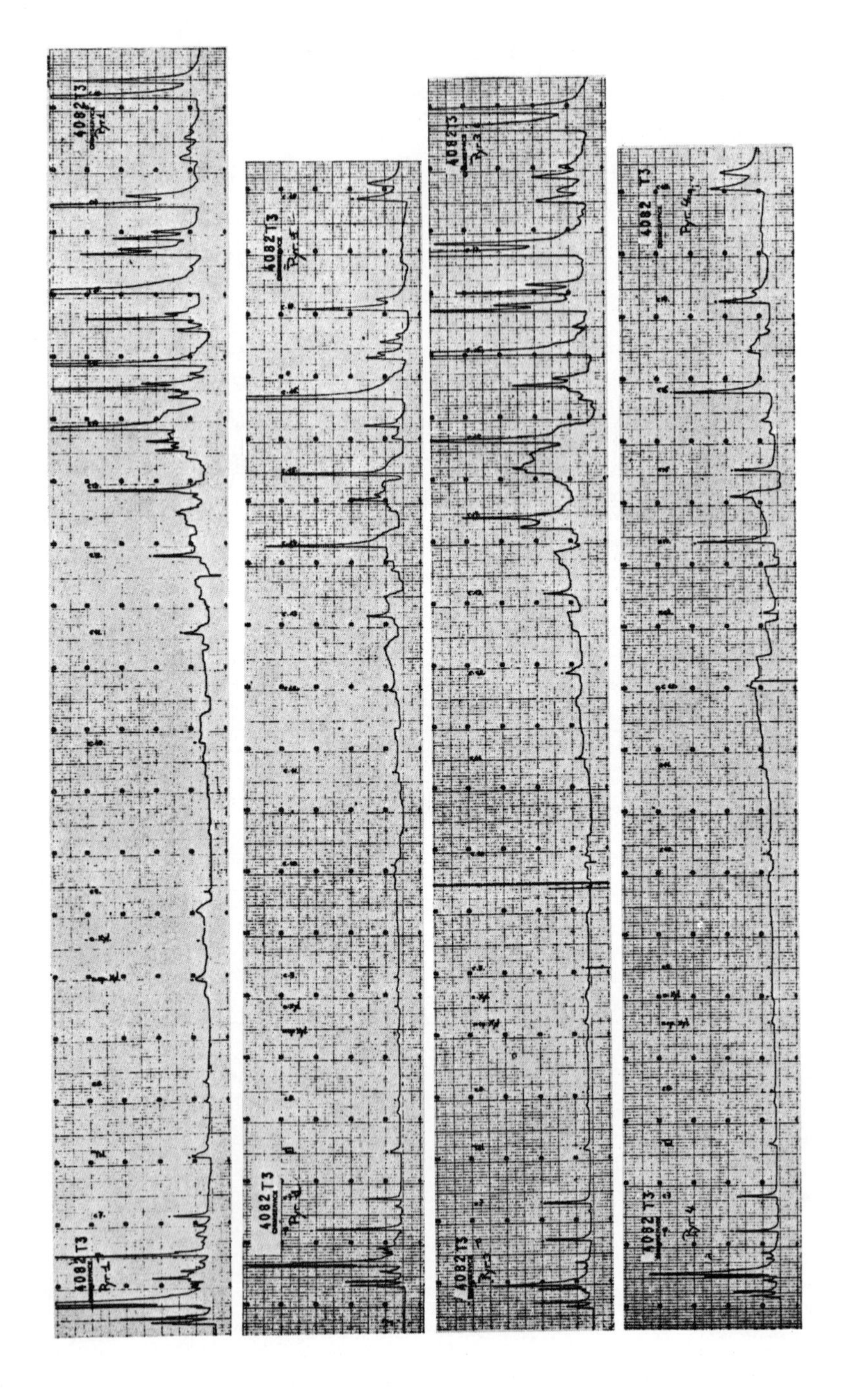

Fig. 6 — SPORE RESIDUES — UPPER DEVONIAN OF LIBYA

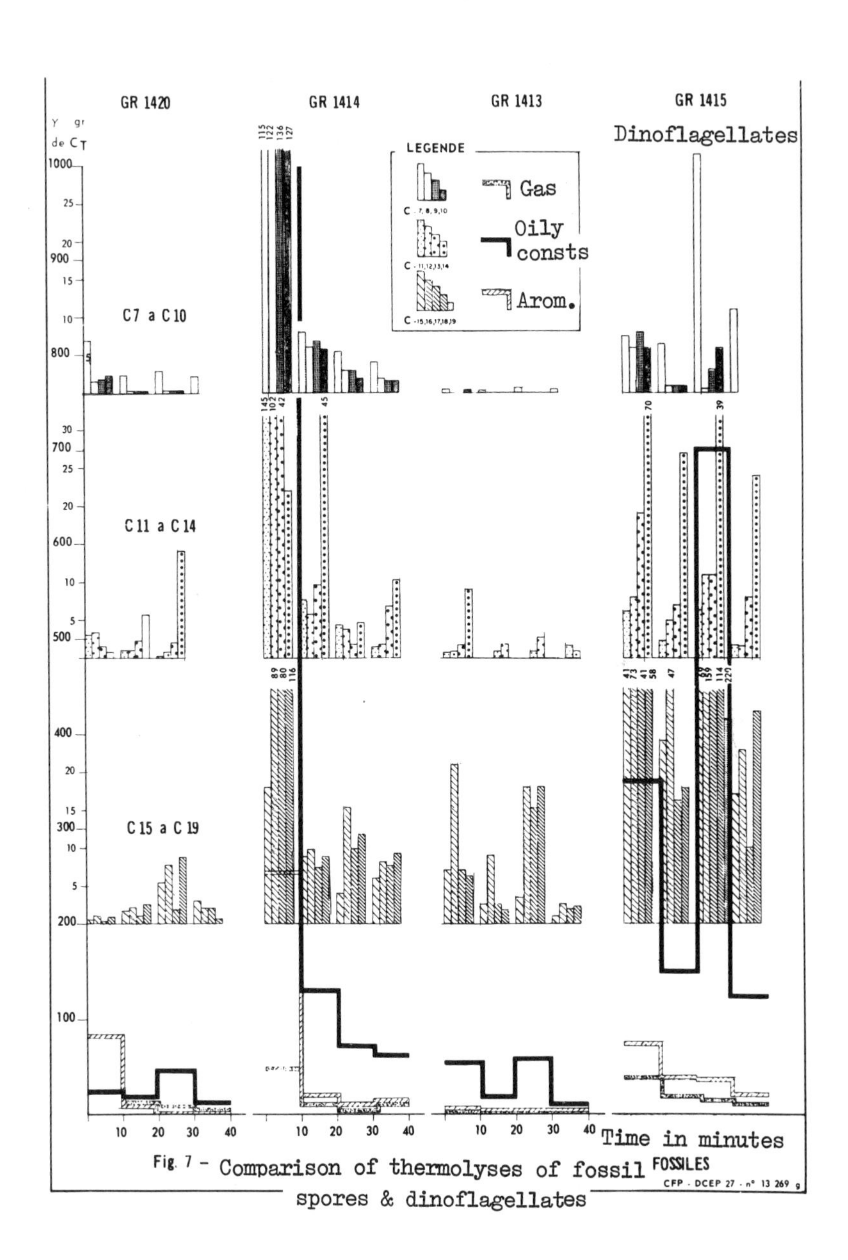

Fig. 7 - Comparison of thermolyses of fossil spores & dinoflagellates

3. Aromatics: Insignificant amounts.

4. Miscellaneous: Absence of components up to C_{14}, but the higher carbon number members show a persistent maximum around the middle of the normal paraffins range. In addition higher members after C_{17} and C_{18} show a maximum that probably represents the isoparaffins that were present in the two previous experiments.

D. Comparison of the results obtained from samples of pollen and spores

Of the 3 fossil materials the second (GR 1414) should be considered individually; this showed a sudden and rich yield of oily constituents and to a lesser extent some monoaromatics, whereas first and third samples had similar low yields with the same maximum for the oily constituents after 20 to 30 minutes of thermolysis. One finds the same exponential decrease of the yield curves for the aromatics and the gas. A similar decrease is found for the light normal-paraffins (C_7 to C_{10}). On the other hand one must note the singular behaviour of the C_{14} terms. The heavier constituents (C_{15} to C_{18}) also have a complex behaviour. All three have a relative maximum during the 3rd thermolysis with C_{16} and C_{18} as the principal constituents. Thus, it is the heavy constituents which are from the late maximum noted in the total yield of oily constituents.

If we examine the results of the fresh pollen thermolysis we find at least two important similarities:

- parallel behaviour of the gaseous phase and monoaromatics, and compared to the fossil material.

- the existence of a much later secondary maximum censed by the heavy paraffins with a predominance of C_{15}.

E. Fossil non-calcareous algae

Several fossil samples were chosen that have an approximate composition similar to sporopollenin (BROOKS and SHAW, 1968) and which are algal.

Tasmanites (GR 398, R 2538): The material used (1.03mg of Tasmanite, C_T = 75%) was selected with the aid of a microscope. It consisted of a monadoid alga similar to the present genus Pachysphaera, which is found from the Silurian and a similar form is possibly present today in the sea (WALL, 1962, PARKE, 1966). The sample studied was extracted from a core of black Silurian shale from a well in the Sahara.

No. Experimental sample	1st thermolysis	2nd thermolysis	3rd thermolysis	4th thermolysis
Sample size	1,03mg	1,03mg	1,03mg	1,03mg
C_T	75 %	75 %	75 %	75 %
TOTAL Gas	53,3	47,2	30,4	14,5
TOTAL Oil	275,3	919,55	45,22	68,5
G/O	0,19	0,05	0,66	21,0
A/O	0,70	0,01	0,10	0,06
TOTAL MONO-ARO	192,4	14,53	4,57	4,3

TABLE 5 - Saharan Silurian Tasmanites - GR 398, R 2538

1. Gas: Low yield which decreases at a linear rate.

2. Oily constituents: High yield during the first two thermolyses, especially during the second. But the characteristics of the two stages are quite different. During the first 10 minutes there is a predominance of the C_{13} components, whilst during the next 10 minutes the yields are rich in the 'even' heavy products such as C_{16} and C_{18}. The next two treatments are similar, although their yields are low and show an increase at the 4th thermolysis for the C_{14} and C_{18} components.

3. Aromatics: The relatively high initial yield decreases suddenly from the 2nd thermolysis and tends towards a level found for the later thermolysis.

4. Miscellaneous: Isomeric phenomenona are notable during the first thermolysis when abundant isoparaffins are produced.

Fig. 8 — PICKED TASMANITES - SILURIAN OF THE SAHARA

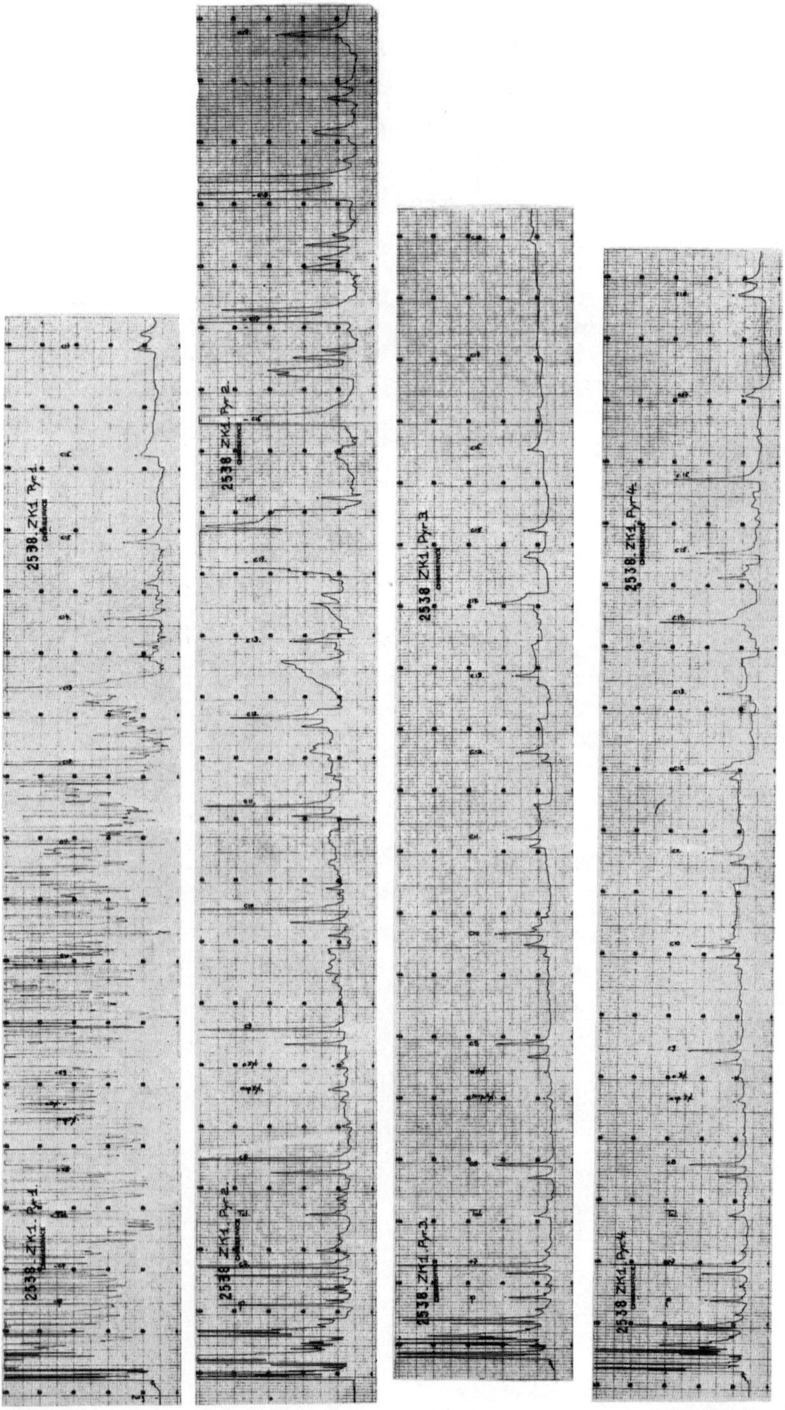

In later thermolysis, this phenomenon disappeared and the persistence of a type of isoparaffins which precedes the normal paraffins C_7-C_{12} are produced.

Cretaceous tasmanite residue from Australia GR 1412 - R 6089: Many tasmanites contain, as well as the abundant spores and pollens, an amorphic algal organic material. Quantity used: 3.6mg - C_T : 75%. The chromatograms obtained are characterised by the absence of the lighter components and the abundance of heavy components.

No. Experimental sample	1st thermolysis	2nd thermolysis	3rd thermolysis	4th thermolysis
Sample size	3,6mg	3,6mg	3,6mg	3,6mg
C_T	75 %	75 %	75 %	75 %
TOTAL GAS	2,7	1	0,9	0,5
TOTAL OIL	1094,10	345,55	2394,09	37,13
G/O	0	0	0	0
A/O	0,005	0	0	0
TOTAL MONO-ARO	4,7	1,2	2,4	0,8

TABLE 6 - Cretaceous tasmanite from Australia - GR 1412, R 6089

1. Gas: Almost non-existent.

2. Oily constituents: The heavy constituents above C_{12} were the most abundant during the 1st thermolysis, but diminish sharply during the 2nd, and again increased strongly at the 3rd thermolysis. At the 4th thermolysis it would appear that only traces of the heavy constituents remain.

3. Aromatics: Low yield with benzene as the principal constituent.

4. Miscellaneous: The bimodal distribution for the C_{12} to C_{13} fractions, etc ... C_{16} and C_{17} give a special interest to the normal-paraffin measurements.

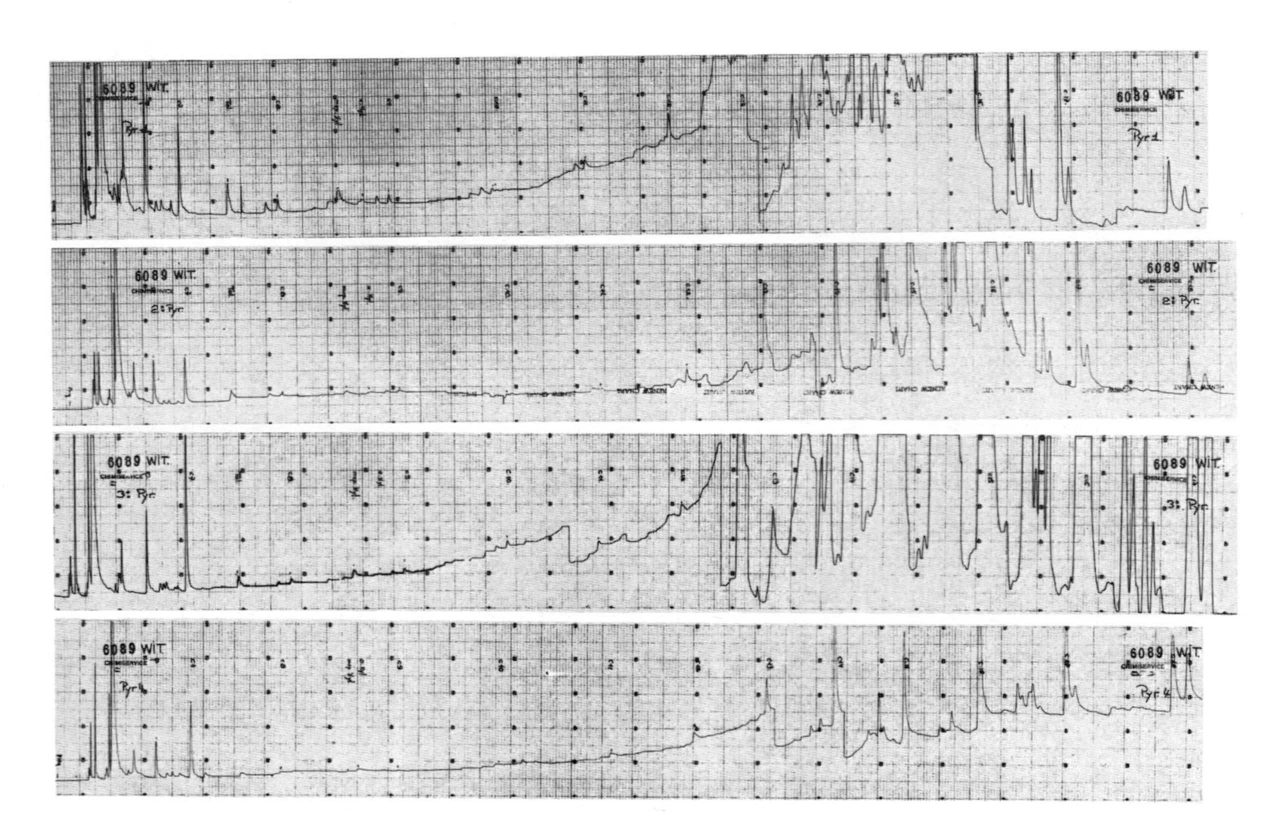

Fig. 9 – TASMANITE RESIDUE - AUSTRALIAN CRETACEOUS

Permian Tasmanite from Tasmania: This is the famous caustobiolithes from the Mersey River which consists essentially of Tasmanites (NEWTON, 1875). The sample was kindly given by D.H. MacCOLL (University of Adelaide), and was studied both as extracted kerogen and also as indigenous rock material.

(a) Kerogen: We define kerogen as the part that remains after removal of all mineral elements (MARCHAND, LIBERT et COMBAZ, 1969). In addition soluble organic material was extracted by percolation with chloroform extraction. Sample quantity studied: 5.99mg - C_T = 72.8%.

No. Experimental sample	1st thermolysis	2nd thermolysis	3rd thermolysis	4th thermolysis
Sample size	5,99mg	5,99mg	5,99mg	5,99mg
C_T	72,8 %	72,8 %	72,8 %	72,8 %
TOTAL GAS	70,6	15,8	19,6	15,2
TOTAL OIL	153,7	37,17	33,11	32,57
G/O	0,46	0,42	0,59	0,47
A/O	0,46	0,38	0,21	0,28
TOTAL MONO-ARO	70,7	14,3	6,9	9,1

TABLE 7 - Tasmanite (kerogen) from Tasmania - A6 - GR 403

No. Experimental sample	1st thermolysis	2nd thermolysis	3rd thermolysis	4th thermolysis
Sample size	10,9mg	10,9mg	10,9mg	10,9mg
C_T	30,8	30,8	30,8	30,8
TOTAL GAS	31,3	18,9	19,3	23,2
TOTAL OIL	61,2	28,3	32,1	32,1
TOTAL MONO-ARO	35,7	17,9	11,9	11,4

TABLE 7b - Tasmanite (Rock) - A6

Fig 10 - TASMANITE - PERMIAN OF TASMANIA

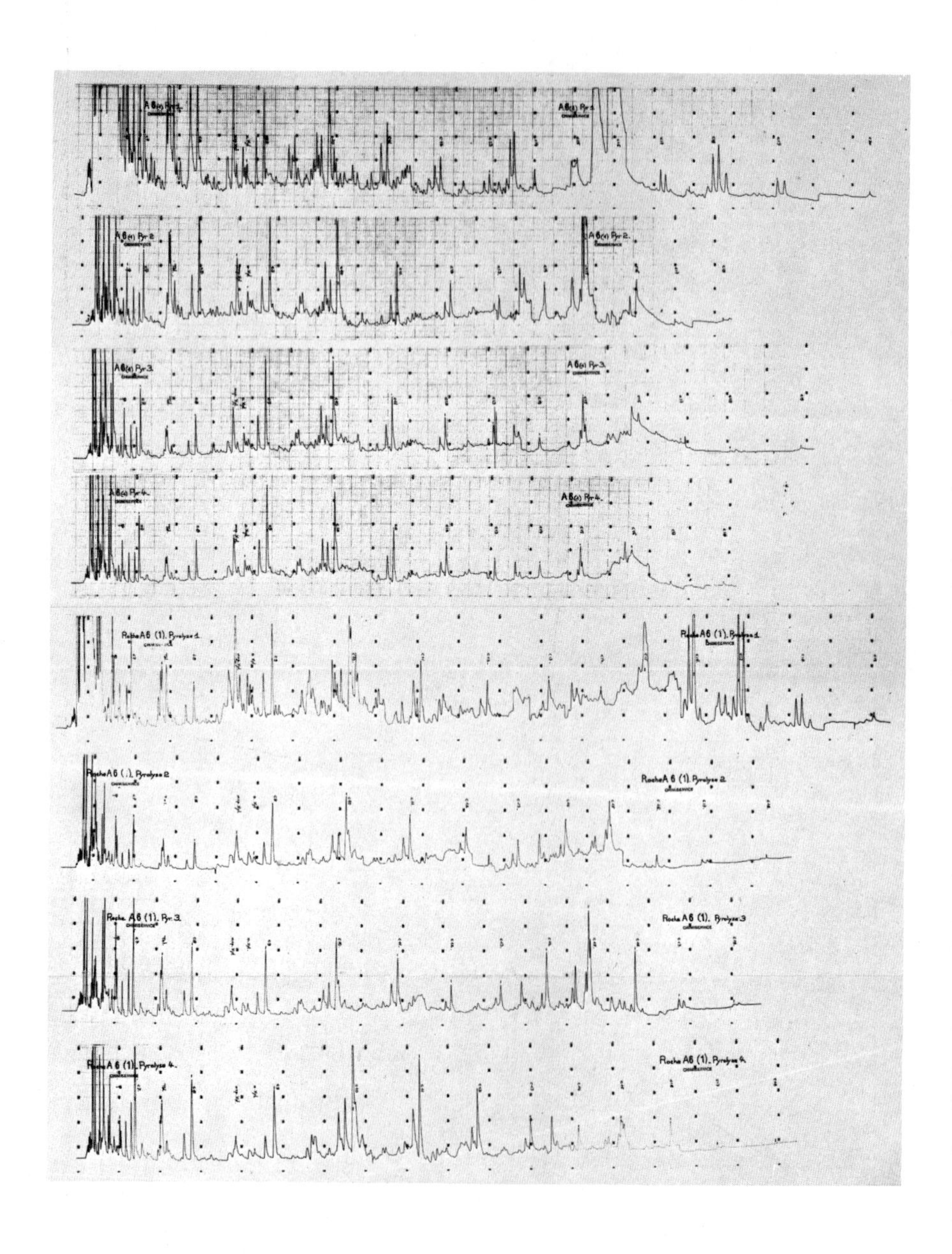

1. Gas: The yield is similar to that of the aromatics, but decreased more slowly and showed small quantitative changes at the 3rd thermolysis.

2. Oily constituents: These are the major products and the yield decreases regularly. The most common constituents are those of the lighter paraffins up to C_{10} with the C_{16} normal paraffin abundant during the 1st thermolysis. The A/H relationship keeps below 1.

3. Aromatics: These show a rapid decrease, whilst the yield in the later stages show a regular value.

4. Miscellaneous: There is little evidence for isomers, and the greater part of the products (C_{15}, C_{16}) are formed during the 1st thermolysis.

(b) Rock containing kerogen: Here the tasmanite sample was simply crushed and solubles extracted with chloroform. The results are similar to those obtained on the extracted kerogen. With the C_{14} to C_{16} fraction of the heavy terms there was an increase in yield at the 3rd thermolysis.

Kukersite: This other caustobiolith from the Upper Ordovician of Estonia consists of the alga called *Gloeocapsamorpha prisca*, and was provided by Professor A. EISENACK (Tubingen).

Kerogen - (cf. MARCHAND and al. 1969): Sample of 3.88mg - C_T : 66%.

No. Experimental sample	1st thermolysis	2nd thermolysis	3rd thermolysis	4th thermolysis
Sample size	3,88mg	3,88mg	3,88mg	3,88mg
C_T	66 %	66 %	66 %	66 %
TOTAL GAS	71,3	41,2	32,1	35,2
TOTAL OIL	97,8	19,54	23,6	21,65
G/O	0,73	2,1	1,36	1,62
A/O	0,93	1,35	1,00	1,20
TOTAL MONO-ARO	91,1	26,3	23,4	25,9

TABLE 8 - Kukersite from the Ordovician of Estonia - GR 401 - A5(1)

Fig 11 KUKERSITE – ORDOVICIAN OF ESTONIA

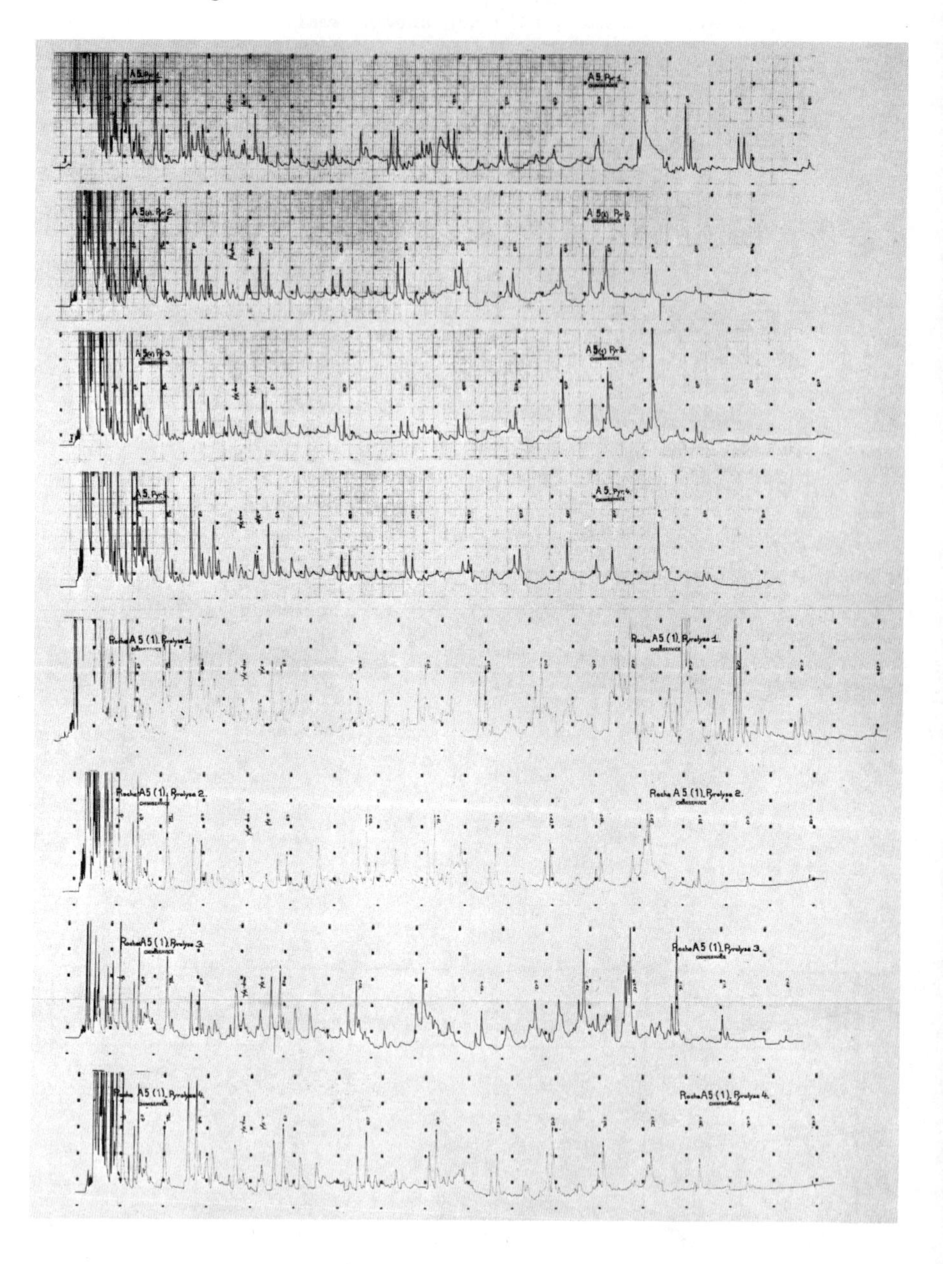

No. Experimental sample	1st thermolysis	2nd thermolysis	3rd thermolysis	4th thermolysis
Sample size	9,45mg	9,45mg	9,45mg	9,45mg
C_T	38,4	38,4	38,4	38,4
TOTAL GAS	104,5	44,7	20,1	31,7
TOTAL OIL	119,5	43,2	23,0	32,8
TOTAL MONO-ARO	65,6	38,7	13,5	25,3

TABLE 8 bis - A5 Rock

1. Gas: Low yield similar to the preceding sample.

2. Oily constituents: Similarities with Tasmanite were found, through with lower yields. Once again it is the C_7 to C_{10} paraffins and C_{16} which are dominant and a relative high yield produced with the 3rd thermolysis. The G/O and A/O relationships are almost unity.

3. Aromatics: These are more abundant than with the tasmanite.

Rock: The rock was crushed and the solubles removed as described previously.

Similar yields of products were obtained as with the demineralised kerogen. A relative high yield was produced with the 3rd thermolysis.

Dinoflagellate rich organic residue (GR 1415 - R 6008): Miscroscopic examination shows numerous *Diconodinium*, *Gonyaulax*, *Odontochitina*, etc ... as well as spores, various pollen and some acritarchs. There was also some algal matter present (COMBAZ, 1967). The colour of the microfossils varies from pale yellow to amber yellow. Sample 0.43mg - C_T : 75%.

Fig. 12 - DINOFLAGELLATE RESIDUES - CRETACEOUS OF AUSTRALIA

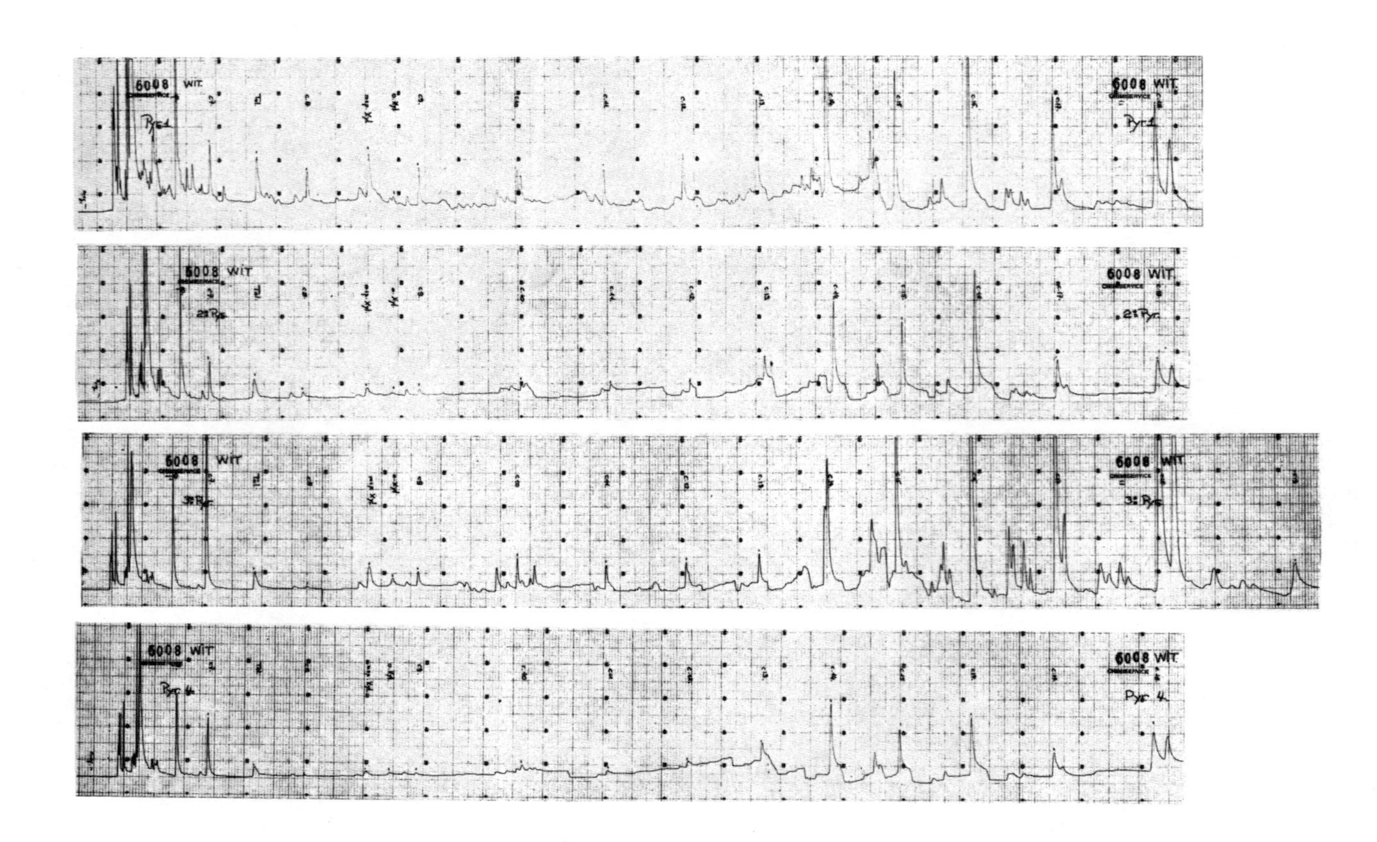

No. Experimental sample	1st thermolysis	2nd thermolysis	3rd thermolysis	4th thermolysis
Sample size	0,43mg	0,43mg	0,43mg	0,43mg
C_T	75 %	75 %	75 %	75 %
TOTAL GAS	35,5	19,3	13	12
TOTAL OIL	348,7	155,1	700,4	124,8
G/O	0,10	0,12	0,02	0,1
A/O	0,19	0,24	0,05	0,16
TOTAL MONO-ARO	68,3	37,6	36	20

TABLE 9 -Dinoflagellates from Cretaceous of Australia - GR 1415 - R 6008

1. Gas: Low with decreasing yields.

2. Oily constituents: High yields especially for the heavier constituents in the C_{13} to C_{18} range and particularly at the 3rd thermolysis. The maximum value (C_{16} component) was shown during the first and second thermolysis and C_{18} was the major component for the last two thermolyses.

3. Aromatics: Low yield showing a regular decrease in value.

F. Comparison of Algal results

The comparison of the yields (Table 10) of these samples underlines the similarity between tasmanite and the low yield kukersite which have their maximum product yield after 10 minutes of thermolysis. The GR 1412 and GR 1415 also show similarities with low yields of the lighter fraction and a contrasting high yield of the heavier fractions, with the highest product yield of both at the 3rd thermolysis. The sorted Tasmanites (GR 398) differ from the other samples in that the high yield of the lighter fraction (1st thermolysis) is separated from that of the heavier fraction (2nd thermolysis).

SAMPLES TYPE		N°	10' C7-C10	10' C11-C19	20' C7-C10	20' C11-C19	30' C7-C10	30' C11-C19	40' C7-C10	40' C11-C19	Total production	
Picked Tasmanites		GR 398	120	155	73	847	28	17	29	40	250	1 059
			275		920		45		69		1 309	
Concentrated Tasmanites		GR 1 412	3	1 091	1	345	4	2 390	2	35	10	3 861
			1 094		346		2 394		37		3 871	
Tasmanite	Kerogen	A6 (1) K	61	93	22	15	18	15	19	14	120	137
			154		37		33		33		257	
Tasmanite	Rock	A6 (1) R	31	30	19	9	18	14	23	9	91	62
			61		28		32		32		153	
Kukersite	Kerogen	A5 (1) K	29	69	11	9	10	14	13	9	63	111
			98		20		24		22		174	
Kukersite	Rock	A5 (1) R	61	59	27	16	9	14	23	9	120	98
			120		43		23		32		218	
Dinoflagell Aust. Cret.		GR 1 415	28	321	9	146	43	657	12	113	92	1 237
			349		155		700		125		1 329	

TABLEAU 10 - COMPARISON OF PRODUCTION

G. Thermodegradation phenomena profile

The first remark that has to be made is that very little of the carbon is transformed into petroleum products (0.2 to 2 o/oo of the total organic carbon).

As regards the various constituents, very little gas is produced, especially from the fossil substances, and the yield seems to obey essentially an exponential decrease rule. The aromatic products which are very abundant from fresh pollen are generally poor in the case of the fossil material. Their yields are much the same as those for the gases. On the other hand, the oily constituents, particularly the heavy fractions, have a higher yield from the fossil material and produce a

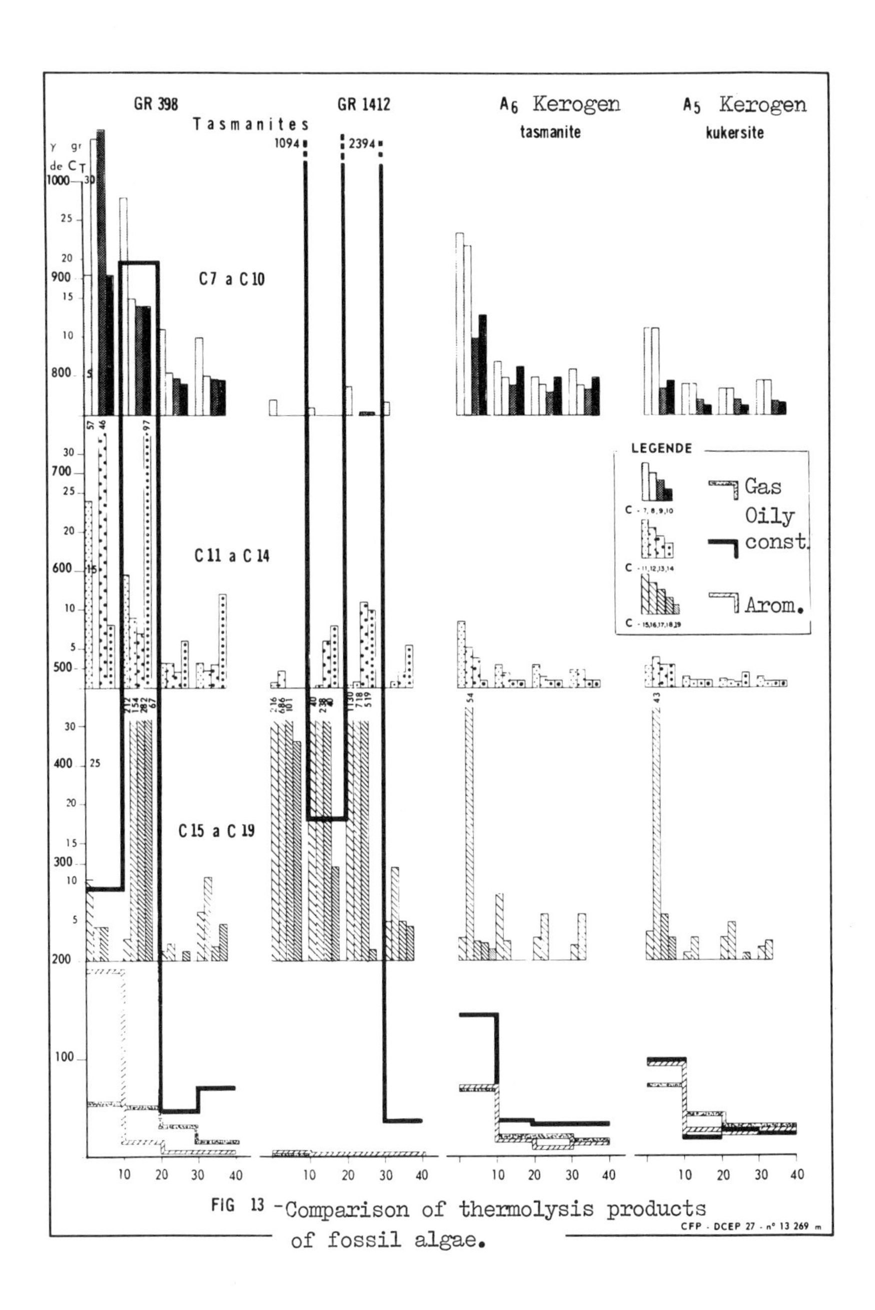

FIG 13 - Comparison of thermolysis products of fossil algae.

complex system of products. Everything seems to happen as if there is a 'superficial chemical state' which can be attained during the 1st thermolysis which liberates them easily, followed by a 'deeper state' producing mainly heavier fractions which is only attained by the 3rd thermolysis for the fossil materials, or the 7th thermolysis for fresh pollen.

Other effects of thermolysis

A. Microscopical examination of thermolysed materials: Little morphological change of the material was produced by the treatment at a constant temperature of 280°C and only a low level of degradation was reached. Only the light absorbing qualities, (ie the colour of the material), were affected to any extent. Fresh pollen, originally colourless, needed 30 minutes heating to produce a pure yellow colour and even after 100 minutes all details were distinguishable but the colour changed to deep yellow with patches of light brown.

These colour change observations are proof of the utility of microscope examination for determining the extent of the diagenesis of surrounding beds.

B. R.P.E. measurements of thermolysed products: We have recently extended our qualitative studies of natural diagenesis by means of paramagnetic electronic resonance (MARCHAND, LIBERT and COMBAZ, 1969), in an attempt to classify the materials according to R.P.E. parameters of the products from our thermolyse experiments. Table 11 shows the paramagnetic electronic susceptibility values ($\chi\rho$) in 10^{10} u.e.m. C.G.S. However, for sample GF 976, the weak signals obliged us to arbitrarily make the strongest signal equal to 100 and record the measurements as relative to it. Its real value was roughly 5.10^{-9} u.e.m. C.G.S. g^{-1}.

The number of measurements obtained were insufficient to plot a χ_ρ curve as a function of heating time. None the less, there is sufficient evidence to indicate an increase of this R.P.E. parameter with relationship to the thermodegradation of the material.

	Pinus pollen - GF 976								Tasmanite A6[1]				Kukersite A5[1]		
Duree en minutes des thermolyses	-	10	30	50	70	100 essai 1	100 essai 2	100 essai 3	-	10	20	40	-	20	40
χ_ρ	0	20	39	46	33	42	45	100	8,9[1]	23	26	25	18	49	46

[1] - Due to lack of sufficient material to make the measurement we have taken this value from MARCHAND, LIBERT and COMBAZ, 1969, p.13.

TABLE 11 - Paramagnetic susceptibilities Values

CONCLUSION

Following from our experiments it seems clear that temperature is an important factor in the alteration of sporopollenin and related algal substances.

The main products produced are hydrocarbons of varying types and quantities according to the source material. The product yield is variable with time of treatment as shown by the various critical peaks. Amongst other things was shown that very little of the total mass of the material plays a part in hydrocarbon production. Microscopical examination of the residual mass shows that even the finest morphological characteristics resist the thermal condition of our experiments.

The Paramagnetic Electronic Resonance (R.P.E.) readings show an increase in the number of paramagnetic centers with increasing diagenetic effect of the thermolysis treatment.

The author has to thank Y. VINCENT (CHIMISERVICE, Marseille) for the thermolysis, and A. MARCHAND (Institut Paul Pascal, Bordeaux) for the R.P.E. without which the paper could not have been written.

REFERENCES

1. ANDREEV, P.F., IVANCOVA, V.V., POLJAKOVA, N.N., SILINA, N.P., 1955, Geologiceskij sbornik, Gostoptekhiz dat, 1, vyp, 83, 171-187 (trad. I.F.P., 1960).

2. BORDENAVE, M., COMBAZ, A. et GIRAUD, A., 1966, 3rd Internat. Cong. Organic Geochemistry, London 1966.

3. BROOKS, J., and SHAW, G., 1968, Nature, 220, 5168, p. 678-679.

4. COMBAZ, A., 1967, Rev. Palaeobotan. Palynol. 1, p.309-321.

5. GIRAUD, A., 1970, Bull. AAPG, Vol. 54, No.3.

6. MARCHAND, A., LIBERT, L., COMBAZ, A., 1969, Revue I.F.P. Vol. XXIV, No.1.

7. PARKE, M., 1966, *Some contemporary studies in Marine Science*, (Harold Barnes, Ed. p. 555-563).

8. SHAW, G., and YEADON, A., 1964, *Grana palynolgica*, *2*, 5, p. 247-252.

9. SHAW, G., and YEADON, A., 1966, *J. Chem. Soc.* p.16-22.

10. WALL, D., 1962, *Geol. Mag.*, *94*, (4), p. 353-362.

Contact between the Spore Cytoplasm and the growing Sporoderm of the Selaginella Megaspore

ANDREAS SIEVERS and BRIGITTE BUCHEN

Abt. Cytologie, Botanisches Institut, Universität Bonn

Abstract. In contrast to the results of FITTING (1900) the cytoplasm of the *Selaginella* megaspore is not separated from the growing sporoderm. Therefore the *Selaginella* megaspore is not an example of cell wall growth without contact with the plasmamembrane. Between the lamellae of the nexine a fibrillar-netlike matrix is situated. In the innermost part of the matrix thin light lines occur parallel to the nexine lamellae. Perhaps these light lines are the pattern for the forming nexine lamellae.

The plasmamembrane is a sharp boundary between the protoplast and the cell wall. Clear contact between both the membrane and the wall is a necessary supposition for wall growth and morphogenesis. The megaspores of *Isoëtes* and *Selaginella* seem to be an exception to this rule. FITTING (1900) concluded, that during the main growth the sporoderm is separated from the protoplast. This is often cited as an example for cell wall growth without contact with the plasmamembrane (BÜNNING 1953, FREY-WYSSLING 1959, ROELOFSEN 1959, HESLOP-HARRISON 1963, SCHNEPF 1969), in spite of the fact that the retraction of spore protoplasts may be a result of stimulation plasmolysis (GEITLER 1938) or even an artefact (PIENIĄŻEK 1938).

In the case of the *Selaginella* megaspore SIEVERS and BUCHEN (1971) demonstrated the contact between the spore cytoplasm and the growing sporoderm. The thin protoplasmatic layer can be separated from the exine by slight plasmolysis (Fig. 1). In electron micrographs of well fixed megaspores (Fig. 2 and 4) one observes that the protoplast contains all the typical organelles of a plant cell. The plasmamembrane lies near the innerst lamellae of the nexine (Fig. 4). In badly fixed spores only the protoplast is retracted from the sporoderm (Fig. 3). In such cases the contact between the plasmamembrane and the inner lamellae of the sporoderm is clearly demonstrated, as often some fragments of the lamellae occur between the retracted protoplast and the artificially loosened nexine (Fig. 3:➤). Probably they are retracted by the adherence to the plasmamembrane.

The exine consists of a lamellated nexine (Fig. 2 and 4: NE) and a spongy sexine (Fig. 2: SE). In the proximal pole a solid layer of sporopollenin exists between the nexine and the sexine (Fig. 2: SS). The elements of the whole exine anastomose except the inner part of the lateral and the distal sexine where a probably artificial, broad cleft (Fig. 1 and 2: ↔) can be observed.

Outside the plasmamembrane and between the lamellae of the nexine a fibrillar-netlike matrix with darkly contrasting granula of different size is situated (Fig. 4). Sometimes in the innerst part of this matrix we observed thin light lines parallel to the nexine lamellae (Fig. 4:→←). The thinnest lamellae also show a thin surrounding electron transparent zone (Fig. 4 a), but not white lines in the center as other exines have (SOUTHWORTH 1966). As the nexine was still growing, in the moment of fixation, it should be possible to imagine that these light lines are the pattern for the forming nexine lamellae.

Supported by the Deutsche Forschungsgemeinschaft.

References

BÜNNING, E., 1953: Entwicklungs- und Bewegungsphysiologie der Pflanze, 3. Aufl., Berlin etc: Springer.
FITTING, H., 1900: Bot. Z. 58, 107.
FREY-WYSSLING, A., 1959: Die pflanzliche Zellwand, Berlin etc.: Springer.
GEITLER, L., 1938: Planta 27, 426.
HESLOP-HARRISON, J., 1963. Symp. Soc. Exp. Biol. 17, 315.
PIENIĄŻEK, S.A., 1938. C. R. Soc. Sci. Lettr. Varsovie, Cl. IV, 31, 211.
ROELOFSEN, P. A., 1959: The plant cell-wall. Handb. Pflanzenanatomie III, 4, 2. Aufl., Berlin: Borntraeger.
SCHNEPF, E., 1969: Sekretion und Exkretion bei Pflanzen. Protoplasmatologia VIII, 8, Wien etc.: Springer.
SIEVERS, A.,and B. BUCHEN, 1971. Protoplasma, in press.
SOUTHWORTH, D., 1966. Grana palynol. 6, 324.

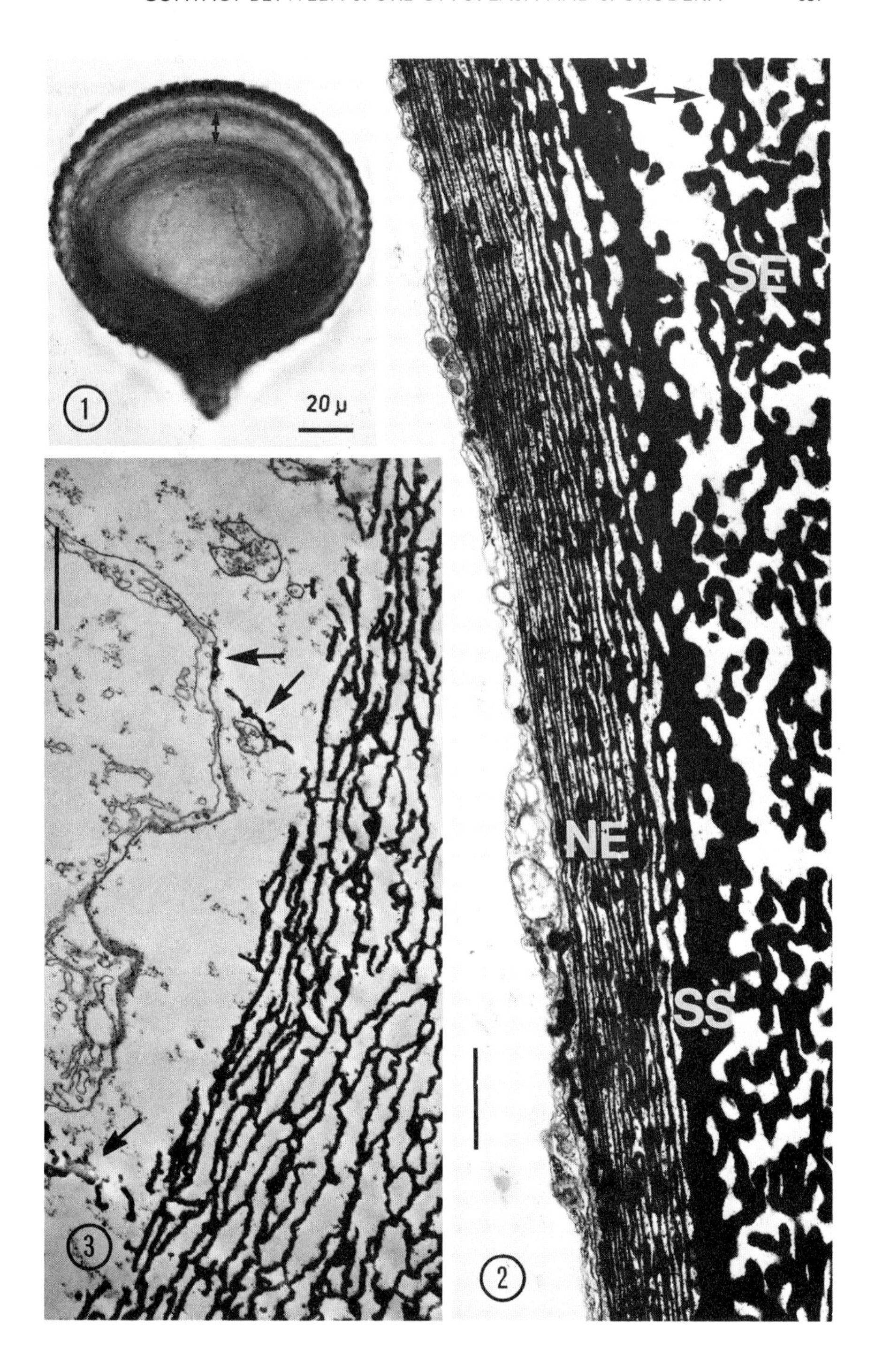
1
20 μ
2
SE
NE
SS
3

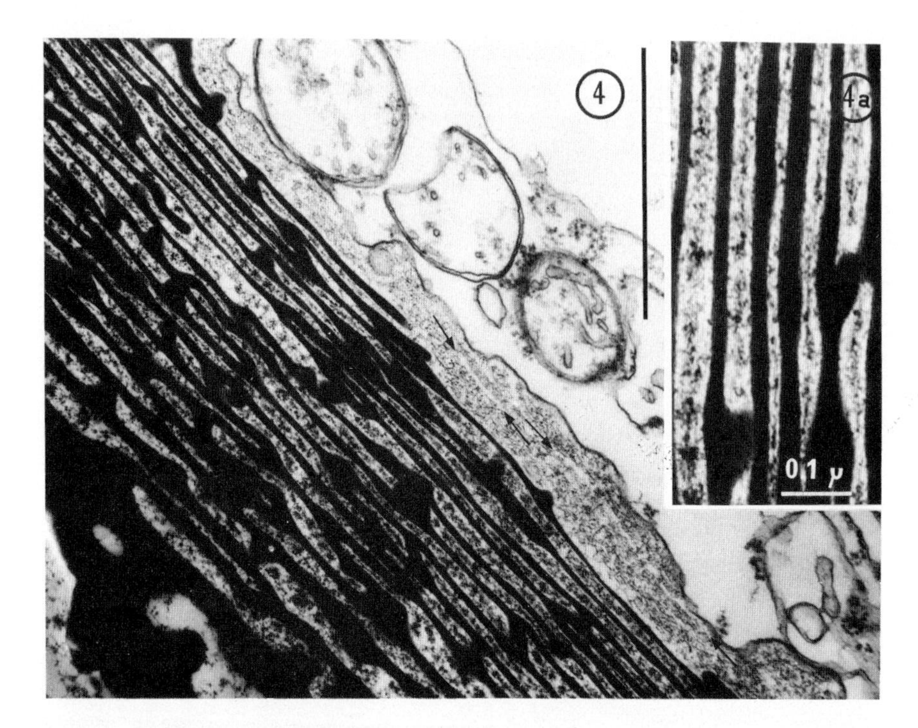

Fig.1. Megaspore after 3 min in 0,08 M sucrose with the retracted protoplast. ↔ : Cleft in the inner part of the sexine.

Fig.2. Section through the more proximal nexine (below) and the more lateral one (above) with a thin protoplasmatic layer containing different cell organelles. NE = nexine, SE = sexine, SS = solid layer of sporopollenin in the proximal exine. ⟷ Cleft. Fix. OsO_4.

Fig.3. Part of a badly fixed spore with retracted protoplast and artificially loosened nexine lamellae. ➤: Fragments of lamellae. Fix. OsO_4.

Fig.4. A fibrillar-netlike matrix with darkly contrasting granula is situated between the anastomosing nexine lamellae. Near the plasmamembrane the matrix shows thin light lines (→←). Fix. OsO_4.

Fig.4a. Electron transparent zone near the thinnest nexine lamellae. Fix. OsO_4.

REVIEW OF THE UV-FLUORESCENCE MICROPHOTOMETRY OF FRESH AND FOSSIL EXINES AND EXOSPORIA

P. van Gijzel

Faculty of Mathematics and Natural Sciences, Catholic University, Nijmegen, The Netherlands

ABSTRACT

UV-fluorescence properties of fresh and fossil pollen and spores of various origin have been investigated by means of microphotometrical methods. The interpretation of the results obtained by means of various measuring techniques are discussed in relation to the problem of sporopollenin. The fluorescence spectra suggest that compounds with a different resistance to geological age, corrosion and coalification must be present in the pollen and spore walls.

Exposure of pollen and spores to UV-radiation under the microscope resulted in a decomposition or alteration of their walls. The exposure caused a "fading effect" of the fluorescence intensities. The fluorescence properties of spectra and fading may be different for well-preserved pollen and corroded or coalified material. Similar differences are found between the sexine and nexine of certain bisaccate pollen types and for fresh spores in various stages of chemical decomposition.

The differences in fluorescence and fading of fresh pollen in various stages of development have been reviewed.

It appears that the UV-fluorescence microphotometry may provide useful information to a better understanding of the chemical character of sporopollenin of various appearance and preservation.

INTRODUCTION

Sporopollenin is one of the most remarkable organic substances in plants. It is extremely resistant to chemical treatment, except to oxidation, which explains the difficulties arising during chemical analysis of its character. Sporopollenin is still not fully understood, although numerous organic compounds in it have been determined. Thus, an exact definition for it cannot yet be given.

The advance in sporopollenin research is based on Zetzsche's work of about forty years ago. He succeeded in making quantitative chemical analysis of fresh and fossil pollen and spore walls of various species. At that time, its resistance was already known - palynology had been born ten years previously - and it must have been a challenge to Zetzsche's genius to apply his new analytical techniques. The insoluble residue in these walls he termed sporopollenin, and he found differing amounts of it in various species. Some years afterwards, Kirchheimer made experiments on the resistance of fresh and fossil material and found that, due to different susceptibilities, not all types of pollen and spores survived equally the geological processes of fossilisation and coalification. But Zetsche and Kirchheimer were in advance of their time, and it took three decades before chemical analysis and resistance investigations on the pollen wall were resumed.

Sporopollenin in fossil palynomorphs can be preserved in rocks for a very long time, even hundreds of millions of years and, according to studies by organic geochemists, probably even in Precambrian rocks and possibly in a certain type of meteorite. The resistance of sporopollenin is in fact the basic hypothesis in palynology. The study of floral changes during geologic time by means of pollen analysis is applied now on a very wide scale, for instance to the dating and correlation of many deposits. Only recently, however, has Kirchheimer's opinion been accepted: this being contrary to the hypothesis of unchangeable exines, which has been generally accepted by palynologists. Recently, several studies have indicated that sporopollenin and other constituents in exines are more or less affected by geological and microbiological processes. Fluorescence microscope investigations showed that the chemical composition of exines and exosporia must have been changed soon after the deposition of the pollen and spores, as was apparent from the changes in their fluorescence colours (van Gijzel, 1961, 1963, 1967). Decay by corrosion and alteration by chemical processes at higher temperature and/or pressure play a more important role in pollen preservation than has been assumed before. If the pollen wall was unable to be fossilised palaeopalynology in its present state would not have existed. Therefore, a knowledge of sporopollenin and its resistance is indeed a basic problem whose importance is increasing with the ever-increasing interest in palynology.

The interest in sporopollenin has developed for other reasons also. Pollen morphological studies show that various plant species may be different in structure and stratification of the pollen wall. Numerous opinions on pollen wall terminology proved the complexity of the composition of the exine. The study of sporopollenin may add a new aspect to pollen morphology (Jonker, this Symposium). Investigations on pollen development by several botanists have provided much new information. In succeeding stages of development, various compounds in the exine, including sporopollenin, are formed. It even appears that material contributing to exine formation is produced from a quite different source in the anther. From all these studies, we know that sporopollenin is present in nearly all pollen and spore walls. It is well known that the sexine and nexine are very different in morphological structure and origin, but the relationships with different contents or composition of sporopollenin is still unknown.

These various aspects of the problem of sporopollenin are subjects of investigation by means of the new method of fluorescence microspectrophotometry. A review of the results will be given below.

For the resistant remains of palynomorphs, the term sporopollenin is used only chemically in this review. Exine and exosporium, both morphological terms, are preferable when a description is given that is based on observations under the microscope.

The development of fluorescence microscopy has much in common with the advance in sporopollenin research. Some decades ago, Max Haitinger, the father of fluorescence microscopy, made his experiments on the fluorescence of plant and animal tissues. He succeeded in improving UV-illumination in the optical microscope, by which the self-fluorescence of cells and even cell-parts could be observed. But much more important was Haitinger's discovery of the phenomenon of 'secondary' fluorescence, i.e. specific colour reactions by means of staining techniques by which many substances could be localised and identified.

As in sporopollenin research, it was only recently that fluorescence microscopy has received a new stimulus. Very sensitive measuring techniques and equipment have been developed, by which it has become possible to record the fluorescence and absorption properties of materials and objects which can be observed visually under the UV-microscope.

The advance in sporopollenin research is an example of the dependance of the study of complex problems on the facilities provided by modern techniques. The use of fluorescence microphotometry and other instrumentation, like the electron microscope and the gas chromatograph is essential to advance the study of sporopollenin. Although fluorescence microphotometry is still under development, in particular measuring ranges and preparation techniques, it has already been applied successfully to palynology (van Gijzel, 1967).

METHODS OF FLUORESCENCE MICROPHOTOMETRY

For a detailed description of these methods, reference should be made to some papers on the special equipment used (van Gijzel, 1966, 1967, 1971a, 1971b). The following points, however, have a special relevance to the study of sporopollenin under the UV-microscope.

Fluorescence microphotometry is very useful for the investigation of microscopic material in situ, and may provide information which cannot be obtained by other means. For instance, chemical differences between fossil palynomorphs from various origins can be observed visually under UV-light, and described objectively by means of the microphotometrical recording of the spectra and other fluorescence properties.

For all recent determinations mentioned in this paper, equipment of the Leitz MPV-system has been used. The excitation of the fluorescence has been derived from long-wave UV at a wavelength of about 365 nm (UV-fluorescence). The spectra have been determined by measuring of the light intensities of the objects at succeeding wavelengths in the range from 400 nm to 700 nm.

Pollen and spores often show a very broad fluorescence spectrum from 400nm to more than 700 nm with a more or less distinct maximum, depending on the state of preservation. This single peak is situated between 460 nm and 600 nm, according to type and fluorescence colour. For instance, young-Holocene *Sphagnum* spores may have a blue colour with a maximum at about 460 nm, compared with certain Gramineae species with an orange fluorescence and a peak at 580 nm.

Fluorescence microphotometry makes special demands on the adjustment and stability of the photometrical equipment and on the preparation of the pollen slides (van Gijzel, 1971a). As mounting medium, distilled water has been used, in order to make the background to the objects (the 'empty picture') as dark as possible.

It is well-known from medical fluorescence microscope studies that organic substances in tissue change in fluorescence colour under microscopical UV-exposure. This 'fading' effect (in German: "Ausbleichung") occurs also in pollen and spore walls and in cuticles, and can give rise to serious difficulties in the determination and interpretation of fluorescence spectra (van Gijzel,1971b). This change in colour, intensity and/or spectrum appears to be irreversible and depends on the period of exposure, wavelength, and on the chemical character of the exposed object. Even in an interval of a few seconds, a change in colour may occur; hence a simultaneous shift in the spectrum. This happens repeatedly in fresh material (see also Willemse, this symposium). The fluorescence intensity may decrease under UV-exposure (negative fading) or increase (positive fading), but it also possible for a large increase to follow a small decrease. For more details of the technique of measuring and the theory of fading, the reader is referred to the above paper.

The amount of fading and the shape of the fading curves appeared to be different for various stages of chemical decomposition of *Lycopodium* spores (see table I). Even when the fluorescence colours of exines and exosporia are similar, their fading properties may differ. After certain treatments, the spores had become dirty yellow or brown in colour. This colour changed after an exposure of one hour, into light yellow or greenish yellow, while the intensity has become very high. This contrasts with fresh untreated material, where the intensity decreases and the spectrum shifts to longer wavelengths.

INTERPRETATION OF THE RESULTS OF FLUORESCENCE DETERMINATIONS

Some difficulties may arise in the interpretation of the fluorescence spectra which have been derived by means of very different measuring techniques.

The measured fluorescence spectrum is a mixture and sum of the spectra of all fluorescent compounds in the exine or exosporium. Organic compounds may have different fluorescence intensities and spectra. The interpretation of such complex compound spectra derived from different palynomorphs is not easy, and is the subject of continuing investigation.

The previously described two peak spectra of fossil pollen and spores (van Gijzel, 1966, 1967) do not differ in principle from the single peak spectra found by means of more advanced

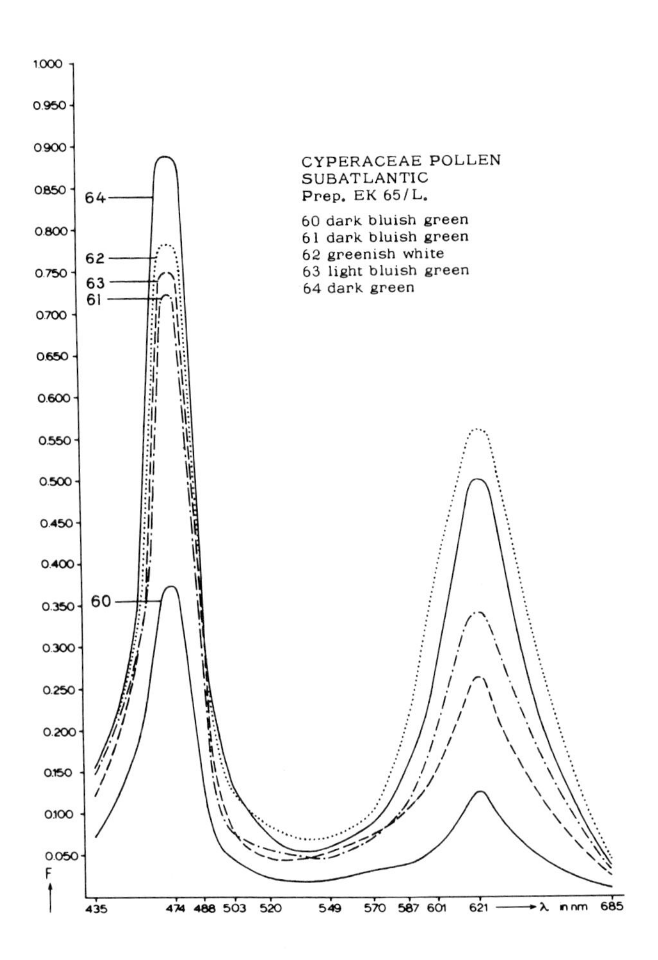

Figure 1. Berek-photometrical spectra of the fluorescence of some Cyperaceae pollen grains from a young Holocene *Sphagnum* peat (after van Gijzel, 1967)

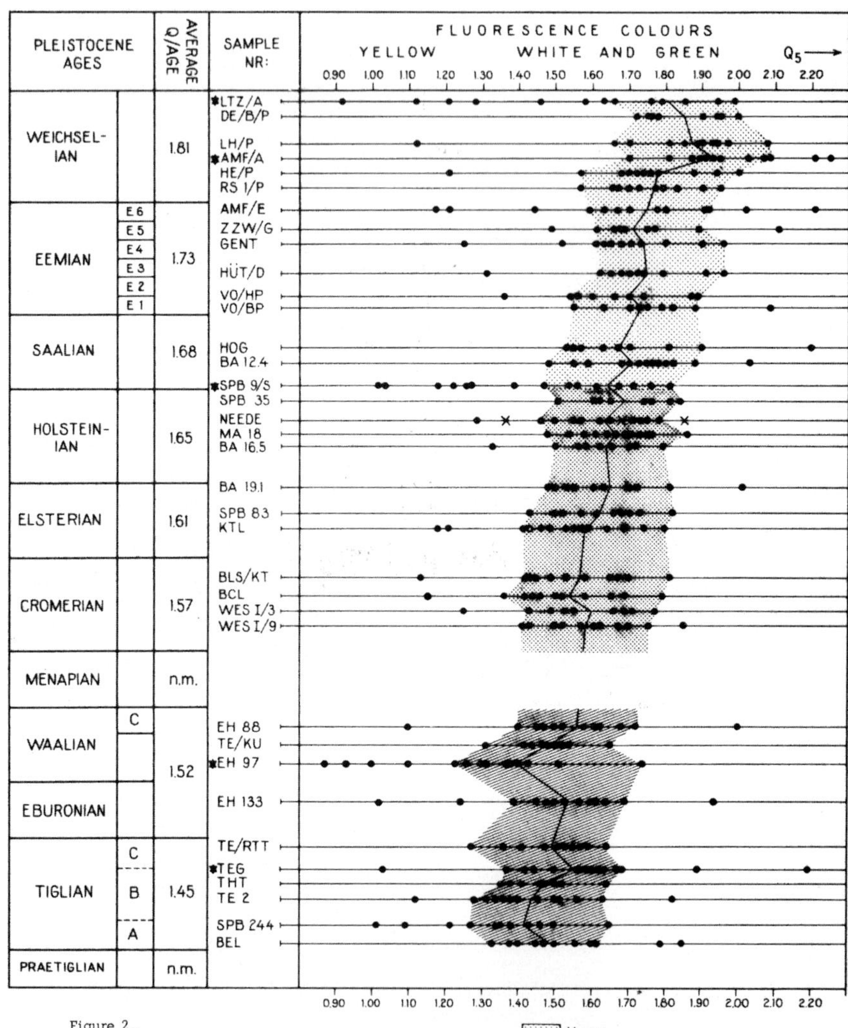

Figure 2.

SPECTRAL RATIO OF PINUS SYLVESTRIS FROM THE PLEISTOCENE
(note the colour shift through time. * = strongly corroded pollen)
(measured with a Berek-photometer)

photo-electrical techniques as the MPV-system. By means of the Berek photometer, two peaks are measured (fig.1), representing the left and right side of one peak. This is due to the fact that the intensities have been compared with and expressed in the fluorescence spectrum of uranyl glass, which is used as a reference light source. By dividing the blue-green intensities by those in the orange-red, the spectral ratio has been obtained. This ratio is equal to the similar value obtained by measurements using newer techniques. The Berek photometer yields incomplete spectra, of which the real maximum cannot be exactly determined. Nevertheless, the spectral ratio of vesiculate pollen grains, of different geological ages, obtained by means of this method (van Gijzel, 1967) gives a reliable impression of the gradual colour change of these pollen types. The use of a more advanced measuring method has no influence on the earlier-made suppositions of several fluorescent compounds in the exine and their gradual decay during geological time.

FLUORESCENCE AND GEOLOGICAL AGE OF PALYNOMORPHS

The fluorescence phenomena of fossil palynomorphs from various geological ages and various types of rocks have been described earlier by the present author (van Gijzel,1961,1963, 1967). From these studies it appeared that the walls of pollen and spores do not remain unchanged during geological time, but they are chemically altered soon after their deposition in the soil. The process of gradual decomposition of various substances in the walls continues during geological time. Easily soluble substances disappear first, followed by a gradual decomposition of the more resistant components in the exines and exosporia.

With increasing geological age, a gradual chanes in the fluorescence colour and spectrum of palynomorphs occurs. For instance, Pleistocene grains of *Pinus sylvestris* show this slow process very clearly (fig.2). The shift in colour is a result of the gradual extinction of fluorescence intensities, which takes place more rapidly for the blue-green than for the orange-red part of the spectrum. Simultaneously a change in the spectral maximum occurs, going to longer wavelengths and covering a range from about 460nm. to 600nm. The yellow-orange colour range covers a longer geological time.

As the fluorescence of a substance is closely related to its chemical composition it may be assumed that the chemical character of palynomorphs has been considerably changed through geological time. Other factors may play a role in this process [see below], but for the Quaternary the influence of temperature and pressure on the colour change may be neglected, as many deposits from that time have been buried under only relatively small sedimentary accumulations. Their rank has never reached the lignite stage.

Various types of fresh and fossil pollen and spores may show different fluorescence properties. This is in agreement with the chemical differences for pollen and spores, as found by Zetzsche and other chemists (see Shaw, this symposium). These differences in fluorescence decrease with increasing age. Even differences in fluorescence properties between various layers of the exines and exosporia may occur [see below].

It may be assumed that the less resistant compounds are responsible for the blue-green intensities in the fluorescence spectra and the more resistant compounds for the orange-red colours.

Several questions arise from the explanation of the fluorescence phenomena. The large spread in the colour distribution of certain pollen samples, as is demonstrated in Fig.2, is difficult to understand. Apparently, grains of the same type can react differently to fossilization. An explanation is also difficult for the differences in the susceptibilities of various types to corrosion, coalification, and other geological agents. From the fluorescence and other studies a large variation in susceptibility has been found. As sporopollenin is considered to be the most resistant substance in the walls of pollen and spores, it is certain that all these fluorescence phenomena must be closely related to the changes in the sporopollenin composition and/or content. The chemical character of sporopollenin is still too poorly understood to explain the fluorescence phenomena exactly.

Nevertheless the fluorescence microscopy of fossil palynomorphs is used in the practice of palynology for age determination of deposits, when the conventional pollen analysis fails as, for instance, in the dating of sediments which have been contaminated by re-worked pollen material. In such deposits pollen produced by the autochthonous vegetation can be distinguished from the re-worked pollen grains by means of the fluorescence properties. The former are more green or yellow in colour and more strongly fluorescent than the latter which are orange, or even brown, and weakly fluorescent. Also, each type of secondary pollen shows a larger variation in fluorescence colour than the autochthonous material. This is due to decay during the re-working.

The fluorescence phenomena are in accordance with the results of staining experiments, Stanley(1965,1966). The effect of stains on exines decreases with increasing age and must also be related to changes occurring in the chemical composition of sporopollenin during geological time.

FLUORESCENCE AND COALIFICATION OF PALYNOMORPHS

When pollen and spores are progressively buried under sedimentary accumulations; their fluorescence colour and spectrum changes due to the increase of pressure and/or temperature with increasing depth of burial. At a higher rank of coal (ie. the degree of coalification) an extinction of the fluorescence occurs, generally, when the palynomorphs become unrecognisable and black in normal illumination (van Gijzel, 1967).

It appears that the change in fluorescence parallels the change in colour with increasing age. Up to a certain stage in the coalification series the fluorescence colour may be determined by time rather than by rank of coal.

With increasing rank the translucency of palynomorphs decreases slowly, but accelerates beyond a certain stage of the coalification series- when they become very rapidly darker and finally lose their structure- remaining simply as black spots. This occurs in the range with fixed carbon content of about 69-78% (Gutjahr,1966). A relationship may exist between the fluorescence and light absorption of palynomorphs (van Gijzel,1967). Both appear to change with increasing depth of burial, probably as a result of temperature increase. The shift in fluorescence, as well as light absorption, has been considered to be a measure of the rank of coal, but up to a certain rank only (F.C. content of about 75%). Both phenomena must be related to changes in the chemical composition of sporopollenin with increasing coalification.

The rate of change in fluorescence with increasing rank is different for various types of palynomorphs: the change will be slower for types with a stronger resistance to coalification.

Fluorescence colours, similar to those of fossil pollen and spores, have been found in organic substances in certain deposits. Dirty yellow, orange and brown colours occur in numerous rocks,such as limestones and sandstones with remains of Molluscs, Foraminifera, or Algae, but also in bituminous rocks and in certain Pre-Cambrian deposits. Such fluorescence colours are even found in the plant remains of the Rhynie Chert from Scotland (van Gijzel, 1971a).

In the exines of bisaccate palynomorphs from medium rank Triassic and Permian rocks the sexine in the wings and around the pollen body is often dark yellow or yellowish orange in colour and much more strongly fluorescent than the dark brown coloured nexine, in which the fluorescence has nearly disappeared. The fading in fluorescence of the sexine is much more than that of the nexine; both increase in intensity under continued UV-exposure. Hence it appears that a larger amount of resistant compounds must be preserved in the sexine than in the nexine.

A corresponding difference occurs in the Palaeocene, lignite rank, exosporia of Triplanisporites sinuosus, where the outer layer becomes light yellow under UV-exposure, while the inner layer remains dark brown, although its intensity increases slightly. In contrast, other spores in the same pollen slide show a light yellow colour without any differentiation due to the layering of the exosporium. Some of them may possess dark orange spots or may even be completely orange. Apparently, in various types of palynomorphs, the sexine and the nexine may possess, in their walls, a different coalification susceptibility. Some species may remain rather strongly fluorescent, while others have more or less completely lost their fluorescence ability.

Coalified palynomorphs may show 'broken' fluorescence peaks and a positive fading under UV exposure, similar to spores of Lycopodium, which have been heat treated to 200°C or more (see below). Again a relationship must exist between the fluorescence properties and the chemical decomposition of the sporopollenin by heating, either by the natural processes of coalification or by treatment in the laboratory.

FLUORESCENCE AND CORROSION OF PALYNOMORPHS

When fossil pollen and spores become corroded their fluorescence colours change more or less into dirty and weak orange and brown hues. When strongly corroded, the grains may be so severely damaged that fluorescence disappears, although they are still recognisable as pollen or spores. The wings of vesiculate pollen may often show less dirty colours than the central bodies, (van Gijzel, 1970). These changes are similar to those in coalified palynomorphs.

Recent investigations on corroded, cryoturbated peat layers from the Weichselian Glacial (U. Pleistocene) by the present author show that the decay of the fluorescent substances is different for various pollen types. When the majority of Pinus sylvestris grains show their original green and white colours, nearly all the Picea pollen (except the smaller type) have changed into orange-brown grains. If present, grains of the small Picea type are similar in colour to Pinus. Their differences in corrosion susceptibility are clearly shown by the fluorescence colours.

Comparable differences in the fluorescence of corroded pollen and spores exists in numerous other types of palynomorphs. Corroded grains of the same type always show a larger spread in their fluorescence colour distribution than non-corroded grains of the same age. Corrosion of algal remains, like *Pediastrum*, and fungal spores results in a similar change in fluorescence, but the rate of change is different from that of spores and pollen grains. In comparison with spores of fungi, algal fluorescence changes much more, but not so much as that of pollen and spores. However, like the more resistant palynomorphs, they differ considerably in fluorescence from the other plant remains, which show very weak brownish fluorescence colours.

It is obvious that corrosion phenomena are also closely related to the chemical properties of sporopollenin.

The fluorescence properties appear to be in agreement with the results of corrosion experiments on various types of fresh pollen and spores, made by Havinga (1964, 1967, see also this Symposium). Types that are more susceptible to corrosion in Havinga's series appeared to change their fluorescence colours more easily than those more resistant to corrosion.

Corrosion may prevent determination of the amount of secondary pollen in a contaminated sediment (van Gijzel,1970). Possibly differences in spectrum and fading may be used to distinguish corroded grains from the reworked material, because their fluorescence colours are, in many cases, very similar. Such an investigation may be useful in pollen analysis for age determination of contaminated deposits which have been more or less corroded.

FLUORESCENCE AND CHEMICAL DECOMPOSITION OF FRESH SPORES

To get an impression of the fluorescence of various organic substances which are present in sporopollenin, a series of measurements has been made on fresh material of *Lycopodium clavatum* that has been treated according to the methods used by Shaw and Yeadon (1966). Ths fluorescence colour of the untreated spores varies slightly between light blue and light green with a rather broad maximum at about 498 to 510 nm. Under UV exposure, the colour changes to greenish yellow with a small shift of the peak to longer wavemengths, while the intensity decreases considerably.

In several stages of treatment, various substances have been released from the spore walls; for every stage the fluorescence spectrum and the fading properties are measured (table I). The removal of contents preceded all the treatments mentioned.

In comparison with the untreated material, the results are as follows:

a) In various steps of the removal of contents only a slight difference in colour occurs: the spectral maximum is narrowed to a distinct peak at 495 nm and the colour becomes more bluish. The fading remains negative, but is slightly different for these steps.

b) After the removal of contents, a small shift in the peak occurs to 510 nm; the colour has become more greenish. The fading is still decreasing at about the same amount, but the shift of the peak after fading is somewhat smaller. As the cytoplasm has a very weak fluorescence (see Willemse, this Symposium), these differences must have been caused by the release of substances from the spore wall.

c) Ozonisation and treatment with KOH for one hour to release all compounds except the cellulose, resulted in a return of the peak from 510 to 497 nm; the colour became more bluish again. After a continued treatment of 16 hours, the peak shifts even further back to 487 nm, which may be related to a higher cellulose content of the walls. The fading remains negative, but increases in %age.

d) Removal of cellulose with 40% phosphoric acid resulted in a shift of the peak from 510 to 490 nm; the colour becomes bluish white and the fading remains negative: whereas, when treated with more concentrated phosphoric acid, the spores become dirty yellow with a flat and 'broken' peak at about 548 to 558 nm; the fading has become positive and very high. After one hour's exposure, the peak is narrowed and has returned to about 512 nm. Similar broad and 'broken' spectra occur also after chemical degradation or heating of the spores. After treatment by acetolysis, a curved maximum may even occur (fig.3).

TREATMENT	FLUORESCENCE COLOUR	MAXIMUM nm	% FADING 2 min.	% FADING 1 hr.	AFTER 1 HOUR FLUORESCENCE COLOUR	AFTER 1 HOUR MAXIMUM nm
Untreated	Blue-green	498-510	0	-22 -50	Greenish-yellow	512-515
Removal of contents by:						
Ether	L. Blue	485	-30			
Ether + alcohol	L. Blue	495	-24			
Ether + alcohol + KOH + water	L. Blue	495	-13			
Contents removed	L. Blue-green	510		-45 -55	L. yellow	508-510
Separate treatments by:						
Alcohol	L. Blue	490	-16 -20	-56 -62	Greenish-yellow	503-506
Concentrated KOH	L. Green	500	-24	-30 -33	Greenish-white	503
Alcohol + KOH + water	L. Blue	493	-8			
Ozonisation and KOH:						
For 1 hour	L. Green	497	-30	-38 -70	Greenish-white	497
For 16 hours	Bluish-white	487-493	-23	-59	Greenish-yellow	493
Ozonisation; no KOH:						
For 1 hour	Greenish-yellow	503	-16	-50 -57	Greenish-yellow	512-151
For 16 hours	L. Blue-green	487-493	-16 -21	-63 -66	Yellowish-green	496-498
Removal of cellulose:						
Phosphoric acid (40%)	Bluish-white	490	-24 -34	-54 -56	Dark yellow	506
Phosphoric acid (85%)	Dirty Yellow	548-558	+20	+215 to +220	Yellowish-white	512-516

<u>Chemical destruction</u>:						
Dilute HNO_3	Dirty Yellow	522-543	-6 -11	+130 to +142	Greenish-yellow	503
Concentrated HNO_3	Dirty Yellow	515-555	+10	+335	L. Yellow-green	505
NaCl + conc. HNO_3	L. Green*	506	-10	+50 to +57	Greenish-yellow	512
<u>Heating</u>:						
100 hours at 200°C, n.e.	Dark Yellow	527-540	+24 to +32	+360	Yellow-green	503
100 hours at 200°C, s.e.	Dark Yellow	540-575	+26 to +50	+140 to +230	Yellow-green	517-525
100 hours at 250°C, n.e.	Yellowish Orange	540-580	+10 to +20	+200	Yellow	505-550
100 hours at 300°C, s.e.	Dark Orange	580-588	+60			
100 hours at 350°C, n.e.	Orange Brown	580-593	+25			
100 hours at 400°C, n.e.	Orange Brown	573-590	+95			
100 hours at 450°C, n.e.	Orange Brown	573-588	0 to +12			
<u>Other separate treatments</u>:						
Br_2 in acetic acid	L. Yellow	515	+50	+180 to +310	Yellow - Ochre	525-540
Sulphur + CS_2	L. Green	500-515	-22	-45 -50	Yellow-green	508
6 normal HCl	L. Green	496-499	-24	-60	Dark Yellow	506-503
18 normal H_2SO_4	Bluish-green	496	-22	-57 -64	Yellow-Ochre	503
30% HF	L. Green	500-510	-20			

Table I: UV fluorescence of fresh <u>Lycopodium clavatum</u> spores after different types of treatment. [All measurements were made using the Leitz MPV-system and an Orthoplan microscope with: Pl 40/0.65 Objective lens, incident illumination, excitation 365 nm (filters UG 1 &BG 38) barrier filter K 430, stabilised high-pressure lamp HBO 100, spores in distilled water. For each treatment: values of fluorescence maximum is average, from 5 - 6 spores.
n.e. = no extraction s.e. = soluble extracted * spores in amorphous mass]

e) With chemical degradation of Lycopodium spores, dirty yellow colours and 'broken' spectra occur; the fading is positive and extremely high, especially when treated with concentrated nitric acid. When different concentrations are used, remarkable fading changes occur even after 2 minutes. When dilute nitric acid is used, the fading curve begins with a decrease, followed by a large increase in intensity. In both cases, however, the position of the peak and its width are similar. In more concentrated nitric acid, the majority of the spores have been decomposed into an amorphous mass. Treatment with NaCl and concentrated nitric acid resulted, however, in the complete destruction of all the spores. The amorphous mass can show a large variation in fluorescence colour.

f) Heating results in dirty yellow, orange or brown colours, and broad maxima without a distinct peak, which is situated (if present) far into the red. A large, positive fading occurs (as much as 50% after 2 minutes of exposure) at which the colour becomes greenish yellow again. At $300^{\circ}C$ or higher, the orange and brown colours appear, greatly resembling the colours of highly coalified palynomorphs.

g) After bromination in acetic acid the spores show a light yellow colour. The shift of the peak is rather small. A large positive fading occurs with a considerable shift of the peak during exposure.

h) Sulphurication results in a change in fluorescence colour and maximum similar to that produced by bromination. The fading is negative with only a small shift of the peak.

i) Treatment with acids, such as hydrochloric acid, sulphuric acid or hydrofluoric acid, as used in the preparation of rock samples, is common practice in palynological laboratories (Faegri & Iversen, 1964). Even with dilute acids, such treatment results in a small shift in fluorescence colour and spectrum of Lycopodium spores. On the other hand, fossil palynomorphs, which have been treated with sulphuric acid or hydrofluoric acid, show a larger change in colour and spectrum.

It appears from the measurements on treated spores that a considerable change in fluorescence properties occurs after heating or treatment with concentrated nitric acid, phosphoric acid or with nitric acid and sodium hydroxide. The spectra have been 'broken' and a high positive fading occurs. Similar fluorescence properties are shown by strongly corroded pollen and spores in rocks. With nearly all other treatments the changes in fluorescence colour and maxima are rather small and the fading is negative and low. A negative fading is found in well preserved fossil exines and exosporia from the youngest deposits (Quaternary or slightly older).

FLUORESCENCE AND POLLEN DEVELOPMENT

The use of secondary fluorescence techniques in combination with polarization microscopy has been applied by Waterkeyn(1970) and by Waterkeyn and Bienfait(this symposium) to the study of callose. They investigated the network of probacula and spine rudiments in the tetrad stage of pollen development in *Ipomoea purpurea*, and *Lilium regale* by means of staining with primuline to activate the secondary fluorescence. In both species a specific fluorescence of the 'Ubisch bodies' occurs. They found also that, after their formation, the sexine and nexine are clearly fluorescent, as opposed to the primexine which shows no fluorescence at all in *Ipomoea*.

Willemse(this symposium) investigated the changes in fluorescence during the development of pollen of *Pinus sylvestris* and *Gasteria verrucosa*. After the first stages of development a shift in colour and spectrum occurs from 500nm. to482nm. in *Pinus* and from 493nm. to 480nm. in *Gasteria*.Differences in fading also occur-this measured after 30 seconds. The intensity and fading are lower for *Gasteria* than for *Pinus*. Willemse concluded that a relation exists between the fluorescence and the changing chemical composition of the exine during development. In the tetrad stage all cellulosic substances are absent and the fluorescence spectrum would be caused by the components of the sexine and nexine 1. As soon as the production of the intine material starts (after the tetrad stage) a decrease of the maximum occurs. The callose appears later and probably causes a further decrease of the maximum. The substances mentioned by Shaw and Yeadon (1966), as well as callose and pecto-callosic materials may be responsible for the pollen wall fluorescence.

Willemse's fluorescence measurements have been made in co-operation with the present author; with the same MPV-equipment. Judging the results of both Willemse's and the present author's requires all the remarks on the methods and interpretation to be taken into account.

DISCUSSION OF THE RESULTS AND CONCLUSIONS

Although we know several of the organic compounds present in sporopollenin, the complete structure remains a problem to solved. The fundamental questions are: "Where, and how many of, these compounds are bound in the layers of pollen and spore walls of different plant species?"; and "Why do these walls react so differently to microbial and geological processes in soils and rocks?".

There is no doubt that the fluorescence phenomena of the walls of pollen and spores are closely related to the chemical composition of the sporopollenin (van Gijzel, 1967). These phenomena cannot be explained, however, unless the above questions are answered. This will only be possible by co-operation of workers from various disciplines. It is clear from this review that fluorescence mirophotometry is able to detect small differences in chemical composition of exines and exosporia. However, the problem of interpretation of the fluorescence analyses must be solved before actual idenification of chemical substances can be made.

A fluorescence spectrum of a pollen wall is determined by:

1) the number and content of fluorescent compounds (ie. the fluorescence is a measure of concentration),

2) the fluorescence ability of every compound, which determines its intensity of fluorescence,

3) the shape (including spectral width and peak) of every

compound,

4) if a rapid fading occurs, as in fresh material, the time needed to measure a spectrum; this can be shortened,

5) properties of the measuring equipment, which can be determined by calibration of the spectra (van Gijzel,1970).

From the fluorescence spectra it is obvious that numerous compounds must be present in sporopollenin, but their identification can be made only when the same material is also investigated by other methods. Fluorescence microphotometry is still at a descriptive stage, and its techniques are under development. Nevertheless, it has provided useful information on the question of sporopollenin.

The gradual shift in the fluorescence spectra, as mentioned in this paper, is determined by a gradual decrease of the blue and green intensities, which occurs more rapidly than the decrease of intensities in the yellow, orange and red part of the spectrum. Apparently, the less resistant substances may possess blue-green colours, while the orange and red colours may be related to the more resistant components of sporopollenin. The final fluorescence colour of a fossil palynomorph is determined by the fluorescence of the most resistant substances - these depend more or less on the results of the combined action of geological processess, of which time and temperature are the most important.

Thus, the differences in fluorescence properties (colour, spectral maximum, and fading) of palynomorphs are related to differences in:

a) decomposition through time, depending on their resistance to corrosion and coalification, (this includes fungal and algal remains)

b) effect of stains on various types,

c) chemical composition, as reported by chemists, of the sporopollenin of various types,

d) chemical composition of the sexine and nexine, related to the differences in sporopollenin composition,

e)chemical decomposition as a result of laboratory treatment, as on _Lycopodium_ spores,

f) chemical composition, which changes during pollen development, when cellulosic and hemi-cellulosic substances and callose subsequently appear.

Even slight differences between specimens of the same type may exist.

The similarity of fluorescence in algal and fungal remains with fluorescence in palynomorphs from corroded deposits may support the opinion that sporopollenin is present in the walls of algae and fungi.

Differences in the fluorescence properties of sexine and nexine, or the exosporium and endosporium, found in coalified palynomorphs and corroded grains depend more or less on the differences in susceptibility to corrosion, coalification or age and may be variable for different types.

In view of the afore- mentioned results, we may expect that by means of the improved techniques of fluorescence microphotometry a contribution will be made toward unravelling some of the mystery still surrounding sporopollenin.

ACKNOWLEDGEMENTS

The author owes special thanks to Dr.J. Brooks, BP Research Centre, for providing numerous samples of chemically treated spores; and is specially grateful for his co-operation in this work. The author is also very grateful to numerous colleagues in palynology and botany, especially to Dr. M.D.Muir, Imperial College, for their encouraging and stimulating support during the preparation of this present paper.

REFERENCES

FAEGRI,K.,and IVERSEN,J., (1964): Textbook of Pollen Analysis, 2nd ed.
GUTJAHR,C.C.M., (1966): Leidse Geol. Meded.,38, 1-29.
van GIJZEL, P.,(1961): Kon.Ned.Akad.Wet.,Proc.,B, 64/1,56-63.
van GIJZEL, P.,(1963): Med. Geol. Sticht., Nw. Ser.,16,25-32.
van GIJZEL, P.,(1966): Leitz Mitt. Wiss. Techn., 3/7, 206-214.
van GIJZEL, P.,(1967): Leidse Geol. Meded., 4o, 264-317.

van GIJZEL, P.,(1970). Med. Afd. Geol. Biol. Lab. Kath. Univ. Nijmegen, 2, 1-9.

van GIJZEL, P., (1971a). Leitz Mitt. Wiss. Techn. V (in press)

van GIJZEL, P., (1971b). Zeitschr. wiss. Mikrosk. mikrosk. Techn., (in press).

HAVINGA, A.J., (1964). Pollen et Spores, 6/2,621-635.

HAVINGA, A.J., (1967). Rev. Palaeobotan. Palynol., 2, 81-98.

SHAW, G., & YEADON, A.,(1966). Journ. Chem. Soc., (C), 16-22.

STANLEY, E.A.,(1965). Nature, 206/4981, 289-291.

STANLEY, E.A., (1966). Marine Geol., 4, 397-408.

WATERKEYN, L., (1970). Grana Palynol., (in press).

Legend to Figure 3 (page 682):

Figure 3. Fluorescence and excitation spectra of fresh Lycopodium spores, measured in suspension by means of a Fluorispec spectrophotometer (after van Gijzel.1967). Spectra not determined microscopically. Spectra are uncorrected for sensitivity of photomultiplier and other properties of the instrument, which explains the difference in the situation of the maximum in comparison with that in Table I.

a) fluorescence spectrum excited at 365 nm, spores treated previously with acetolysis and KOH

a') excitation spectrum belonging to the fluorescence at 445 nm, spores treated similarly to a (above)

b) fluorescence spectrum, excited at 365 nm, spores treated with KOH only

b') excitation spectrum,belonging to the fluorescence at 445 nm, spores treated similarly to b (above)

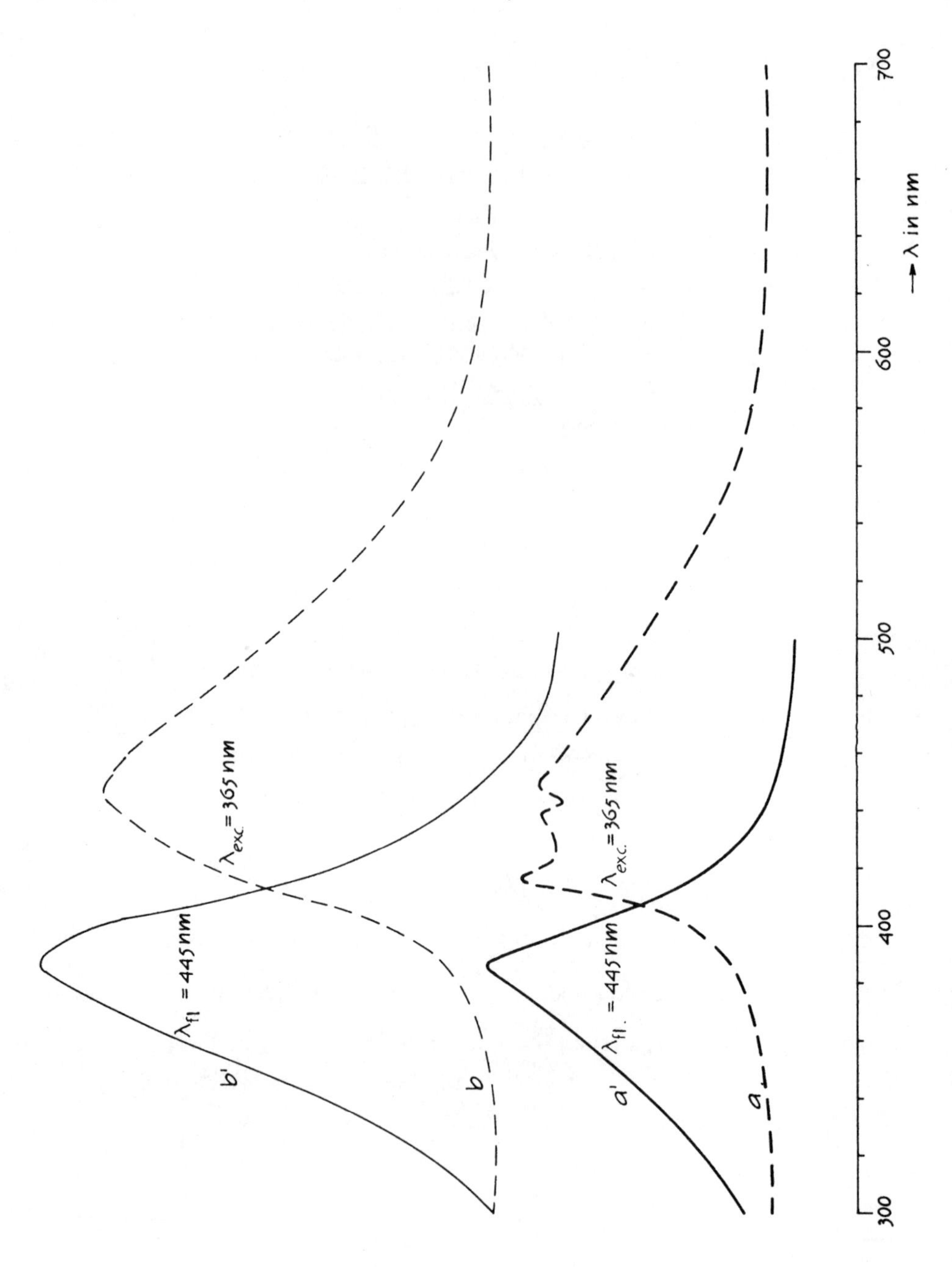
b'
b
a'
a
$\lambda_{fl} = 445$ nm
$\lambda_{exc} = 365$ nm
$\lambda_{fl.} = 445$ nm
$\lambda_{exc.} = 365$ nm
300
400
500
600
700
→ λ in nm

Figure 3

DISCUSSION ON DR. M. MUIR & MR. P. GRANT's and DR. P. VAN GIJZEL's PAPERS.

VAN GIJZEL: Confusion might arise between the use of cathodoluminescence and fluorescence for what is really the same thing. Why not use the term electron fluorescence?

MUIR: Cathodoluminescence is the proper term for luminescence stimulated by an electron beam. Electrofluorescence, easily confused with the proposed "electron fluorescence," is an entirely different phenomenon.

HESLOP-HARRISON: Yes, cathodoluminescence is a physicist's term, not a biologists. It cannot be used interchangeably for fluorescence. I would like to make one or two comments on Dr. van Gijzel's very interesting paper. Some spores have coloured exines, some being bright red (not to be confused with exines coloured by the presence of carotenoids, etc.). These exines do not fluoresce very much, and the quenchimg of the fluorescence is apparently related to the presence of the pigment (*Impatiens sultanii* provides a good example of this). This leads me to suggest that we should be careful when drawing chemical conclusions about sporopollenin from fluorescence properties. Further, the various solvents employed in preparation are almost certain to have an effect on the fluorescence colours depending upon their extraction of various uncombined carotenoids etc.which remain trapped in the interstices of the sporopollenin.

VAN GIJZEL: Yes, this is probably true for the investigation of fresh material, but in the case of fossil material, the problem is entirely different, and the effects of different preparation methods on fluorescence have been examined. While many difficulties remain, I think we are approaching the problem in the right way.

SHAW: I thought Dr. van Gijzel's work might have perhaps satisfied Professor Rowley. The presence of cellulose seemed to be quite clear.

CORRIEA: You have shown slides that showed the changes in fluorescence as a function of the reflectivity of vitrinite. Now, it appears that above a reflectivity value of 0.40% (more or less) fluorescence does not vary. You explained how the relatively great age of these samples explained the absence of variation in fluorescence. I believe that this explanation should perhaps be re-thought because if the age of the sample is greater than that where fluorescence disappears, or in the region of 0.40%, a good thermal marker, of vitrinite then fluorescence cannot be used,ie. fluorescence methods should only be used when the degree of evolution of the organic matter is small.

VAN GIJZEL: The relationship between fluorescence changes and coalification is a difficult problem, but with a more thorough investigation, then we can come to some conclusions. Fluorescence studies should always be carried out in combination with other work, and then the results of the combined work are extremely useful.

CALDICOTT: In order to fluoresce a substance has first to absorb UV (or IR) and then re-emit. Can you tell me what substance is likely to absorb the UV?

VAN GIJZEL: If you take a pure mineral it does not fluoresce. However, if it contains impurities or structural defects then it fluoresces with a colour characteristic of the impurity or the defect. It is possible that trace elements will stimulate fluorescence in organic materials too. But in fluorescence in general, when an atom is irradiated with UV an electron will move from one energy level to another with the absorption of a photon. Later, the electron will jump back from its meta-stable state to its original lower energy state and in the process a photon may be emitted. This emission is generally of longer wavelength than the irradiating UV. This is the case for inorganic materials.
In organic material parts of molecules, and even whole molecules may be in varying meta-stable states. It is therefore difficult to give a simple illustration of an excited organic compound.

CALDICOTT: Dr. van Gijzel described a blue fluorescence like cellulose. Is there a characteristic fluorescence for aromatic compounds, and is there any inorganic material present in sporopollenin?

SHAW: Ozonolysis tends to destroy aromatic constituents. But there is quite a considerable inorganic constituent to sporopollenin (up to 1%).
Has Dr. van Gijzel tested pure cellulose for fluorescence, and was it the same as repored above?

VAN GIJZEL: Yes.

ROWLEY: About so-called carbonisation of pollen walls: what does this do to the fluorescence colours of sporopollenin?

VAN GIJZEL: Well, this raises the question of how long we call the substance sporopollenin. I think that during carbonisation various compounds are lost; and what we are trying to do with our fluorescence studies is to try to investigate the spectra of each of these compounds and to seek any relationships that may exist.

SHAW: It seems to me that you should have a collection of

standard compounds.

WHITEHEAD: I think that we should not look at the problem of carbonisation of sporopollenin except in the context of the diagenisis of petroleum. Both lead to the formation of aromatics (of low free energy). The presence of O and N etc. suggests that sporopollenin may act as an ion-exchange resin and will absorb inorganic elements too.

HESLOP-HARRISON: This refers back to the question of cellulse. I was suprised that you were able to get fluorescence from cellulose, and particularly from callose because callose has practically no auto-fluorescence.

VAN GIJZEL: Callose has very faint, but measurable fluorescence.

HESLOP-HARRISON: The absorption of cellulose acetate is extremely small.

VAN GIJZEL: The fluorescence of cellulose is small.

ORBELL: Perhaps when you irradiate spores for long periods with UV., you may possibly create a reaction between them and the proteinaceous medium in which they are examined.

VAN GIJZEL: Yes fading is a sort of photochemical reaction, but I do not, at the moment, know very much about it.

MUIR: Yes it might be very informative to make a chemical analysis on the medium in which the spores are kept, and to try and find out what is going in and what is coming out during irradiation.

SUMMARY AND GENERAL DISCUSSION

F.P. Jonker, Botanisch Museum en Herbarium,
Utrecht, the Netherlands.

In my lectures in the University of Utrecht, the Netherlands, I formerly used to tell my students that classical pollen analysis is based on the property of the pollen wall to remain unchanged during centuries, if protected against oxidation. This opens a way to reconstruct former vegetations, to build up a fine stratigraphy, and to date layers as well as prehistoric objects which were met in these layers. It does not apply to Quaternary pollen grains and layers only, the resistance of the pollen or spore wall is so strong that it is applicable from the Precambrian onwards. And then I used to add: now you will ask me on what this resistance is based, or, in other words, which resistant chemical compound is building up those walls. I can give two answers to that question.

The first answer is: I don't know. The second answer is: the pollen wall consists mainly of sporopollenin. Both answers, I must admit, were exactly the same though the second makes a much better impression and sounds much more scientific. Nowadays I cannot tell this story any longer. Especially in the Netherlands we know, after the publications by VAN GIJZEL, that the pollen wall does not remain unchanged during hundreds or thousands of centuries. The chemical composition starts to change soon after the deposition of the pollen grain as the changing autofluorescence shows. It might be wondered now, however, whether the pollen morphology, i.e. the morphological features of a pollen wall, remains unchanged indeed during millions of years, whereas the chemical composition and physical properties are changing with increasing geological age.

With regard to this question, one has to investigate whether a morphological change occurs, by comparing and measuring recent exines and those of the same species of an older geological age. As far as your present reporter knows, hardly any research along this line has been carried out. In this Symposium, Professor ERDTMAN presented some data pointing to the direction that the sexine, particularly its outermost part in certain tectate pollen grains, is more resistant than the nexine of the same grain. Up till now, however, a dating according to morphological decomposition, or morphological change in another sense, is impossible.

With regard to the investigation of the statement that chemical composition and, in relation to this, physical properties are changing with increasing geological age, one is dependent on the present knowledge of the chemistry of sporopollenin. Especially after having attended this Symposium, one cannot maintain that nothing is known regarding the properties, its chemical and physical characters, the ontogeny, the deposition and the fossilisation of sporopollenin.

In the immature pollen sac (locule) of an angiospermous anther the sporogenous tissue shows a rapid mitosis and cellular division until the tissue connection severs and unconnected pollen mother-cells come into existence. Then meiosis starts, giving rise to four haploid cells which, in the beginning are set out in a tetrad. The cell walls of the young pollen grains (= microspores) are formed in the same way as normal cellulose walls, and they form the intine of the mature pollen grain. During this process the cells of the tapetum - the inner covering of the endothecium, i.e. the inner part of the pollen sac wall - expand, release, and envelop the tetrad. These tapetum cells are chiefly responsible for the formation of the resistant outer pollen wall (exine).

A comparable process takes place in the formation of pollen grains of Gymnosperms and spores of Pteridophytes; we only call the cellulose wall in the latter endosporium, and the resistant outer wall exosporium.

The exine shows a number of layers of which the innermost one is optically homogeneous whereas the outer ones show a heterogeneous stratification and ornamentation.

The outermost elements forming an ornamentation pattern may consist of pillar-like elements, which are crowned by an apical swollen part or not. The swollen parts may be united to a roof supported, consequently, by pillars. In accordance with the recent paper by REITSMA (1970) I call this roof tectum, a pillar which is not supporting a tectum and not crowned by a swollen part a baculum, a pillar which supports a tectum or is crowned by a swollen part a columella, and the swollen part a caput. Columella and caput together form a pilum. Outside the tectum a varying ornamentation may occur.

In this symposium Mme. CERCEAU showed beautiful "Stereoscan" photographs from which became apparent that in the Umbelliferous genus *Peucedanum* a series of transitions exists, from free pila to a homogeneous tectum.

This short exposition already demonstrated that we need a rather detailed terminology to describe pollen grains, but as the fathers of pollen morphology each used their own terminology an almost chaotic confusion arose.

A homogeneous layer between the bases of the bacula or columellae and the intine, especially, gave rise to controversies as its outer part - the foot layer - shows a staining capacity which is more in accordance with that of the already discussed elements outside of it than with that of the inner part of the homogeneous layer. Does it belong, on the basis of its staining capacity, to the outer heterogeneous layer or, on the basis of its homogeneity, to the inner one. REITSMA (1970) recently published suggestions towards a unification of exine terminology and in accordance with his views I shall call the inner homogeneous layer of the exine, including the foot layer, the nexine, and the outer heterogeneous part, from the bases of the bacula or columellae outwards, the sexine. REITSMA also proposes to indicate the different sublayers which might be distinguished - either in nexine or in sexine - by figures (numbers). In my opinion this proposal makes sense, avoids the use of too many terms and, moreover, is convenient to non-pollen morphologists who have to mention nexine - or sexine - stratification. I may add to this, finally, that the complicated exine stratification as outlined here, occurs especially in Angiospermous pollen grains.

In Gymnospermous pollen grains or in spores it may be different or less complicated which also might have significance regarding the sporopollenin problem. Mme. CERCEAU and collaborators demonstrated, in this symposium, the possibility of breaking acetolysed pollen grains, i.e. exines, by ultrasonic vibration. This enabled them to study transverse sections of the exine and its different layers by means of scanning electron microscopy (CERCEAU *et al.*, 1970). This new technique offers especially good results with ellipsoid pollen grains.

I touch upon pollenmorphological terminology as we may need it in discussion the localization of sporopollenin in the exine. I wish to state here that sporopollenin research demands a pollenmorphological basis and, consequently, a uniform, standardized, concise terminology. Pollen morphologists have to realise that the latter is not only true for pollen morphology itself but the more for the sake of those investigating the chemical composition and physical properties of the different parts of the pollen wall.

As the intine appears to be an ordinary cell wall mainly consisting of cellulose associated with some other polysaccharide material (hemicellulose, pectin), the sporopollenin must be localized in the exine.

We all know the statement in FAEGRI and IVERSEN'S standard

work (1964): "The exine is formed by one of the most extraordinary resistant materials known in the organic world" but this does not offer us definitions, neither for exine nor for sporopollenin.

A definition for exine is a mere pollenmorphological matter and when we again quote REITSMA (1970) it is the part of the pollen wall outside the intine. The intine is the cell wall proper (cell membrane), the inner layer of a pollen wall. With regard to the definition of sporopollenin we may follow G. SHAW (1970) with some slight modifications: sporopollenin is the name for the highly resistant chemical substance or substances in the exine of the pollen wall and the exosporium of a large number of spore walls.

This definition implies that other chemical compounds occur in the exine. I mention carotenoids and carotenoid esters and some hydrocarbons, but this falls outside this definition. In his symposium paper CHALONER expressed his concept to restrict the term sporopollenin taxonomically, i.e. to apply it to spores and pollen grains of vascular plants only, and not to extend it to resistant algal and fungal cells. The present reporter does not favour this. In his opinion the term exine needs a morphological definition and the term sporopollenin a chemical one which is independent of botanical taxonomy.

Now a number of questions arise which are partly answered

in this symposium and by previous work:

1. Is the sporopollenin uniformly distributed over the different layers of the exine or do the more resistant outermost layers contain a higher percentage than the nexine, according to ERDTMAN'S observation that these outermost layers are more resistant.

2. Is sporopollenin formation and sporopollenin deposition restricted to the action of tapetum cells or do the immature pollen grains produce sporopollenin as well?

3. What is known about the chemistry of sporopollenin? Is the sporopollenin in, e.g., the nexine identical with regard to its chemical composition, to that of the outer layers? Is sporopollenin from spore walls identical to that of pollen walls? Is the sporopollenin of the different types of pollen grains indentical? Is sporopollenin to be regarded as a single chemical compound or is it composed of two or more compounds?

4. Are the differences between the properties of different exines - e.g., a different resistance to corrosion and to both chemical biological decay, a different autofluorescence, a different staining capacity both with regard to different layers of the exine - caused by differences in chemical composition of its sporopollenin, or different proportional

composition of the compounds together forming the sporopollenin, or different proportional quantities of sporopollenin in the different exines or different layers?

5. Does sporopollenin in a chemical sense occur in other wall structures than those of pollen grains and spores of Pterido-phytes? Does it occur e.g. in resistant algal cysts or fungal spores? What is, in this connection, the geological age of sporopollenin? Does it occur outside the plant kingdom?

6. What happens to sporopollenin during the division and germination of pollen grains or spores?

And now the answers to these questions as given by previous work and during this symposium:

The first question referred to the distribution of sporo-pollenin over the different layers of the exine. From previous work by HAVINGA (1967) we know the differences in resistance to corrosion in different pollen types are in accordance with the difference in resistance to oxidation and also with differences in sporopollenin content. Professor ERDTMAN reported that the sexine and particularly its outermost parts in certain tectate pollen grains is more resistant that the nexine of the same grains. We might gather from his statement perhaps that at least in those grains the sporopollenin content increases in centrifugal direction. Professor POTONIE, in his symposium seminar, gave examples of Palaeophytic sporomorphs of which the inner walls were more resistant than the outer layer. This might be the case both in spores and in pollen grains producing, after cell division in the male gametophyte, free germs provided with cilia (antherozoide). It might be in accordance with the germination polarity which shows a reversal in those taxa in which the male nuclei are transported by a pollen tube, as outlined by CHALONER in his symposium paper. Pollen grains (and spores) producing free germs germinate proximally whereas pollen grains producing a pollen tube germinate distally. In that case both reversal of germination polarity and reversal of the chief resistance from inner wall to outer wall would be related to pollen tube formation.

A noteworthy fact is that the reversal of germination polarity does not occur in megaspores and embryo sacs. ROWLEY postulated the thesis that in *Epilobium* pollen the sporopollenin in the sexine is not the same as that in the nexine since the resistance to chemical treatments is different. I wonder whether a difference in sporopollenin content might be responsible as well.

Judith H. FORD reported in a paper exhibited at the symposium, that in the mature pollen tetrads of the Epacridaceae the sexine has a high sporopollenin content, and the nexine has a high lignin content. These and other more or less unusual features are explained with reference to the special pollen ontogeny in this family.

BROOKS and SHAW who originally also deduced a lignin component in Lycopodium spores and Pinus pollen on the basis of a formation of phenolic acid after fusion of sporopollenin and potassium hydroxide, however, withdrew this suggestion later on (1968). They remarked that there is now perhaps no need to postulate "lignin-like material in sporopollenin", as phenolic acids are obtainable from polymerized carotenoids.

Both exine and exosporium form, consequently, during the transport a highly resistant protecting coat around the precious interior.

In spores of Equisetum the exosporium has not, in the first place, a protecting function. It splits into two pairs of hygroscopic mobile elaters. In accordance to this the sporopollenin percentage in Equisetum spores is very low. It is, according to KWIATKOWSKY and LUBLINER-MIANOWSKA (1957) 1.8% whereas it varies in other spores or pollen grains from 2.4% in Narcissus to 29% in Lycopodium.

Yet data regarding the distribution of sporopollenin or different sporopollenins over the different layers of the exine are still scarce and meager. Future fluorescence-microscopical research to observe and compare the fluorescence of the various layers of an exine may be very valuable (though very difficult as well) especially when the results are compared with evidences supplied by transmission electron-microscopy, by scanning electron microscope techniques and by the new technique of cathodoluminescence. In this symposium M.D. MUIR and P.R. GRANT made a start in this respect, by describing the differences in properties of endosporia and exosporia of ferns. Future research may benefit from co-operation with pollen morphologists like the greater part of problems touched upon in this symposium will benefit from team work.

My second question referred to sporopollenin formation. From HESLOP-HARRISON'S former work (see review 1964) it became known that, during the pre-meiotic phase of the pollen mother-cells, the formation of a special mother-cell wall within the original cellulose wall of the mother cell starts.

It consists of callose and thickens during prophase. During the metaphase, when the microspores are formed, a synthesis of sporopollenin in the mitochondria of the tapetum protoplasts starts. This sporopollenin is during later stages transported to the walls of the tapetum cells on which it is found in special bodies of various shape (plaques, Ubisch bodies, sporopollenin orbicules, cupules, particles).

Then autolysis of the tapetum cell-walls begins. At the end of meiosis we find a tetrad of microspores without interconnections within the callose wall. In that stage individual spore walls are formed; that first spore wall is called by HESLOP-HARRISON the primexine.

The ornamentation of the future exine is already preformed in this primexine. The next stage is the release of spore individuals in the pollen sac by a dissolution of the callose wall. Immediately after, both deposit of nexine sporopollenin and disruption of the primexine starts. The intine formation begins and this is followed by the final dissolution of the tapetum cells to a mass in which the immature pollen grains are embedded until the mass dries up against the wall of the pollen sac. At the same time sexine formation is in progress. New data and new insights regarding this process have been reported in a number of papers during this symposium. HESLOP-HARRISON himself reported on experiments in which young stages are centrifuged. This highly influences and damages cleavage, resulting in, e.g., a sporopollenin sexine deposit on a colpus, or the formation of small anucleate grains without apertures.

WILLEMSE reported that, in *Pinus*, the cytoplasm of the microspore starts the production of the sporopollenin of the nexine when the tetrad is still enclosed by the callose wall, i.e. before the formation of the intine. The fluorescence of the sporopollenin of the nexine during sporogenesis is quite interesting as it shows a single peak spectrum of increasing intensity and a shifting maximum.

WATERKEYN and BIENFAIT reported that the primexine shows no fluorescence at all, in *Ipomoea*. The Ubisch bodies on the tapetal cell walls show fluorescence, in later stages the nexine layers exhibit fluorescence as well, and so do the sexine elements during their development.

DICKINSON drew our attention to the differences in primexine formation in Gymnosperms and Angiosperms, whereas no principal differences exist with regard to the formation of Ubisch bodies on the tapetum cell-walls in both groups. Especially during nexine formation differences between Gymnosperms and Angiosperms exist regarding the deposition of sporopollenin. He, moreover, showed interesting details with regard to the formation of the bladders (sacs, wings) of *Pinus* pollen.

We may answer my second question in this sense that sporopollenin formation occurs both in the tapetum cells and in the cytoplasm of the young spores. Nexine sporopollenin is chiefly or totally produced in the microspores; sexine sporopollenin is deposited by the dissolution of tapetum cells in a pattern which was already preformed at the primexine stage.

The role of the diploid or even polyploid tapetum was amply discussed by ECHLIN who also stated that, during the first stages of pollen development, sporopollenin is formed within the cytoplasm of the haploid pollen grain. He especially discussed the formation of the Ubisch bodies and the transport and deposition of their sporopollenin to the pollen wall. In *Tradescantia* no sporopollenin formation should occur in the tapetum.

ROWLEY reported on finer details of this deposition in connection with both the properties of the microspore surface and the submicroscopic structure of the nexine before the intine is present. By means of tracer research he comes to the conclusion that the exine is not, in the first place, an impermeable barrier but rather more a molecular sieve. His already mentioned statement that, in Epilobium, the sporopollenin in nexine and sexine are different is perhaps to be carried back to the difference in origin, viz. from the microspore cytoplasm and from the tapetum.

ROWLEY'S data were gathered in co-operation with a number of collaborators from the U.S.A., Sweden and elsewhere. I will speak here a word in remembrance of the late Botjah PRIJANTO, a promising Indonesian botanist and palynologist, who died in 1969, rather soon after his stay in Europe (Edinburgh, Solna and Utrecht) by a car accident.

My third question referred to the chemistry of sporopollenin of which, after the very important classical work by ZETZSCHE, new data became available by the work of G. SHAW and co-workers. SHAW and YEADON (1966) published molecular formulae of the sporopollenin of a number of spores and pollen grains, showing considerable variation.

In this symposium SHAW discussed early work on the chemical degradation of sporopollenin extended by new data regarding the incorporation of labelled carotenoids into sporopollenin of Lilium and some fungi. His former collaborator J. BROOKS added some observations regarding properties of sporopollenin in sediments, compared to changes in the structure of the exine when heated. He reported on (1) the sporopollenin/cellulose ratio, (2) the carbon 12/carbon 13 ratio which is dependent on the environment, and (3) the ratio of intine to exine which varied through geological time. He was able to determine environmental differences in the case of two fossil *Tasmanites* species on the basis of the carbon 12/carbon 13 ratio. DODD and EBERT reported on their investigations on the presence of free radicals on the spore surface of *Osmunda* and *Lycopodium* by electron spin resonance, using charcoal as a model.

So far I can answer my question by the statement that the chemical composition of the spore wall of both different spore types and different pollen types are not the same: quite a considerable variation exists. Sporopollenin of the nexine is, in all probability, different from that of the sexine. Fluorescence microscopy appears to be a useful technique in investigating this problem and VAN GIJZEL (1967) published a number of fluorescence spectra of fresh spores and pollen grains, all showing two peaks. This might suggest that sporopollenin consists of two components, though other explanations are perhaps possible, and a composition of more than two components might be possible as well. From the data reported before and during this symposium I originally deduced that one of the peaks might reveal the nexine sporopollenin the other the sexine sporopollenin. From the shift of fluorescence with increasing geological age and rank of coal, and from the more rapid decrease of one of the peaks compared with that of the other, we might deduce that the less resistant sporopollenin of the nexine is revealed by the more rapid extinction of one of the peaks. The slower extinction of the other peak then reveals, consequently, the more resistant sexine sporopollenin. This, in turn would be in agreement with WILLEMSE's observation that the fluorescence spectrum of the young nexine is a single peak spectrum.

In his symposium paper, however, VAN GIJZEL ascribed the two peaks in the spectra, as published in his former work, to inadequate measuring techniques. After having improved the latter, his former two peak spectra appeared to belong to the single peak type too. In the one peak spectra, a decrease in blue-green and an increase in orange-red intensities occurs, similar to that in the former two peak spectra. Studies on fresh _Lycopodium_ spores in various stages of chemical decomposition showed that the one peak spectra consist of a mixture of spectra. The final fluorescence of fossil material will be determined by the spectra of the most resistant and most strongly fluorescent compounds.

However, it is important to find methods and techniques which enable the separation and comparison of the fluorescence spectra of the nexine and the sexine. The varying chemical and physical properties of sporopollenin are the subject of investigation both into degradation by biologic activity and degradation by oxidation, heating, decay, and corrosion, i.e. during every chemical and physical degradation which may occur during geological time or coalification. VAN GIJZEL reported that in more strongly coalified deposits, the sexine of bisaccate pollen may differ considerably in fluorescence properties from the nexine, due to their different coalification susceptibility.

FAEGRI drew attention to the fact that micro-organisms do not attack or damage the exine, though he leaves the possibility that micro-organisms might exist that possess an enzyme system which enables them to degrade sporopollenin. Experiments by HAVINGA (1966, and reported in this symposium point to the direction that attack of sporopollenin by micro-organisms in the soil is successful only after previous degradation of the sporopollenin by oxidation. He described the different, presently known deterioration types.

From experiments in which different pollen grains and spores were buried in different types of soil it appeared that the different pollen and spore types showed a different resistance to corrosion, oxidation and subsequent microbial attack. Moreover, different types of deterioration exist and occur, dependent both on the pollen or spore type, and on the type of soil, and also as especially postulated in this symposium, on the factor time, i.e. duration of burial. It appeared that the percentage of undamaged grains changed considerably during a prolonged burial.

ELSIK compared the change in exine pattern caused by chemical and physical degradation to that caused by microbial attack. He found also that the type of degradation pattern is dependent from the pollen type as well, which might point to differences in sporopollenin chemistry. He reported on corrosion types in fossil spores from the Cretaceous onwards, and concluded that both his perforate and his non-perforate scar type have been produced by micro-organisms.

Non perforate scars degrade only one layer of the exine or exosporium.

My fourth question has been treated by me already for the

greater part.

I believe that differences in resistance, differences in autofluorescence, and differences in staining capacity are caused by a different composition, chiefly chemical composition of the sporopollenin. With regard to the different layers of the exine we may state that nexine sporopollenin is characterized by another chemical composition and another origin compared to sexine sporopollenin.

This again offers an interesting question to pollen morphologists. In the beginning of my report I mentioned that the nexine consists of at least two layers. The outermost, known as foot layer, we call nexine I. It is characterized by a staining capacity which is more in accordance to that of the sexine than to that of nexine II. This suggests perhaps that only nexine II is formed by the immature microspores whereas the foot layer = nexine I has possible been formed by the action of the tapetum cells like the sexine. In that case we come to the statement that the foot layer morphologically belongs to the nexine though both its origin and the chemical composition of its sporopollenin might agree to those of the sexine.

Regarding my fifth question we may gather from this symposium that the occurrence of sporopollenin is not restricted to Spermatophytes and Pteridophytes. It also occurs in fungal spores. It has been demonstrated in fossil algae of Upper-

Devonian age as well as in pollen and spores and dinoflagellates from the Carboniferous onwards.

The oldest plant life is known from the Praecambrium Onverwacht Series in Transvaal, South-Africa, originally dated 3200 million years before present and according to more recent datings 3400-3700 million years BP, and from the Fig Tree Series of Barberton, Transvaal, 3000-3100 million years BP. BROOKS reporeted that he found insoluble organic matter in sediments up to 3700 million years old, that might show a common identity with modern sporopollenin, i.e. compounds characteristic of the former presence of sporopollenin (see also BROOKS and SHAW, 1968).

If these conclusions are correct, i.e. if sporopollenin occurred as early as the older Praecambrium, then either sporopollenin has been produced by very primitive organisms, or we must assume that under the much more reducing atmospheric conditions of the early Praecambrium an abiogenic origin of sporopollenin was possible. In this connection it is striking that SHAW (1970) isolated these insoluble substances from carbonaceous chondrites, viz. the Orgueil meteorite and the Murray meteorite. This points either to the occurrence of comparable early life forms on the moon (UREY, 1962, 1965) or elsewhere in our solar system, or in the universe, as CLAUS and B. NAGY (1961, 1962) postulated, or to atmospheric conditions comparable to those existing during our early Praecambrium in other planetary bodies.

SHAW (1970) and co-workers favour the thesis that the most plausible explanation of these facts is that life on Earth had an extra-terrestrial origin.

This possibility had already been suggested by LEIBNIZ in the very beginning of the 18th century.

KREMP (1968) suggested, however, that a "pre-life" form, a coacervate which coagulated into a cyst similar to that of blue-green algae (Cyanophyta), may have provided the organic matter that is preserved in the objects described by CLAUS and B. NAGY (1961, 1962), TIMOFEJEV (1963), and others.

Besides the already mentioned research on heating of sporopollenin by BROOKS, a number of biochemical and physico-chemical and partly statistical papers were presented. DOUGLAS, DUNGWORTH, POWELL and McCORMICK reported on lipid components in *Lycopodium* spores and *Pinus* pollen compared with Carboniferous spores. Lipid extracts were analysed and identified, the remaining sporopollenin residues subjected to high temperature and high pressure hydrogenolysis. CORREIA and PENIGUEL reported on their statistical computer studies on the diagenetic changes of sporopollenin, dependent on depth and on temperature belonging to that depth and, to a minor degree, on lithology and geologic age. This study not only proved to be of importance to the sporopollenin problem but to oil exploitation as well. COMBAZ compared the properties of sporopollenin to related resistant substances, again with regard to heating. He also paid attention to morphological changes which proved to be less distinct.

COOPER and MURCHISON, finally, discussed the petrology and geochemistry of sporonite. They use this term for the organic substance or maceral found in Palaeophytic coals. It is formed by coalification of sporopollenin and its properties are dependent on the rank of coal.

Also in this case the properties showed considerable variation, even in single coals.

All these new and modern approaches to the problem of the nature of sporopollenin may not only help palynologists to understand the properties on which their branch of science is based, but even might be a further step to our knowledge of the origin and evolution of life on Earth.

My sixth and last question referring to the behaviour of sporopollenin during mitosis and germination of spores and pollen grains has not been discussed during this symposium though Professor FAEGRI touched upon this problem in his paper. He especially wondered what happens to the empty exine which remains on an angiospermous stigma after the development of the pollen tube.

Up till now we know nothing of it. Also in this case, perhaps, fluroescence microscopy might be helpful, next to morphological research and microchemical analyses.

REFERENCES

CERCEAU, M.Th., M. HIDEUX, L. MARCEAU and F. ROLAND, C.R. Acad. Sc. Paris 270, 1970, p. 66-69.

CLAUS, G. and B. NAGY, Nature 192, 1961, p. 594-496.

CLAUS, G. and B. NAGY, Phys. Soc. Amer. New Bull. 15, 1962, p. 15-19.

BROOKS, J. and G. SHAW, Nature 219, 1968, p. 532-533.

BROOKS, J. and G. SHAW, Nature 220, 1968, p. 678-679.

BROOKS, J. and G. SHAW, Grana Palynologica 8, 1968, p. 227-234.

FAEGRI, K. and Johs. IVERSEN, Textbook of Pollen Analysis, 2nd rev. ed., Oxford, 1964, 237 pp. (p.5).

HAVINGA, A.J., Review of Palaeobotany and Palynology 2, 1967, p. 81-98.

HESLOP-HARRISON, J. in LINSKENS, H.F. (ed.). Pollen Physiology and Fertilization, Amsterdam 1964, p. 39-47.

KREMP, G.O.W., J. British Interplanetary Soc. 21, 1968, p. 99-112.

KWIATKOWSKI, A. and K. LUBLINER-MIANOWSKA, Acta Societ. Botanicorum Poloniae 26, 1957, p. 501-514.

REITSMA, Tj., Review of Palaeobotany and Palynology 10, 1970, p. 39-60.

SHAW, G., Sporopollenin in "Phytochemical Phylogeny" Chapter 3, 1970, p. 31-57 (Academic Press, London).

SHAW, G., and A. YEADON, Journ. Chem. Soc. C, 1966, p. 16-22.

TIMOFEJEV, B.V., Grana Palynologica 4, 1963, p. 92-99.

UREY, H.C., Nature 193, 1962, p. 1119-1123.

UREY, H.C., Science 147, 1965, p. 1262.

VAN GIJZEL, P., Leidse Geologische Mededelingen 40, 1967,

INDEX